DESIGN AND CONSTRUCTION OF TUNNELS

PIETRO LUNARDI

DESIGN AND CONSTRUCTION OF TUNNELS

Analysis of controlled deformation in rocks and soils (ADECO-RS)

Springer

Pietro Lunardi
Rocksoil SPA
Piazza S. Marco 1
20121 Milano
Italy

Originally published in Italian “Progetto e costruzione di gallerie – Analisi delle deformazioni controllate nelle Rocce e nei suoli”, Hoepli 2006, ISBN 88-203-3427-5

Translation by James Davis B.A. (hons).

ISBN 978-3-540-73874-9 e-ISBN 978-3-540-73875-6

DOI 10.1007/978-3-540-73875-6

Library of Congress Control Number: 2007936432

Typesetting and production: LE-TEX Jelonek, Schmidt & Vöckler GbR, Leipzig, Germany
Cover design: WMXDesign GmbH, Heidelberg

Printed on acid-free paper

9 8 7 6 5 4 3 2 1

springer.com

About the Author

Pietro Lunardi

A civil engineer in the field of transport, he is one of the greatest experts in world on the design and construction of underground works, the creator of highly innovative solutions: the cellular arch, developed for the construction of the Porta Venezia station on the Milan Railway Link Line for which he was nominated "Man of the Year in the construction field" by the United States journal "*Engineering News-Record*"; shells of improved ground using jet-grouting techniques; full face mechanical precutting, face reinforcement using fibre glass structural elements; he devised and developed the revolutionary new approach to design and construction, known by the acronym ADECO-RS, described in detail in this book, which for the first time has made it possible to construct tunnels even in the most difficult geological-geotechnical and stress strain conditions with reliable forecasting of construction times and costs.

A former university lecturer in "Soil and rock improvement" in the Faculty of Engineering of the University of Florence and in the "Defence and conservation of the soil" in the Faculty of Engineering of the University of Parma, he has filled many institutional roles including that of Minister of Infrastructures and Transport for five years in the second Berlusconi government (2001-2006).

The author of more than 130 publications he has held more than 40 national and international conferences on the subjects of tunnelling and geo-engineering.

Contents

Preface

To those who believed...

Geological hazard and the lack of appropriate survey, design and construction instruments for tackling those terrains we call "difficult", with good prospects of success, have always made the design and construction of underground works a risky affair, which could not therefore be faced with the same degree of accuracy as other civil engineering works. As a consequence they have always occupied a subordinate position with respect to similar surface constructions and in the past they were only resorted to when the latter seemed impractical or of little use.

However, decisive progress made in the field of geological surveys, the availability of powerful computers for making calculations and above all the introduction of excavation technologies that are effective in all types of ground have created the conditions for a qualitative quantum leap forward. The last formidable negative factor to be overcome to achieve that transparency in this field, which has until now been the prerogative of traditional surface works, remains the absence of a modern and universally valid design approach, capable, that is, of integrating and exploiting the new capabilities and of guiding the design engineer through the stages of design and construction. In fact even today the answer to the apparently obvious and banal question, "What does the design and construction of an underground work consist of?", would find many design engineers in disagreement not only on the form but also on contents of design. And this is not surprising because this type of problem has always been addressed in a very indeterminate fashion. Until not very long ago the inadequacy of the available knowledge and means meant that the design of an underground construction had to be improvised during tunnel advance. As a consequence, the design of such a construction was merely a question of identifying the geometry of the route and some of the tunnel section types, while the means of excavation, intervention to stabilise the tunnel and which linings to use were largely decided during construction as the tunnel advanced.

The practice of "observing" the response of the ground to excavation in order to devise appropriate countermeasures to stabilise a tunnel in the short and the medium term has therefore always lain at the basis of underground construction. In the last century some engineers sought to develop design and construction "methods" around this practice and although they were based on incorrect scientific theory, they nevertheless constituted significant progress at the time. This brought them great success at first, and despite many clamorous failures, they have managed to survive and flourish, assisted by a lack of alternative ideas caused by an unexplainable, lazy, and far too common tendency to conform. These methods, led by the NATM, were not only found to be inadequate in really difficult geotechnical and geomechanical conditions, but they also appear very much behind the times, because they cannot, by their very nature, furnish solutions which will enable construction to be planned in any way, in terms of finance and schedules, an undoubtedly essential requirement for transparent and prudent management of resources in modern societies.

This is the context in which, a little more than ten years ago, the presentation of the approach based on the Analysis of Controlled Deformation in Rocks and Soils (ADECO-RS) was met with great general interest, mixed with a degree of scepticism. I and my research team had developed it over a long period of theoretical and experimental research conducted outside traditional lines. It finally recognised how important the three dimensional nature of statics and the dynamics of tunnel excavation was and by taking this to its ultimate consequences and by appropriately exploiting the new technologies, it seemed to hold the promise of that long awaited quantum leap forward. It would for the first time enable an underground work to be designed before construction commenced with all the consequent advantages in terms of planning, construction costs and schedules.

Since then the validity of the approach has been tested on the construction of more than 300 km of tunnel and at least 150 km of this was under very difficult stress-strain conditions. These have been fully discussed as the occasion arose in conferences and publications in which it has been demonstrated beyond any doubt that we know how to transform our promises into reality. The approach had in fact made it possible to predict times and costs for the construction of underground works with a fair degree of precision (proportional to the knowledge of the geology acquired beforehand), minimising unforeseen events and eliminating tunnel advance problems, which were previously encountered under the same ground and overburden conditions. It seemed to have become finally possible to make a reliable estimate of the cost benefit ratio for an underground project, a fundamental parameter in the decision making progress of selecting design alternatives.

We are therefore on the right track; however, I do feel that much investigation and study is still necessary. The purpose of this book is not just to illustrate the basic concepts of the approach as fully and exhaustively as possible and to show how, by following its principles, underground works can be designed and constructed with a reliability and accuracy never attained before. Its purpose is above all to furnish the scientific community with a useful reference text around which all may work together to improve the ADECO-RS approach or even to go beyond it.

Pietro Lunardi

A note to the reader

I was concerned in writing this book to make it as easy and pleasurable to read as possible, despite the very technical and highly specialised nature of the contents. I therefore drew on the experience I had acquired in past years as a university lecturer, trying throughout the book to imagine the curiosity and desire for greater explanation that might arise in my readers. It was by trying to respond to this curiosity, which sometimes even led me touch on to subjects apparently quite distant from those being dealt with, that I felt I was often able to make the explanation more straightforward and to stimulate the attention of my readers, even on the more complex concepts.

The outcome is a book with two sets of contents, one for odd numbered pages on which the central theme of the book unfolds and one for even pages, which can be read independently of the text. It is on these pages that I have sought to satisfy the reader's desire for greater explanation with observations and extra detail.

Thanks

To complete this book, which collects together experiences from forty years of working on numerous construction sites, in universities and other professional environments, I wish to sincerely thank those who have believed in me over these long years, teaching me and advising me. They include Angelo Palleschi from Capistrello, one of the many tunnel miners who have helped me during those long hours spent on tunnel construction sites, Angelo Farsura, an enlightened entrepreneur who gave me the chance in the 1960's and 1970's to follow the works on site for the Gran Sasso tunnel, one of the most complex and fascinating projects of the last fifty years and so many other people who I obviously cannot mention here, but to whom I am bound, through my memories, by gratitude and friendship.

Finally, I wish to say a special thank you to those who have worked most closely with me, Renzo Bindi, Giovanna Cassani and Alessandro Focaracci, who, with their shrewd engineering sense, have helped me to develop this new approach to the design and construction of underground works. It is an approach, which I hope will serve as a useful guide for young engineers who wish to study and implement these works, which differ from other civil engineering works because of the extreme and continuous variation in the geological, geotechnical and stress-strain conditions in which the design engineer is obliged to operate.

Pietro Lunardi

From the research to ADECO-RS

There are no tunnels which are easy or difficult because of the overburden or the ground to be tunnelled, but only stress-strain situations in the ground in which it is, or is not possible to control the stability of the excavation, which will depend on our knowledge of the pre-existing natural equilibriums, on a correct approach to the design and on the availability of adequate means for excavation and stabilisation.

CHAPTER 1

The dynamics of tunnel advance

1.1 The basic concepts

Anyone who sets out to construct underground works, finds themselves having to tackle and solve a particularly complex civil engineering problem, because it is far more difficult to determine the basic design specifications for underground works in advance than it is for surface constructions (Fig. 1.1).

It is not, as with surface constructions, a question of gradually assembling materials (steel, reinforced concrete, etc.) with well known strength and deformation properties to build a structure which, when subject to predictable loads, finds its future equilibrium in the desired final configuration. On the contrary, one has to intervene in a pre-existing equilibrium and proceed in some way to a "planned disturbance" of it in conditions that are only known approximately.

Another peculiarity of underground works, well known to design and construction engineers, but not always given sufficient weight, is that very often, the stage at which the structure is subject to most stress is not the final stage, when the tunnel is finished and subject to external loads predicted at the design stage, but the intermediate construction stage. This is a very much more delicate moment because

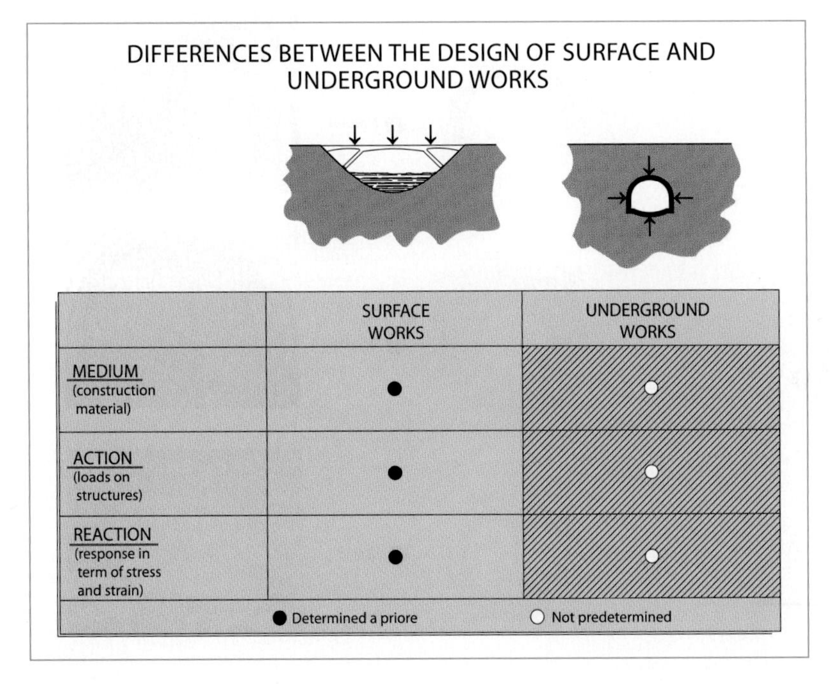

Fig. 1.1

The arch effect

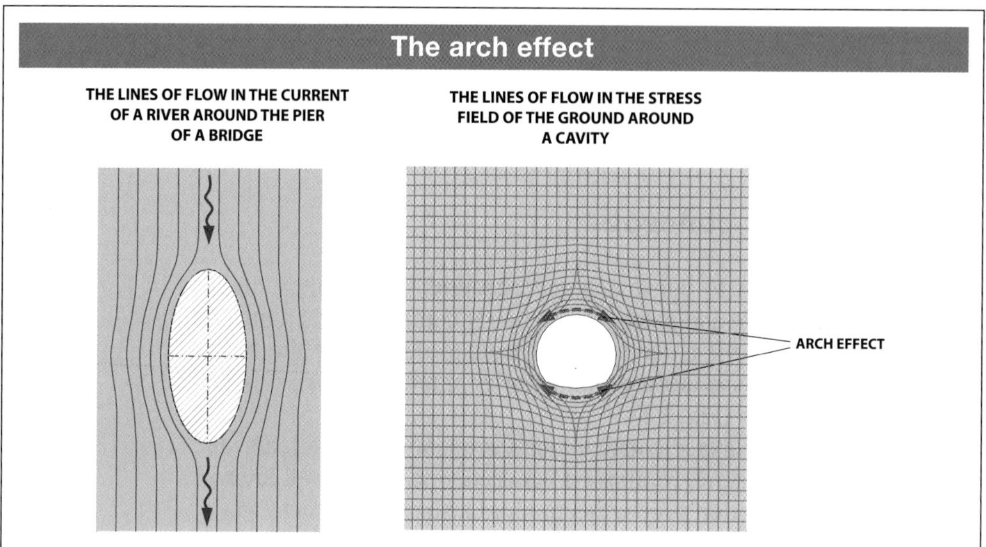

In a similar fashion to the lines of flow in the current of a river, which are deviated by the pier of a bridge and increase in speed as they run around it, the flow lines of the stress field in a rock mass are deviated by the opening of a cavity and are channelled around it to create a zone of increased stress around the walls of the excavation. The channelling of the flow of stresses around the cavity is termed an **arch effect**.
The arch effect ensures that the cavity is stable and will last over time.

THE FORMATION OF AN ARCH EFFECT IS SIGNALLED BY THE DEFORMATION RESPONSE OF THE ROCK MASS TO EXCAVATION

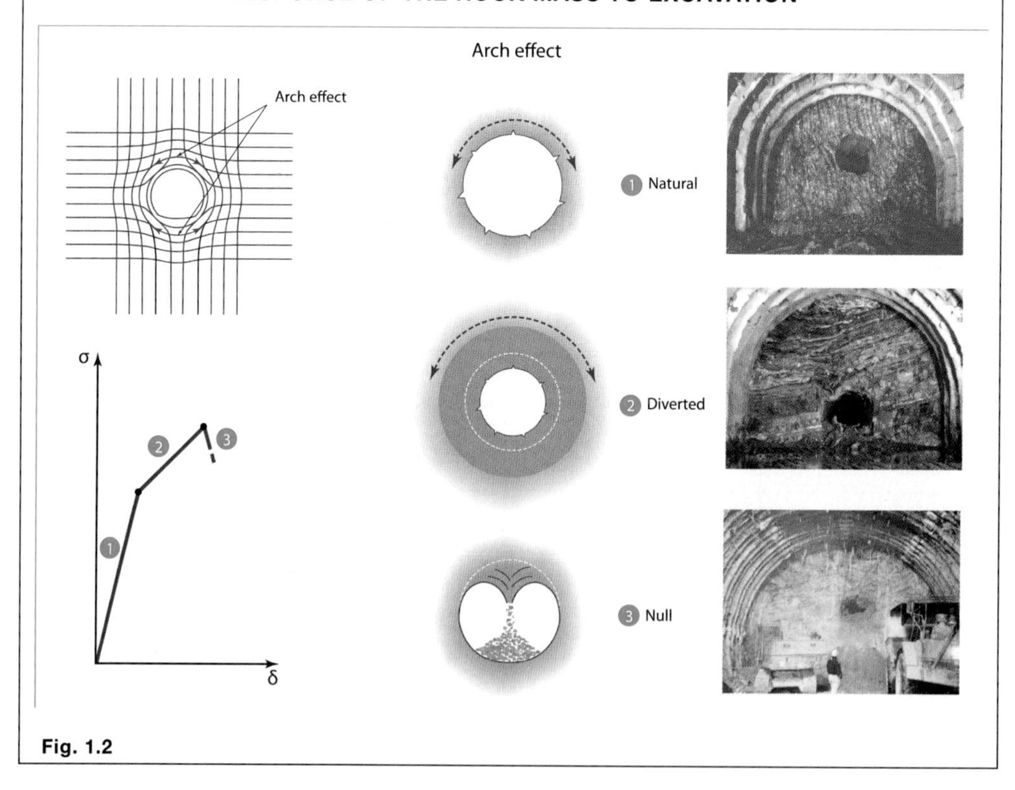

Fig. 1.2

the effects of the disturbance caused by excavation have not yet been completely confined by the final lining at this stage, when the pre-existing stresses in the rock mass are being deviated by the opening of the cavity and channelled around it (**arch effect**) to create zones of increased stress on the walls of the excavation.

The particular delicacy of this intermediate stage becomes clear if one considers that it is precisely on the correct channelling of stresses around the cavity that the integrity and life of a tunnel depends. Channelling can be produced, depending on the size of the stresses in play and the strength and deformation properties of the ground, as follows (Fig. 1.2):

1. close to the profile of the excavation;
2. far from the profile of the excavation;
3. not at all.

The first case occurs when the ground around the cavity withstands the deviated stress flow around the cavity well, responding elastically in terms of strength and deformation.

The second case occurs when the ground around the cavity is unable to withstand the deviated stress flow and responds anelastically, plasticising and deforming in proportion to the volume of ground involved in the plasticisation phenomenon. The latter, which often causes an increase in the volume of the ground affected, propagates radially and deviates the channelling of the stresses outwards into the rock mass until the triaxial stress state is compatible with the strength properties of the ground. In this situation, the arch effect is formed far from walls of the excavation and the ground around it, which has been disturbed, is only able to contribute to the final statics with its own residual strength and will give rise to deformation, which is often sufficient to compromise the safety of the excavation.

The third case occurs when the ground around the cavity is completely unable to withstand the deviated stress flow and responds in the failure range producing the collapse of the cavity.

It follows from this analysis of these three situations that:

- an arch effect only occurs *by natural means* in the first case;
- an arch effect by natural means is only produced effectively in the second case if the ground is "helped" with appropriate intervention to stabilise it;
- in the third case, since an arch effect cannot be produced naturally, it must be produced *by artificial means,* by acting appropriately on the ground before it is excavated.

The first and most important task of a tunnel design engineer is to determine if and how an arch effect can be triggered when a tunnel is excavated and then to ensure that it is formed by calibrating

The medium

If we simplify to the maximum, we can say that there are three main mediums in nature: sand, clay and rock, which have three different natural **consistencies**:

- the consistency of sand, which has its effect above all in terms of friction, giving rise to **loose** type behaviour;
- the consistency of clay, which has its effect above all in terms of cohesion, giving rise to **cohesive** type behaviour;
- the consistency of rock, which has its effect in terms of cohesion and friction, with markedly higher values than in the case of sand and clay giving rise to **rock** type behaviour.

It is the **natural consistency** of the medium which determines local differences in the earth's crust.

Morphology characteristic of the consistency of sand

Morphology characteristic of the consistency of clay

Morphology characteristic of the consistency of rock

In its natural state, the medium appears with the characteristics of its own type of consistency, however, when tackled underground, where it is subject to stresses which increase with depth, it has a consistency which varies as a function of the entity and anisotropy of the stress tensor (**acquired consistency**).
The manner in which the consistency of the medium varies as a function of its stress state is studied by means of triaxial tests on samples and is described by the intrinsic curve and stress-strain diagrams.

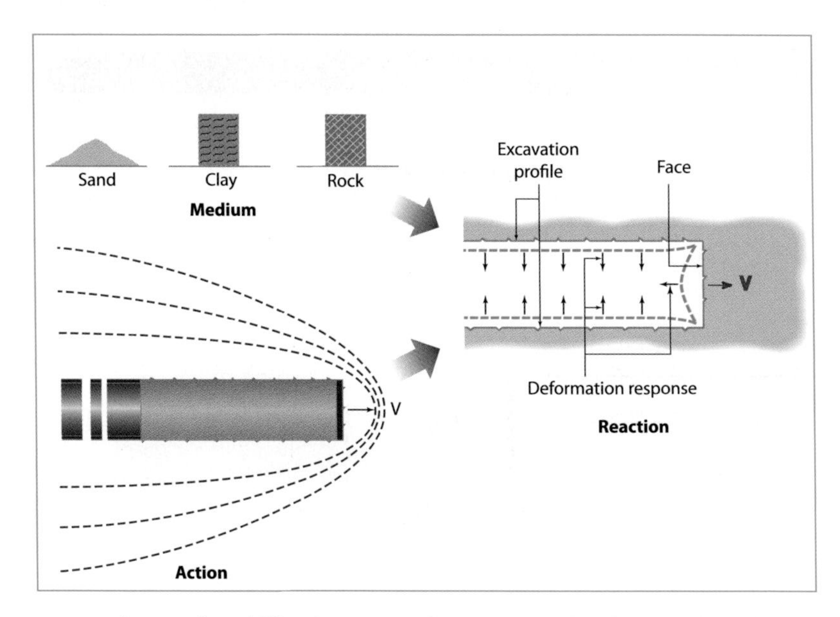

Fig. 1.3

excavation and stabilisation operations appropriately as a function of different stress-strain conditions.

To achieve this, a design engineer must have a knowledge of the following (Fig. 1.3):

- the *medium* in which operations take place;
- the *action* taken to excavate;
- the expected *reaction* to excavation.

1.2 The medium

The **medium**, and that is the ground, which is in practice the actual "construction material" of a tunnel, is extremely anomalous when compared to traditional materials used in civil engineering: it is discontinuous, unhomogeneous and anisotropic. *On the surface*, its characteris-

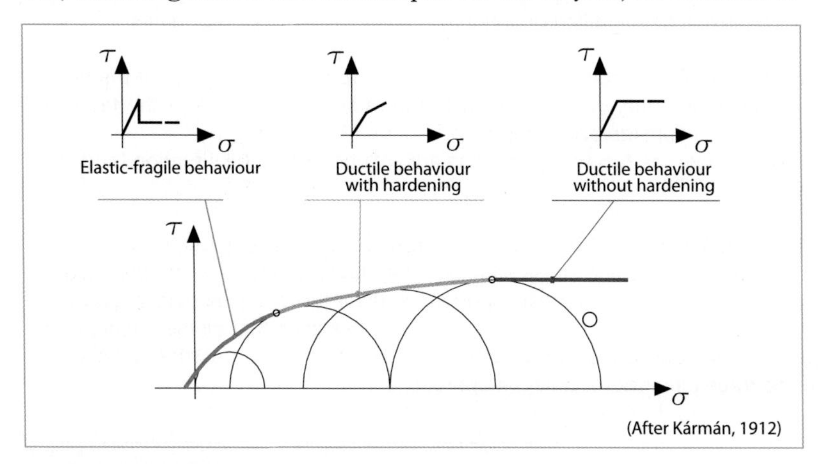

Fig. 1.4 *The same material can reach failure with different types of behaviour according to the stress range*

The characteristic zones

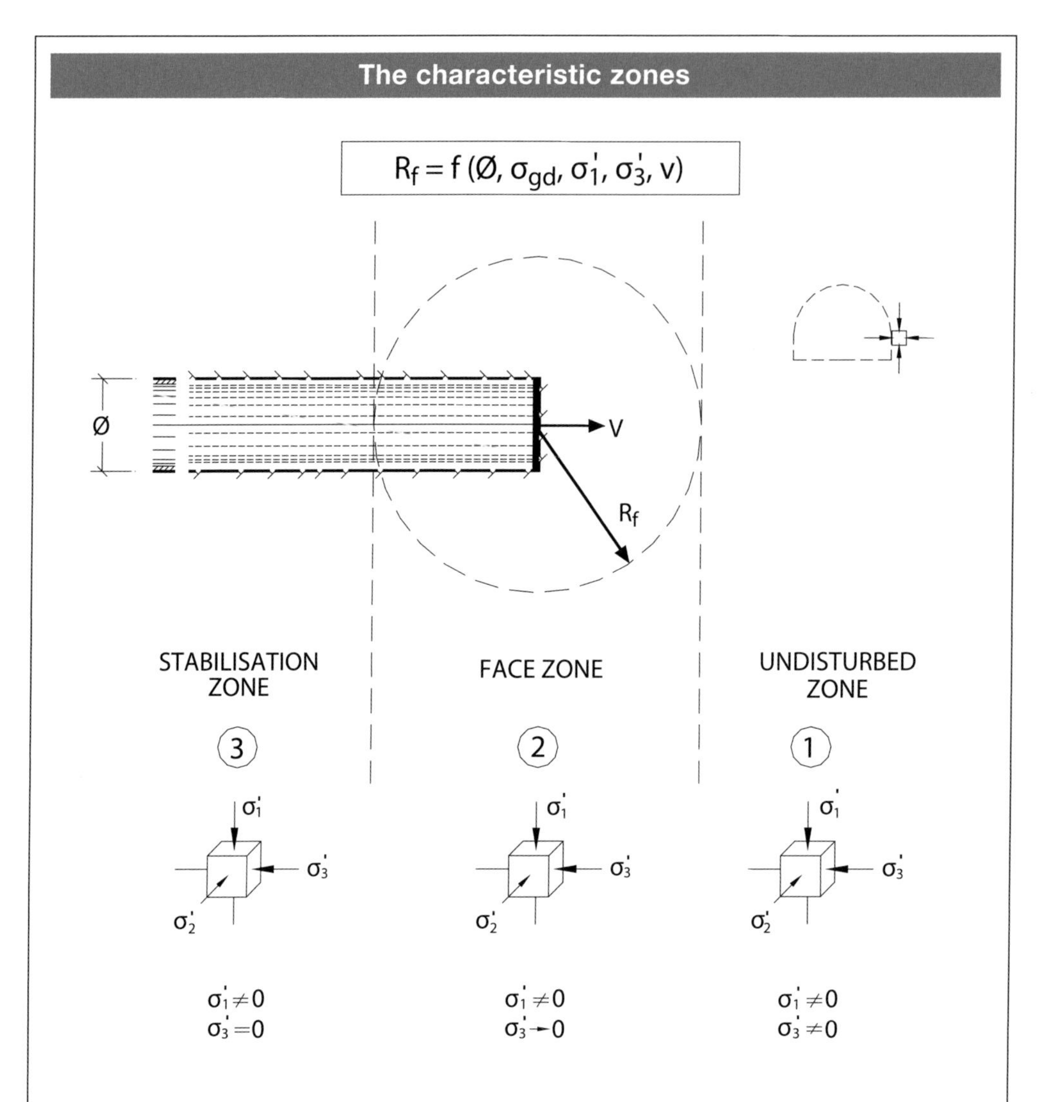

Three characteristic zones can be identified during tunnel advance in an unlined tunnel.

1. an undisturbed zone, where the rock mass is not yet affected by the passage of the face;
2. a tunnel face or transition zone, corresponding to the radius of influence of the face, in which its presence has a considerable effect;
3. a stabilisation zone, where the face no longer has any influence and the situation tends to stabilise (if possible).

It is important to observe that in passing from the undisturbed zone to the stabilisation zone, the medium passes from a triaxial to a plane stress state and that the face zone is where this transition takes place. Consequently, this is the most important zone for the design engineer. It is here that the action of excavation disturbs the medium and it is on this zone that all the attention of the design engineer must be focused for proper study of a tunnel. It is not possible to achieve this without employing three dimensional analysis approaches.

tics vary but this depends solely on its own intrinsic nature (natural consistency), which conditions the morphology of the earth's crust, *while at depth* its characteristics also change as a function of the stress states it is subject to (acquired consistency) and this conditions its response to excavation [1] (Fig. 1.4).

1.3 The action

The **action** is that whole set of operations performed to excavate the ground. It is seen in the advance of the face through the medium. It is therefore a *distinctly dynamic phenomenon*: the advance of a tunnel may be imagined as a disk (the face) that proceeds through the rock mass with a *velocity* **V**, leaving an empty space behind it. It produces a *disturbance in the medium*, both in a longitudinal and transverse direction, which upsets the original stress states (Fig. 1.5).

Within this disturbed zone, the *original field of stresses*, which can be described by a network of flow lines, *is deviated* by the presence of the excavation (Fig. 1.2) and concentrates in proximity to it, producing increased stress, or, to be more precise, an increase in the stress deviator. The size of this increase determines the amplitude of the disturbed zone for each medium (within which the ground suffers a loss of geomechanical properties with a possible consequent increase in volume) and, as a result, the behaviour of the cavity in relation to the strength of the rock mass σ_{gd}.

The size of the disturbed zone in proximity to the face is defined by the *radius of influence of the face* R_f, which identifies the area on which

Fig. 1.5

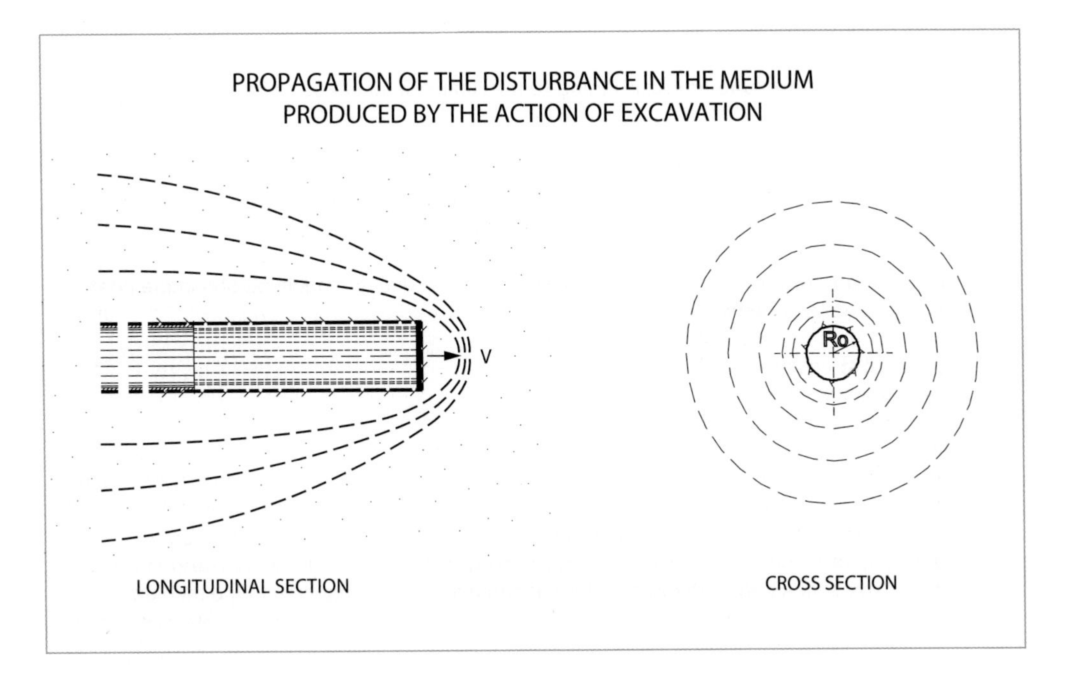

The type and the development of the deformation response (reaction)

The deformation response of the medium to excavation manifests in different forms depending on the range in which it occurs and these can be described with simple diagrams. For example:

a ***solid load*** response, primarily when the failure occurs in a medium generally subject to stress in the elastic range, which is localised and produced mainly as a result of gravity, when the strength of the medium is exceeded along pre-existing discontinuity surfaces;	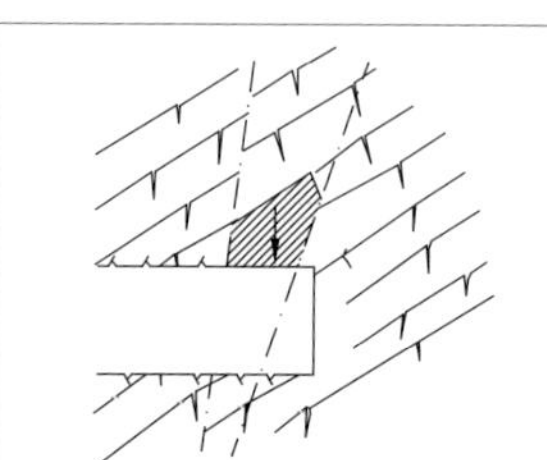	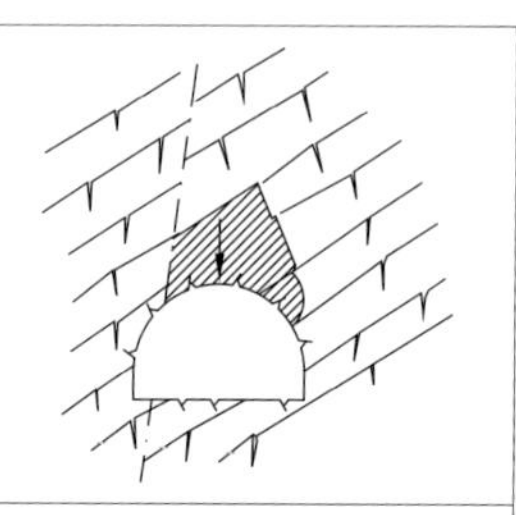
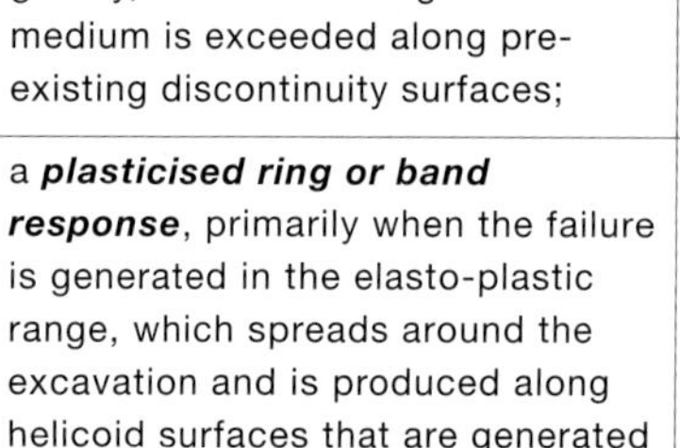 a ***plasticised ring or band response***, primarily when the failure is generated in the elasto-plastic range, which spreads around the excavation and is produced along helicoid surfaces that are generated inside the medium after it has plasticised.	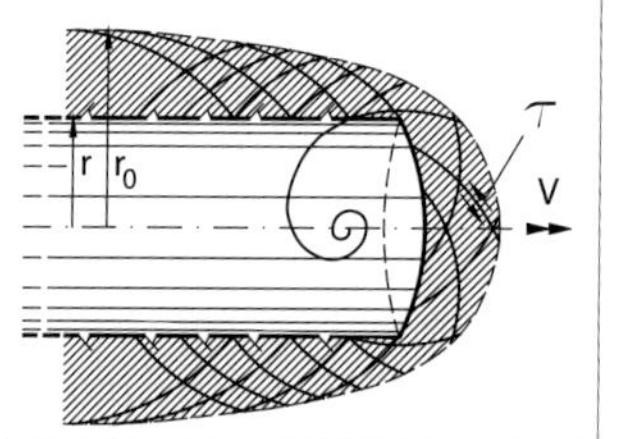	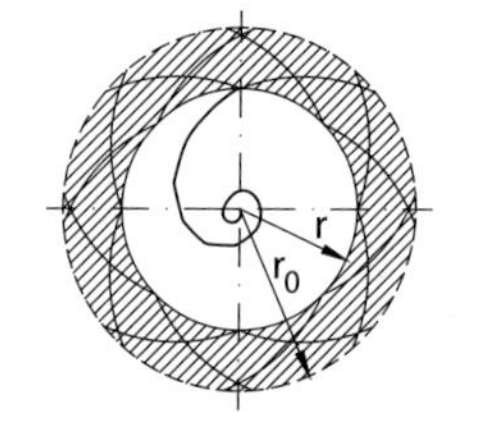

Let us now consider the three characteristic zones illustrated on page 8 and examine how the stress and deformation situation evolves in each of them.

1. **Undisturbed** zone characterised by:
   - natural stress field;
   - triaxial stress state at all points;
   - nil deformation.

2. **Face or transition zone** (corresponding to the radius of influence of the face Rf), characterised by:
   - disturbed stress field (variation in the stress state);
   - the stress state evolves from triaxial to biaxial (increase in the stress deviator);
   - increasing, immediate and negligible deformation if in the elastic range, deferred and large deformation if in the elasto-plastic range.

3. **Stabilisation zone** for deformation phenomena (if the design specifications implemented in the face zone were correct) characterised by:
   - equilibrium of the stress field restored;
   - biaxial stress state;
   - plane deformation state;
   - deformation phenomena at an end or ending.

Experimental measurements indicate that no less than 30% of the total convergence deformation produced in a given section of tunnel develops before the face arrives. It follows that the ground ahead of the face is the first to deform and that it is only after it has deformed that convergence of the cavity is produced. It also follows that the convergence measurements taken inside the tunnel only represent a part of the total deformation phenomenon that affects the medium.

the design engineer must focus his attention and within which the passage from a triaxial to a plane stress state occurs (the face or transition zone); proper study of a tunnel therefore requires three dimensional methods of calculation and not just plane methods.

1.4 The reaction

The **reaction** is the **deformation response** of the medium to the action of excavation. It is generated ahead of the face within the area that is disturbed, following the generation of greater stress in the medium around the cavity. It depends on the medium and its stress state (consistency) and on the way in which face advance is effected (action). It may determine the intrusion of material into the tunnel across the theoretical profile of the excavation. Intrusion is frequently synonymous with instability of the tunnel walls.

Three basic situations may arise (Fig. 1.6).

If on passing from a triaxial to a plane stress state during tunnel advance, the progressive decrease in the confinement pressure at the face ($\sigma_3 = 0$) produces stress in the elastic range ahead of the face, then the wall that is freed by excavation (the face) remains *stable* with *limited and absolutely negligible deformation*. In this case the channelling of stresses around the cavity (an "arch effect") is produced by natural means close to the profile of the excavation.

If, on the other hand, the progressive decrease in the stress state at the face ($\sigma_3 = 0$) produces stress in the elasto-plastic range in the ground ahead of the face, then *the reaction is also important* and the wall that is freed by excavation, the face, will deform in an elasto-plastic manner towards the interior of the cavity and give rise to a condition of *short term stability*. This means that in the absence of intervention, plasticisation is triggered, which by propagating radially and

Fig. 1.6

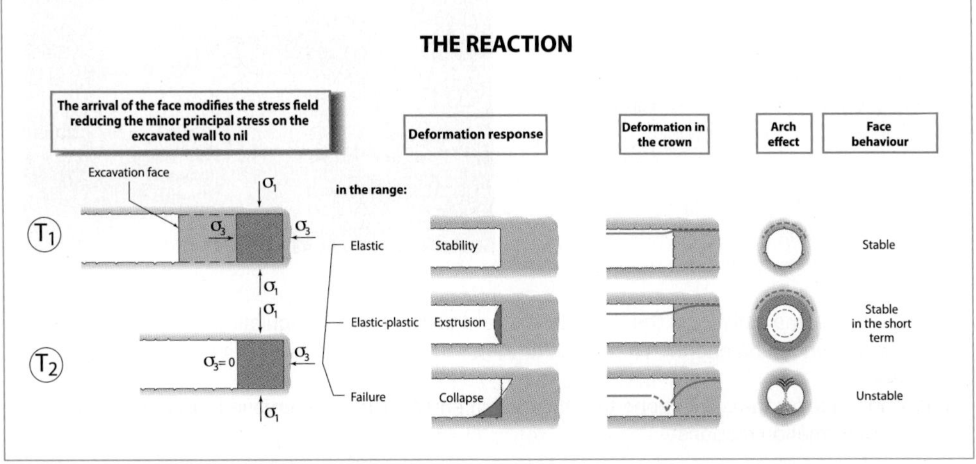

The three fundamental situations of stability

The behaviour of the medium at the face as a result of being 'deconfined' depends above all on its acquired consistency.

If the ***consistency*** is that ***of rock*** then rock type behaviour and therefore a stable face situation results.

If the ***consistency*** is that ***of clay*** (cohesive type behaviour), the face and the perimeter of the cavity deforms plastically intruding into the tunnel giving rise to a stable face in the short term situation.

If the ***consistency*** is that ***of sand*** (loose type behaviour) an unstable face situation results.

As we will see, the stability of the face plays a very decisive role in regulating and controlling deformation phenomena and therefore also for the short and long term stability of an underground construction.
It is in the face (or transition) zone that the design engineer must intervene to regulate and control the deformation response.

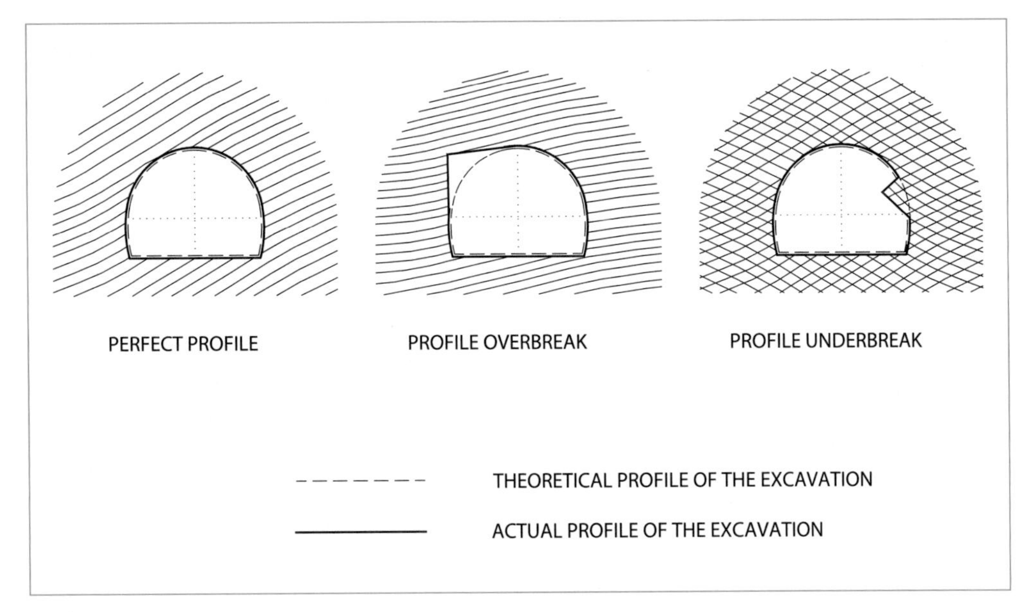

Fig. 1.7

longitudinally from the walls of the excavation, produces a shift of the "arch effect" away from the tunnel further into the rock mass. This movement away from the theoretical profile of the tunnel can only be controlled by intervention to stabilise the ground.

If, finally, the progressive decrease in the confinement pressure at the face ($\sigma_3 = 0$) produces stress in the failure range in the ground ahead of the face, then the *deformation response is unacceptable* and a condition of ***instability*** exists in the ground ahead of the face, which makes the formation of an "arch effect" impossible: this occurs in non cohesive or loose ground and an "arch effect" must be produced in it artificially since it cannot occur by natural means.

It therefore follows that it is important from a statics viewpoint to avoid over-break and to keep to the theoretical profile of the tunnel, especially in fractured and stratified rock masses. Accidental over-break, caused mostly by the geological structure of the ground, helps to shift the arch effect away from the walls of the cavity and this decreases the stability of a tunnel (Fig. 1.7).

However, the most important conclusion to be drawn is that the formation of an arch effect and its position with respect to the cavity (on which we know that the long and short term stability of a tunnel depend) are signalled by the quality and the size of the "deformation response" of the medium to the action of excavation.

The next chapter illustrates the evidence accumulated over the last twenty five years from research study on the relationships between changes in the stress state in the medium induced by tunnel advance and the consequent deformation response of the tunnel.

CHAPTER 2

The deformation response of the medium to excavation

2.1 The experimental and theoretical research

In the previous chapter, we carefully examined the statics and dynamics of tunnel advance to arrive at two important considerations:

1. the short and long term stability of an underground cavity is closely connected with the formation of an *arch effect*, which must therefore be the primary object of study for a tunnel designer;
2. the formation of an arch effect and its position with respect to the cavity are signalled by the "deformation response" of the medium to the action of excavation, in terms of both size and type.

When reasoning over these two important considerations around thirty years ago (it was 1975), we felt the need to conduct in-depth studies on the relations between the stress state in the medium, induced by tunnel advance (action), and the consequent deformation response (reaction). These studies were conducted as part of theoretical and experimental research, which, although still in progress, has already furnished important and very useful indications.

It was developed in three stages.

The **first research stage** was dedicated above all to systematic observation of the stress-strain behaviour of a wide range of tunnels during construction. Particular attention was paid to the behaviour of the face and not just that of the cavity as is normally done. Very soon, the complexity of the deformation response, what we were studying, became clear as did the consequent need to identify *new terms of reference* in order to be able to define it fully (Fig. 2.1):

- the **advance core**: the volume of ground that lies ahead of the face, virtually cylindrical in shape, with the height and diameter of the cylinder the same size as the diameter of the tunnel;
- **extrusion**: the primary component of the deformation response of the medium to the action of excavation that develops largely inside the advance core. It depends on the strength and deformation properties of the core and on the original stress field to which it is subject. It manifests on the surface of the face along the longitudinal axis of the tunnel and its geometry is either more or less axial symmetric (bellying of the face) or that of gravitational churning (rotation of the face);
- **preconvergence of the cavity**: convergence of the theoretical profile of the tunnel ahead of the face, strictly dependent on the relationship between the strength and deformation properties of the advance core and its original stress state.

Design and construction approaches for underground works: geomechanical classifications

Two main types of approach have been followed to date in the design and construction of underground works: one is mainly empirical, the other theoretical. Some authors working on the first type have proposed systems to assist design engineers in the design of tunnel stabilisation and lining works, which are based on geomechanical classifications.
Some of the most well known of these are those produced by Bieniawski (R.M.R. System) [2] and by Barton (Q System) [3]. Both identify geomechanical classes on the basis of a series of geomechanical and geostructural parameters. Stabilisation works which determine tunnel section designs are associated with each class. In theory it is possible to immediately select the most appropriate tunnel section type to ensure the long and short term stability of a tunnel by extrapolating the necessary parameters from core samples and direct measurements at the face.
Unfortunately, as the authors themselves have sometimes complained, an extremely distorted use of this type of classification has been made, as people have tried to use them as the basis for complete design and construction methods and not as a simple support tool for tunnel designers which the creators of the systems intended. It is interesting in this respect to quote some thoughts taken from an article written by Bieniawski and published in the July 1988 edition of the journal *Tunnels & Tunnelling* [4]: "When used correctly and for the purpose for which they were intended, rock mass classifications can be a powerful aid design. When abused, they can do more harm than good. ... Rock mass classifications are not to be taken as a substitute for engineering design. ... There are instances when rock mass classifications simply do not work."
When used for purposes other than those for which they were designed, geomechanical classifications, and as a consequence those design and construction methods that are based on them, such as the NATM, suffer from considerable shortcomings.
They are difficult to apply in the domains of soft rocks, flysch and soils, give insufficient consideration to the effects of natural stress states and the dimensions and geometry of an excavation on the deformation behaviour of a tunnel and fail to take account of new constructions systems. These constitute objective limitations, which make design and construction methods that are based on them inevitably incomplete and not universally valid.

THE FIRST PART OF THE TABLE CONCEIVED BY BIENIAWSKI TO EVALUATE THE R.M.R. INDEX									
PARAMETER			RANGE OF VALUES						
1	Strenght of intact rock material	Point-load strength index (mpa)	> 10	4 – 10	2 – 4	1 – 2	For this low range - uniaxial compressive test is preferred		
		Uniaxial compressive strength (mpa)	> 250	100 – 250	50 – 100	25 – 50	5 – 25	1 – 5	< 1
	Rating		15	12	7	4	2	1	0
2	Drill core quantity RQD (%)		90 – 100	75 – 90	50 – 75	25 – 50	25		
	Rating		20	17	13	8	3		
3	Spacing of discontinuities		> 2 m	0.6 – 2.0 m	200 – 600 mm	60 – 200 mm	< 60 mm		
	Rating		20	15	10	8	5		
4	Condition of discontinuities		Very rough surfaces Not continuous No separation Unweathered wall rock	Slightly rough surfaces Separation < 1 mm Slightly weathered walls	Slightly rough surfaces Separation < 1 mm Highly weathered walls	Slickensided surfaces or Gouge < 5 mm thick or Separation 1–5 mm Continuous	Soft gouge > 5 mm thick or Separation 1–5 mm Continuous		
	Rating		30	25	20	10	0		
5	Ground water	Inflow per 10 m tunnel length (L/min)	None	< 10	10 – 25	25 – 125	> 125		
		Ratio Joint water pressure / Mayor principal stress	0	< 0.1	0.1 – 0.2	0.2 – 0.5	> 0.5		
		General conditions	Completely dry	Damp	Wet	Dripping	Flowing		
	Rating		15	10	7	4	0		

Subsequently, during the **second research stage**, detailed analysis - above all in terms of timing – was performed on instability phenomena observed during the construction of at least 400 km of tunnel in an extremely wide range of ground types and stress-strain conditions. The aim was to seek a connection between the stress-strain behaviour of the core-face (extrusion and preconvergence) and that of the cavity (convergence).

Once we had established that the deformation response as a whole (extrusion, preconvergence and convergence) is systematically conditioned by the rigidity of the core of ground at the face as a function of the stress state acting on it (which is therefore the real cause of it), at a third stage, **the third research stage**, we worked to discover to what extent the deformation response of the cavity (convergence) could be *controlled* by acting on the rigidity of that core.

To do this, the stress strain behaviour of the advance core, systematically compared to that of the cavity, was analysed both in the absence and the presence of intervention to protect and to reinforce the advance core.

2.1.1 The first research stage

The first research stage (systematic observation of the deformation behaviour of the core-face) was conducted by using instruments and visual observation to monitor the stability and deformation behaviour of the advance core and walls of tunnels, with particular attention paid to the following phenomena (Fig. 2.1):

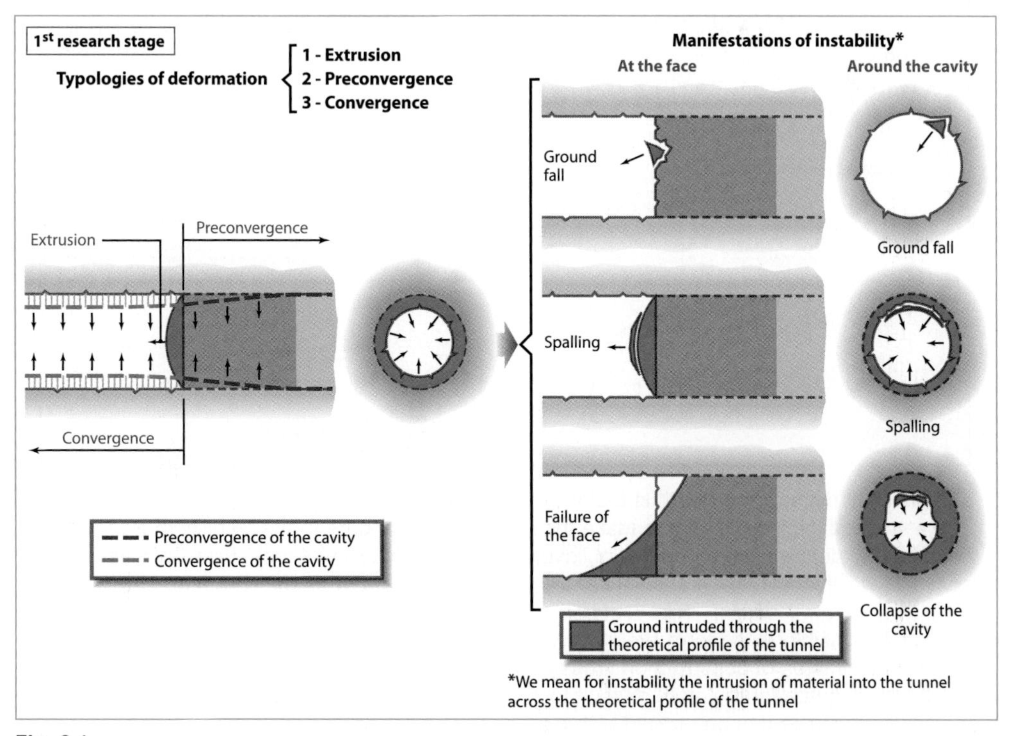

Fig. 2.1

The new Austrian method (NATM)

The New Austrian Tunnelling Method (NATM), developed between 1957 and 1965 by Pacher and Rabcewicz [5], which laid claim to the technological innovations of the *Sprayed Concrete Lining* or *SCL Method*, is a design and construction philosophy based on purely observational criteria. The starting point is a system for classifying rock masses based on a qualitative description of the conditions these present when an underground opening is made. The geomechanical parameters of the design, the excavation system (full or partial face) and the tunnel section type are associated to each rock type on an empirical basis and the final dimensions are in any case decided during construction on the basis of cavity convergence measurements.

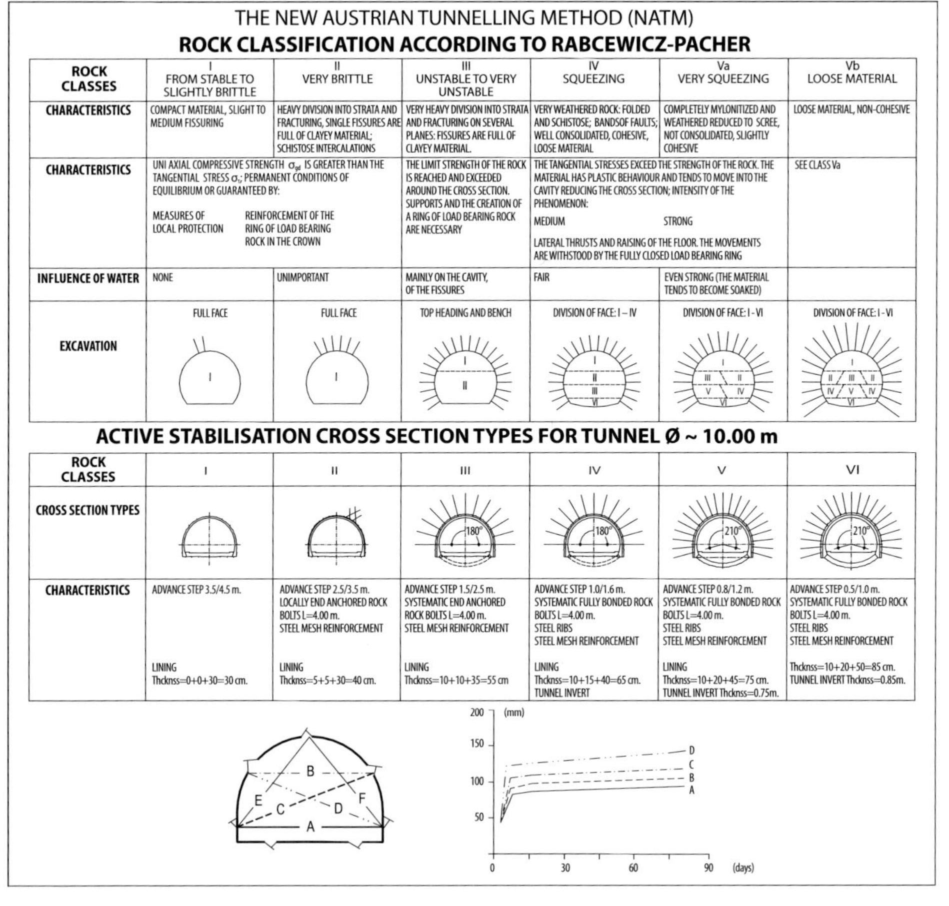

THE NEW AUSTRIAN TUNNELLING METHOD (NATM)

ROCK CLASSIFICATION ACCORDING TO RABCEWICZ-PACHER

ROCK CLASSES	I FROM STABLE TO SLIGHTLY BRITTLE	II VERY BRITTLE	III UNSTABLE TO VERY UNSTABLE	IV SQUEEZING	Va VERY SQUEEZING	Vb LOOSE MATERIAL
CHARACTERISTICS	COMPACT MATERIAL, SLIGHT TO MEDIUM FISSURING	HEAVY DIVISION INTO STRATA AND FRACTURING, SINGLE FISSURES ARE FULL OF CLAYEY MATERIAL; SCHISTOSE INTERCALATIONS	VERY HEAVY DIVISION INTO STRATA AND FRACTURING ON SEVERAL PLANES: FISSURES ARE FULL OF CLAYEY MATERIAL.	VERY WEATHERED ROCK: FOLDED AND SCHISTOSE; BANDSOF FAULTS; WELL CONSOLIDATED, COHESIVE, LOOSE MATERIAL	COMPLETELY MYLONITIZED AND WEATHERED REDUCED TO SCREE, NOT CONSOLIDATED, SLIGHTLY COHESIVE	LOOSE MATERIAL, NON-COHESIVE
CHARACTERISTICS	UNI AXIAL COMPRESSIVE STRENGTH σ_{gd} IS GREATER THAN THE TANGENTIAL STRESS σ_t; PERMANENT CONDITIONS OF EQUILIBRIUM OR GUARANTEED BY: MEASURES OF LOCAL PROTECTION / REINFORCEMENT OF THE RING OF LOAD BEARING ROCK IN THE CROWN		THE LIMIT STRENGTH OF THE ROCK IS REACHED AND EXCEEDED AROUND THE CROSS SECTION. SUPPORTS AND THE CREATION OF A RING OF LOAD BEARING ROCK ARE NECESSARY	THE TANGENTIAL STRESSES EXCEED THE STRENGTH OF THE ROCK. THE MATERIAL HAS PLASTIC BEHAVIOUR AND TENDS TO MOVE INTO THE CAVITY REDUCING THE CROSS SECTION; INTENSITY OF THE PHENOMENON: MEDIUM / STRONG. LATERAL THRUSTS AND RAISING OF THE FLOOR. THE MOVEMENTS ARE WITHSTOOD BY THE FULLY CLOSED LOAD BEARING RING		SEE CLASS Va
INFLUENCE OF WATER	NONE	UNIMPORTANT	MAINLY ON THE CAVITY, OF THE FISSURES	FAIR	EVEN STRONG (THE MATERIAL TENDS TO BECOME SOAKED)	
EXCAVATION	FULL FACE	FULL FACE	TOP HEADING AND BENCH	DIVISION OF FACE: I – IV	DIVISION OF FACE: I - VI	DIVISION OF FACE: I - VI

ACTIVE STABILISATION CROSS SECTION TYPES FOR TUNNEL Ø ~ 10.00 m

ROCK CLASSES	I	II	III	IV	V	VI
CROSS SECTION TYPES			180°	180°	210°	210°
CHARACTERISTICS	ADVANCE STEP 3.5/4.5 m. LINING Thcknss=0+0+30=30 cm.	ADVANCE STEP 2.5/3.5 m. LOCALLY END ANCHORED ROCK BOLTS L=4.00 m. STEEL MESH REINFORCEMENT. LINING Thcknss=5+5+30=40 cm.	ADVANCE STEP 1.5/2.5 m. SYSTEMATIC END ANCHORED ROCK BOLTS L=4.00 m. STEEL MESH REINFORCEMENT. LINING Thcknss=10+10+35=55 cm	ADVANCE STEP 1.0/1.6 m. SYSTEMATIC FULLY BONDED ROCK BOLTS L=4.00 m. STEEL RIBS STEEL MESH REINFORCEMENT. LINING Thcknss=10+15+40=65 cm. TUNNEL INVERT	ADVANCE STEP 0.8/1.2 m. SYSTEMATIC FULLY BONDED ROCK BOLTS L=4.00 m. STEEL RIBS STEEL MESH REINFORCEMENT. LINING Thcknss=10+20+45=75 cm. TUNNEL INVERT Thcknss=0.75m.	ADVANCE STEP 0.5/1.0 m. SYSTEMATIC FULLY BONDED ROCK BOLTS L=4.00 m. STEEL RIBS STEEL MESH REINFORCEMENT. Thcknss=10+20+50=85 cm. TUNNEL INVERT Thcknss=0.85m.

The principal merit of the NATM is that it explained the importance of using reinforcement and active stabilisation instruments to make the rock mass contribute to the stability of a tunnel by adapting the deformation properties of linings to the deformability of rock mass. Its main shortcomings, which now make it obsolete and incompatible with new tendencies, are:

- it is impossible to perform preliminary design of construction with it in enough detail to allow estimates of construction times and costs to be made that are sufficiently reliable;
- it is inadequate for the more difficult terrains and stress-strain conditions, which it erroneously presumes can be tackled with partial face tunnel advance;
- practical implementation suffers from too much subjectivity, the result of being based on what are essentially qualitative parameters.

a) *extrusion at the face,* which can manifest with either a more or less axial symmetric geometry (bellying of the face) or a gravitational churning geometry (rotation of the face), depending on the type of material and the existing stress state;
b) *preconvergence of the cavity,* understood as convergence of the theoretical profile of the tunnel ahead of the face, strictly dependent on the relationship between the strength and deformation properties of the advance core and its original stress state.
c) *convergence of the cavity* (which manifests as a decrease in the size of the theoretical cross section of the excavation after the passage of the face).

In addition to the systematic implementation of the well known measurements of cavity convergence at the walls of the tunnel or of the ground inside the rock mass, new types of experimental monitoring were studied, developed and implemented to achieve this, which enabled the deformation response of the medium to be studied in detail for a given section of tunnel *before, during and after* the arrival of the face, with particular attention paid to the face zone itself.

They consisted of *pre-convergence measurements*, performed from the surface using multi-point instruments to measure deformation (extensometers), which were inserted vertically into the ground above the crown and the springline of the tunnel to be constructed whenever the morphology of the ground and the depth of the overburden permitted.

In most cases preconvergence measurements were accompanied by measurements of advance core *extrusion* performed by inserting *a sliding micrometer* horizontally into the face, supplemented with line of sight measurements targeted on marks positioned on the face.

Systematic visual observation performed inside the cavity enabled the following *manifestations of instability* (instability is intended as occurring whenever material intrudes into an excavation beyond the theoretical profile) located on the face or around it to be associated with the types of deformation mentioned above:
a) rock fall, spalling and failure of the face in the core-face system;
b) rock fall, spalling and collapse of the cavity in the roof and tunnel wall zone.

2.1.2 The second research stage

Once the different types of deformation and manifestations of instability that occur on the *core at the face* and on the *roof and walls of a tunnel* had been identified, we asked ourselves whether observation of the former might in some way give us an indication of what the type and size of the latter might be. The second stage of the research thus commenced [to seek *possible connections* between the deformation of the core-face (→ extrusion and preconvergence) and that of the cavity (→ convergence)]. It was performed by studying, observing and monitoring deformation at the face and in the cavity, with particular attention paid to its magnitude and the chronological sequence of it in relation to the systems, stages and rates of construction adopted at different times.

It is essential to first briefly illustrate the observations we made on a few of the tunnels we ourselves designed before presenting the results of this experimental stage.

More on the NATM

In 1963 Rabcewicz changed the name from "the sprayed concrete lining method" to "the new Austrian tunnelling method" to claim the merits of this fascinating technological progress in the use of steel ribs, sprayed concrete and rock bolts in the construction of tunnels for "his country of origin". Subsequently he sought, together with other engineers (Pacher, Müller, etc.), to set the method in a scientific framework by formulating a series of principles which were not always found to be correct.

Thanks to the undoubted marketing abilities of the protagonists and the apparent simplicity of the basic concepts, the NATM enjoyed great popularity in most parts of the world between 1970 and 1990, despite the clear and severe shortcomings of the method for driving tunnels under stress-strain conditions that differ from the ordinary.

It has recently been recognised that the protagonists of the NATM had attributed a construction practice to themselves (the "sprayed concrete lining") which had been described in the technical literature since 1920, taking the key arguments from material already placed in the public domain by the international scientific community.

Directives [6] recently published by the UK Institution of Civil Engineers states that:

"The use of sprayed concrete support for a tunnel is often erroneously referred to as NATM. In view of this, and to avoid any confusion, this guide will generally use the description 'sprayed concrete linings (SCL)'".

The former professor of engineering in "Tunnel Construction" at the Polytechnic of Zurich, Kalman Kovári, who has demonstrated the lack of a scientific basis for the NATM in various scientific papers [7] [8], has observed that:

"this choice is a direct consequence of the pseudo-scientific character of the NATM".

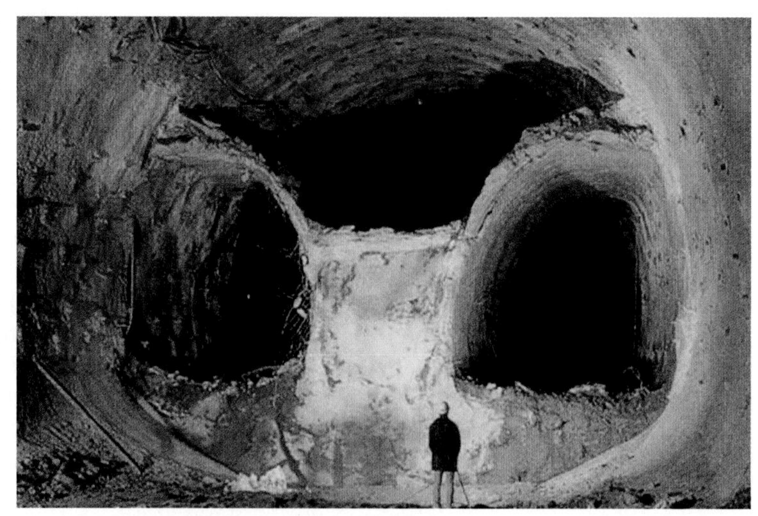

Typical tunnel advance following complete division of the face, advised by the NATM in "squeezing" or loose grounds

2.1.2.1 The example of the Frejus motorway tunnel (1975)

Ninety five percent of the length of the *Frejus* motorway tunnel (13 km long with overburdens of up to 1,700 m) passes through a metamorphous formation of schistose crystalline limestone, that is lithologically homogeneous along the tunnel alignment (Fig. 2.2).

The design of the tunnel was performed on the basis of a geological and geomechanical survey conducted from an adjacent railway tunnel (completed in 1860) and from service tunnels. The strength and deformation tests performed on samples of the schistose crystalline limestone gave the following average geotechnical results:

- angle of friction: 35°;
- cohesion: 30 kg/cm^2 (= 3 MPa);
- elastic modulus: 100,000 kg/cm^2 (= 10,000 MPa).

No forecasts of deformation behaviour were made in the original design for the tunnel (1975), because it was not standard practice to make them at that time. Account was taken of what was known of Sommeiller's experience during the construction of the adjacent railway tunnel about a century before and full face advance was decided with immediate stabilisation of the ring of rock around the cavity with end-anchored, tensioned rock bolts to a depth of 4.5 m supplemented with shotcrete. The final lining was in concrete, with an average thickness of 70 cm, cast afterwards to complete the tunnel.

The study and monitoring of deformation behaviour, which we conducted systematically, constituted the most significant part of the campaign of observations and measurements performed during construction. The purpose was to keep the response of the rock mass to the stabilisation intervention performed under control in the exceptional circumstance of driving a tunnel for the first time though a homogeneous rock mass (crystalline limestone) subject to a stress field which increased and varied as a function of the overburden (0 –1,700 m) [9].

Up to an overburden of 500 m, the rock mass remained stressed within the elastic range and the tunnel manifested *stable face* behaviour, with negligible deformation and limited instability at the face and in the cavity, consisting exclusively of rock fall.

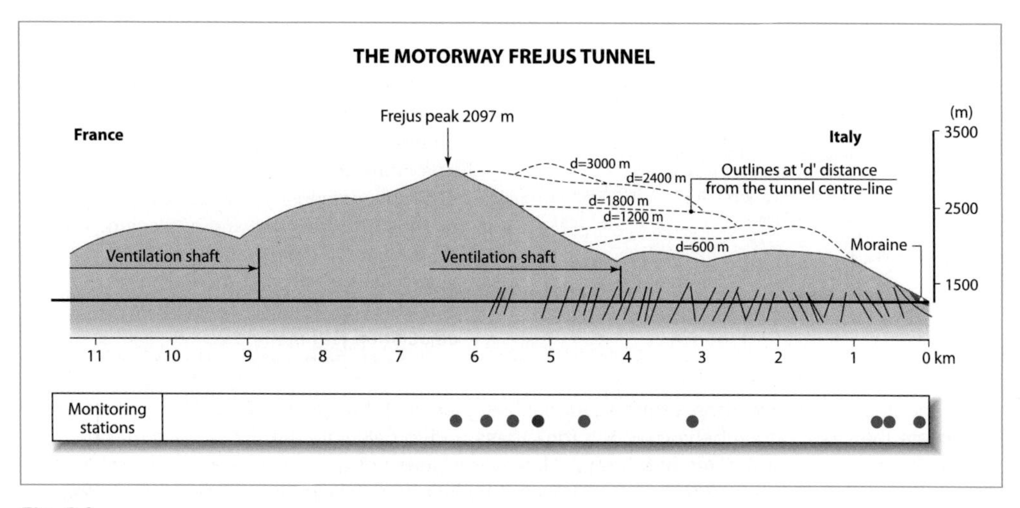

Fig. 2.2

Design and construction approaches for underground works: the theoretical approach

A distinctly *theoretical* approach has been used as an alternative to *empirical* approaches based on geomechanical classifications, which has produced interesting developments, even if it has so far been limited to the design phase only of underground works. It uses mathematics to describe the stress-strain behaviour of the ground and the lining structures as far as is possible in order to decide the dimensions of the latter. The most important results which this type of approach has generated are:

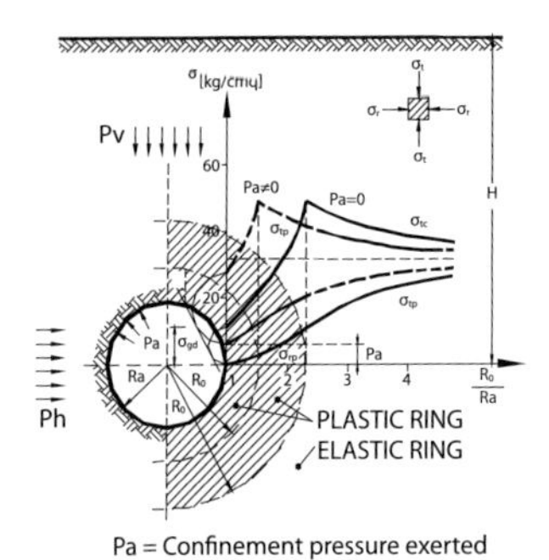

Pa = Confinement pressure exerted by the lining

definition of the concept of the confinement pressure of the cavity, with which the design engineer is able to control the extent to which the plasticised zone around the tunnel develops (plastic radius Rp), as demonstrated analytically for the first time by Kastner in 1962 [10];

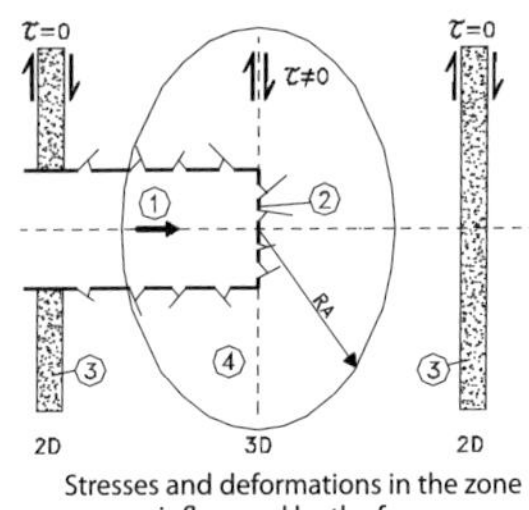

Stresses and deformations in the zone influenced by the face

① : direction of tunnel advance
② : face
③ : zones where the stress state is considered two dimensional
④ : zones where the stress state is three dimensional
RA : radius of action of the face

recognition that the problem of calculating the size of a tunnel lining is a completely three dimensional problem and that it is not acceptable to overlook this, especially when tunnel advance is through terrains in which stress states are high in relation to the strength and deformability of the ground [11];

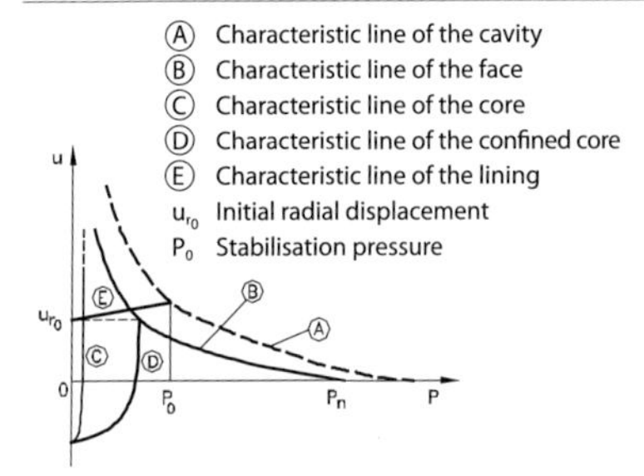

the demonstration that the pressure exerted by the surrounding rock mass on the stabilisation and lining structures of a tunnel is not predetermined, but depends, amongst other factors, on the method of excavation and of placing the structures themselves (as can be seen from the "characteristic line" [11] and the "convergence-confinement" calculation [12] methods).

The theoretical approach has furnished tunnel designers with the calculation tools they need to evaluate the stress-strain behaviour of a rock mass and to calculate the dimensions of the stabilisation and lining structures of a tunnel. However it does not give adequate consideration to the construction stages and therefore it does not constitute a fully integrated method of design and construction.

As the overburden and consequently the stress state increased, the rock mass entered the elasto-plastic range and tunnel *behaviour* became *stable face in the short term*, with convergence of the cavity measurable in the order of decimetres (diametrical convergence 10 – 20 cm). The band of bolt reinforced rock contributed effectively towards the statics of the tunnel, limiting convergence and preventing the appearance of consequent instability.

Assisted by the good quality of the rock, tunnel advance proceeded at rates of 200 m per month until work halted temporarily at metre 5,173 for the summer holiday break, in a zone of homogeneous rock with an overburden of approximately 1,200 m.

Convergence measurement station No. 6, installed immediately 1 metre back from the face (metre 5,172), recorded maximum deformation of approximately 10 cm, after advance had halted for 15 days (Fig. 2.3). It was undoubtedly *fluage* deformation (with a constant load) only, since the face had remained at a complete halt during that time.

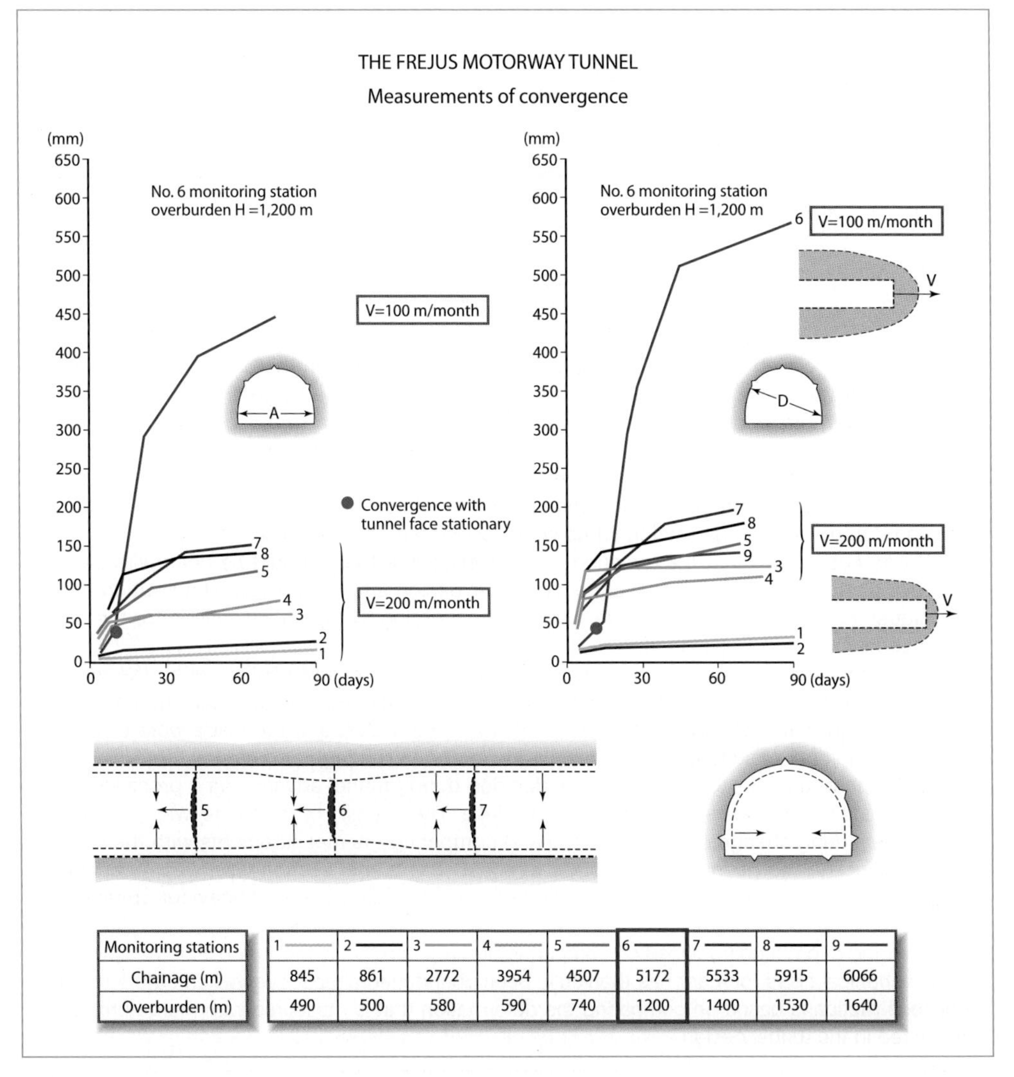

Monitoring stations	1	2	3	4	5	6	7	8	9
Chainage (m)	845	861	2772	3954	4507	5172	5533	5915	6066
Overburden (m)	490	500	580	590	740	1200	1400	1530	1640

Fig. 2.3

The system typically used in Italy in the 1970's for the design and construction of tunnels

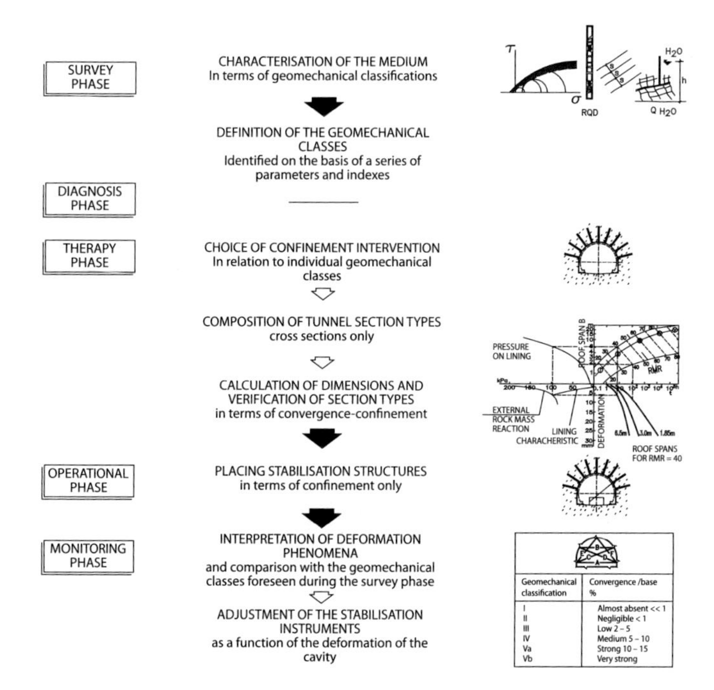

- As recommended by the NATM, the design was conducted on the basis of the results of a geological survey campaign, often very limited, and of a characterisation of the medium for the purpose of subsequent geomechanical classification.
- Each section of tunnel was placed in classes on the basis of the geostructural characteristics of the rock mass to be excavated and the geomechanical characteristics, to which the values of determined geomechanical parameters, determined convergence values and determined tunnel section types were associated on an empirical basis.
- It was considered that the question of the statics of a tunnel could be dealt with clearly as a plane problem; as a consequence tunnel design employed radial stabilisation techniques only (ribs, bolts, shotcrete, concrete) according to the specifications for the geomechanical class in question.
- The dimensions of tunnel cross sections were calculated and verified using the wedge thrust method proposed by Rabcewicz or the hyperstatic reaction method, employing the well known calculation programmes STRESS and STRUDL and the loads were obtained from empirical or semi-empirical calculation methods (Terzaghi, Kommerel, etc.).
- The type of tunnel section to use was decided during tunnel advance as it proceeded on the basis of geomechanical surveys of the face. The stability of the tunnel was monitored by measuring changes in convergence and comparing the measurements with the values specified for the geomechanical class forecast. The tunnel section type was redesigned each time it was necessary, as a function of the actual deformation behaviour observed.

This method of proceeding based on comparing deformation that is measured with the specified geomechanical class forecast has thankfully been abandoned almost everywhere. Modern tunnel construction is based on comparing the deformation that is measured with the deformation predicted in the tunnel design.

When advance recommenced, diametric convergence for that cross section increased very sharply to values never before encountered, reaching 60 cm after 3 months, while further ahead it returned to normal values (diametrical convergence of 20 cm) after 20 or 30 metres as tunnel advance continued.

It is important to point out here that before work stopped, the tunnel had been reinforced up to one metre from the face with more than 30 radial rock bolts per linear metre of tunnel, but there had been no intervention on the advance core at the face. Once tunnel advance resumed, stabilisation procedures around the cavity continued with the same intensity and at the same rate as before.

The conclusion drawn was that while tunnel advance was halted, the core of ground at the face had had time to extrude into the elasto-plastic range since it had not been assisted by any reinforcement and this had triggered a phenomenon of relaxation of the stresses in the rock mass around it (preconvergence) due to creep. This in turn had then caused the considerable increase in convergence of the cavity compared to normal values.

2.1.2.2 The example of the "Santo Stefano" tunnel (1984)

The *Santo Stefano* tunnel is on the new route of the twin track Genoa-Ventimiglia railway line located on the section between S. Lorenzo al Mare and Ospedaletti.

It runs through the Helminthoid flysch formation characteristic of western Liguria. It consists of clayey and clayey-arenaceous schists with thin banks of folded and intensely fractured sandstones and marly limestones. The clay schist component is heavily laminated. The passage between the H2 unit and the more marly-limestone-like H1 unit of the formation (Fig. 2.4) is marked by an extremely tectonised transition zone. The tunnel overburden in that zone varies between 40 m and 80 m.

Strength tests carried out in the laboratory on samples gave angle of friction values varying between 20° and 24° and cohesion from 15 kg/cm^2 (= 1.5 MPa) to 0.

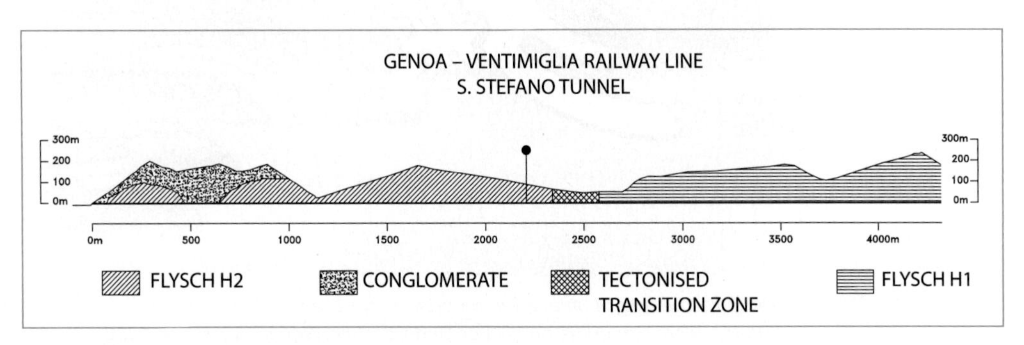

Fig. 2.4

Fundamental requirements

What requirements should a tunnel design and construction method have to meet to be considered consistent with reality and universally valid?
It should certainly have to satisfy at least the following four fundamental requirements:

1. the method must be valid and applicable in all types of ground and under all stress conditions;
2. the method must provide design and construction instruments that are able to solve the problems of different statics conditions in all types of ground;
3. the method must distinctly separate the design stage from the construction stage, although allowing for refinements to be made to the specifications during construction;
4. the method must allow projects to be reliably planned in terms of construction schedules and costs.

None of the design and construction approaches which have been employed to date have been able to meet these four fundamental requirements. Even the NATM, which for many years was considered, rightly or wrongly, the industry standard, has not furnished satisfactory answers and in some ways has been actually misleading.
In order to formulate a modern and universally valid approach, the problem of the design and construction of underground works must be radically reformulated. In order to achieve this, one must clear one's mind of preconceived ideas and assumptions that govern the existence of an underground cavity and examine the phenomena that occur during the excavation of a cavity in terms of cause and effect.

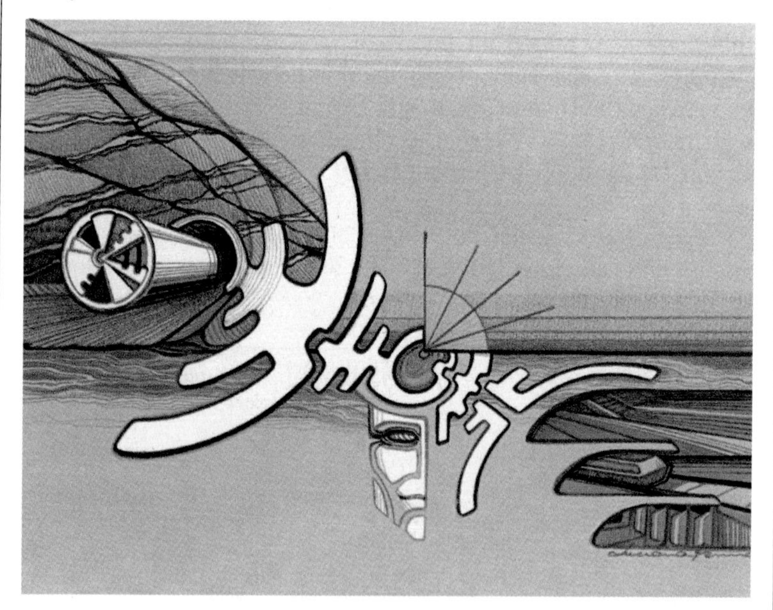

Work by the artist Luciana Penna created for the manifesto of the International Congress on "Major Underground Works", Florence 1986

In this case too no forecasts of tunnel deformation behaviour were made when work started in 1982.

The original design was for full face advance with steel ribs and shotcrete for the primary lining and a thick ring of concrete (up to 110 cm) for the final lining.

During excavation it was found that as long as tunnel advance proceeded under elastic conditions, deformation at the face and in the tunnel was negligible and there were practically no manifestations of localised instability (*stable face* behaviour). When the tunnel entered a zone affected by residual stress states of tectonic origin, where ground conditions were in the elasto-plastic range, manifestations of deformation began to cause some difficulty and this was also because sizeable asymmetrical thrusts appeared caused by rigid masses dispersed in the plastic matrix. As this occurred, layers of material were seen to break off at the face, a sure sign of extrusion-type movement typical of a *stable face in the short term* situation, while convergence reached decimetric values.

At a certain moment, since the stress state of the rock mass had obviously developed into the failure range, the entire face failed (*unstable face* situation), followed within a few hours by the collapse of the cavity with diametric convergence of more than 2 m even in the part already stabilised with steel ribs and shotcrete for a considerable stretch of more than 30 m back from the face (Fig. 2.5).

Fig. 2.5
Collapse of the cavity (S.Stefano Tunnel)

It should be noted at this point, that the type of ground encountered in the three stress-strain situations mentioned was essentially the same and that the one instance of tunnel collapse which occurred with convergence measurable in metres even in a part of the tunnel where the primary lining had already been placed, only happened when there was no longer any contribution from a rigid core at the face.

The research

Having considered that:

① THE SHORT AND LONG TERM STABILITY OF A CAVITY IS CLOSELY CONNECTED WITH THE FORMATION OF AN ARCH EFFECT, WHICH MUST THEREFORE BE THE PRIMARY OBJECT OF STUDY FOR A TUNNEL DESIGNER

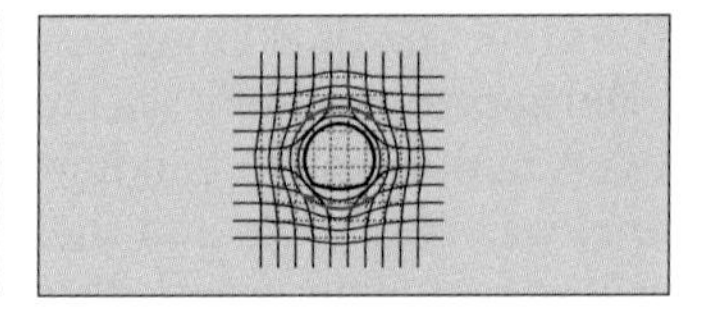

② THE FORMATION OF AN "ARCH EFFECT" AND ITS POSITION WITH RESPECT TO THE CAVITY ARE SIGNALLED BY THE "DEFORMATION RESPONSE" OF THE MEDIUM TO THE ACTION OF EXCAVATION, IN TERMS OF SIZE AND TYPE

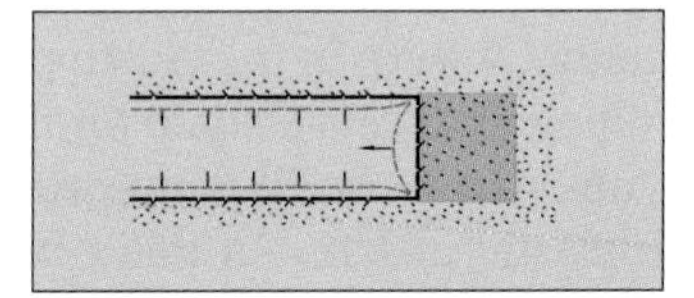

experimental and theoretical research was started in 1975 with the aim of throwing light on the relationships between modifications to the stress state of the medium induced by tunnel advance (the action) and the consequent deformation response (the reaction).

The research was conducted in three stages:

① SYSTEMATIC OBSERVATION OF THE DEFORMATION BEHAVIOUR OF THE CORE-FACE AND NOT JUST OF THE CAVITY

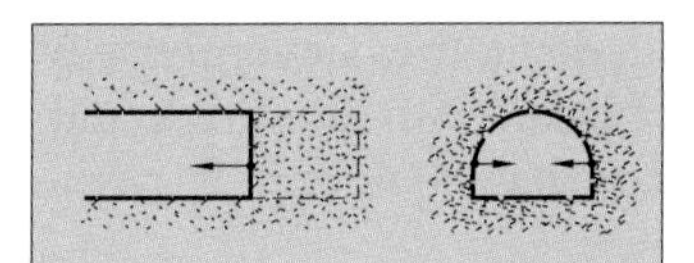

② VERIFICATION OF THE EXISTENCE OF CONNECTIONS BETWEEN THE DEFORMATION BEHAVIOUR OF THE CORE-FACE AND OF THE CAVITY

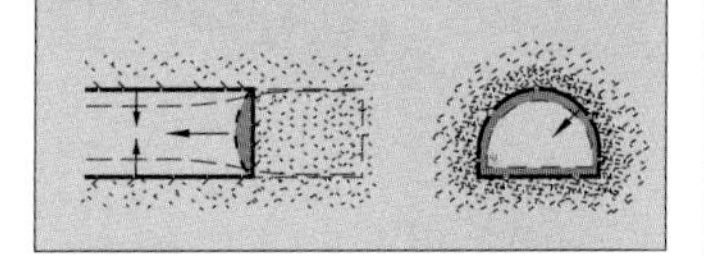

③ VERIFICATION OF HOW THE DEFORMATION RESPONSE OF THE CAVITY CAN BE CONTROLLED BY ADJUSTING THE RIGIDITY OF THE CORE-FACE

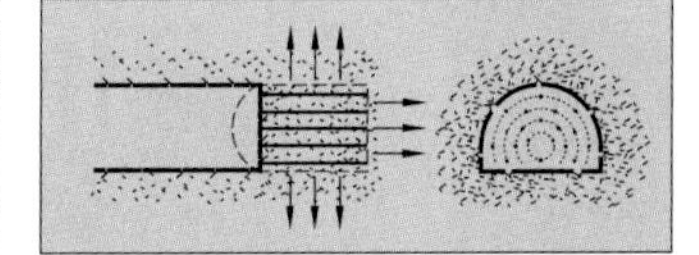

and, although it is still in progress, it has already furnished important and very useful results.

2.1.2.3 The S. Elia tunnel (1985)

The *S. Elia* tunnel, which forms part of the Messina-Palermo motorway, enters a flyschoid series of sedimentary origin which alternates between clayey-arenaceous and marly-arenaceous, after passing through rather coarse detritus material on the eastern side. The change occurs across a brief transition zone consisting of heavily fractured rock (Fig. 2.6).

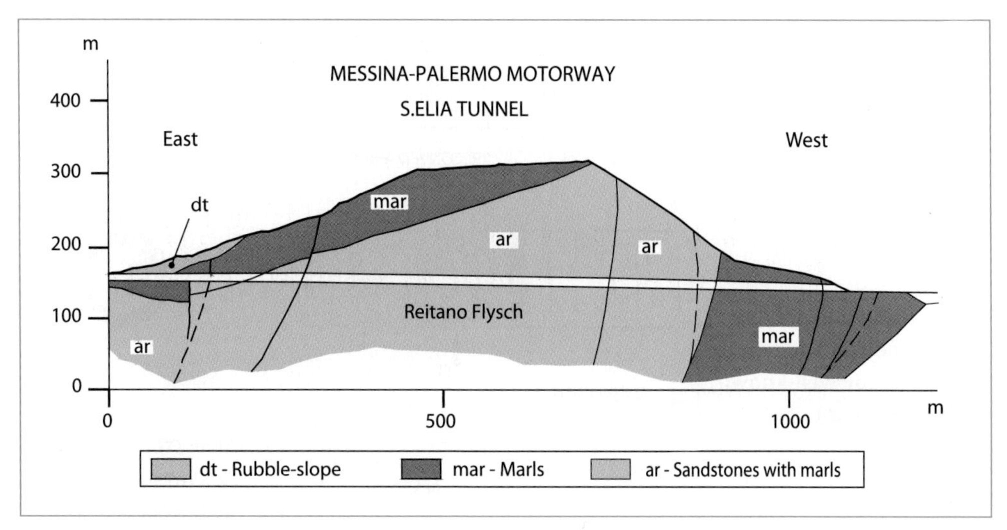

Fig. 2.6

The conventional design was for half face advance after ground improvement in advance consisting of jet-grouting around the tunnel in the detritus zones and the placing of ribs and shotcrete in the more consistent rock. The final lining, in concrete, with a tunnel invert to close it was placed afterwards.

The tunnel passed through the detritus material with no problems thanks to the ground reinforcement in advance and then entered the apparently better formation without any particular complications. The face appeared stable in the short term with limited deformation phenomena. On penetrating further into the transition zone, a water table under pressure was intercepted which rapidly produced instability at the face. A few hours after the failure of the face, the perimeter of the half cross section, already lined with shotcrete and ribs, failed for some tens of metres back from the face with radial convergence of more than one metre (Fig. 2.7) [13].

1ST RESEARCH STAGE – A new framework of reference

The need to identify a *new framework of reference* to define the deformation response of the rock mass to the action of excavation in its full complexity emerged clearly from the outset in the first research stage.

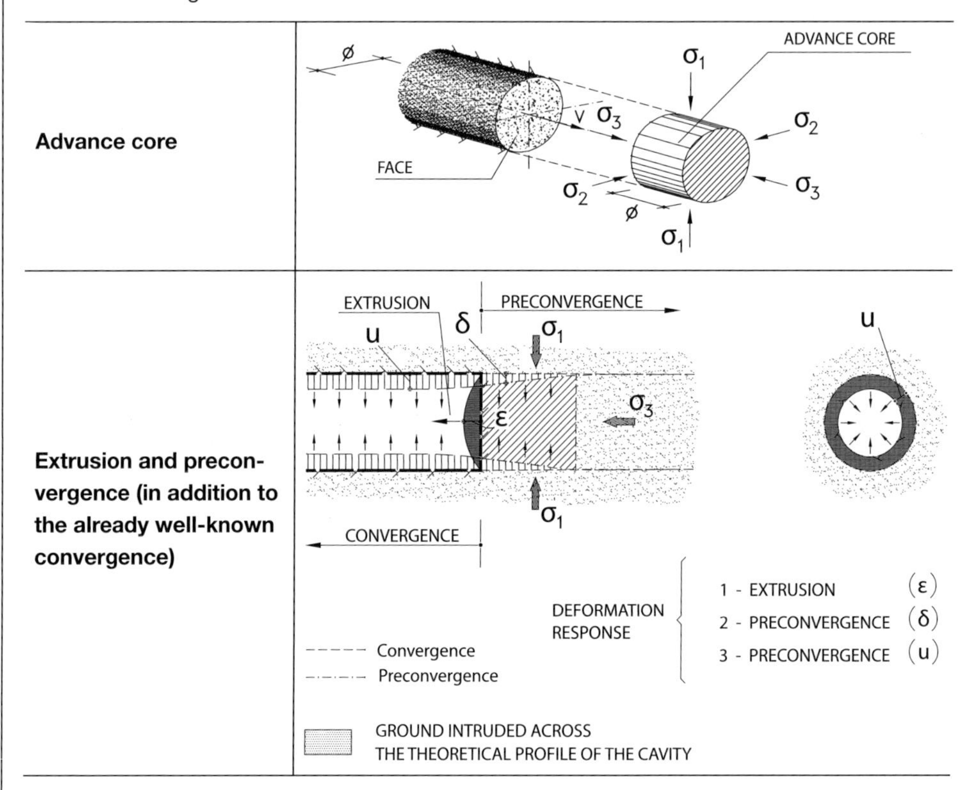

Manifestations of instability

Observations made in the tunnel allowed the resulting manifestations of instability to be associated with each type of deformation:

Location	Manifestations of instability		
Core-face	*Rock fall*	*Spalling*	*Failure of the face*
Around the cavity	*Rock fall*	*Spalling*	*Collapse of the cavity*

Fig. 2.7
Collapse of the cavity (S. Elia Tunnel)

2.1.2.4 The example of the "Tasso" tunnel (1988)

The *Tasso* tunnel is one of a series of tunnels excavated towards the end of the 1980's for the new "Direttissima", high speed Rome to Florence railway line. The area in which the tunnel is located lies in the lake basin of the *Valdarno Superiore* and consists of silty sands and sandy silts interbedded with silty clay levels containing sandy lentils and levels saturated with water (Fig. 2.8).

The original State Railways design involved *half face* advance and stabilisation of the walls with steel ribs and shotcrete. The ribs were anchored at the feet with sub-horizontal tie bars and given a foundation of micropiles or columns of ground improved by jet-grouting. Initially excavation proceeded under *stable face in the short term* conditions with no appreciable deformation phenomena either at the face or in the tunnel.

As the overburden increased, and therefore also the stress state of the medium, and given also the poor geomechanical characteristics of the ground, conditions of *stable face in the short term* rapidly changed to those of an *unstable face*.

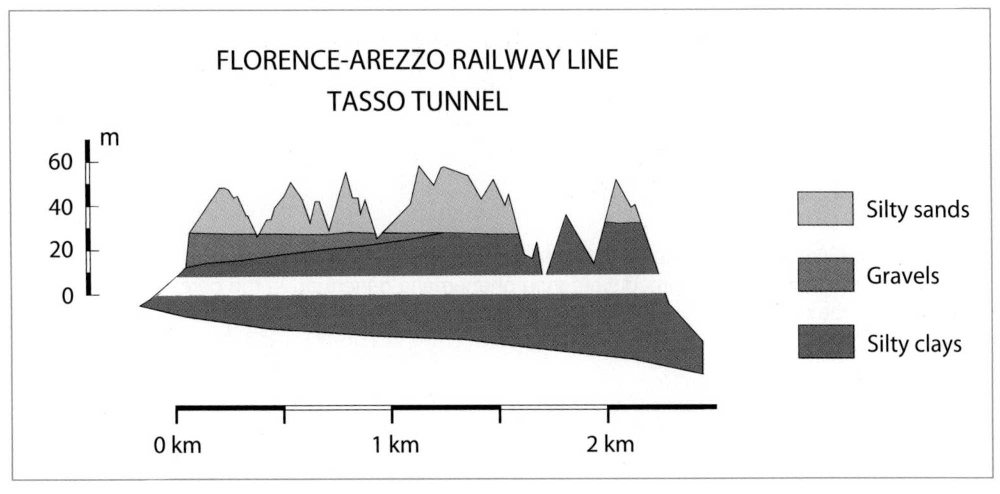

Fig. 2.8

2ND RESEARCH STAGE – Connections

The second research stage consisted of verifying the existence of **connections** between the deformation behaviour of the core-face system (determined by the lesser or greater development of extrusion and preconvergence phenomena) and that of the cavity (convergence). To achieve this, core-face and cavity deformation events were observed and monitored with particular attention paid to the size and chronological sequence as well as to their relationship to the method of advance employed.
The following emerged from these observations and monitoring activities:

- other conditions remaining unchanged, the size of convergence in a given section of tunnel is proportional to the size of extrusion and preconvergence phenomena generated previously in the same section (and therefore to the rigidity of the core-face on which these depend);
- normally the collapse of the cavity follows the failure of the face.

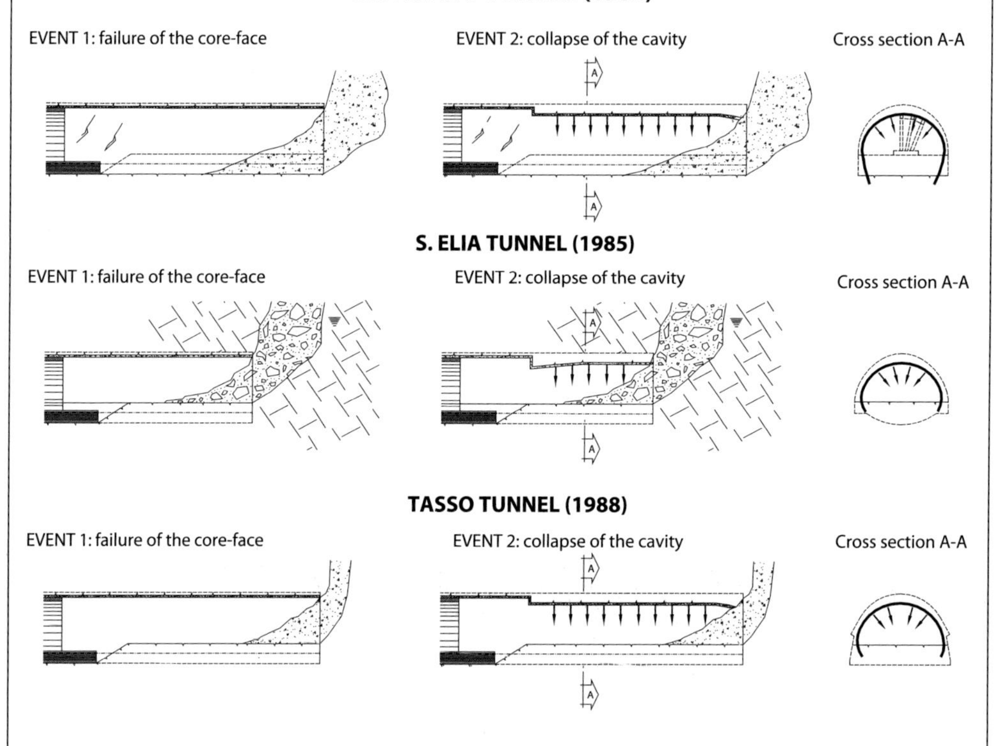

Important confirmation was therefore found for the hypothesised existence of close connections between extrusion and preconvergence phenomena generated in the core-face and convergence of the cavity.

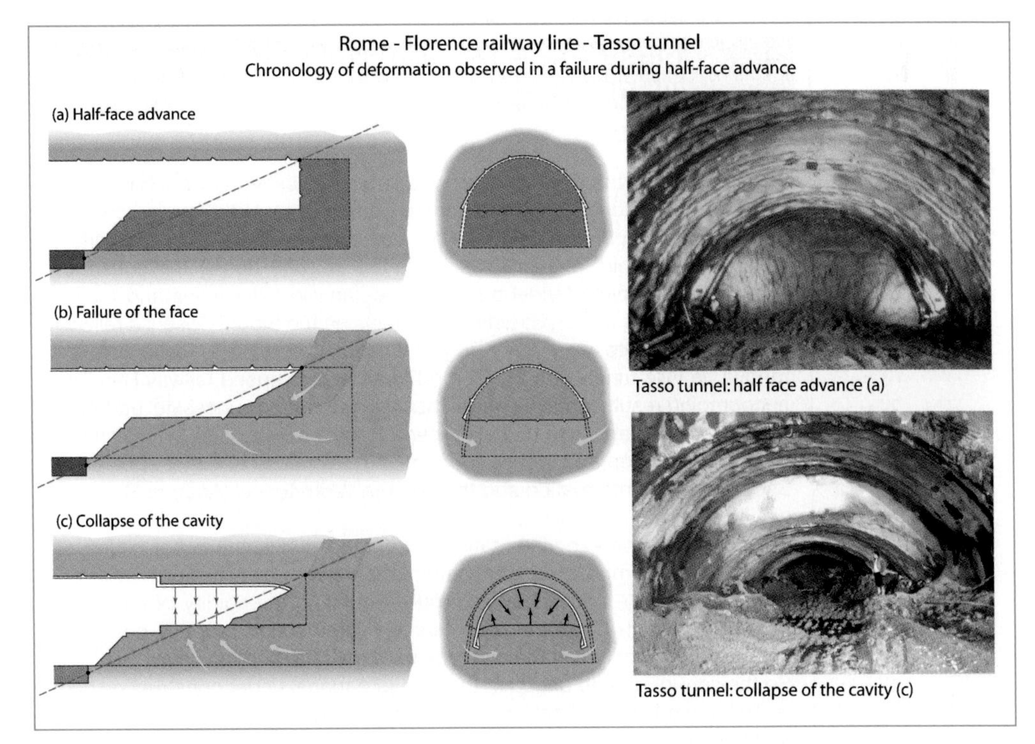

Fig. 2.9

After the failure of the face, despite half-face advance, approximately 30-40 m of tunnel already excavated and protected with ribs and shotcrete also collapsed during the course of one single night with convergence in the order of 3–4 m (Fig. 2.9).

2.1.2.5 The results of the second research stage

Study and analysis of the cases illustrated and other similar cases, which it would take too long to describe here, produced many ideas of great interest. The following points appeared clearly from the Frejus study:

- when advancing through ground in elasto-plastic conditions it is very important to keep excavations rates high and constant in order to avoid giving the core time to deform. This prevents extrusion and preconvergence, which constitute the starting point of subsequent convergence of the cavity, from being triggered.

What also emerged from the other experiences cited and similar cases was that:

- the failure of the core is generally followed by the collapse of the cavity and it is very rare for the latter to be preceded by the former.

3RD RESEARCH STAGE – the face-core, a new instrument for control and stabilisation

The purpose of the third research stage was to ascertain to what extent it was possible to control the deformation response of the cavity (convergence) by acting on the rigidity of the core-face system. From a historical viewpoint, the third research stage started in 1983 during the construction of the Campiolo tunnel on the Udine-Tarvisio railway line and then it continued during the construction of tunnels on the Sibari–Cosenza railway line and the Talleto, Caprenne, Poggio Orlandi, Crepacuore, Tasso and Terranova Le Ville tunnels on the Rome-Florence high speed railway line. The encouraging results obtained were then verified during successful attempts to salvage the San Vitale tunnel (Caserta Foggia railway line), on which excavation had been abandoned for some time because of huge, apparently uncontrollable deformation and then on the Vasto tunnel (Ancona-Bari railway line).
The results of the experimentation conducted during the construction of these tunnels demonstrated that deformation of the cavity can be controlled and substantially reduced by artificially regulating the rigidity of the core-face system with adequate reinforcement intervention properly designed and distributed between the core ahead of the face and the cavity.
So the core of ground ahead of the face of a tunnel under construction can really be used as an effective instrument for stabilising a cavity in the short and long term.

The core-face in clayey silts of the Tasso tunnel after treatment with fibre glass reinforcement.

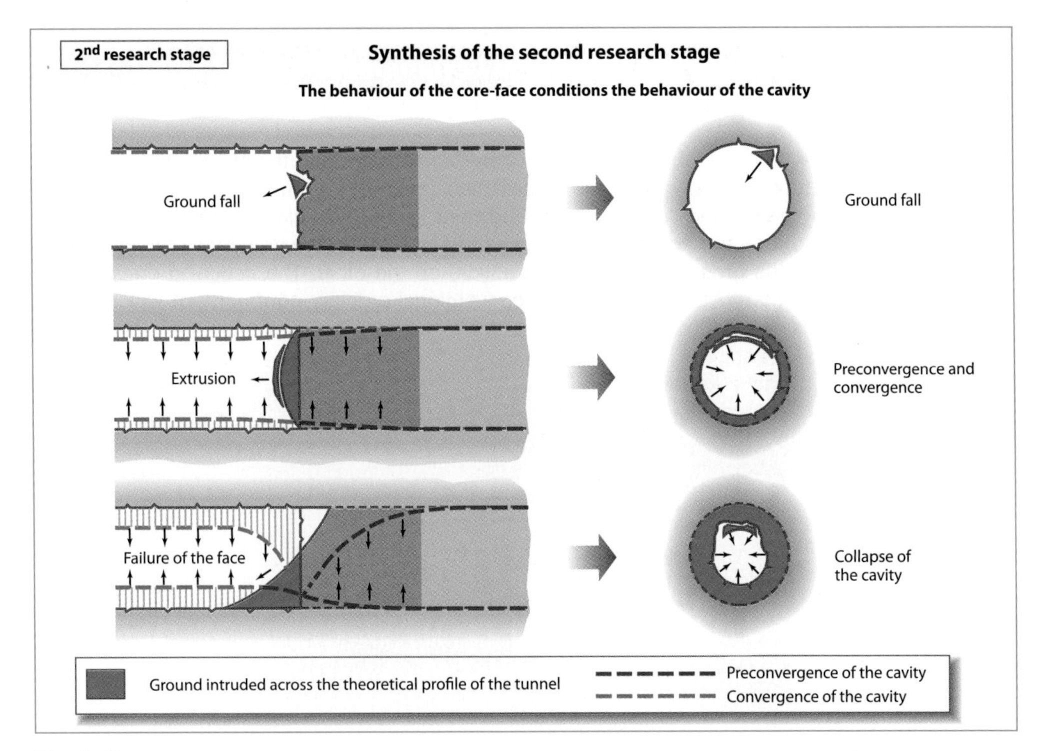

Fig. 2.10

The following was therefore clear from the second research stage (Fig. 2.10):

1. there is a close connection between extrusion of the advance core at the face and preconvergence and convergence of the cavity;
2. there are close connections between the failure of the core-face and the collapse of the cavity even if it has already been stabilised;
3. chronologically deformation in the cavity normally follows and is dependent on deformation in the core of ground at the face.

It was also clear that it was necessary for an arch effect, which as we know conditions the stability of a tunnel, to have already been triggered ahead of the face in order to be able to continue to function in a determined cross section after the face has already passed ahead of it.

2.1.3 The third research stage

The results of the second research stage reinforced the impression, which we already had, that the deformation of the advance core of a tunnel was the true cause of the whole deformation process in all its components (extrusion, preconvergence and convergence) and that

Action for preconfinement and confinement of the cavity

How was it possible to adjust the rigidity of the core-face system and therefore verify whether this actually made it possible to control the deformation response (*convergence*) of the cavity? It was achieved by generating "cavity **preconfinement**" action designed to release the core from excess stress *(protective action)* on the one hand and to conserve or improve the strength and deformability of the ground (reinforcement action) on the other. The action to preconfine the cavity is thus termed to distinguish it from conventional ordinary **confinement** which acts around the cavity inside the tunnel.

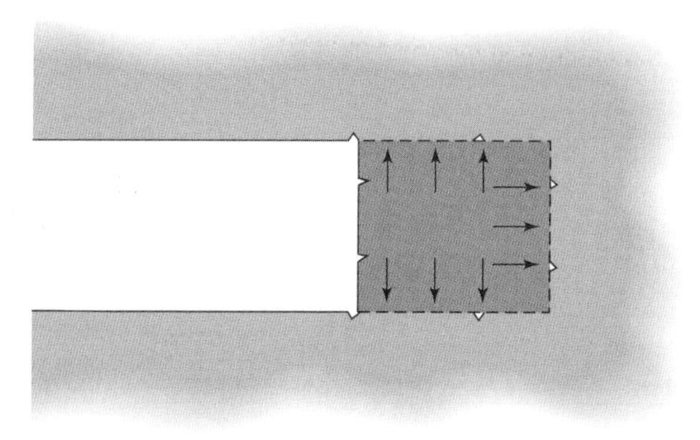

Preconfinement action

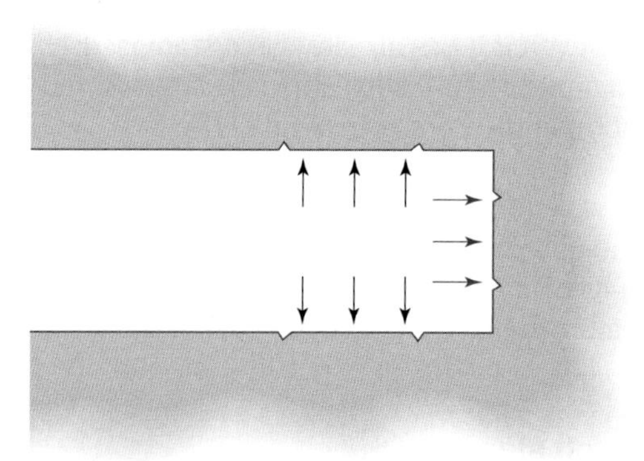

Confinement action

Fig. 2.11

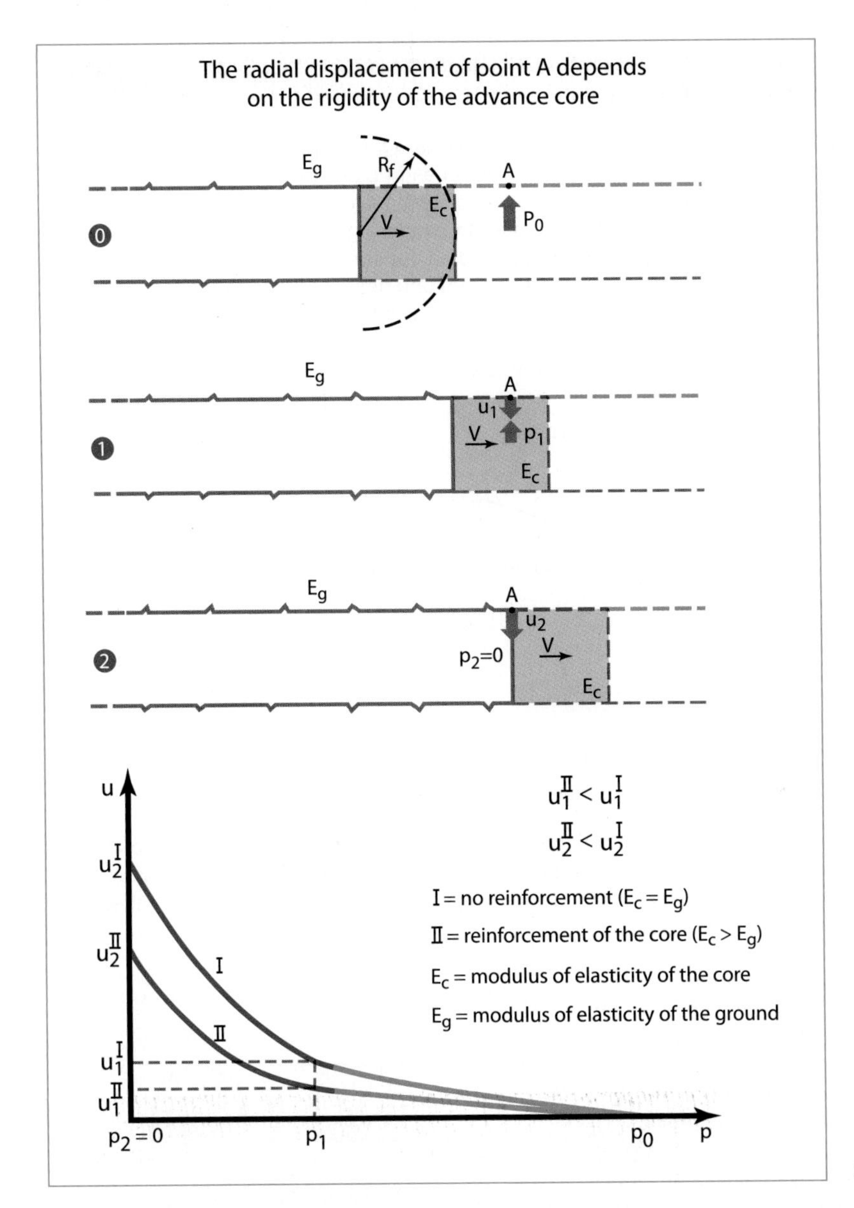

Fig. 2.12

as a consequence, the rigidity of the core played a determining role in the stability of a tunnel both in the long and short term.

If one takes a point A, located on the profile of the crown of a tunnel still to be excavated, it is in fact quite clear that its radial movement u, (preconvergence) as the face approaches it, depends largely on the strength and deformation properties of the ground inside the profile of the future tunnel.

If the course of its radial movement is plotted on a graph p-u (where p is the confinement pressure exerted radially on A), it can be seen (Fig. 2.12) that while the face is still distant (distance from A greater

Conservative intervention

Preconfinement of a cavity can be achieved using various types of intervention depending on the type of ground (natural consistency), the stress states in play and the presence of water. They are all relatively recent types of intervention which, because of the action that they perform ahead of the face (designed to prevent the rock mass from relaxing and to conserve the principal minor stress σ_3 at values greater than zero), are defined as **conservative**.
The "conservative" action exerted by this intervention can be described on the Mohr plane as a line, termed, as one might expect, the *conservative line*, which represents the limit below which the principal minor pressure σ_3 must not be allowed to fall, in order to maintain the intrinsic curve of the ground as close as possible to the peak values and not lose control of the deformation response of the rock mass.

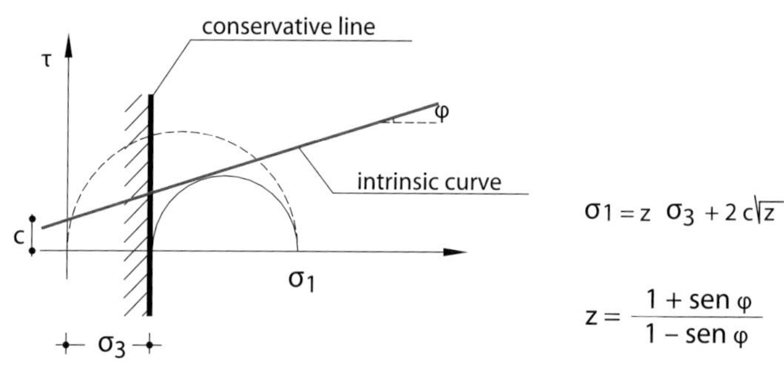

This intervention, which complements ordinary conventional confinement of the face and the cavity [insofar as their effectiveness is dependent on the continuity and regularity of the passage from action to preconfine the cavity (ahead of the face) and that to confine the cavity (after the passage of the face)] can be divided into:

- *conservative protective intervention,* when it channels stresses around the outside of the face performing, as the term says, a protective action which conserves the natural strength and deformation characteristics of the ground (e.g. shells of improved ground formed by means of sub-horizontal jet-grouting, shells of fibre reinforced shotcrete or concrete created in advance by either mechanical pre-cutting or pretunnel techniques);
- *conservative reinforcement intervention*, when it acts directly on the consistency of the advance core, improving its natural strength and deformation characteristics by means of appropriate reinforcement techniques (e.g. reinforcement of the core using structural fibre glass elements).

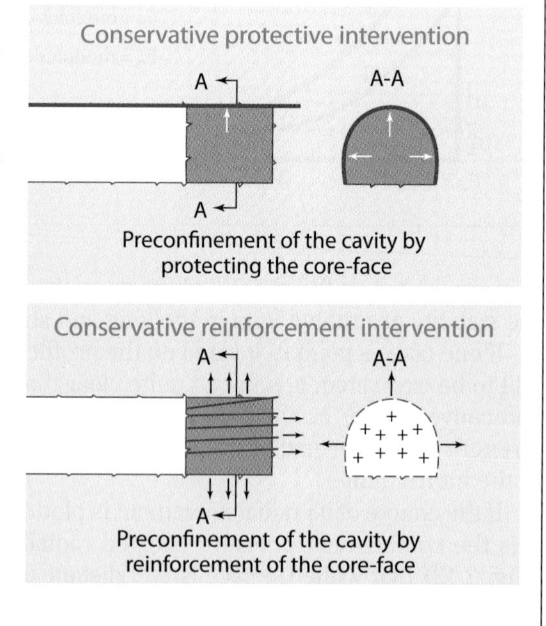

than the radius of influence of the face R_f), the stress condition of A remains unchanged (radial confinement pressure p_0 = original pressure).

As the face approaches, however, the thickness of the core between A and the face narrows and the radial confinement pressure p also diminishes as a consequence. The point A will then start to move radially towards the inside of the future cavity. How far it moves depends, as we have said, and as is quite obvious, on the stresses present and on the deformation properties of the core, which determines its equilibrium, and not just on the geomechanical characteristics of the surrounding ground.

After the passage of the face, on the other hand, the radial movement of A will continue in either the elastic or elasto-plastic range, as a function of the pre-existing stress states, the characteristics of the ground around the tunnel and the radial confinement pressure exerted by stabilisation structures (preliminary lining, final lining) on which the equilibrium of point A depends.

The qualitative graph in Fig. 2.12 shows the course of the deformation to which A is subject, other conditions remaining unchanged, for a deformable advance core (curve I) and for a rigid advance core (curve II). Obviously the radial deformation to which point A is subjected as the radial confinement pressure p decreases, before the passage of the face, is less for a rigid core than for a deformable core.

It would also appear probable that even after the passage of the face, and therefore without the confinement exerted by the advance core, the curves I and II would remain quite separate and that the movement of point A would be determined by its previous stress-strain history. It follows that the deformability of the advance core constitutes the factor most capable of conditioning the deformation response of the ground to excavation and must therefore be considered the true cause of it.

Now, if the deformability of the advance core is the true cause of the deformation response of the ground to excavation, it seems logical to hypothesise the possibility of making use of the core as a *new instrument for controlling* it, by acting on its rigidity with appropriate intervention.

We therefore worked on the possibility of *controlling the rigidity of the advance core* in order to ascertain to what extent this would make it possible to *control the deformation response of the cavity.*

To do this, new technologies and new types of intervention had to be researched and developed that would act on the core and protect it from excessive stress (protective techniques) and/or conserve or improve its strength and deformation properties (reinforcement techniques). These particular types of intervention are termed **conservation intervention** or alternatively **cavity preconfinement intervention** to distinguish them from *ordinary confinement,* which only acts on the ground surrounding the cavity after the passage of the face (Fig. 2.11) [14].

The Campiolo Tunnel (1983)

The construction of the Campiolo Tunnel on the twin track Udine-Tarvisio railway line required an initial section of 170 m of tunnel to be driven from the Udine portal through rubble slopes, with boulders, some of large dimension, dispersed in a sandy silty matrix, under an overburden varying from 0 m to 70 m. Until then, tunnel advance through this type of material had always presented considerable difficulty in terms of face stability and was generally tackled by top heading and bench type advance on several faces or using a core (nose) at the face with spiles and ribs. While it did not at all exclude rock fall, this manner of proceeding did not even guarantee the safety of tunnel workers. However, the alternative was to improve the ground ahead of the face using conventional grout injections, but this was often costly and produced uncertain results since it was impossible to control how the mix spread through the ground.
It was in this context that in 1983 the idea of experimenting with a new tunnel advance technology was proposed. It had been studied as part of the preparatory work for the third research stage to develop "conservative" intervention capable of exerting controlled cavity preconfinement action, by acting appropriately on the rigidity of the advance core. This technology, termed *horizontal jet-grouting,* was tried for the first time in the world on the Campiolo tunnel. It was used to create a shell of improved ground ahead of the face around the profile of the future excavation, which was strong enough to guarantee the necessary protection for the ground which constitutes the advance core and therefore the formation of an arch effect very close to the profile of the excavation. The geometry of the treatment, which on that particular historic occasion was performed half-face, is shown in the figure below.

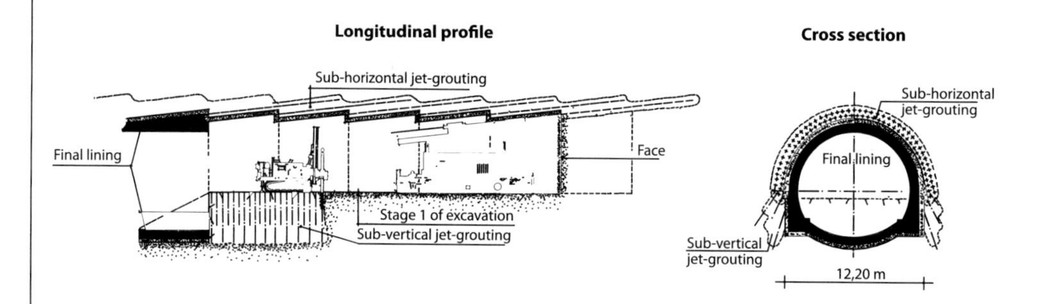

The results achieved were encouraging from both a technical and an economical viewpoint [15]. Not only did they guarantee extremely fast advance rates for the geotechnical conditions (an average of 1.7 m/day of finished tunnel), but the new technology enabled excellent programming of the works and great operational safety. The deformation measured was practically nil in terms of both extrusion and convergence, and this confirmed the validity of the experimental hypotheses based on the third stage of the research. One reason why the choice of the jet-grouting system to improve the ground was so successful was that it also solved problems of pollution and permeation, dependent on the granulometry of the material treated, typically encountered with conventional cement grouting methods. At a distance of twenty years, it has demonstrated that it is also very long lasting.

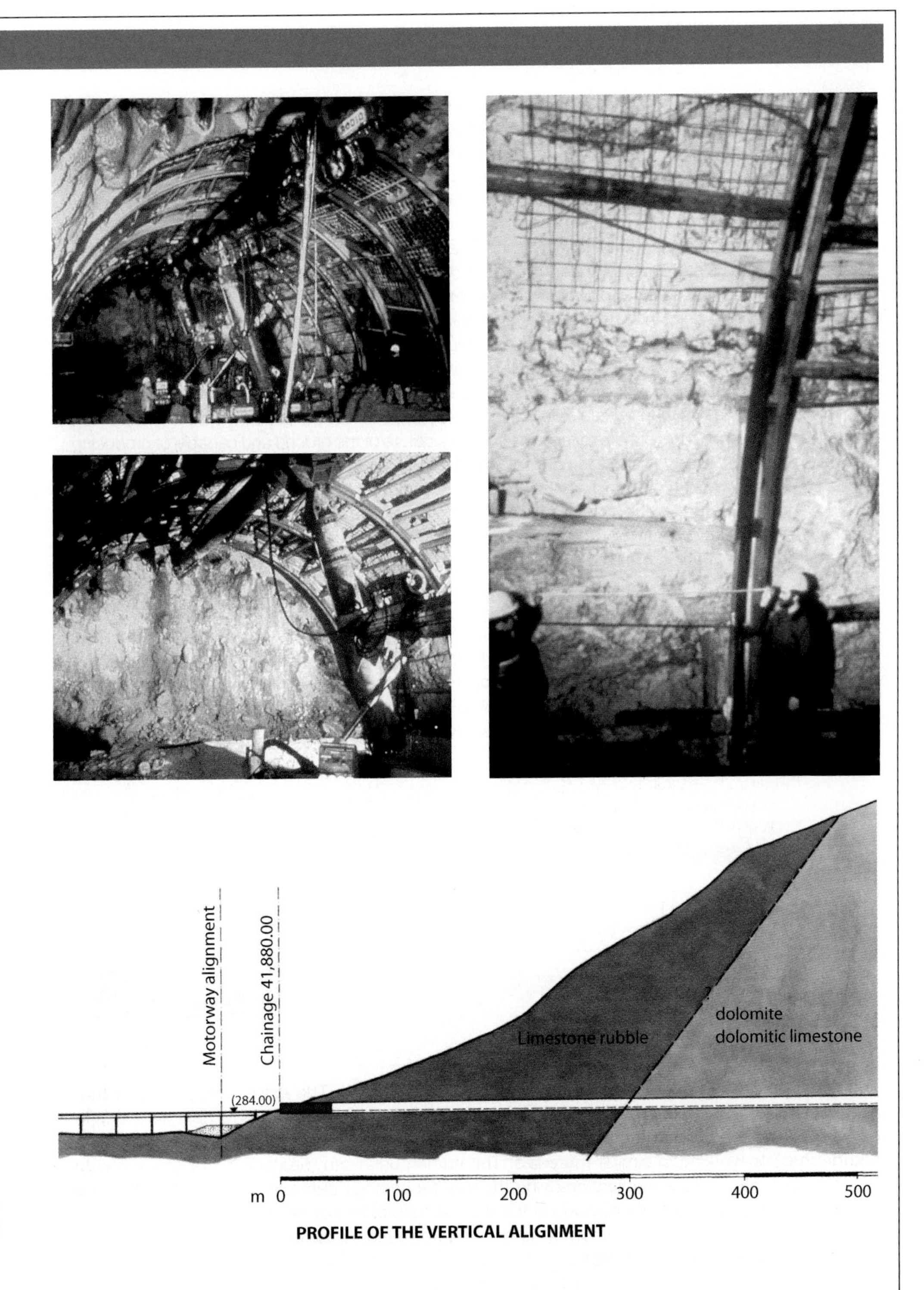
Motorway alignment
Chainage 41,880.00
(284.00)
Limestone rubble
dolomite
dolomitic limestone
m 0 100 200 300 400 500
PROFILE OF THE VERTICAL ALIGNMENT

The tunnels on the Sibari – Cosenza railway line (1985)

The project to construct a new route for the single track Sibari-Cosenza railway line, between the S. Marco Roggiano and Mongrassano – Cervicati stations required the construction of four 10 m diameter tunnels for a total length of approximately 4,000 m The ground to be tunnelled had maximum overburdens of around 100 m and consisted of grey-azure clays belonging to the Calabrian-Pliocene formation characterised by poor strength and deformability characteristics. The calculations performed to assess the behaviour of the tunnels during excavation in the absence of intervention predicted serious face and cavity instability. In similar situations, tunnel advance using conventional systems had always created a series of difficulties in planning the works, which were conducted under precarious safety conditions with frequent incidents, making any forecast of construction times and costs uncertain. In fact enormous and irreversible deformation phenomena was triggered, impossible to avoid with those systems, and led to frequent tunnel collapses.
After the positive experience with *horizontal jet-grouting* on the Campiolo tunnel, it seemed clear that the only way of working successfully would be to prevent deformation phenomena by intervening before the arrival of the face with appropriate stabilisation systems. To achieve this, a new type of intervention needed to be designed, appropriate to the fairly cohesive nature of the ground and capable of producing continuous preconfinement action even during and after the passage of the face.
It was precisely at that time that *"predecoupage mecanique"* had been experimented with half face advance in France for the construction of a section of tunnel in clays on the Lille metro line. It consisted of making a cut of predetermined thickness and length ahead of the face around the profile of the extrados of the tunnel. The cut was then immediately filled with shotcrete. A shell or 'prelining' was thus created before the advance core was excavated, thereby preventing the confinement pressure it provided from reducing to zero. The problem at this point was to develop the French idea to adapt it to the Calabrian clays, which were much softer than those at Lille. In order to avoid the delicate problems posed by excavating down from half to full face in poor quality grounds, the technology was redesigned to allow it to be applied full-face [17]. This required, amongst other things, careful study to perfect the geometry of the precut shells and appropriate modifications to ensure that they rested on a sound reliable foundation.

Longitudinal profile

Mechanical precut
3.50 m
0.50 m
th. = 0.14 m
Final lining
Face
≤ 1.5 Ø
Tunnel invert

Cross section

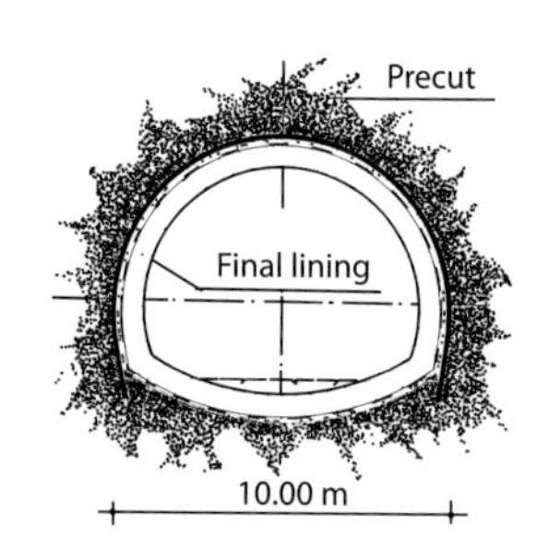

The effectiveness of the new technology was clear immediately. The extreme regularity of tunnel advance rates, which stabilised at average speeds of around 3,00 m/day, despite the difficult ground, allowed construction operations, times and costs to be planned in a manner never before possible working in similar materials. The numerous *in situ* measurements performed to study the behaviour of the excavation using the new technology demonstrated its effectiveness. The movements, which usually develop in the ground before the passage of the face, were practically eliminated, while the maximum gradient of deformation was reached during the excavation of the tunnel invert, which when cast resulted in the immediate stabilisation of the cavity. Since the ground had conserved most of its original characteristics, thanks to the preconfinement, even the long term stress acting on the lining was considerably reduced compared to the stresses that would have been expected with conventional tunnel advance.

The "Vasto" tunnel (Ancona-Bari railway line), 1991, ground: clayey silts, overburden <50 m). Top: reinforcement of the advance core with fibre glass reinforcement; bottom: excavation for tunnel advance

The new ideas were then tried out on the construction of tunnels under very difficult stress-strain conditions. One particularly difficult engineering project, and therefore also very significant in this respect, was that of the Vasto tunnel.

2.1.3.1 The Vasto tunnel (1991)

The alignment of the tunnel, part of the new Ancona to Bari railway line, runs for approximately 6,200 metres under the hills on which the village of Vasto (Pescara) lies.

From a geological viewpoint, the lower and middle part of this hillside consists of a complex of mainly silty, clayey ground, grey in colour, stratified with thin sandy interbedding, while the top part consists of a bank of conglomerates, cemented to varying degrees with an overlying layer of sandy-silty ground, yellowish brown in colour.

The tunnel runs entirely through the lower clay formation with the exception of the initial sections near the portals. At the depth of the tunnel the ground is saturated with water and extremely sensitive to disturbance.

2.1.3.1.1 A brief history of the excavation

Work began in 1984 at the North portal and continued slowly with repeated and serious incidents until April 1990 (Fig. 2.13) [16].

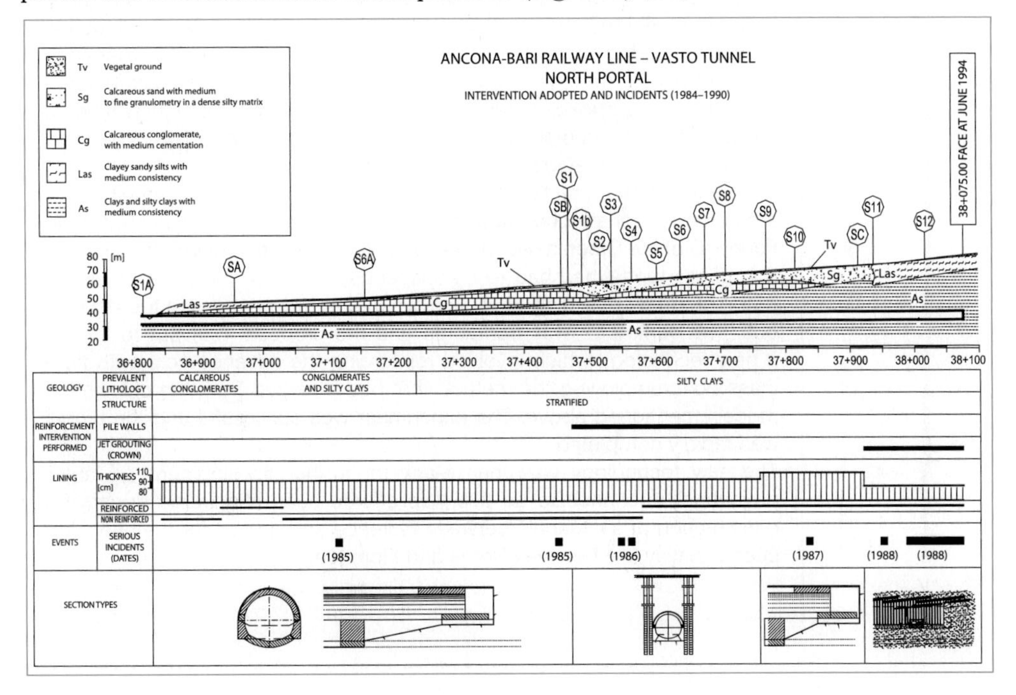

Fig. 2.13

The birth of fibre glass reinforcement of the advance core

Not only was full face mechanical precutting technology used for the first time in the world during the construction of the tunnels on the Sibari-Cosenza railway line, but the foundations were also laid for the development of another type of conservative intervention, which was to become an important part of the history of tunnelling in years to come. This was the reinforcement of the advance core using fibre glass structural elements.
The idea had started to take form after it was decided to bolt the face to counter spalling phenomena which was produced without fail on the face when construction work halted at weekends [18] during the construction of the tunnels mentioned above (it was 1984).
The operation performed on that occasion consisted of simply inserting a certain number of 24 mm Ø steel bolts, 4 m in length, into the face on which a small cross had been fitted on one of the ends as a handle. They were pushed in using an excavator bucket. The operation which was completed with a thin layer of shotcrete turned out to be very effective and when work started again on Monday, it was sufficient to pull out the bolts, gripping them by the cross, to resume excavation without any problems.
We were convinced of the importance of the rigidity of the advance core for controlling tunnel deformation phenomena, a conviction which had grown with positive experimental observations of the stress-strain behaviour of the tunnels already constructed using operations for the pure protection of the core, the horizontal jet-grouting on the Campiolo tunnel and mechanical precutting on the Sibari-Cosenza railway line (described on the preceding pages). We therefore started to study the possibility of increasing this rigidity artificially to reach the desired value. To reinforce the advance core with special systematic bolting, capable of guaranteeing significant increases in the strength and deformability of the cores treated, while not hindering demolition during excavation, seemed worth studying to achieve that goal. Having selected fibre glass from those examined as the most suitable material for this purpose, potential operational methods were then studied for putting the technology into practice. Their effectiveness was tested on original calculation models run on computers.
Experimentation in the field then became necessary. The opportunity soon arose for the construction of a short water tunnel, 5 m in diameter, for the floodway of the Citronia river at Salsomaggiore Terme. Work on the excavation of the tunnel had had to be suspended because the great instability of the ground, consisting of scaly clays, made tunnel advance too dangerous. The contractor was contacted and the proposal was put to try out a new technology which involved reinforcing the advance core with fibre glass bolts employing the method that had emerged from the calculation models mentioned above. The experiment was successful and the tunnel was rapidly completed.
The new technology was then ready for further development on larger projects. The chance to do this was offered soon afterwards during the construction of six tunnels between Florence and Arezzo for the new high speed railway line between Rome and Florence.

Fig. 2.14
Failure of the core-face because of extrusion (Vasto Tunnel)

The original design involved half face excavation, protected immediately with a temporary lining of shotcrete, steel ribs and steel mesh reinforcement. The final lining in reinforced concrete, one metre thick, was cast immediately behind the face, always close to the advance core. The side walls of the tunnel were cast subsequently for underpinning and the casting of the tunnel invert completed the work.

After the first serious incident, an attempt was made to resume tunnel advance employing several methods, which all, however, proved to be completely inadequate and finally a disastrous cave-in occurred at chainage km 38+075 under an overburden of 38 m, which involved the face (Fig. 2.14) and then a section of approximately 40 m down from it. It produced deformation of enormous entity (greater than one metre) in the final lining and it was impossible to continue working.

At this point I was called in to find a solution that would allow the halted works to resume and to complete the tunnel. I tackled the far from simple problem by employing a new advance method for the remaining part of the tunnel. The principles of the method were based on controlling deformation by stiffening the core at the face and therefore by action to preconfine the tunnel.

2.1.3.1.2 The survey phase

Before commencing the new design, it was felt wise to acquire more detailed geotechnical knowledge of the material to be excavated.

The ground in the lower clayey formation was classified as clayey silts or silty clays ranging from medium to highly plastic and impermeable, markedly susceptible to swelling when soaked in water.

Although direct and triaxial cell shear tests provided rather a widely scattered range of values for cohesion and angle of friction, they did give very low average values for strength.

Some *triaxial cell extrusion* tests (see section 3.1.2) were then used to model tunnel advance in the laboratory under the actual stress conditions of the ground *in situ*. These, together with simple finite element

The tunnels on the Rome–Florence high speed railway line

If reinforcement of the advance core using fibre glass structures is considered in the same way as any other bolting operation to be performed occasionally along short sections of tunnel as a counter measure against ground falling in at the face, it is difficult to establish with certainty when this practice first started.

If, however, reinforcement of the advance core with fibre glass structures is considered more correctly as a true and genuine construction technology to be applied systematically under medium to extremely difficult stress-strain conditions to achieve full control of deformation phenomena (and of consequent surface subsidence, when necessary), then the introduction of this practice definitely started in 1985 when it was experimented for the first time in the world, after the successful result on the construction of the water tunnel at Salsomaggiore Terme (see the previous focus box) during construction on tunnels for the new high speed railway line between Rome and Florence on the Florence-Arezzo section (Talleto, Caprenne, Tasso, Terranova Le Ville, Crepacuore and Poggio Orlandi tunnels), where the poor quality of the geological formations (sandy/clayey silts and lacustrian deposits, often under the water table) caused considerable problems, requiring work to stop and the tunnels to be redesigned.

It was decided, thanks to the enterprise and farsightedness of the State Railway management and the contractors (Ferrocemento S.p.A. and Fondedile S.p.A.) and the trust they placed in myself the consulting engineer, to try out the new technology which I had proposed. It had the effect of turning the whole line being constructed between Florence and Arezzo into one large experimental tunnel construction site.

After reinforcement of the advance core with fibre glass tubes, excavation was performed full face giving the face a concave shape to favour the natural formation of a longitudinal arch effect. The tunnel invert and side walls were systematically placed at a maximum distance from the face of 1.5 times the diameter of the tunnel [19].

The Tasso tunnel which collapsed during half face advance (NATM)

The Tasso tunnel during full face advance after reinforcing the advance core with fibre glass tubes

The positive results achieved on the Rome-Florence line during the construction of more than 11 km of tunnel under objectively difficult conditions, confirmed the complete reliability of the full face advance principles in the presence of a rigid core in soft ground and rapidly established the success of this technology of reinforcing the advance core using fibre glass and of conservative cavity preconfinement techniques in general.

mathematical models, also made it possible to calibrate the geomechanical parameters (c, φ, E) for use in the subsequent diagnosis and therapy phases. Direct simulation of the triaxial cell extrusion tests available (integrated with both consolidated and non consolidated triaxial cell shear tests) were used to determine the following ranges of variation for the main geomechanical parameters:

c_u = undrained cohesion = 0.15 – 0.4 Mpa (= 1.5 – 4 Kg/cm^2)
c' = drained cohesion = 0 – 0.2 MPa (= 0 – 2 Kg/cm^2)
φ' = drained angle of friction = 18° – 24°
E = Young's elastic modulus = 50 – 500 Mpa (= 500 – 5,000 Kg/cm^2).

2.1.3.1.3 The diagnosis phase

The input for this phase was the geological, geotechnical, geomechanical and hydrogeological knowledge obtained using theoretical and experimental methods to process the data from *in situ* surveys and laboratory tests performed on the ground in question. It was used to make forecasts of the stress-strain behaviour of the face and the cavity in the absence of intervention to stabilise the tunnel. The purpose was to divide the tunnel into sections, each with uniform deformation behaviour in terms of the three basic stress-strain conditions that might occur.

The diagnosis study then continued with an analysis of the shear mechanisms and instability kinematics that would be produced following the development of deformation and concluded with an estimate of the extent of the unstable zones and the size of the loads mobilised which, however, are not dealt with here.

2.1.3.1.4 Assessment of the stress-strain behaviour

Assessment of the stress-strain behaviour along the tunnel alignment was performed using two different procedures, both of which valid for low, medium and high stress states (Fig. 7.1): the first, more immediate, is based on **characteristic line** theory (calculated using analytical or numerical methods according to the situation); the other, rather lengthier, is based on **triaxial cell extrusion tests**, mentioned in the section on the survey phase.

In the case of the Vasto tunnel, apart from short sections near the portals, both procedures forecast unstable face behaviour with considerable extrusion and, as a consequence, also preconvergence and convergence (over 100 cm radially). These values are sufficient to produce serious manifestations of instability, such as the failure of the face and, as a consequence, the collapse of the cavity.

2.1.3.1.5 The therapy phase

The forecasts produced in the diagnosis phase were used as a basis for deciding on the type of action to exert (preconfinement or simple confinement) and on the intervention

The first studies and experimentation of fibre glass reinforcement of the advance core

It is interesting to illustrate the main characteristics of the first experiments with reinforcement of the advance core at Arezzo on the construction sites of the Rome to Florence high speed railway line.
The following parameters were employed:

- Length of each reinforced section: L = 15 m
- Resistant cross section of the fibre glass elements: ø = 60/40 mm
- Intensity of the reinforcement: I = 0.35 – 0.51 elements/m²
- Overlap between reinforced sections: S = 5 m

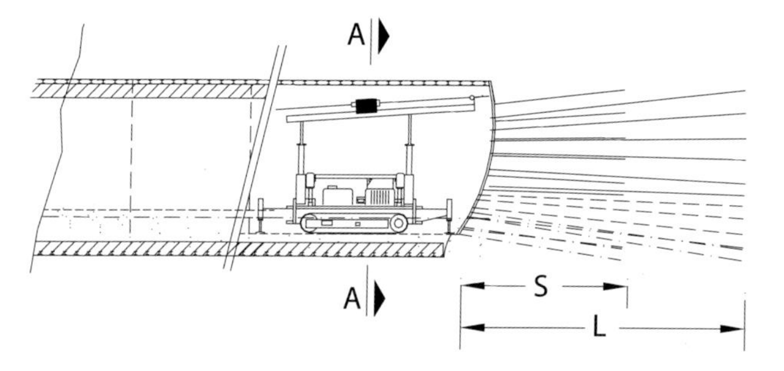

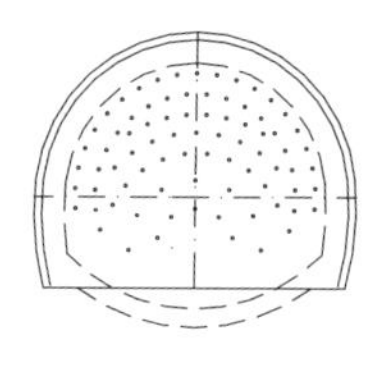

Excavation was performed full face and the tunnel invert and side walls were systematically placed at a maximum distance from the face of 1.5 times the diameter of the tunnel. The tunnel was given a concave shape to favour the formation of a natural longitudinal arch effect.
The figure below illustrates the work cycle adopted for the excavation of the Caprenne and Talleto tunnels in those sections where face reinforcement was accompanied by mechanical precutting.

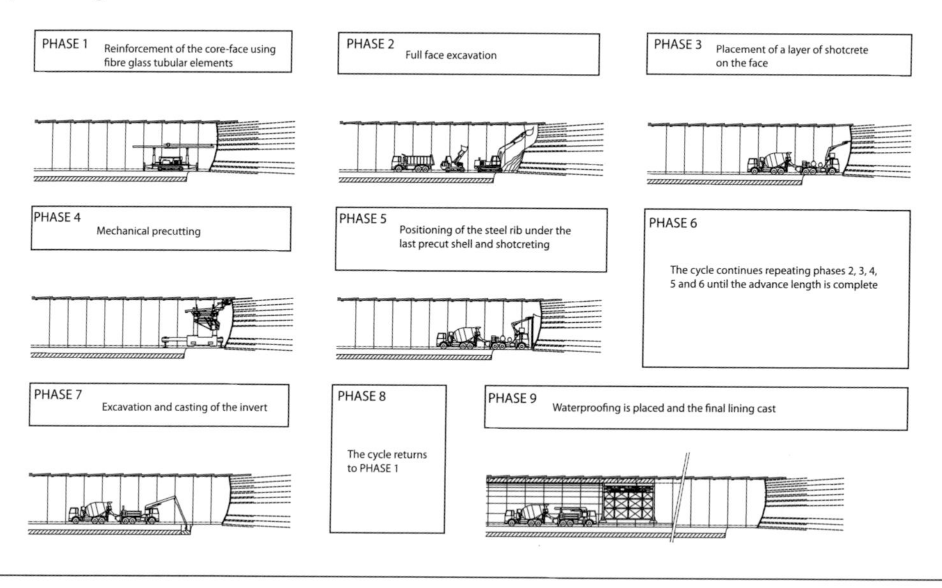

required, in the context of the predicted behaviour category, to achieve complete stabilisation of the tunnel.

Given the characteristics of the ground still to be tunnelled (including the Southern portal to be opened under land slip conditions) and the results of the diagnosis phase, which forecast face unstable behaviour for the whole length of the underground alignment (stress in the failure range, zero arch effect, typical manifestations of instability: failure of the face, collapse of the cavity), it was decided to stabilise the tunnel using preconfinement with decisive intervention ahead of the face to guarantee the formation of an artificial arch effect, again, ahead of the face itself.

More specifically, full face advance was decided after first adopting mixed conservation techniques to create a preconfinement effect by acting both around the core (protective action) and on it directly (reinforcement action).

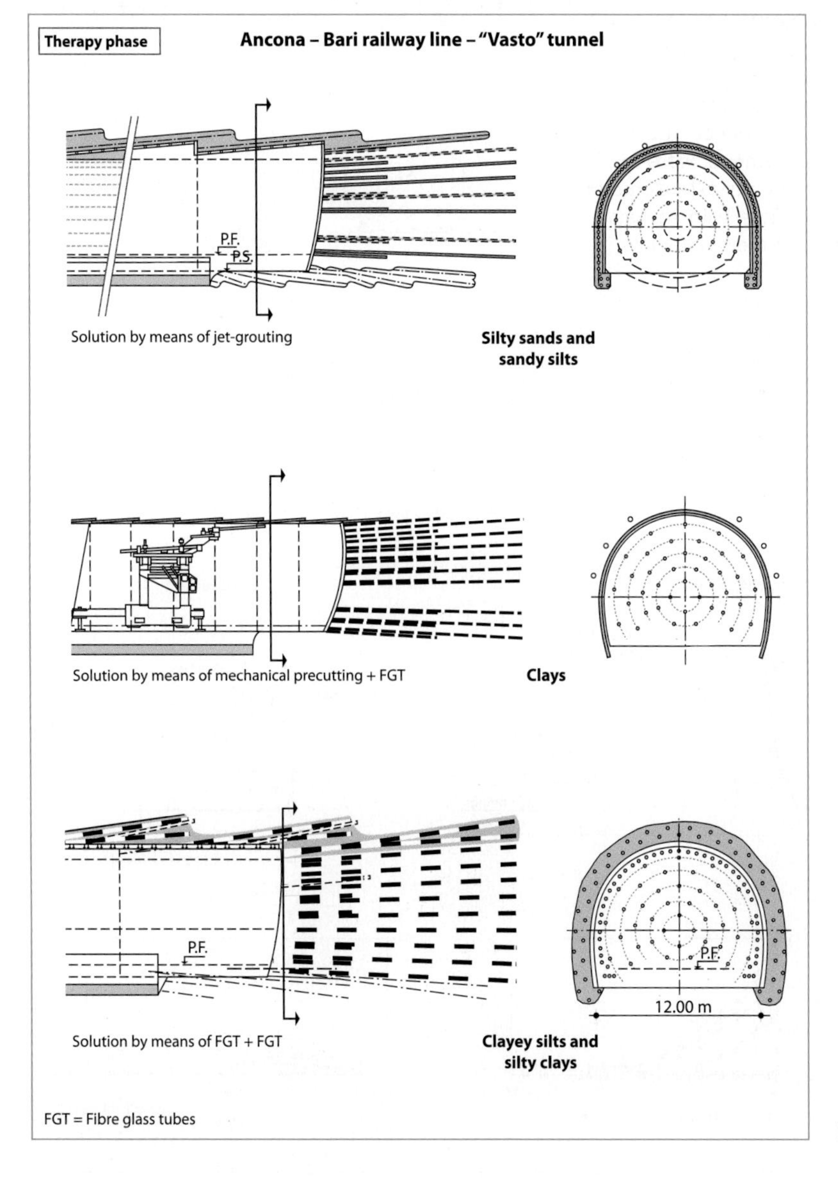

Fig. 2.15

Tests, monitoring and measurements carried out during experimentation of fibre glass reinforcement of the advance core

Experimentation of fibre glass reinforcement of the advance core conducted on the Rome-Florence railway line required detailed knowledge of the characteristics of both the interaction between the fibre glass elements and the ground surrounding them and also of the effects of reinforcing the core on the stress-strain behaviour of the tunnel ahead of the face under different conditions of reinforcement. A whole series of *in situ* deformation and 'pull-out' tests were first dedicated to studying the former aspect (the results are given in the figure on the left [20]), while for the second aspect, work started on developing extrusion tests, which were then to become widespread in tunnelling as an accompaniment to more traditional convergence measurements. These measurements were made by inserting an incremental extensometer 15 m long into the face with measurement points fitted at 1 metre intervals.

The results of the extrusion, preconvergence and convergence measurements taken, significantly increased our theoretical knowledge of the stress-strain behaviour of a tunnel at the face and confirmed the effectiveness of the new technology in controlling deformation behaviour.

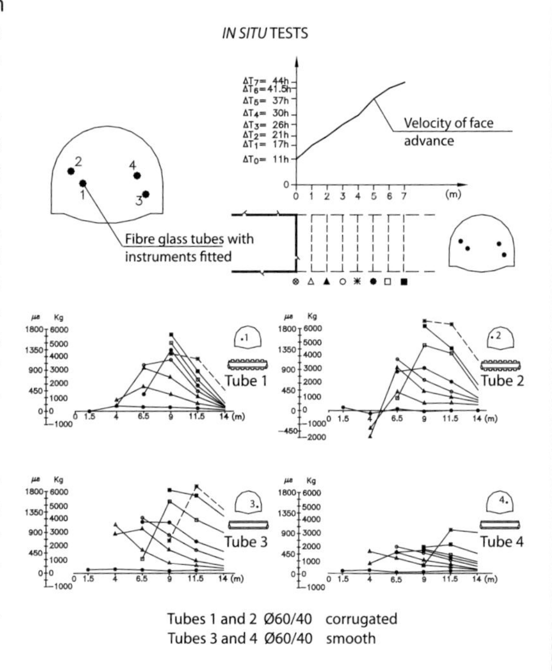

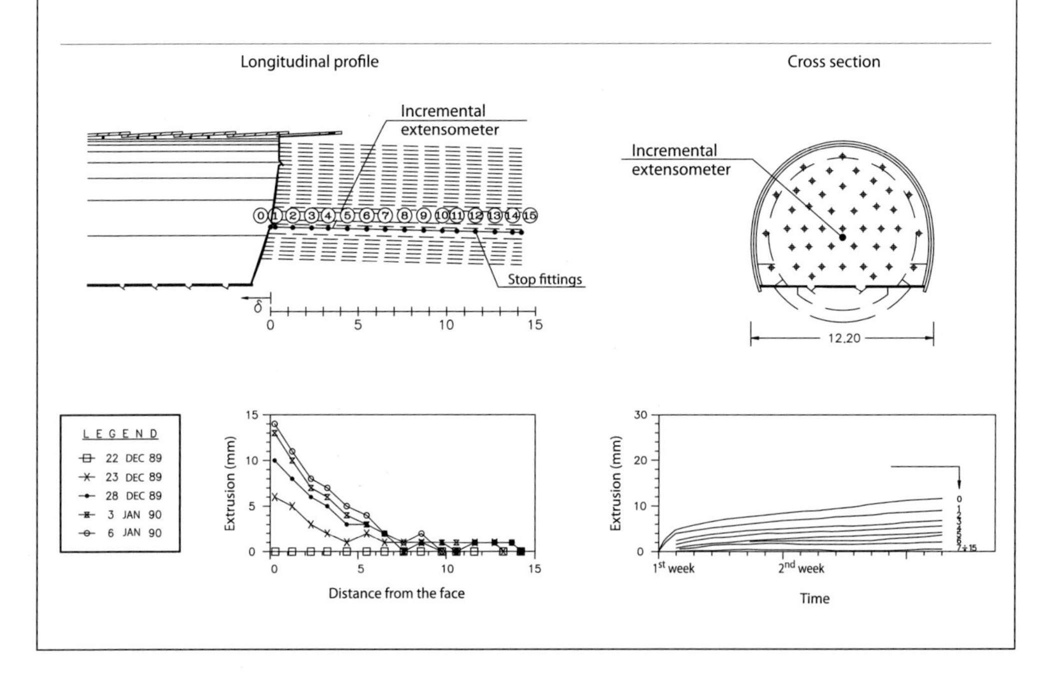

Three alternative tunnel section types were then designed to be adopted according to the homogeneity and consistency of the ground encountered during tunnel advance (Fig. 2.15).

The only difference between them was the type of treatment (prereinforcement-pre-confinement) to be employed ahead of the face around the future cavity, while the method of ground reinforcement employed in the advance core was the same for all three.

The choice of technique to be employed around the core was strictly dependent on the nature and the acquired consistency of the ground to be tunnelled.

Horizontal jet-grouting was used in granular ground or ground with poor cohesion, characterised by weak shear strength.

The most appropriate technology to create strong shells ahead of the face in cohesive, compact and homogeneous ground, able to guarantee the mobilisation of an "arch effect", is that of mechanical precutting.

In grounds where the values for undrained shear strength and cohesion make the use of this technology inadvisable a band of improved ground ahead of the face around the core and the cavity can be obtained using "*claquage*" injections performed using fibre glass tubes specially fitted with valves.

All three tunnel section types were completed with a preliminary confinement in the tunnel consisting of steel ribs and shotcrete, closed with the tunnel invert and subsequent placing of the final lining in concrete.

Once the tunnel section type had been selected, the specifications for the reinforcement of the advance core using fibre glass tubes were decided in terms of the number of tubes to be inserted, the length and the geometry of the pattern on the face.

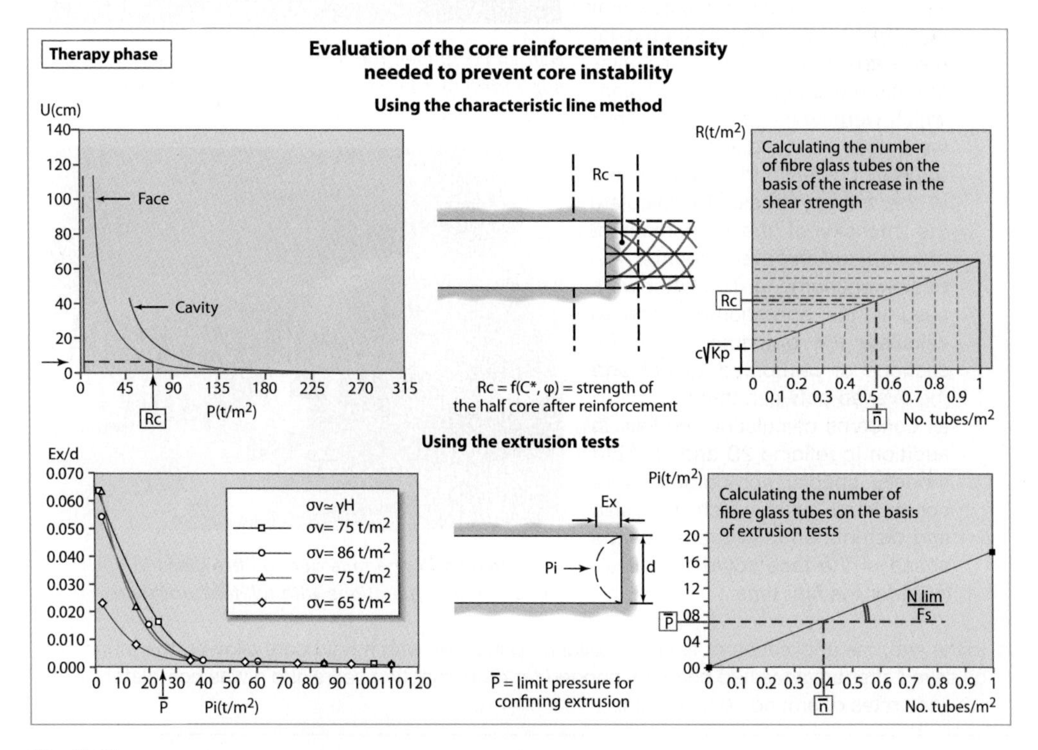

Fig. 2.16

The San Vitale Tunnel (1991) – 1

Since it was first experimented fibre glass reinforcement of the advance core technology has undergone significant development of the design instruments, the materials, the types of implementation and the operational technologies. The construction of the San Vitale Tunnel (Caserta-Foggia State Railway line) in 1991, performed in the scaly clays formation where conventional methods (partial-face advance and preliminary stabilisation using ribs, radial bolts and shotcrete) had failed completely [21], represented a fundamental step forward for each of these aspects. The difficulty of the undertaking not only constituted a rigorous test and a chance to fully explore the new technology, but it was also an opportunity to develop new types of implementation, new types of laboratory and *in situ* measurements and 3D calculation models capable of correctly investigating the effect of intervention in the core-face zone on the stress-strain behaviour of tunnels and on the size of the loads on linings in the long and short term.

- as concerns the first of these, the FGT + FGT type of implementation (valved fibre glass reinforcement injected ahead of the face around the cavity + fibre glass reinforcement cemented in the core) was introduced to replace PT + FGT (mechanical precutting around the cavity and fibre glass reinforcement cemented in the face) which was not well suited to the ground because it was difficult to guarantee the continuity of the precut shell;
- as concerns measurements, new equipment was designed for triaxial cell extrusion tests and for systematic measurement of extrusion, which were very useful. The former were used in the diagnosis phase to predict behaviour categories and in the therapy phase to calculate the intensity of the reinforcement required to effectively counter extrusion phenomena. The latter were used in the operational phase to calculate the optimum length for sections of reinforced tunnel and the overlap between them;
- as concerns calculation models, in addition to refining 2D and 3D FEM models, special nomograms were constructed with which the size and distribution of preconvergence ahead of the face could be calculated for the first time.

San Vitale Tunnel: the half cross section face supported by I-beams and tie rods after tunnel advance was abandoned

San Vitale Face: the face during fibre glass reinforcement operations with full face advance

In the end, the application of full face advance principles with a rigid core allowed the tunnel to be saved after advance had been abandoned for more that two years and it was completed with advance rates of around 50 m/month.

As with the approach adopted in the diagnosis phase for predicting cavity behaviour, two different procedures were adopted to decide the number of fibre glass tubes (Fig. 2.16).

The first procedure was based on the characteristic line method, taking into account, in a simplified manner, the effect of reinforcing the advance core when calculating the corresponding characteristic line.

The second procedure for the design of the advance core reinforcement was based on interpretation of the extrusion curves obtained from triaxial extrusion tests. Having first identified the minimum confinement pressure on the curve P_i required to stabilise the face (defined as the borderline pressure between the "elastic" and the "elasto-plastic" parts of the extrusion curve), experimental diagrams of the type shown in the same Fig. 2.16 were used to calculate the number of tubes required to guarantee the safety of the face with the desired safety coefficient.

Both approaches (extrusion tests and characteristic lines) furnished comparable results, which confirmed the conceptual similarity they have in common.

2.1.3.1.6 The operational phase

Work resumed almost simultaneously in 1992 on both portals, at the North portal to repair the collapsed section of tunnel and at the South portal to begin tunnel advance. Average advance rates working 7 days per week were approximately 50 m per month of finished tunnel (Fig. 2.17).

Fig. 2.17 *Reinforcement of the core-face (Vasto Tunnel)*

The charts in Fig. 2.18 give a comparison of average monthly advance rates with convergence measurements during the same period. The distinct tendency of the latter to be indirectly proportional to the

The San Vitale Tunnel (1991) – 2

The San Vitale Tunnel deserves further consideration because, as it was monitored constantly and extensively, it constituted a true and genuine laboratory for the full scale study of the deformation response of the medium to the action of excavation. This was studied as a function of the intensity and type of the preconfinement intervention in terms of: extrusion of the core and preconvergence and convergence of the cavity; pressures on preliminary and final linings; the stress state in the linings.

The tunnel, which had been driven half face according to conventional principles (NATM) in scaly clays of extremely poor quality in relation to the stress states in play, recorded convergence of a magnitude of metres and uncontainable pressure on the linings, which obliged site work to stop. When work resumed with full face advance after first protecting and stiffening the core-face, convergence values of less than a decimetre were recorded and average radial pressure on the final lining was of the order of only 5 bar (see the figure below).

This, together with other similar cases, constituted very firm confirmation that the beneficial effect produced ahead of the face by this type of intervention can be conserved after the passage of the face, with extremely significant results, if it is appropriately maintained by means of suitable confinement of the cavity.

S. VITALE TUNNEL

MEASUREMENTS OF STRESS WITHIN THE FINAL LINING

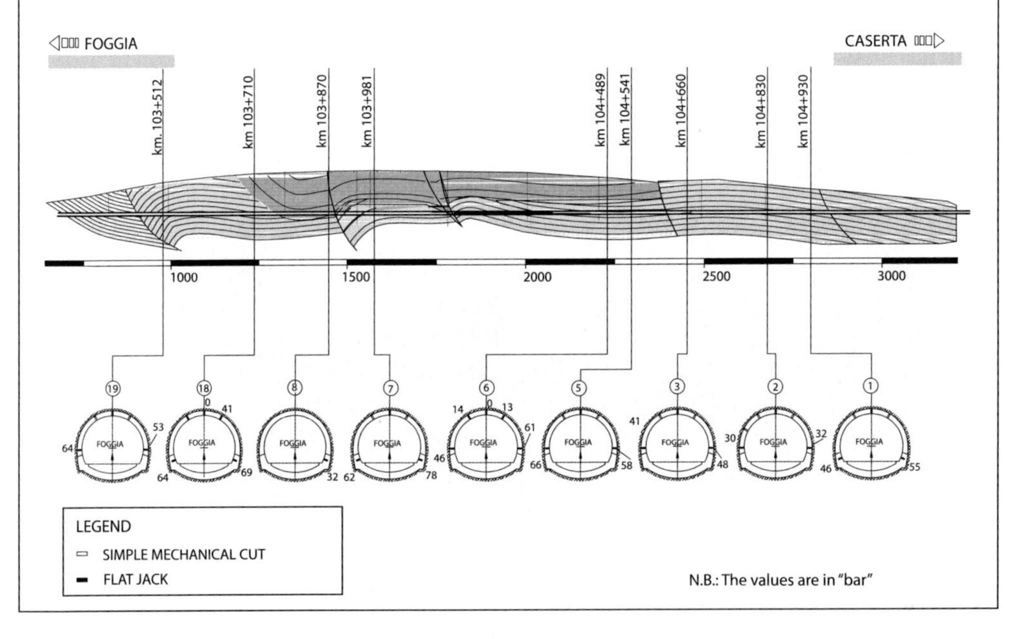

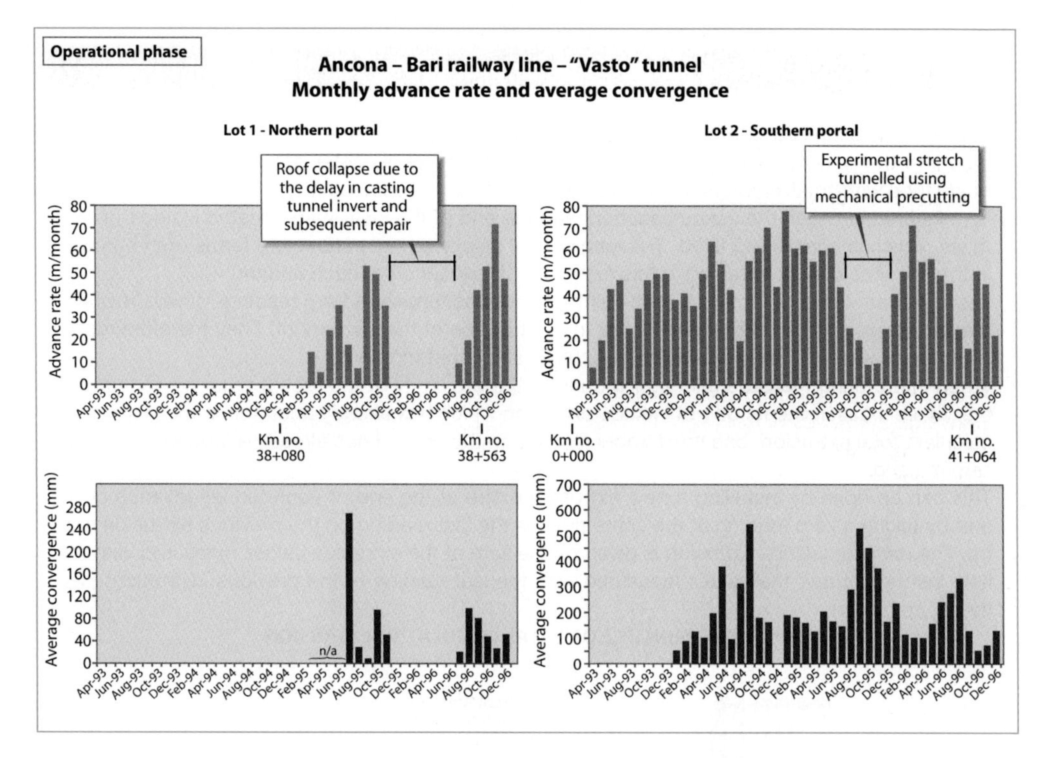

Fig. 2.18

former is particularly significant and confirms the hypothesis that the less time the core is given to deform, the less extrusion and preconvergence is triggered and convergence, which depends on that extrusion and preconvergence, is also more limited as a consequence.

2.1.3.1.7 The monitoring phase during construction

The monitoring phase began at the same time as excavation resumed and consisted of interpreting the deformation response of the medium to excavation in order to optimise and calibrate intervention to stabilise the tunnel.

In addition to the normal measurements of convergence and pressure, systematic and simultaneous extrusion and convergence measurements were also taken in the Vasto tunnel. These were new and of particular interest, especially considering the results that they have furnished to date.

The results of these measurements are summarised in the graphs given in Fig. 2.19, which show the simultaneous course of extrusion and convergence for a complete work cycle.

Examination of the graphs shows that following face advance and as the depth of the reinforced part of the advance core gradually reduces from an initial 15 m to only 5 m (with a consequent reduction in its

Extrusion measurements – 1

Extrusion measurements are obtained by fitting an extrusion metre of the sliding micrometer type into the advance core with a length of 2–3 times the diameter of tunnel. It consists of a guide tube with small anchors fitted externally to act as reference points. The change in distance is measured between the various anchors and the end of the instrument located inside the rock mass, which is considered fixed. The longitudinal displacement in absolute terms of the ground in the advance core is obtained in this fashion for the location of each anchor.
It is important to keep in mind that the extrusion values furnished from readings *always relate to the zero reading* taken immediately after the installation of the instrument. They therefore represent the *increase in extrusion* that has occurred since that moment.
When interpreting measurements for ground in the proximity of the face it must be considered that these measurements do not include extrusion which has already occurred in the ground. To calculate total extrusion, one must know how much the ground has already extruded *before the zero reading.*
This can be done by installing a new extrusion metre at the end of each tunnel advance cycle and by taking a zero reading at the same time as the last reading on the previous extrusion metre. The *total cumulative extrusion* is given by the sum of the extrusion values measured with the new instrument and the values measured with the last reading of the previous instrument (see the figure below).

THE PRINCIPLE OF TOTAL CUMULATIVE EXTRUSION

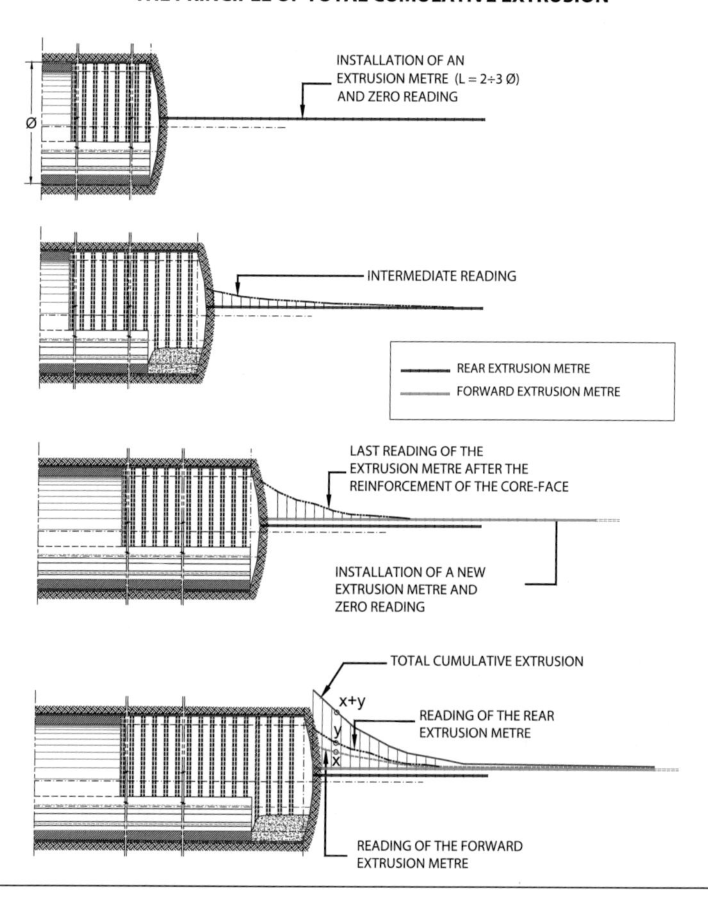

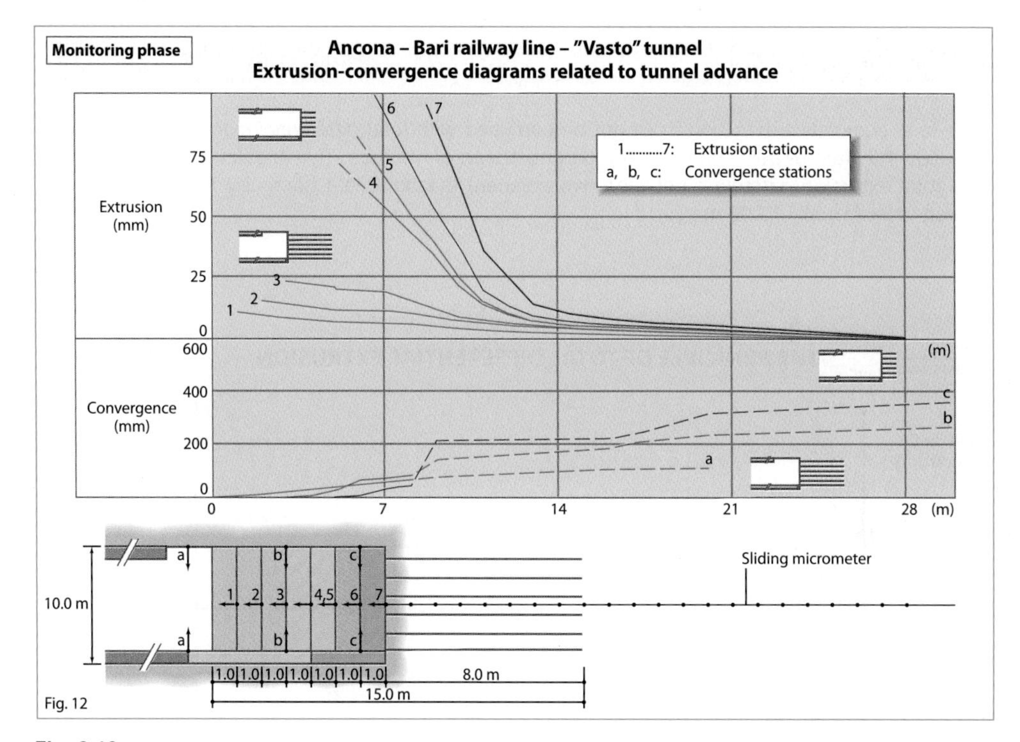

Fig. 2.19

average rigidity), a deformation response of the core itself (extrusion) and of the cavity back from the face (convergence) develops and this gradually moves from the elastic range towards the elasto-plastic range. The convergence curves in particular start with an initial tendency typical of a situation which stabilises rapidly (maximum values of the order of 10 cm are produced after maximum values for extrusion of less than 2.5 cm) and then gradually change to record values which indicate an increasing inability of deformation to come to a halt. For example, when the length of the reinforced core falls to just 5 m, extrusion of the order of 10 cm develops, which gives rise to convergence four times greater than that measured at the beginning of the work cycle.

In this context, then, combined reading of extrusion and convergence of the cavity is an extremely important indicator for the design engineer to use in order to be able to establish the moment at which advance must halt to carry out another cycle of reinforcement and restore the minimum depth of reinforced core required to maintain the ground, if not actually in the elastic range, at least out of the failure range.

Extrusion measurements – 2

The total cumulative extrusion must not be confused with *total differential extrusion*, which expresses the state of deformation in the ground ahead of the face. This is calculated by dividing the total extrusion difference between two adjacent measurement bases by the distance between them (usually 1,000 mm).

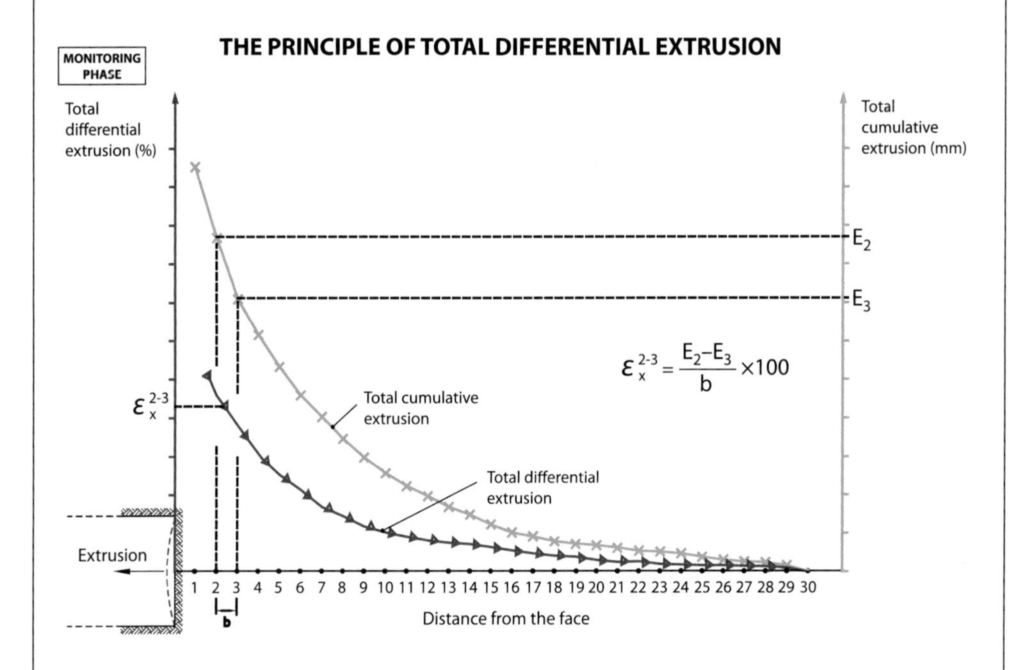

By comparing the deformation value obtained with the limit, elastic and failure values characteristic of the ground (found by conducting triaxial extrusion tests in the laboratory), the real state of deformation in the ground can be assessed with precision and more specifically whether it is stressed in elastic range, in the elastic-plastic or in the failure range.
Knowing the distribution of the values for total differential extrusion in the ground ahead of the face (advance core) means being able to reliably evaluate the size of the band of ground stressed in the elastic range, in the elastic plastic or in the failure range ahead of the face, and therefore being able to regulate the depth of reinforcement intervention and to evaluate the safety of the core-face in terms of stability.

2.1.3.2 Results of the third research stage

The study and experimentation conducted on the Vasto tunnel enabled us to see that there is both a close connection between deformation that occurs inside the advance core of a tunnel (extrusion) and that which develops subsequently around the cavity after the passage of the face, (convergence) and also (Fig. 2.20, the results of the third research stage), that deformation of a cavity can be controlled and considerably reduced by artificially controlling the deformability, and therefore the rigidity, of the advance core (confinement of extrusion). This is possible by employing appropriate stabilisation techniques carefully dimensioned and balanced between the advance core and the cavity, as a function of the strength and deformability of the medium in relation to the stress conditions.

More specifically, if a medium is stressed in the *elasto-plastic range*:

- if the stress state is low in relation to the characteristics of the medium, then it may be sufficient to intervene on the cavity only, using radial action with no longitudinal intervention in the advance core at all;

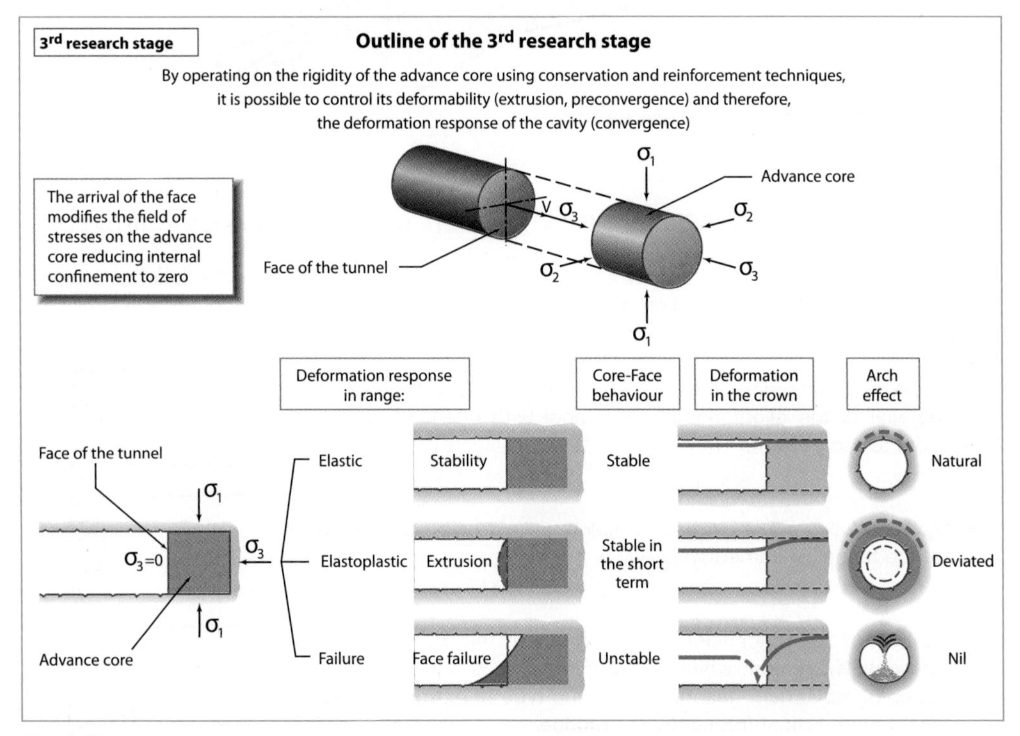

Fig. 2.20

The types of extrusion

The study of extrusion phenomena was perfected by analysing maps of displacement at given points measured topographically on a large number of faces in different types of ground and under a very wide variety of stress-strain conditions.

MAP OF DISPLACEMENT AT POINTS ON THE WALL OF A FACE

(over 10 days with no face advance)

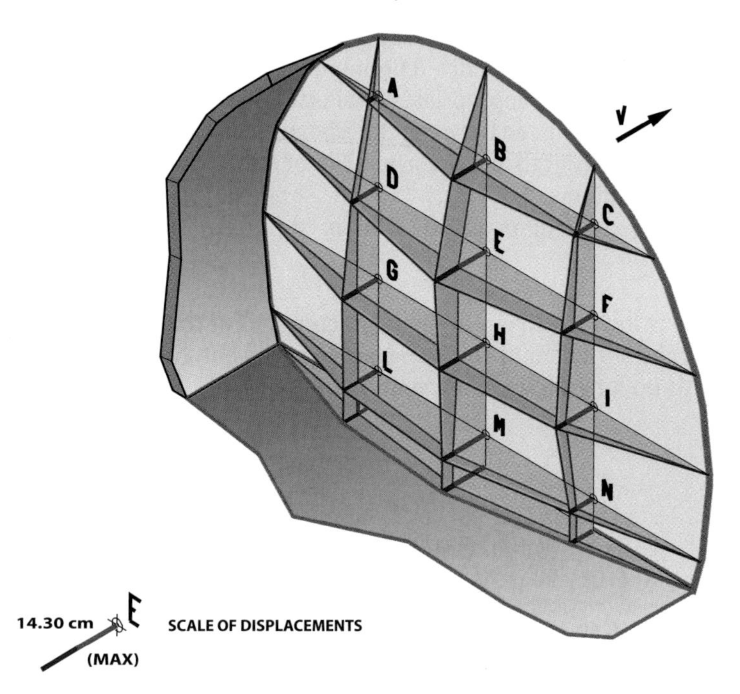

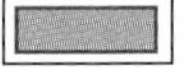 **DISPLACEMENT AFTER 0 – 5 DAYS** 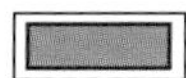 **DISPLACEMENT AFTER 6 – 10 DAYS**

It was established from these studies that the advance core extrudes from the face according to tree basic types of deformation depending on the type of material involved and the stress it is subject to.

- *cylindrical extrusion*: the face presents translational movement parallel to the alignment of the tunnel with intensity increasing from top to bottom;
- *spherical dome-like extrusion*: maximum extrusion at the centre of the tunnel or immediately below it;
- *combined* cylindrical dome-like extrusion: the wall of the face presents movement which is a combination of the two previous types of movement described. It is the type of extrusion which is encountered most frequently in practice.

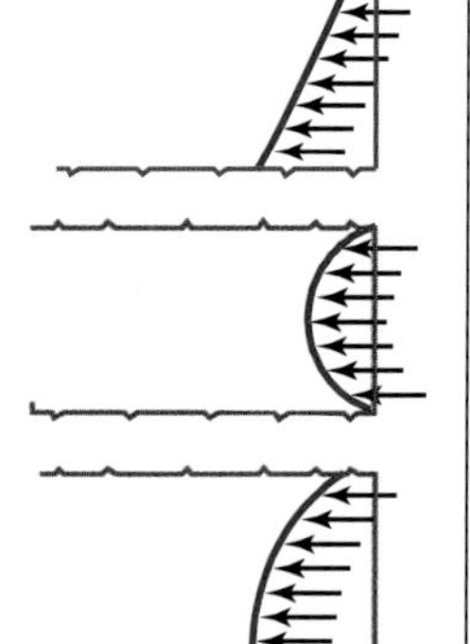

- if, however, the stress state is high, then on the contrary, it will be necessary to act above all on the core reinforcing it with longitudinal intervention and calibrating radial measures after the passage of the face appropriately.

If a medium is stressed *in the failure range*, it is of primary importance to stiffen the advance core with preconfinement of the cavity, which may be supplemented with appropriate action to confine the cavity after the passage of the face. In these cases, experience (and that described in the previous paragraphs is particularly significant) advises (Fig. 2.21):

- working ahead of the face on the *form* and *volume* of the advance core by creating a protective crown of improved ground around it.

This method was used effectively on the construction of the Vasto tunnel to advance through particularly difficult ground.

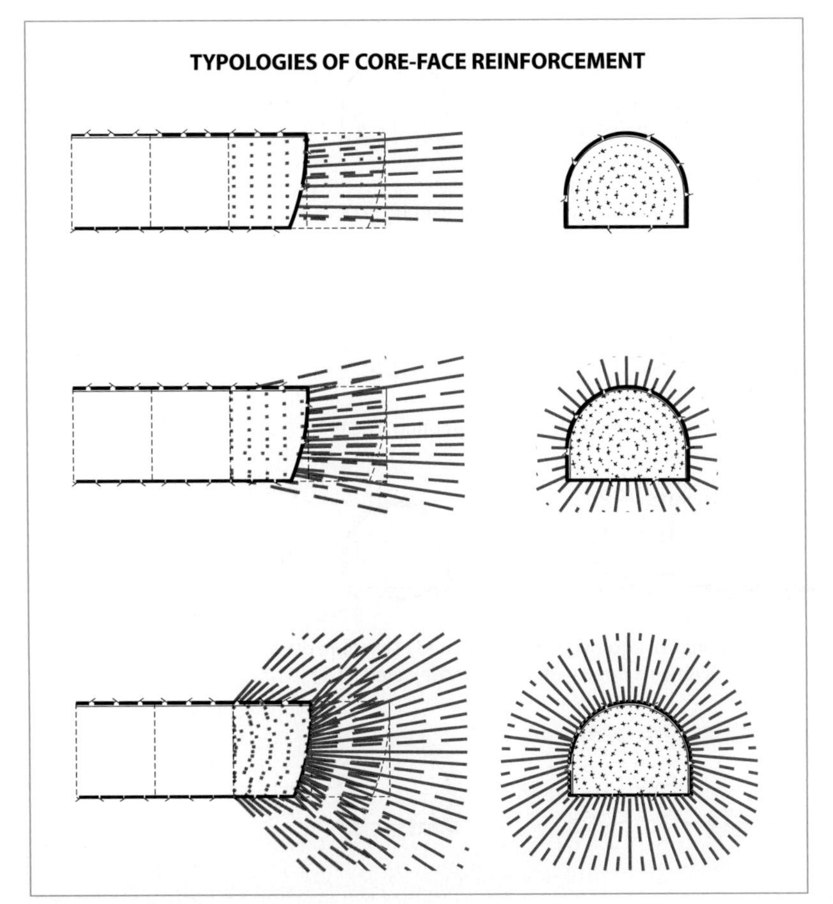

Fig. 2.21

Fibre glass reinforcement of the advance core

Preconfinement of the tunnel by reinforcement of the advance core using fibre glass elements (VTR) has been performed to date using four principal methods developed by myself, to be used according to the type of ground and the stress-strain conditions to be tackled:

1. (**FGT**): simple reinforcement of the advance core using fibre glass reinforcement (indirect conservative intervention);
2. (**PT + FGT**): fibre glass reinforcement of the core and protection of the core at the same time by means of "pre-arches" in shotcrete created using mechanical precutting (mixed conservative intervention);
3. (**HJG + FGT**): fibre glass reinforcement of the core and protection of the core at the same time by means of "pre-arches" of improved ground performed using sub-horizontal jet-grouting (mixed conservative intervention);
4. (**FGT + FGT**): fibre glass reinforcement of the core and protection of the core at the same time by means of "pre-arches" of improved ground around the tunnel using fibre glass elements, fitted with valves, placed in advance and injected with grout (mixed conservative intervention).

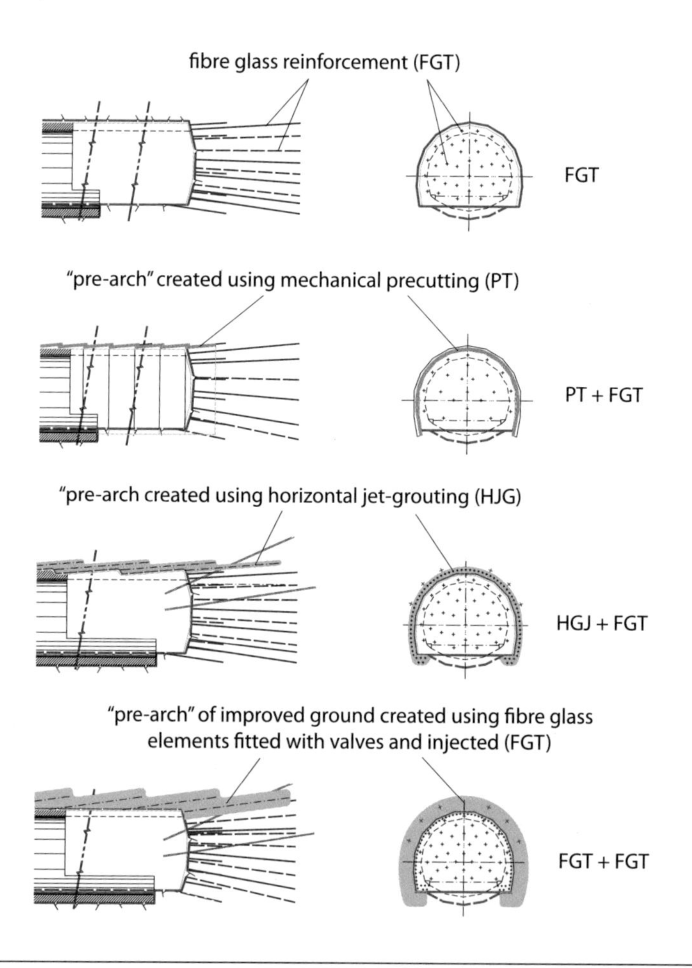

If that is found to be insufficient then it is necessary to:

- perform further radial ground reinforcement around the cavity of an entity sufficient to absorb residual convergence that the core, although stiffened, is not able to prevent by itself.

In the latter case, the balance of intervention between the core and the cavity decided at the design stage, can be adjusted or "fine tuned" during construction.

2.2 The advance core as a stabilisation instrument

The results of the research can be very briefly summarised as follows:

- during **the first research stage** three *fundamental types of deformation* (extrusion at the face, preconvergence and convergence) and consequent manifestations of instability (rock fall, spalling, failure of the face and collapse of the cavity) were identified;
- in the **second research stage** experimental confirmation was obtained showing that all the deformation behaviour (extrusion at the face, preconvergence and convergence) and the consequent visible manifestations of instability visible inside the cavity (rock fall, spalling, failure of the face and collapse of the cavity) depend directly or indirectly on the *rigidity of the advance core*;
- in the **third research stage** experimentation was performed on how it was possible to use the *advance core as a stabilisation instrument*, by intervening artificially on the rigidity of the core itself to regulate deformation of the cavity.

The research also achieved the following results:

- it confirmed that the deformation response of the medium to the action of excavation must be the principal question with which a tunnel designer is concerned, because, amongst other things, it indicates the triggering and position of an arch effect in relation to the profile of a tunnel, or, in other words, the level of stability reached by a tunnel;
- it showed that the deformation response begins ahead of the face in the advance core and develops backwards from it along the cavity and that it is not only convergence, but consists of extrusion, preconvergence and convergence. Convergence is only the last stage of a very complex stress-strain process;
- it clearly indicated the existence of a direct connection between the deformation response of the core-face and that of the cavity, in the sense that the latter is a direct consequence of the former, underlining the importance of monitoring and controlling the deformation response of the core-face and not just that of the cavity;
- it demonstrated that it is possible to control deformation of the advance core (extrusion, preconvergence) and as a consequence also control deformation of the cavity (convergence) by acting on the rigidity of the core with measures to protect and reinforce it.

In conclusion, the results of the research allowed the advance core to be seen as a *new stabilisation tool* for the cavity in the short and long term: an instrument whose strength and deformability play a determining role because it is able to condition that aspect which should concern tunnel designers more than any other: the *triggering of an arch effect in the ground when the face arrives.*

■ 2.3 The advance core as a point of reference for tunnel specifications

If the advance core is an effective tool for long and short term stabilisation of a cavity, capable of conditioning its behaviour when the face arrives, then tunnel designers should focus above all on stress-strain phenomena in the core-face, which is to say on its stability, in order to be able to draw up designs capable of guaranteeing the long and short term stability of a tunnel.

It follows that the stability of the core-face can be assumed as a point of reference for standardising tunnel specifications, with the advantage that it is an indicator that conserves its validity in all types of ground and in all statics conditions.

From this viewpoint, the three fundamental stress-strain conditions of the core-face already described in section 1.4 (see also Figs. 1.6 and 2.20) also identify three possible types of cavity behaviour (Fig. 2.22):

- stable core-face behaviour (behaviour category A);
- stable core-face in the short term behaviour (behaviour category B);
- unstable core-face behaviour (behaviour category C).

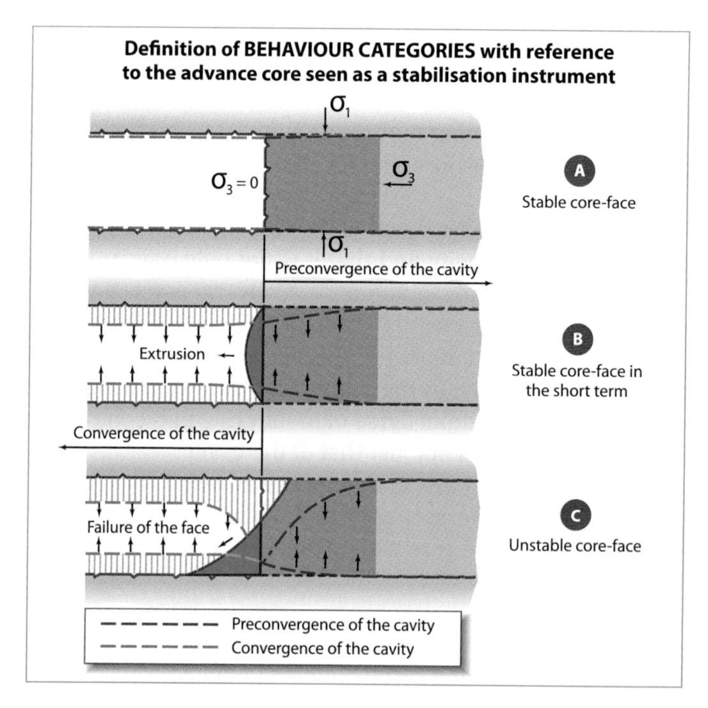

Fig. 2.22

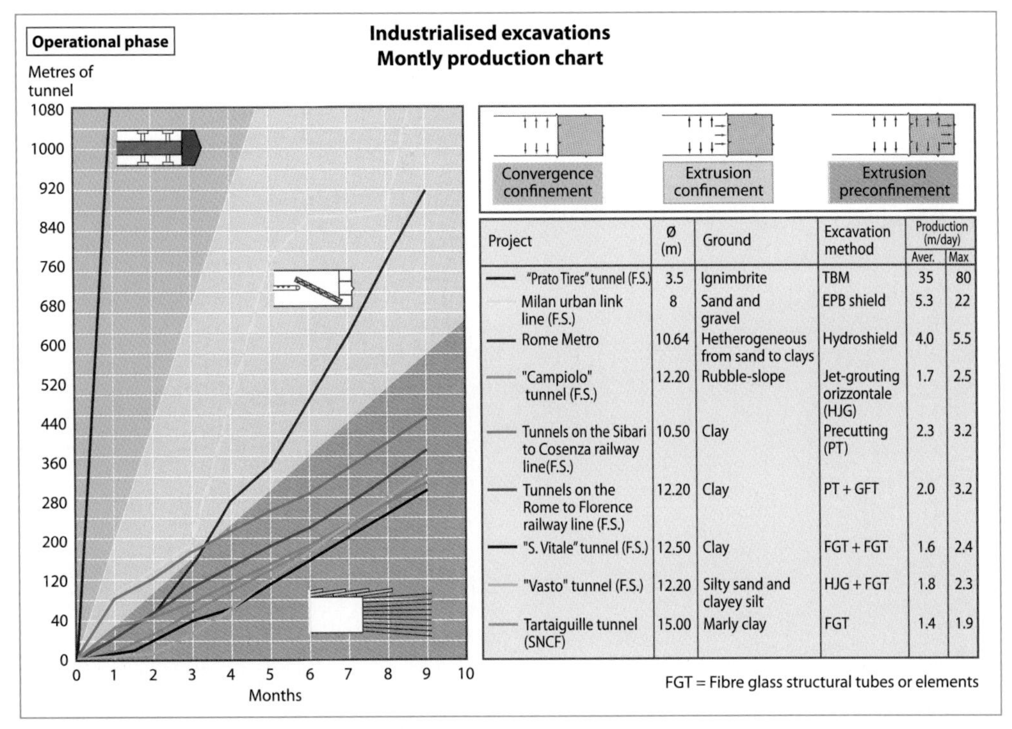

Project	Ø (m)	Ground	Excavation method	Production (m/day) Aver.	Production (m/day) Max
"Prato Tires" tunnel (F.S.)	3.5	Ignimbrite	TBM	35	80
Milan urban link line (F.S.)	8	Sand and gravel	EPB shield	5.3	22
Rome Metro	10.64	Hetherogeneous from sand to clays	Hydroshield	4.0	5.5
"Campiolo" tunnel (F.S.)	12.20	Rubble-slope	Jet-grouting orizzontale (HJG)	1.7	2.5
Tunnels on the Sibari to Cosenza railway line(F.S.)	10.50	Clay	Precutting (PT)	2.3	3.2
Tunnels on the Rome to Florence railway line (F.S.)	12.20	Clay	PT + GFT	2.0	3.2
"S. Vitale" tunnel (F.S.)	12.50	Clay	FGT + FGT	1.6	2.4
"Vasto" tunnel (F.S.)	12.20	Silty sand and clayey silt	HJG + FGT	1.8	2.3
Tartaiguille tunnel (SNCF)	15.00	Marly clay	FGT	1.4	1.9

Fig. 2.23

Where there is stable core-face behaviour (A), the overall stability of the tunnel is practically guaranteed even in the absence of stabilisation works.

The results of the research in situations (B) and (C) indicate that preconfinement intervention must be adopted to prevent instability of the face, and therefore of the cavity, and to try and restore stable core-face conditions (A). It must be properly balanced between the face and the cavity and of an intensity appropriate to the real stress conditions in relation to the strength and deformability of the medium.

The application of these concepts in design practice has resulted in numerous significant successes. Figure 2.23 groups together advance rates achieved during the excavation of tunnels designed and constructed in Italy and France over the last ten years, under an extremely wide range of different geological conditions and stress states. What strikes here is not only the high average advance rates maintained in relation to the type of ground involved, but above all the linearity of the rates achieved, even under the most difficult stress-strain conditions, an indicator of industrial type construction performed with regular work cycles and without hitches (see the appendices A, B and C).

At this point it seemed both necessary and urgent to take the extraordinary knowledge acquired to its extreme conclusion and to develop a design and construction approach which adhered more closely to reality than those in common use.

CHAPTER 3

Analysis of the deformation response according to the ADECO-RS approach

3.1 Experimental and theoretical studies

In order to take the new knowledge acquired during the research illustrated in chapter two to its extreme consequences and to develop a design and construction approach which fits reality more closely than those commonly in use today, it was essential to start a new programme of both theoretical and experimentally research to add to and deepen the knowledge we had already acquired of the relationship between the stress-strain behaviour of the advance core and the cavity.

The ADECO-RS approach (acronym in the Italian language for Analysis of Controlled Deformation in Rocks and Soils) (Fig. 3.1), sprang from these studies and were also the crowning result of them. By observing that:

- the phenomena that accompany the excavation of a tunnel can be interpreted in terms of a process of cause and effect (action and reaction);
- normally in these processes the cause must first be fully identified if the effect is to be controlled effectively;
- the only way to achieve full identification of the cause is through in-depth analysis of the effect;

this approach focuses on the effect itself (deformation response of the rock mass) both ahead of and back from the face and by firstly *analysing* its genesis and development with

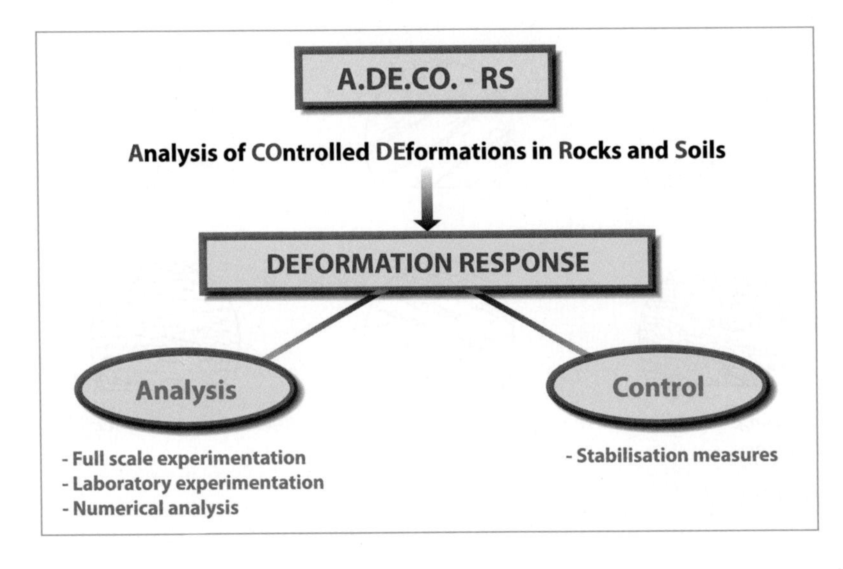

Fig. 3.1

Information sheets for classifying tunnel faces

Full scale experimentation was employed to analyse the stress-strain behaviour of the advance core **in terms of its stability** and this was compared systematically with that of the cavity using an observational approach, which in actual reality involved classifying more than a thousand working faces using special sheets. They were designed as follows (see the example given in Fig. 3.2): the top was reserved for general information (project, client, contractor, date, chainage, overburden), below which space was left for a 10 x 15 cm colour photograph of the face, while the remaining space was divided into two parts, one over the other.
The upper part was for geological (name of the formation, description of the lithology, type of structure, hydrogeological condition) and geomechanical information (on both the matrix and the discontinuities) which could be of both a qualitative (e.g. consistency, weathering) and a quantitative (compressive strength, cohesion, etc.) nature. The geomechanical characterisation of the discontinuities was based on ISRM (International Society for Rock Mechanics) recommendations and the description considered the type, position, spacing, persistence and hydrogeology of joints for each family. There was a space on the right for an illustrative diagram of the face, which is always very useful for highlighting salient characteristics.
The lower part was dedicated to excavation behaviour and included information on the method of excavation (explosives, hammer, ripper, etc.), on the type location and intensity of deformation phenomena (extrusion, convergence), behaviour category (A, B, C), stabilisation intervention and the advance rate achieved.

The systematic compilation of these information sheets made it possible to create an *archive of tunnel faces* of inestimable value. This instrument was found to be irreplaceable for a large number of studies and investigations into the stress-strain response of a tunnel to excavation, as determined by the geological and geomechanical context and the excavation and stabilisation methods employed.

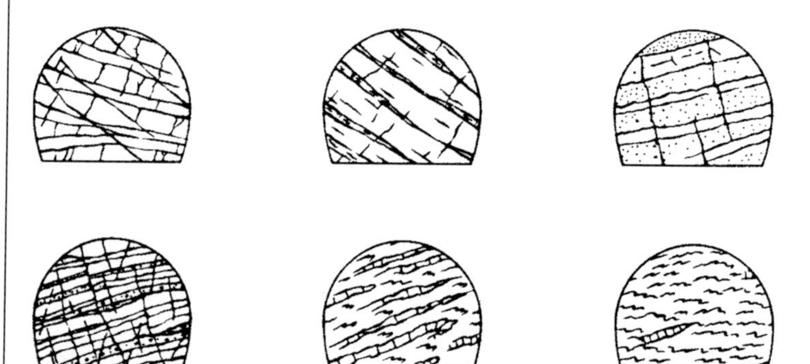

Examples of illustrative diagrams of tunnel faces in different types of rock mass

full scale and laboratory experimentation and numerical calculation instruments, both focusing on the advance core behaviour, it identifies the cause in the deformability of the ground ahead of the face.

By then *controlling* the deformability of the ground ahead of the face (the advance core) using adequate stabilisation tools, it is found that it is possible to thereby control the deformation response of the rock mass, which is incontrovertible evidence that it is the true cause of the process under examination.

3.1.1 Full scale experimentation

A series of observations and measurements both *in situ* and in the laboratory was used to study the stress-strain behaviour of the advance core in terms of stability and deformation

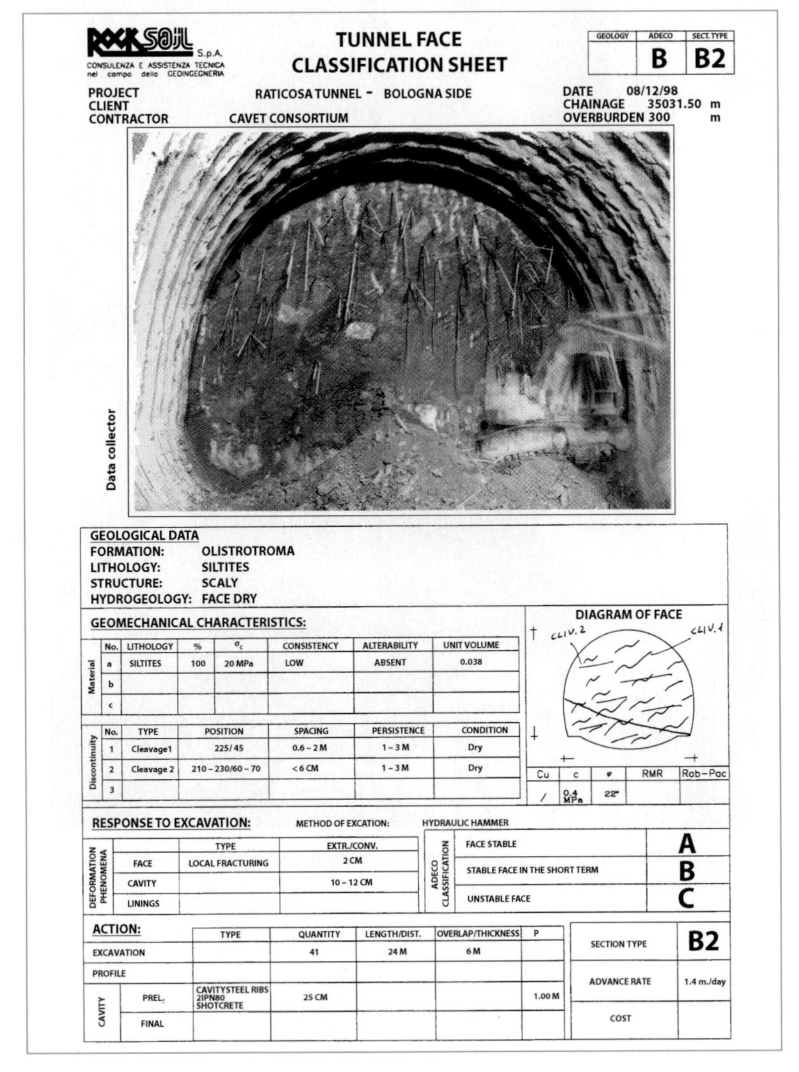

ROCKSOIL S.p.A.
CONSULENZA E ASSISTENZA TECNICA nel campo della GEOINGEGNERIA

TUNNEL FACE CLASSIFICATION SHEET

GEOLOGY	ADECO	SECT. TYPE
	B	B2

PROJECT RATICOSA TUNNEL - BOLOGNA SIDE
CLIENT
CONTRACTOR CAVET CONSORTIUM

DATE 08/12/98
CHAINAGE 35031.50 m
OVERBURDEN 300 m

Data collector

GEOLOGICAL DATA
FORMATION: OLISTROTROMA
LITHOLOGY: SILTITES
STRUCTURE: SCALY
HYDROGEOLOGY: FACE DRY

GEOMECHANICAL CHARACTERISTICS:

Material	No.	LITHOLOGY	%	σ_c	CONSISTENCY	ALTERABILITY	UNIT VOLUME
	a	SILTITES	100	20 MPa	LOW	ABSENT	0.038
	b						
	c						

Discontinuity	No.	TYPE	POSITION	SPACING	PERSISTENCE	CONDITION
	1	Cleavage1	225/45	0.6 – 2 M	1 – 3 M	Dry
	2	Cleavage 2	210 – 230/60 – 70	< 6 CM	1 – 3 M	Dry
	3					

DIAGRAM OF FACE

CLIV. 2
CLIV. 1

Cu	c	φ	RMR	Rob–Pac
/	0.4 MPa	22°		

RESPONSE TO EXCAVATION: METHOD OF EXCATION: HYDRAULIC HAMMER

DEFORMATION PHENOMENA	TYPE	EXTR./CONV.
FACE	LOCAL FRACTURING	2 CM
CAVITY		10 – 12 CM
LININGS		

ADECO CLASSIFICATION	
FACE STABLE	A
STABLE FACE IN THE SHORT TERM	B
UNSTABLE FACE	C

ACTION:

		TYPE	QUANTITY	LENGTH/DIST.	OVERLAP/THICKNESS	P
EXCAVATION			41	24 M	6 M	
PROFILE						
CAVITY	PREL.	CAVITYSTEEL RIBS 2IPN80 SHOTCRETE	25 CM			1.00 M
CAVITY	FINAL					

SECTION TYPE	B2
ADVANCE RATE	1.4 m./day
COST	

Fig. 3.2

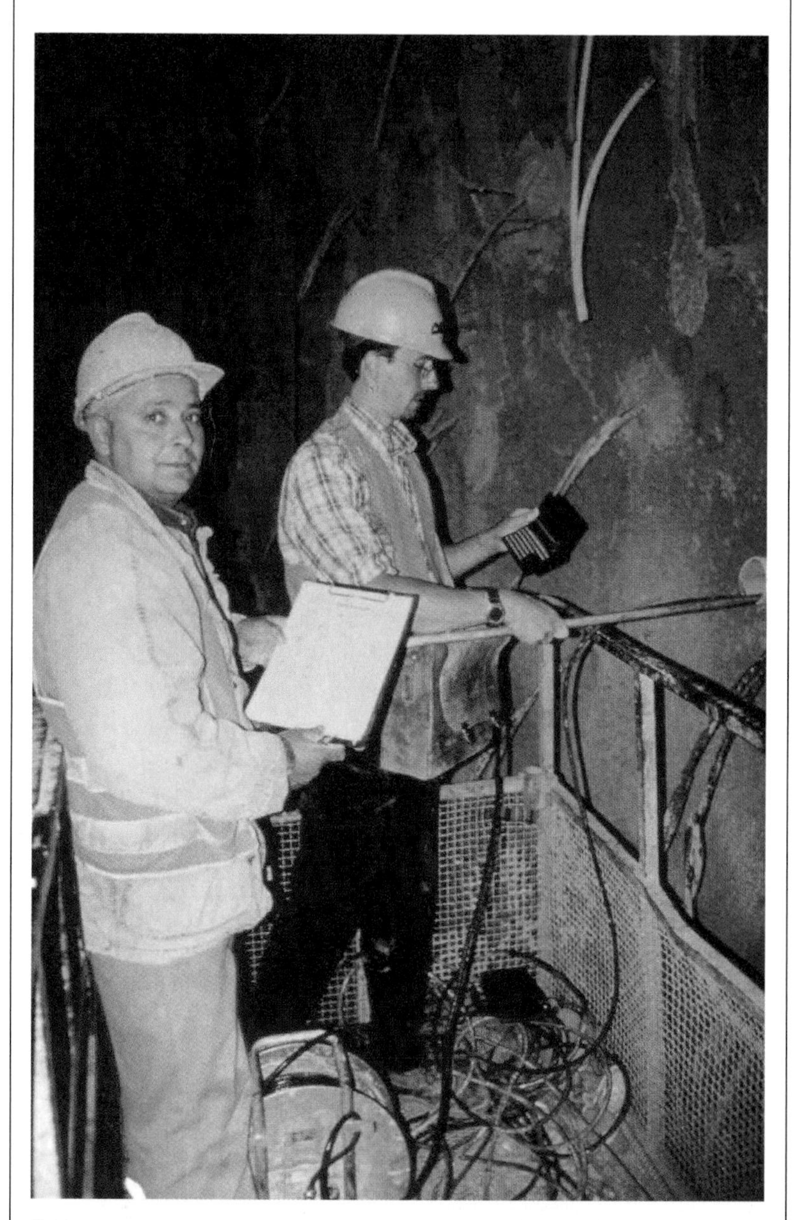

Taking core-face extrusion measurements using a sliding micrometer

and to compare it systematically to that of the cavity, both in the absence and the presence of protective and reinforcement intervention.

Its behaviour in terms of **stability** was analysed using an observational approach with more than one thousand tunnel faces classified and the principal data for them summarised on special sheets (see Fig. 3.2).

In **deformation** terms, on the other hand, the advance core was studied by systematic measurement of (Fig. 3.3):

- extrusion, obtained by fitting the advance core with a horizontal extrusion meters (sliding micrometer type) of a length equal to 2 –3 times the diameter of the excavation. These give longitudinal deformation in absolute terms of the ground in the advance core, both as a function of time (static phase, face halted) and as a function of face advance (dynamic phase);
- topographical plottings of movements in absolute terms of the face, by means of optical targets, taken with the face at a halt;
- preconvergence taken from the surface, whenever the morphology of the terrain and the size of the overburden allowed, by setting up multi-point extensometers, inserted vertically into the ground

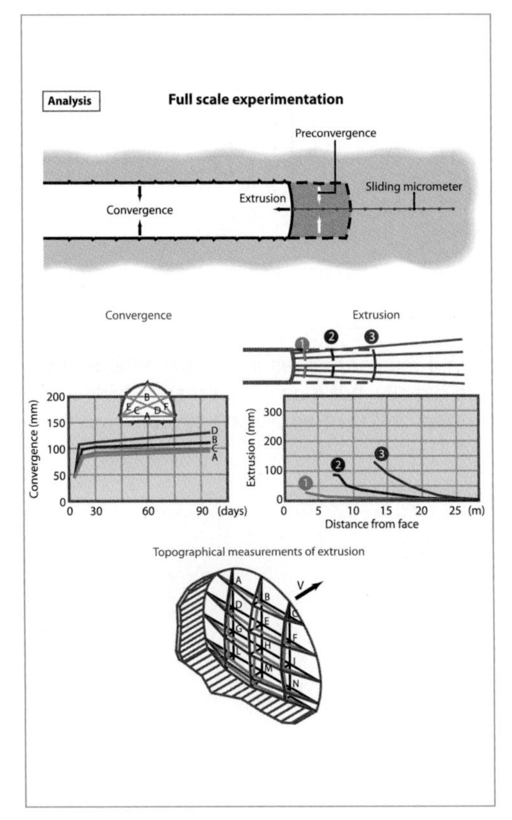

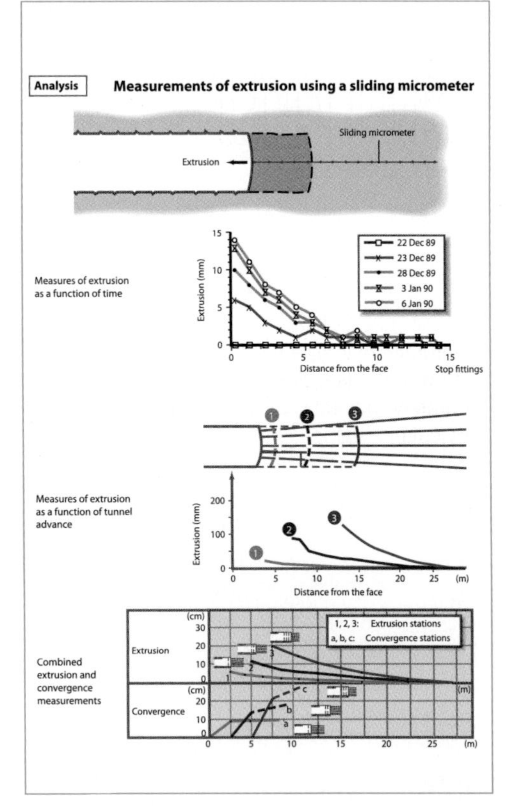

Fig. 3.3

Preconvergence measurements from the surface

While the size and development of the visible part of the radial deformation of the walls of an excavation that is produced back from the face, commonly termed convergence, can always be known by measuring it directly using, for example, simple tape distometers, knowledge of the invisible part of that deformation, that which develops ahead of the face and which is therefore termed “preconvergence”, is quite a recent conquest.
The first measurements of preconvergence were taken a little more than 15 years ago by placing multi-base rod extensometers vertically over the crown of a tunnel to be excavated from the surface and sufficiently ahead of the arrival of the tunnel face.

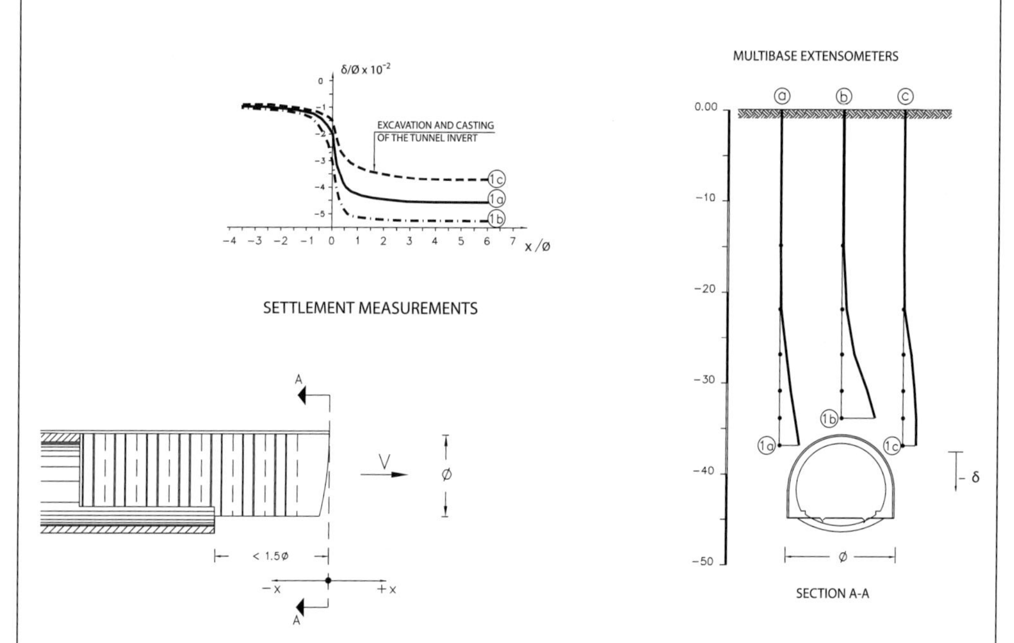

Naturally this type of measurement can only be taken if the overburden is shallow. Where this is not the case, preconvergence cannot be measured directly, but it can be assessed fairly accurately indirectly by measuring advance core extrusion and then employing preconvergence calculation tables.

Fig. 3.4
Installation of a sliding micrometer to measure core-face extrusion

before the arrival of the face above the crown and sides of the tunnel being driven [19].

These measurements were naturally always accompanied by traditional measurements, such as measures of convergence and of stress on linings.

Full scale experimentation achieved the following:

- it confirmed, by the construction of special extrusion-convergence diagrams (Fig. 3.3), both the existence of a close correlation between the magnitude of the extrusion allowed in the advance core and that of convergence that manifests after the passage of the face and also that these both decrease as the rigidity of the core is increased;
- it established that the advance core extrudes through the wall of the face (surface extrusion) with three basic types of deformation (cylindrical, dome-like spherical, combined) depending on the material involved and the stress state it is subject to;
- the calculation of preconvergence in absolute terms, using simple volumetric calculations that can be easily performed using tables, even when it is not possible to measure it directly from the surface (Fig. 3.5);
- it was verified that as preconfinement action of the cavity increases and the band of plasticised ground around the tunnel decreases as a consequence, there is subsequently a proportionally smaller load on the preliminary and final linings.

Estimating preconvergence using preconvergence calculation tables

By very careful study of advance-core extrusion measurement diagrams produced in tunnels and by reasoning in terms of the simple balance of the volumes in play (volume of ground intruding radially ahead of the face across the theoretical profile of the cavity = volume of ground extruding longitudinally across the theoretical wall constituted by the tunnel face), it was possible to calculate the correlation of extrusion itself with preconvergence of the tunnel. The correlation was then perfected on the basis of the three types of extrusion identified by mapping the displacements of points measured topographically on tunnel faces (see the sheet *The types of extrusion*), mapping which was performed on a very large number of different types of ground and under a great variety of stress-strain conditions.
The result of processing all these observations was the development of a summary in the form of *preconvergence calculation tables* which made it possible to reconstruct, for the first time, the entire deformation history of a cross section of a tunnel (which as is known, consists of preconvergence + convergence and not just of convergence, the only component that could be measured before now), even when it is impossible to take extensometer measurements from the surface.

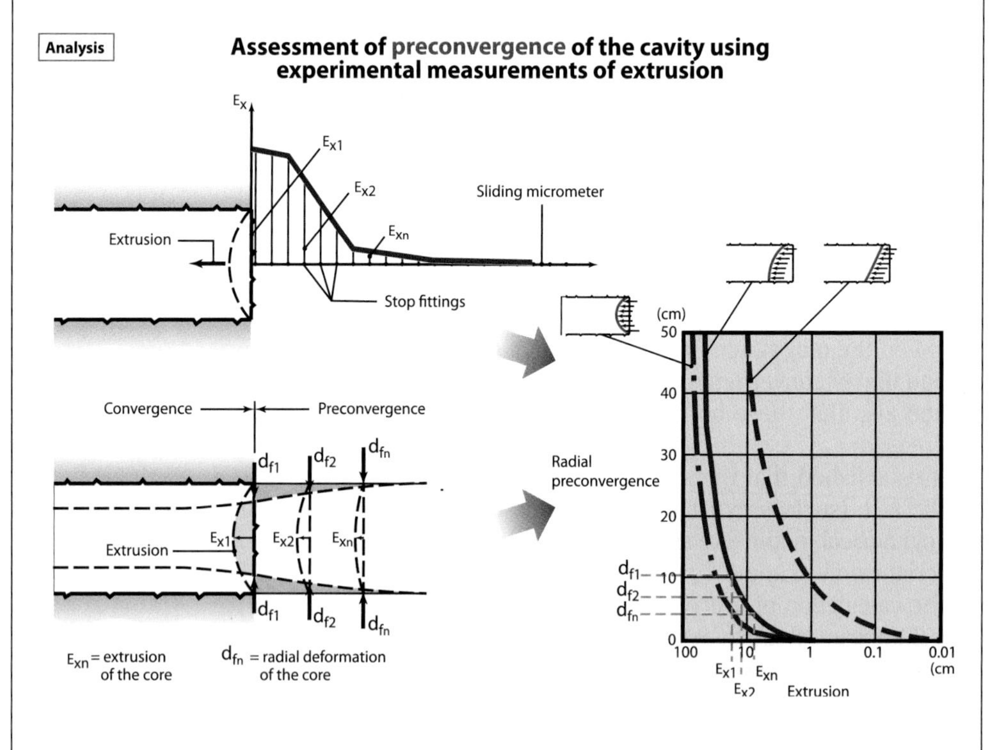

Fig. 3.5

3.1.2 Laboratory experimentation

Laboratory experimentation was also carried out in parallel with full scale experiments in order to fully analyse all aspects of face extrusion in tunnels.

The phenomenon had already been studied by Broms and Bennemark in 1967 (Fig. 3.6) but only in terms of stress trigger thresholds. They had defined a stability coefficient $N_s = f(\sigma_1, \sigma_3, c_u)$, with which it seemed to be possible to predict whether extrusion in a tunnel would be triggered or not [25], [26].

a - Broms and Bennermark's extrusion test

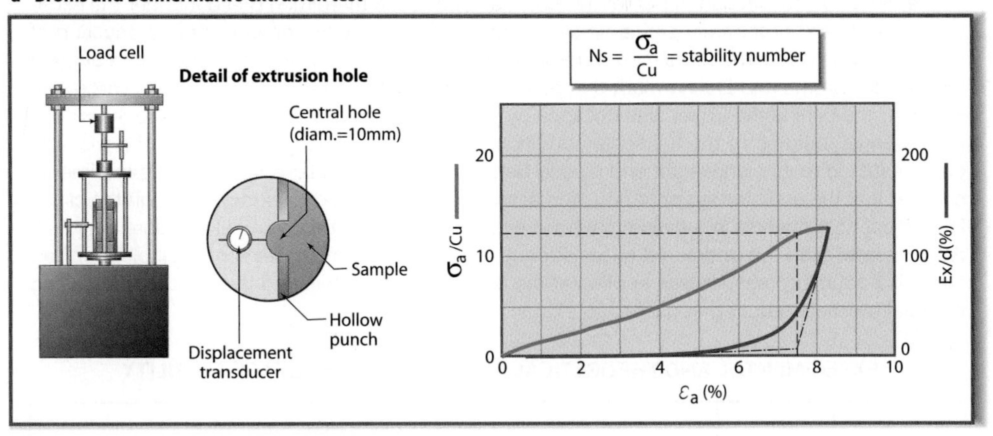

Fig. 3.6

This method had been applied in the 1980's to study and design a series of tunnels that were then being constructed on the new Sibari-Cosenza and Florence-Arezzo railway lines. Although it was found to be reliable for pure clays, it seemed to be of little use at the design stage for predicting the size of extrusion in terms of the dynamics of deformation, as the authors themselves pointed out.

New experimental models capable of reproducing the dynamics of face behaviour more accurately were designed to overcome these limits and to provide preliminary indications on the magnitude of the action required to control extrusion phenomena. Two newly designed tests were therefore studied and developed:

- the triaxial cell extrusion test;
- the centrifugal extrusion test.

In the **triaxial cell extrusion test** (Fig. 3.7), a sample of ground is inserted into a cell and the original stress state σ_o of the rock mass is recreated. The pressure of a fluid is then used to also reproduce

The stability of the face: experimental and theoretical studies – 1

The problem of predicting the stability of a tunnel face in terms of extrusion has been examined scientifically by numerous authors using different types of approach. The table that follows, which is certainly not exhaustive, summarises the most significant contributions on which we initially based our studies.

The problem was first addressed experimentally by Broms and Bennermark following a serious accident that occurred on 20th November 1964 at Edsadalen, near Stockholm, in a tunnel driven through clayey ground. Ninety minutes after a circular hole was opened, 1.98 m in diameter, at the base of a support wall 10.70 m high, the face extruded disastrously even though the surface had been inspected just an hour before and the clay had been hard and dry. The first sign of failure, noticed a few seconds before the event by one of the three men present under the excavation, consisted of a barely perceptible movement of the exposed ground. Before any kind of alarm could be given, the three men were buried under 47 m^3 of clayey material and one of them lost his life. Broms and Bennermark thought they would increase their knowledge of the failure mechanisms behind the Edsadalen collapse by performing a series of experiments to reproduce the essential features of the full scale conditions in the laboratory, similar to that of a mass of clay subjected to its own weight and placed behind a vertical circular opening. The experiments showed that the undisturbed or remodelled samples of clay started to extrude through the circular vertical hole opened in a cylindrical hollow punch, when the total vertical pressure was 6–8 times greater than the undrained shear strength of the ground at the level of the hole. Furthermore, if a counter pressure was applied at the level of the hole, this had the effect of increasing the total pressure required to cause the extrusion.

EXPERIMENTAL AND THEORETICAL STUDIES OF TUNNEL FACE STABILITY

AUTHOR	METHOD USED	FIELD OF APPLICATION										PARAMETERS USED						
		GEOMETRY OF THE CAVITY		OVERBURDENS			GROUND											
		SHAPE	$\varnothing_{max}$ EXPERIMENTED [m]	SHALLOW 10	20	DEEP 30 DEEPER [m]	PURELY COHESIVE c → 0		0 ← φ WITH FRICTION	STRATIFIED	WATER TABLE	C_U	C'	φ'	E	σ_t	F_S	N
BROMS-BENNERMARK (1967) ATTEWELL-BODEN (1971)	EXPERIMENTAL	O	0.02 – 0.04	*			*					*				*		*
LOMBARDI (1974) PANET-GUENOT (1982)	ELASTO-PLASTIC CALCULATION	O			*	*	*	*	*				*	*	*	*		
TAMEZ (1984)		O	8	*		*	*				*	*	*	*		*	*	
ELLSTEIN (1986)	CONE OF GROUND	O	4–6	*			*				*	*				*	*	*
TAMEZ-CORNEJO (1988)		O		*		*	*	*	*	*	*	*	*	*		*	*	
LECA-PANET (1988)	LIMIT ANALYSIS	O		*		*	*	*	*			*	*	*		*		
CHAMBON-CORTE' (1990)	LOGARITHMIC SPIRAL	O	4–5	*		*			*			*	*	*				

b - Triaxial extrusion test

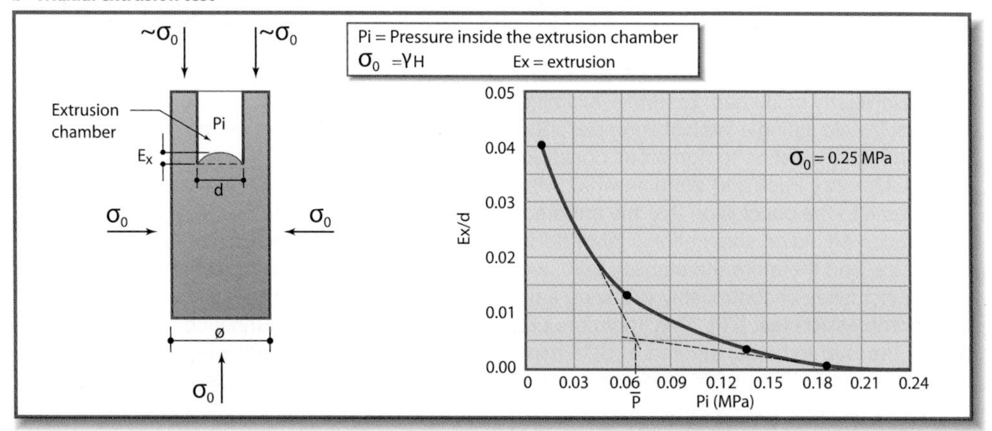

Fig. 3.7 *Apparatus for triaxial extrusion test.*

the stress state inside a special cylindrical volume, termed an "extrusion chamber", which is cut out from inside the test sample before the test. The chamber is coaxial to the sample and simulates the face zone of a tunnel.

By maintaining the stress state around the sample constant and gradually reducing the pressure P_i of the fluid inside the extrusion chamber, a realistic simulation is obtained of the gradual decrease in stress produced in the medium for a given cross-section of tunnel, as the face approaches and a forecast of extrusion at the face as a function of time or as a function of the decrease in internal confinement pressure P_i is therefore obtained. The curves generated are similar to those in Fig. 3.7 and in all those cases in which it is reasonable to

The stability of the face: experimental and theoretical studies – 2

In 1968 Attewell and Boden started experimental work in the engineering geology laboratories of the University of Durham to further examine the hypotheses that had guided Broms and Bennermark in calculating a 'stability number'. The value of between 6 and 8 that they had proposed seemed high in relation to numerous construction experiences reported by researchers such as Kuesel, Moretto, Peck and Ward, in which where instability occurred at lower values.
They came to the conclusion that the apparent incongruity was basically due to a problem of interpretation based on conventional laboratory approaches to the stress-strain principle whereby the failure, and therefore the stability ratio, were defined by the yield load (beyond which deformation increases uncontrollably following a tiny increase), obtained from the stress-strain curve for a compression test [26]. They therefore proposed that under actual tunnel construction conditions the load to be considered to determine a stability number, was that which produced the maximum acceleration of the extrusion movement, corresponding to 4.5 times the undrained shear strength of the clayey material.
Attewell and Boden processed the results of their laboratory tests statistically for mineral composition, the index properties and the undrained shear strength of the ground and found that there is no single value for the ratio applicable to all cohesive soil conditions, but that it can vary as a function of the physical properties of the clay found *in situ* (for the liquidity index above all).
Although the experiments on extrusion through the face of a tunnel conducted by Broms and Bennermark and continued by Attwell and Boden, suffer from all the problems typical of studies performed in the laboratory (effect of small scale, impossibility of working on completely undisturbed samples, difficulty in reproducing *in situ* conditions faithfully, the results cannot be generalised to all types of ground), they furnish a guide for the first time that is sufficiently reliable for engineering predictions of the phenomenon in purely cohesive soils.

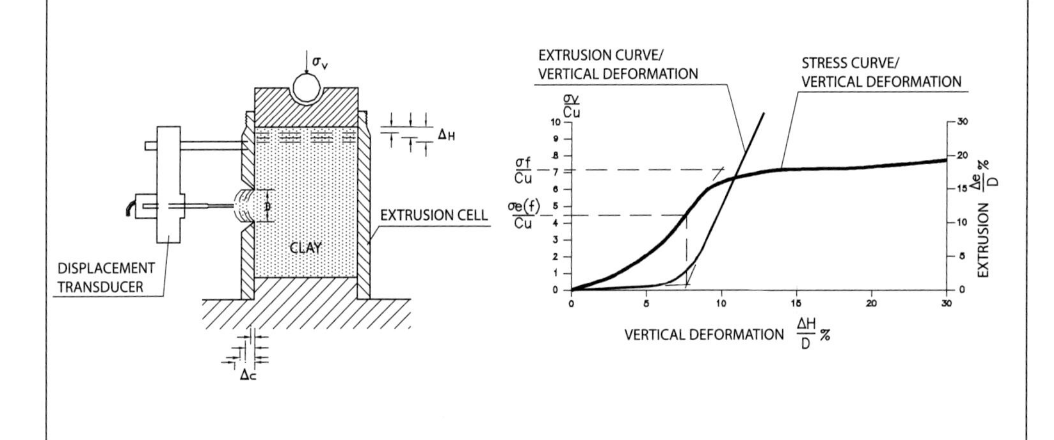

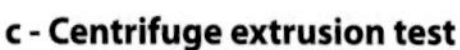

c - Centrifuge extrusion test

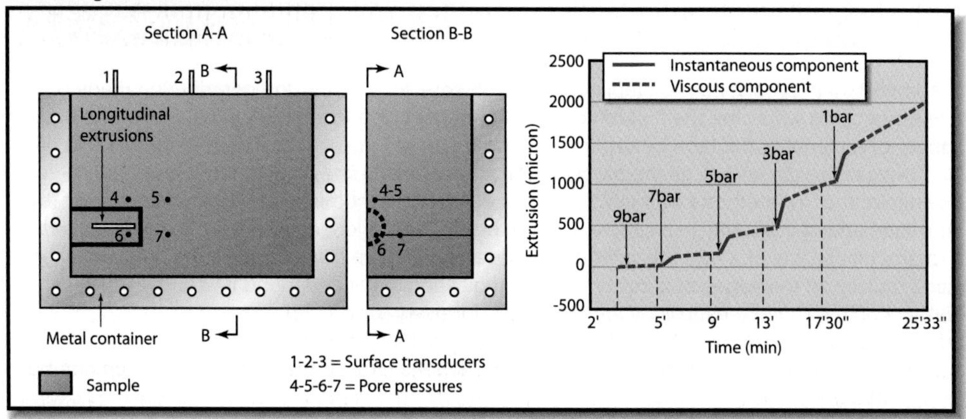

Fig. 3.8

overlook the effect of gravity on extrusion (medium to deep overburdens), they can be immediately used in the design stage for calculating the preconfinement pressure required to guarantee a given rigidity of the core and, as a consequence, the control of preconvergence.

Some considerations can be formulated from the numerous extrusion tests performed in triaxial cells:

1. given the modest dimensions of the sample, these tests apply mainly to the matrix of the ground, which must be mainly clayey;
2. if there are irregularities in the ground (schists or scaliness) the test will only work if these are of negligible size relative to that of the sample;
3. the more homogeneous the ground the more probability there is of the results being applicable to the full scale situation.

Centrifuge extrusion tests (Fig. 3.8) [21] were developed and performed for those cases in which gravity has a significant influence on extrusion. Use of these is limited to a few specific cases because of their high cost and complexity.

Special markers and transducers for measuring deformation and pore pressure are inserted in the sample of ground, which is then placed in a special box with a transparent wall. After having cut out the opening of the tunnel in this, a steel tube is inserted representing a first approximation of the preliminary lining, the lining and the tunnel invert. The cell thus obtained is filled with a fluid kept under appropriate pressure. The natural geostatic pressure is then recreated in the centrifuge and once this is reached the pressure in the cell is reduced to simulate the excavation of ground at the face.

The results obtained (Fig. 3.8 gives those for a centrifuge extrusion test performed on a reconstituted sample of ground) show that extrusion at the face manifests rapidly at the point in time when the pressure is released and develops with increasing speed as relaxation of the core progresses. The figure shows both the instant and the viscous extrusion separately for each step of pressure release. It is easy to see that the latter accounts for 50% of total extrusion at the end of the test.

The stability of the face: experimental and theoretical studies – 3

It is important in predicting the stability of a tunnel face to consider the contribution made by a few researchers who were the first to convincingly schematise the complex problem of the equilibrium of a tunnel at the face by using a pseudo three dimensional calculation approach in the elasto-plastic range known as the *theory of the characteristic lines.* This theory is in fact still today one of the few methods with which it is possible to predict the stress-strain response of the core-face and of the cavity in a simple manner, but with sufficient accuracy for many purposes. A characteristic line of a cavity is intended as meaning a curve that connects the pressure exercised radially on the edges of a cavity with the radial convergence around it.
Three principal characteristic lines are calculated for every tunnel [34]:

- the core-face characteristic line from which the behaviour of the tunnel at the face can be predicted;
- the characteristic line of the cavity at the face, which takes account of the three dimensional effect of the stresses close to it and which can be used, by correlating it with that of the core-face using a simple graph, to calculate the preconvergence that the tunnel has already been subject to in the tunnel cross section at the face;
- the characteristic line of the cavity, valid for any cross section sufficiently distant from the face, for which the deformation state can be considered flat.

Once the characteristic lines for the core-face and for the cavity at the face are plotted on a graph, the point at which they intersect identifies the equilibrium stress-strain situation at the face of a tunnel in the absence of stabilisation intervention. Four different situations can be distinguished corresponding to four different cases of stability [35]:

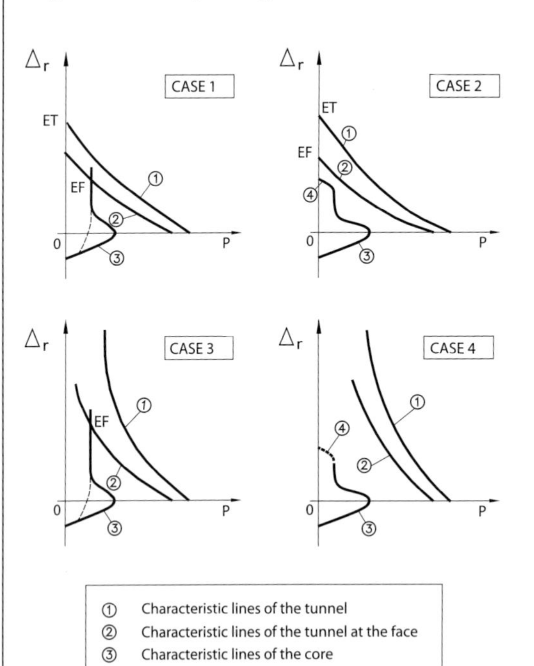

① Characteristic lines of the tunnel
② Characteristic lines of the tunnel at the face
③ Characteristic lines of the core
④ Failure of the core
EF Equilibrium at the face
ET Equilibrium of the tunnel

Case 1: both the face and the cavity are stable by themselves.

Case 2: the cavity is stable, but the face isn't (possible in theory, but rare, although it is seen in reality).

Case 3: stability is only guaranteed close to the face.

Case 4: neither the face nor the cavity are stable.

Experiments which reproduce the phenomenon of advance core extrusion in the laboratory were fundamental, together with the results of full scale measurements, for the correct weighting of the geomechanical strength and deformation parameters (c, ø, E) in the mathematical models used in the theoretical part of our studies on controlling deformation responses.

3.2 Numerical analyses

The complexity of the mechanisms that come into operation ahead of the face and the initial difficulty in identifying objective criteria for predicting the stress-strain behaviour of the advance core, which go beyond intuition and experimental findings, required an effort to produce an organic and unified interpretation of the numerous aspects investigated which in turn would provide a general theoretical framework capable of overcoming the limits of current theories.

To do this, our analysis of the deformation response continued using theoretical tools. Three different approaches were employed:

- initially we tried to make use of existing calculation theories updating them where necessary;
- we then sought to solve the problems using axial-symmetric finite element and finite difference mathematical models;
- finally, we resorted to three dimensional mathematical modelling.

3.2.1 Studies using analytical approaches

To start with, we tried to solve the problem by modifying and bringing existing analytical calculation approaches up-to-date. We sought in particular to introduce the concept of the core and reinforcing it into some of the classic formulas used for designing tunnels, for example the Convergence-Confinement Method [11] and the Characteristic Line Method [12], this being one of the few in which the concept of the core appears explicitly.

Both of these methods allowed us to simulate the effects of reinforcing the core and to reproduce some of our experimental results, including the resulting reduction of the radius of plasticisation R_p and decrease in deformation in the face zone.

However since both these methods separate the calculation of the stress-strain situation at the face from that at a distance from it (see Fig. 3.9), they cannot hold in memory the effects of what has happened ahead of the face for use in the equations that are valid for the zone down from the face. Consequently they are not able to interpret and represent the phenomena correctly as a whole.

This applies in particular to the decrease in the radius of plasticisation R_p and the consequent reduction in the deformation of the cavity (convergence) and in the loads acting on the preliminary and final linings. These phenomena were not detected in the results obtained using the two methods of analysis considered, while they have been systematically observed from experimental measurements [24].

We therefore concluded that these approaches, although useful in the diagnosis phase for defining the behaviour of the ground when excavated in the absence of

The stability of the face: experimental and theoretical studies – 4

If appropriate counter measures are not adopted in cases of face instability phenomena before the failure process sets in, then this progresses and affects increasing volumes of material. In the end a new state of equilibrium will be formed at the cost, however, of the failure of the face along the plane of maximum shear and the formation of a raise or a chimney in the crown of the tunnel.

Various researchers have taken this observation as a starting point to propose approaches based on failure calculation and formulations that are valid for both cohesive soils and ground with friction properties to predict the stability of a tunnel. In practice it is assumed that there is a portion of ground inside the core-face and above the tunnel which is detached along a certain longitudinal section. The shape of it, defined with varying degrees of accuracy, depends on the method of calculation employed (shapes that are closest to real failure surfaces can only be investigated by using a computer).

These methods generally produce rather high values, compared to other types of approach, for the confinement pressures required for face stability as a consequence of the large volumes of ground set in motion and of the simplification of the three dimensional effects. Nevertheless in the case of tunnels under soils subject to cohesionless landslide, where the failure mechanism is that of a near vertical chimney which intersects the surface (subsidence), they interpret the physical reality fairly faithfully. However, they have the disadvantage that they take only very approximate account of the dome effect that may be produced in the ground or don't even consider it at all and this is a disadvantage. Consequently these methods should be used with caution for tunnels located at a depth of more than twice their diameter.

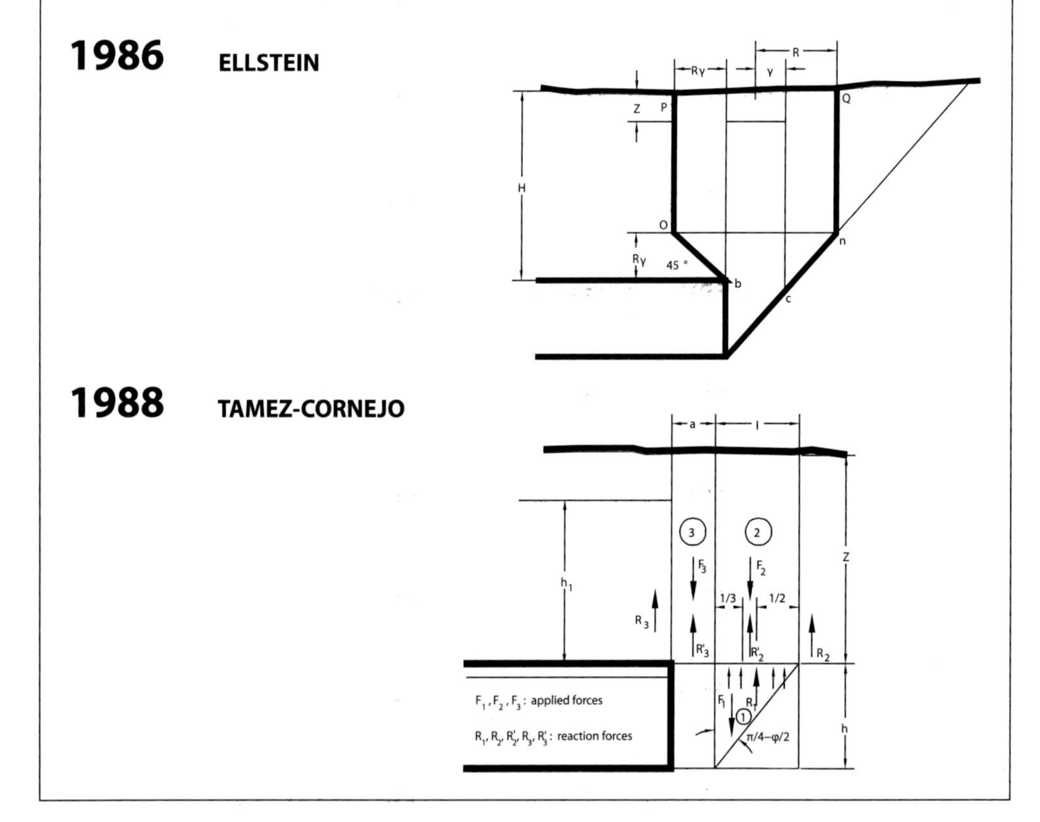

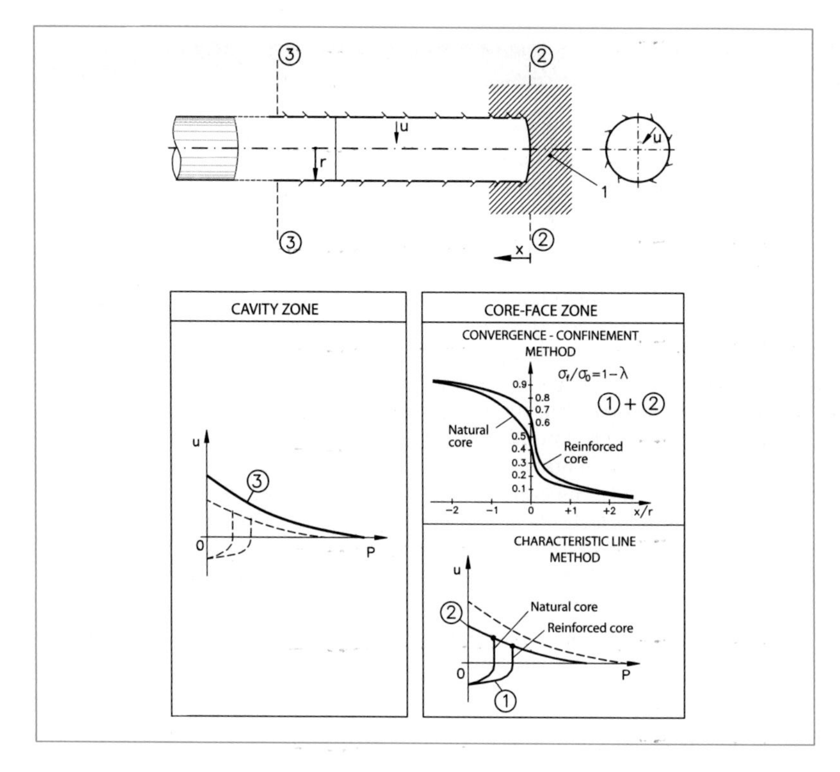

Fig. 3.9

preconfinement intervention, they are not equally useful in the therapy phase where preconfinement is employed. This is because they cannot predict deformation of the cavity with sufficient accuracy, nor can they be used to calculate the dimensions of the preliminary and final linings correctly.

It was therefore decided to abandon these types of approach and to take the path of numerical models (finite elements or finite differences), which are able to take continuous account of the whole stress-strain history of the medium around the excavation, passing from the area ahead of the face to that behind it.

3.2.2 Studies using numerical approaches on axial symmetrical models

The effect of reinforcing the core was therefore investigated using finite element and finite difference numerical models. We began using **axial symmetric** (Fig. 3.10) type models since they are easier to manage than three dimensional models.

Although we were unable to overcome some of the intrinsic limitations of these approaches (tunnel perfectly circular, uniform stress states in the ground surrounding it, the impossibility of considering

The stability of the face: experimental and theoretical studies – 5

The approach proposed by Chambon and Corté in 1990 [33] is a logarithmic spiral method. These methods study the equilibrium of a sliver of ground, equipped with both friction and cohesion which, as it comes away from the face, slides along a failure surface given by a logarithmic spiral which intersects the lower generatrix of the tunnel. This shape of the failure surface is in fact very similar to those observed in reality. The application of the method studied by Chambon-Corté is two dimensional, but it is possible to develop a three dimensional version which takes account of the real shape of the failure surface. Chambon and Corté furnish solutions for cases of tunnels at depth and for tunnels close to the surface, for which account can be taken of the potential presence of surface loads. The results, which were compared with experimental results obtained by the same authors on tests conducted on centrifuge models, appear sufficiently accurate to identify the order of magnitude of the limit pressure to be exerted on the face to prevent failure. Naturally the results of the calculation obtained from the three dimensional study over estimate the degree of pressure really required to ensure face stability, because in this case the calculation overlooks the contribution of the cohesion acting along the lateral faces of the real three dimensional failure surface. The theoretical analysis not only considers the influence of the different geotechnical characteristics of the ground (angle of friction and cohesion) on face stability, but also appears able to furnish design engineers with reasonable indications of the form of the failure mechanism and its critical depth, beyond which the failure of the face can be produced without affecting the surface.

1990 CHAMBON - CORTÉ

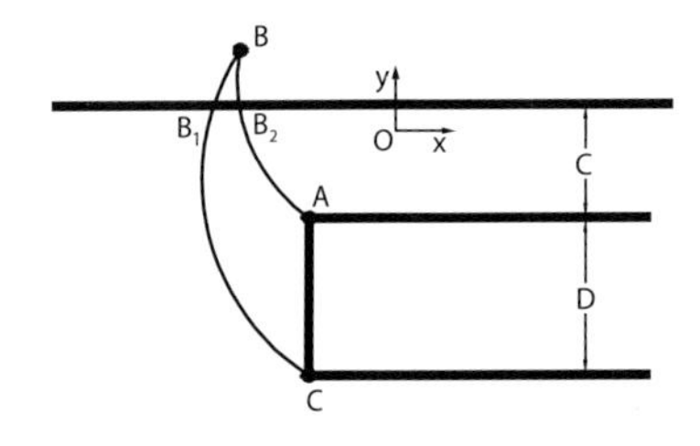

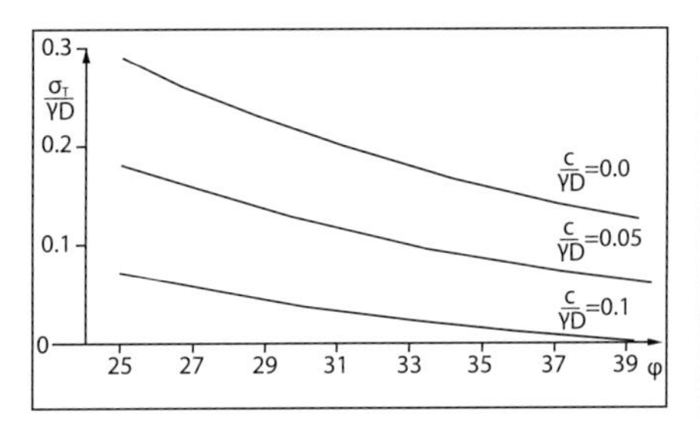

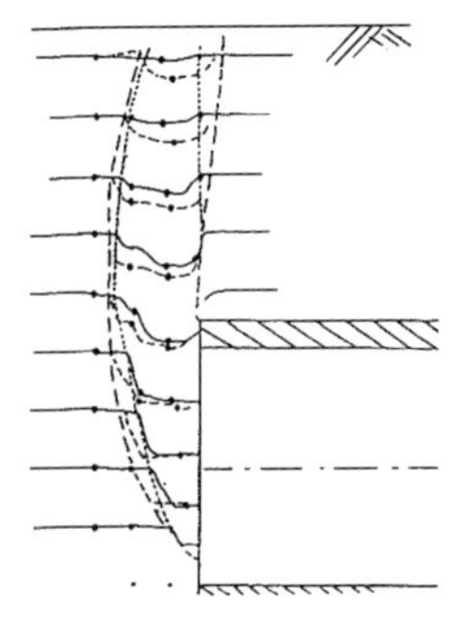

Development of the failure surface for a tunnel with a shallow overburden, obtained from a centrifuge test. [32]

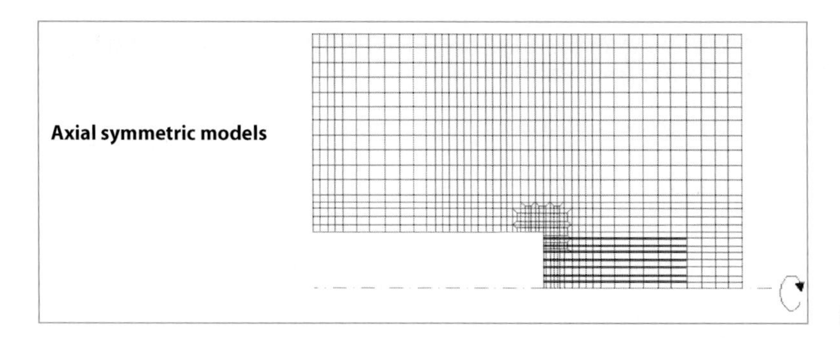

Fig. 3.10

any linings other than closed rings and therefore of simulating real construction cycles), the use of these models did nevertheless show that ground reinforcement of the core produces a different distribution of stresses ahead of the face and around the tunnel. This finally confirmed from the use of calculation too, that reinforcement of the core results in a reduction in the size of the band of plasticised ground and in all deformation around the tunnel both ahead of the face and down from it (not only extrusion and preconvergence but also convergence) (Fig. 3.11). In addition, analyses conducted using axial symmetric numerical models showed that it is not possible to control extrusion and preconvergence by varying the rigidity of tunnel linings and/or the

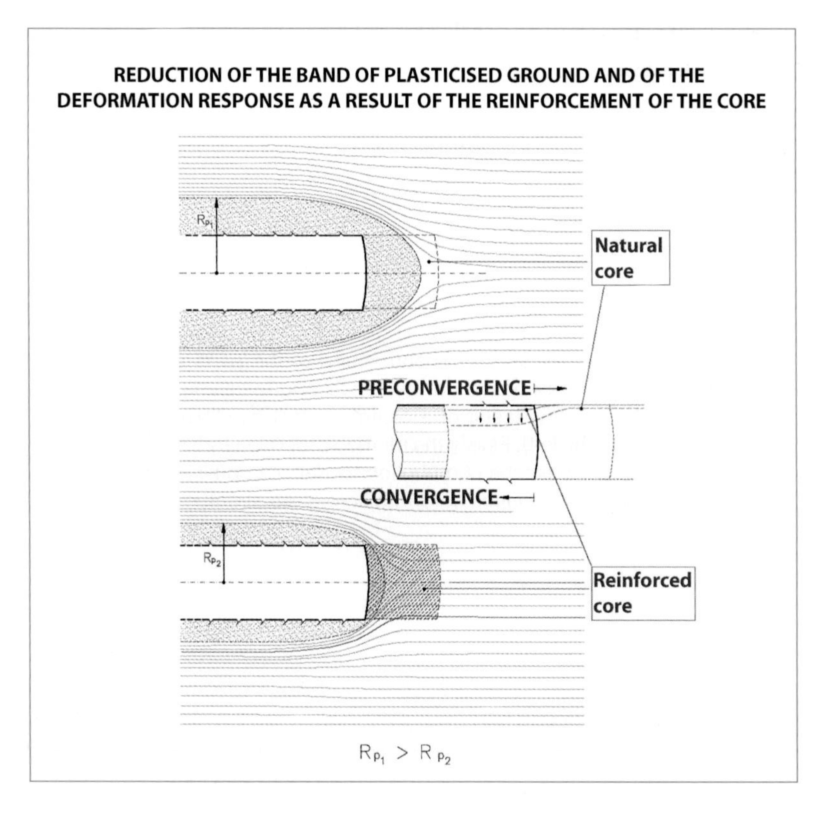

Fig. 3.11

The stability of the face: experimental and theoretical studies – 6

An interesting study conducted between 1989 and 1990 for a degree thesis [37] at the Faculty of Civil Engineering at the University of Florence was also based on logarithmic spiral methods. The purpose of the study was to develop a computer programme based on those used to verify the land stability of slopes which could be used to:

1. identify, from the infinite possibilities, the logarithmic spiral shaped slip surface with the minimum safety coefficient;
2. consequently, predict the core-face stability situation during excavation in the absence of stabilisation intervention;
3. assess the variation in that situation as a function of the type, geometry and intensity of possible stabilisation intervention using fibre glass structural elements applied to the advance core.

The basic scheme considered, within which the programme operated, consisted of hypothesising that the advance core was affected by a failure surface with the shape of a logarithmic spiral along the longitudinal section and that the mass of ground bearing on the volume of ground coming away, again along the longitudinal section, had the form of a directrix parabola.

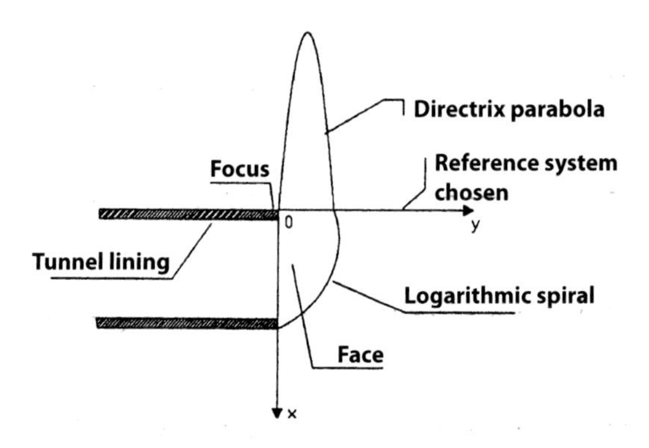

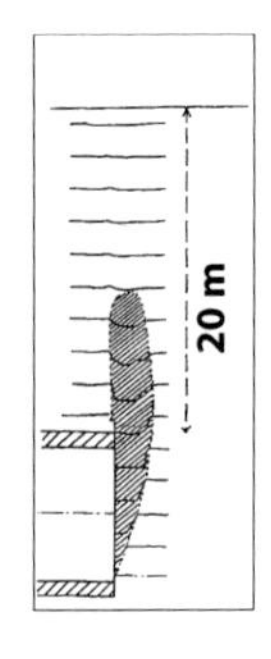

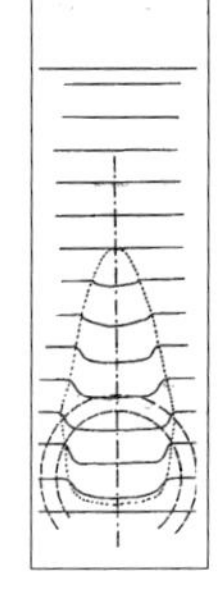

These choices for the forms and shape of the parts of the ground coming away were based on theoretical and experimental considerations [32] [38] (the result of a centrifuge test by Chambon-Corté in the figure on the left). Finally the method proposed by Anthoine A. [39] was used to assess the contribution to stability obtained by the reinforcement of the advance core with fibre glass structural elements.

distance from the face at which they are placed alone. In other words, they demonstrated that it is impossible to remedy what has already happened ahead of the face with confinement action only [36].

Although axial-symmetric models displayed a reasonable ability to simulate tunnel advance in the presence of advance core reinforcement and furnished better results than those obtained using analytical methods, they did not display an equal ability to predict the loads acting on the preliminary and final linings, which with this type of model would be more or less the same as those that would result in the absence of advance core reinforcement, other conditions remaining the same.

This contrasts, as has already been said, with observations that have been made during experimental research, which have been confirmed many times in practice during construction. This is because it is impossible with these types of models to take account of the gravitational effects produced by the plasticised ground around the tunnel and of the real construction cycles involved in placing the preliminary and final linings.

3.2.3 Studies using numerical approaches on 3D models

Resort had to be made to three dimensional numerical modelling to overcome the difficulties encountered with axial symmetric numerical models (Fig. 3.12). With this method the real geometry of a tunnel could be introduced into the calculations, which is no longer simply circular as in the case of the *convergence-confinement method*, or the *characteristic line method* and (finite element or finite difference) axial symmetric analysis. It is also possible to consider stress states in the ground that are not of the hydrostatic type, which take due account of gravitational loads and which also calculate the effects which the various construction stages have on the statics of a tunnel by simulating the real geometry of lining structures and the sequence and the distance

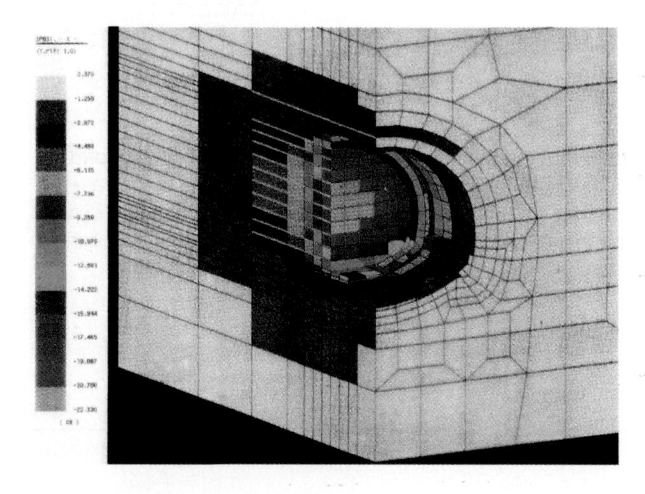

Fig. 3.12

The stability of the face: experimental and theoretical studies – 7

The approach proposed by Panet and Leca in 1988 [31] examines the problem of face stability for a circular tunnel located at shallow depth and excavated using a drilling machine capable of exerting continuous confinement pressure on the face, both in purely cohesive and friction soils. It is assumed that the tunnel is lined for the whole of its length up to a distance P from the face with a perfectly rigid ring lining, while a uniform confinement pressure σ_T is exerted on the wall of the face. Face stability conditions are assessed by using the upper and lower limit theorems of plasticity theory which provide an estimate against and in favour of safety respectively. In order to define the upper and lower limits of the load systems which could potentially be withstood by the statically and plastically admissible load systems and the kinematically compatible failure mechanisms are considered. The combination of the approximate solution obtained using the lower limit theorem and the equally approximate solution obtained using the upper limit theorem provide a picture of the face stability conditions. This combination can be depicted graphically on a plane for purely cohesive grounds (Tresca criterion), while depiction is more difficult for grounds with friction and cohesion (Mohr-Coulomb criterions) because it must occur in a space.

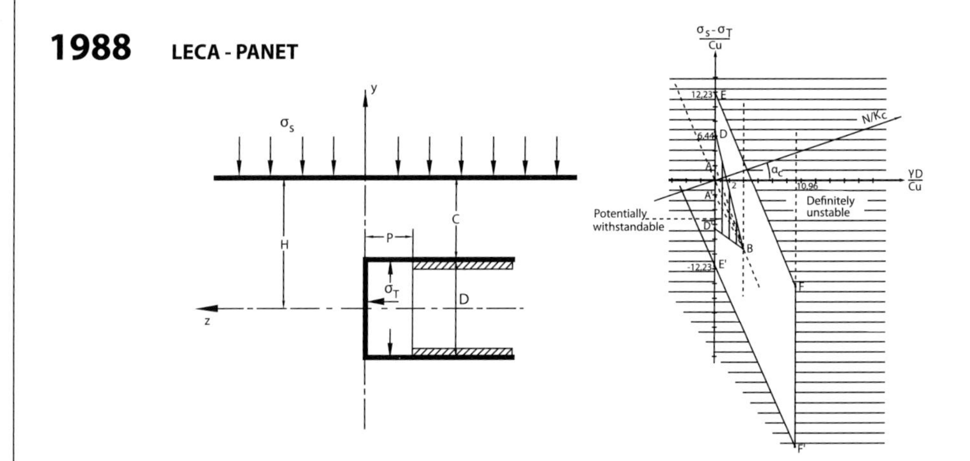

The gap between the boundaries obtained by applying the two theorems is rather wide, partly because of the low number of load systems and failure mechanisms considered. In most cases, however, those researchers who have addressed the problem of tunnel face stability agree in reporting a good match between the experimental findings and the solutions obtained by applying the lower limit theorem of plasticity theory.

from the face in which they are placed. It was therefore possible, as is illustrated below, to employ numerical calculation to study how the distribution of extrusion movements at the face and the consequent different failure mechanisms vary as a function of the distance at which the tunnel invert is placed.

The results obtained from 3D models generally agree well with experimental observations, both as far as deformation is concerned (extrusion, preconvergence and convergence) and with regard to stresses on linings, which are lower when the advance core is reinforced, as found in the experimental research.

3.3 Results of the experimental and theoretical analyses of the deformation response

Experimental and theoretical analysis of the deformation response using the advance core *as the key to interpreting long and short term deformation phenomena in tunnels* enabled us to identify, with certainty, the strength and deformability of the advance core as the true cause of the whole deformation process (extrusion, preconvergence and convergence). It also confirmed, beyond any reasonable doubt, that it is possible to control deformation of the advance core (extrusion, preconvergence) by stiffening it with protective and reinforcement techniques and as a consequence to also control the deformation response of the cavity (convergence) and the size of the long and short term loads on the tunnel lining.

Consequently, if the strength and deformability of the advance core constitute the true cause of the deformation response of the ground to excavation, it is possible to consider it as a new tool for controlling that response: an instrument whose strength and deformation properties play a determining role in the long and short term stability of the cavity.

CHAPTER 4

Control of the deformation response according to the ADECO-RS approach

The experimental and numerical research on the deformation response of the ground shows that the true cause of the entire stress-strain process (extrusion, preconvergence and convergence) that is triggered when a tunnel is excavated lies in the deformability of the advance core. It follows therefore that in order to solve all types of stress-strain problems, but above all those found under difficult ground conditions, one must act first on the advance core by regulating its rigidity appropriately. In terms of forces this means employing preconfinement and not just confinement action, where preconfinement is defined as any active action which favours the formation of an arch effect in the ground ahead of the face.

It is for this reason that it is necessary to achieve complete control of the deformation response in the ground (Fig. 4.1):

1. ahead of the face, by regulating the rigidity of the advance core using appropriate preconfinement techniques;
2. back from the face in the cavity, by controlling the manner in which the advance core itself extrudes using tunnel confinement techniques capable of providing immediate continuous active confinement of the cavity close to the face.

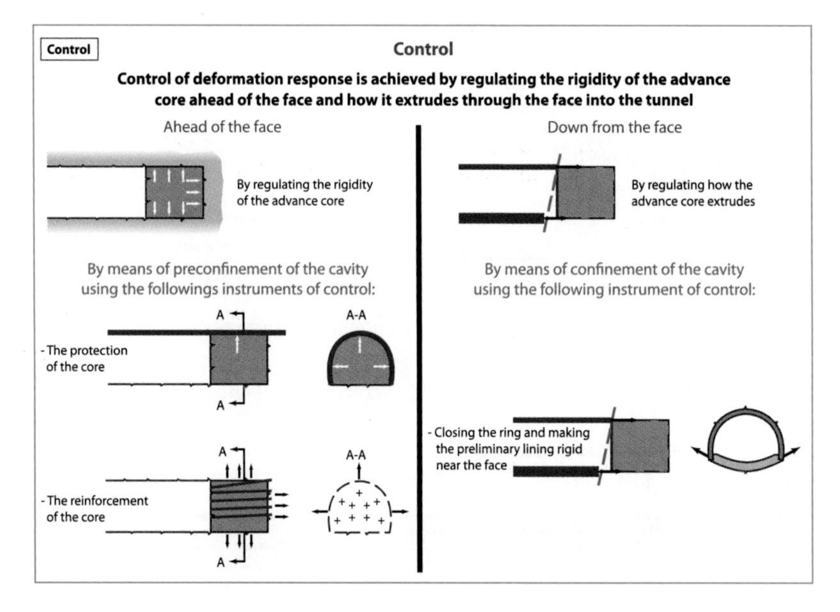

Fig. 4.1

Some reflections on the use of characteristic lines – 1

As we have already seen in the focus box, *The stability of the face: experimental and theoretical studies – 3*, the *theory of characteristic lines* makes it possible to predict the long and short term stress-strain response of the core-face and of the cavity in a simple manner, but with sufficient accuracy for many purposes. It must, however, be made clear that because of the lack of connection between the two formulas used to calculate the characteristic line that is valid at the face and the characteristic line that is valid for any section that lies outside the range of influence of the face, this theory is not always able to accurately account for the effects of preconfinement of the cavity on the stress-strain response of a tunnel.

Let us examine, for example, the case in which preconfinement is launched ahead of the face around the cavity (sub-horizontal jet-grouting, fibre glass structural elements distributed around the perimeter of a tunnel, etc.). The effect of this type of intervention can still be studied using characteristic lines, on condition, however, that a ring of material with its own geometrical dimensions and mechanical properties is introduced into the mathematical formulas for the two lines (that which is valid at the face and that which is valid at a distance from it). The figure below shows the results of such a study and it highlights how the characteristic lines of the tunnel change appreciably in the absence or in the presence of annular preconfinement of the cavity.

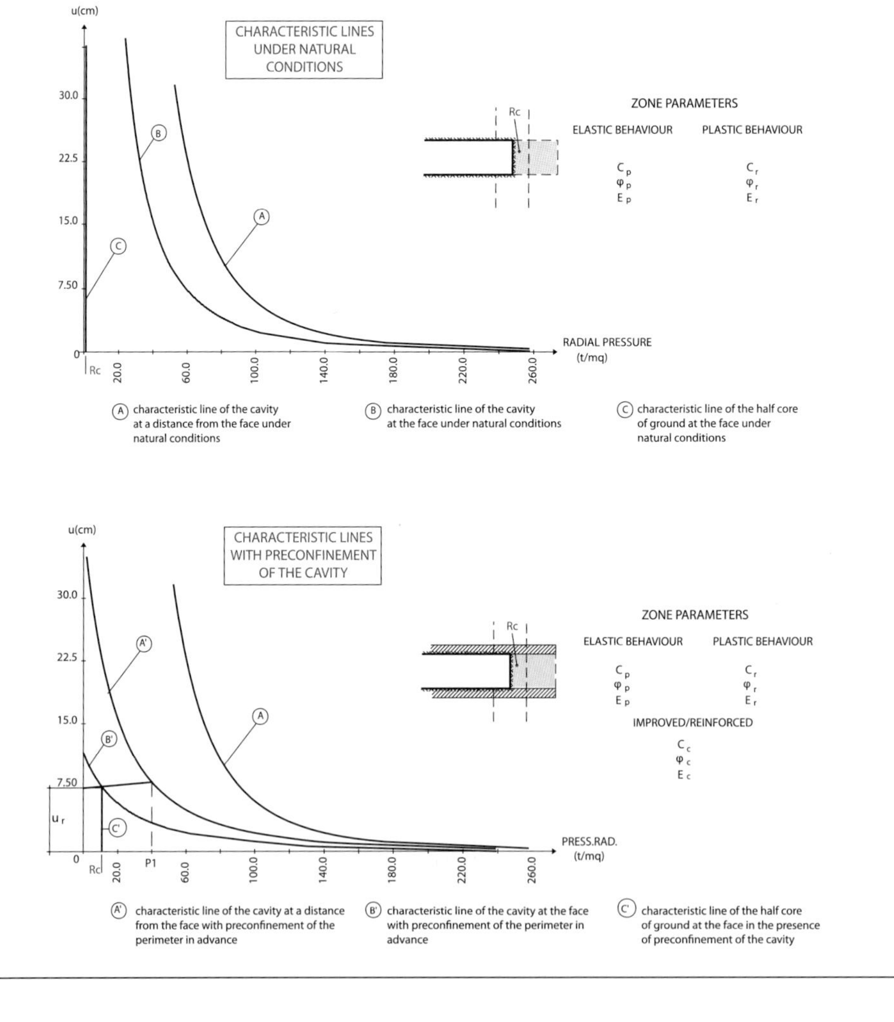

4.1 Control ahead of the face

In order to regulate the rigidity of the advance core and to thereby create the right conditions for complete control of the deformation response of the ground and therefore, in the final analysis, for complete stabilisation of the tunnel in the long and short term, the ADECO-RS approach proposes, as will be seen, numerous types of intervention that have been fully illustrated in numerous articles, some of which are listed in the bibliography.

All these types of intervention can be divided into two single categories (Fig. 4.2):

- *protective intervention*, when the intervention channels stresses around the advance core to perform a protective function that ensures that the natural strength and deformation properties of the core are conserved (e.g. shells of improved ground by means of sub-horizontal jet-grouting, shells of fibre reinforced shotcrete or concrete ahead of the face by means of mechanical precutting);
- *reinforcement intervention*, when the intervention acts directly on the consistency of the advance core to improve its natural strength and deformation properties by means of appropriate ground reinforcement techniques (e.g. ground reinforcement of the core using fibre glass structural elements).

Although these types of intervention for controlling the deformation response ahead of the face have a rather limited field of application in relation to the nature of the ground when considered individually, if they are considered as a whole, they are able to guarantee solutions for all possible geotechnical conditions. Naturally, in extreme stress-strain conditions, there is no reason why two or more types of intervention cannot be used at the same time to obtain a mixed action of protection and reinforcement (Fig. 4.3).

4.2 Control in the tunnel back from the face

As opposed to techniques based on conventional principles of tunnel advance, which ignore the cause of deformation and allow the core to deform, but then require the installation of flexible linings to absorb deformation which has already been triggered (a practice which in really difficult stress-strain conditions frequently turns out to be inadequate), the application of these new concepts in tunnel advance in the presence of a rigid core, characteristic of the ADECO-RS approach, requires the use of equally rigid linings as an essential condition, if the advantage obtained by reinforcing the core ahead of the face is not to be lost behind it. It is also just as important that maximum care and attention be paid to ensure that the *continuity of action in the passage from preconfinement to confinement occurs as gradually and as uniformly as possible, never forgetting that the cause of the whole deformation process that must be controlled lies in the strength and deformability of the advance core.*

Furthermore, numerical analyses performed on computers show extremely clearly that (Figs. 4.4 and 4.5):

1. when extrusion is produced, it occurs through a theoretical surface, termed the **extrusion surface**, which extends from the point of contact between the ground and the leading edge of the preliminary lining to the point of contact between that same ground and the leading edge of the tunnel invert;

Some reflections on the use of characteristic lines – 2

When preconfinement of the cavity has been performed by stiffening the advance core with fibre glass structural elements, jet-grouting or other means, the effect cannot be fully studied using the characteristic lines of the tunnel calculated with the analytical formulas currently in use. The characteristic lines of a cavity obtained in this way are in fact the same as those calculated in the absence of core stiffening. Only the line for the half core changes as a consequence of that stiffening. It follows that the value for the final pressure acting on the lining is greater than that calculated in the absence of reinforcement. This clearly conflicts with all the experimental evidence (see paragraphs 2.1.2.5 and 2.1.3), which clearly shows that stiffening of the advance core has a significant effect on all the characteristic lines of the cavity and not just on the line for the core. More specifically, the experimental evidence shows that the real characteristic lines for the cavity, both close to and at a distance from the face, are decidedly flatter than those calculated using the formulas mentioned above and that the values for convergence and the final pressure on the lining are significantly lower. Consequently, when the face core has been reinforced, the characteristic lines cannot represent the reality accurately unless they are calculated on a computer using axial symmetric or three dimensional models in the non linear field.

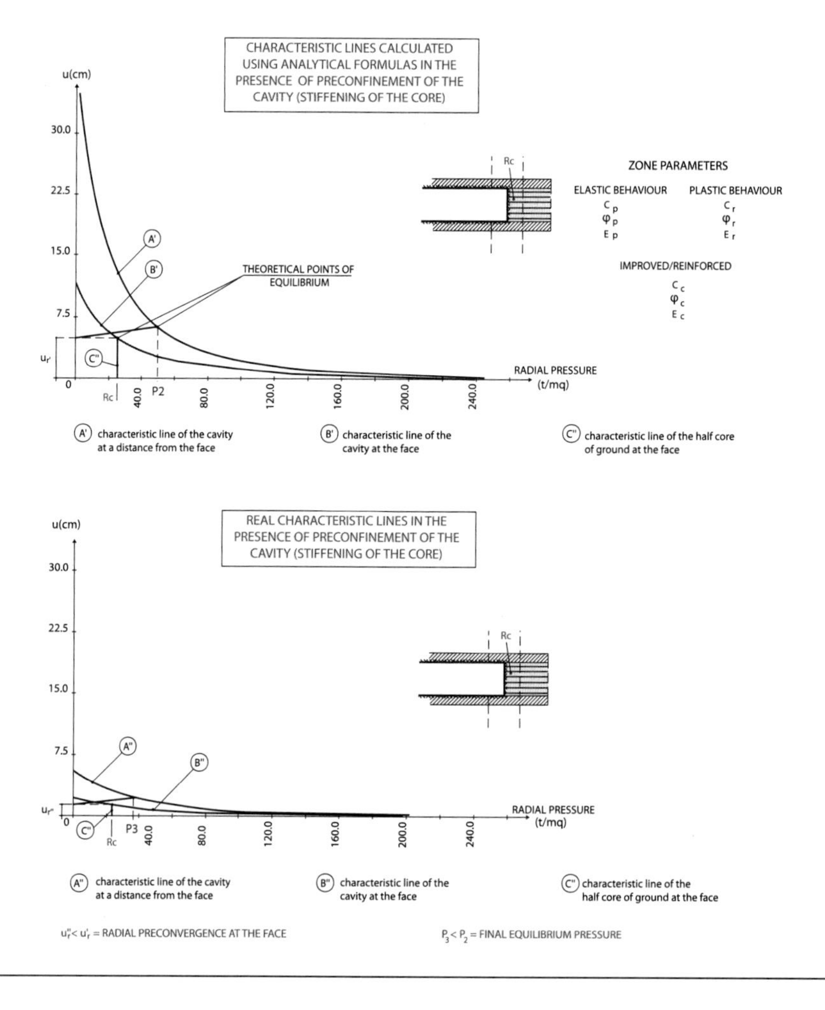

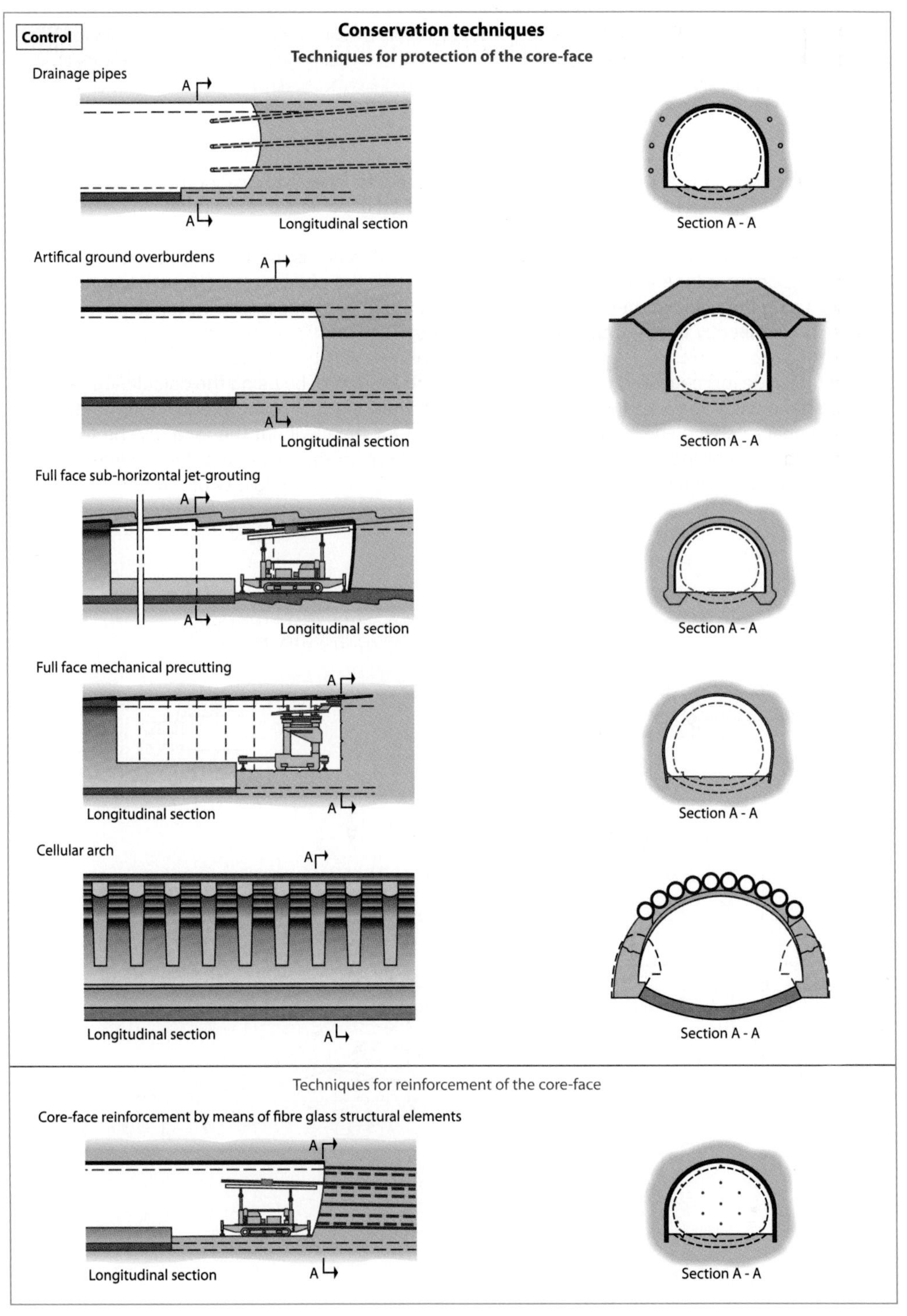
Control
Conservation techniques
Techniques for protection of the core-face
Drainage pipes
A
A
Longitudinal section
Section A - A
Artifical ground overburdens
A
A
Longitudinal section
Section A - A
Full face sub-horizontal jet-grouting
A
A
Longitudinal section
Section A - A
Full face mechanical precutting
A
Longitudinal section
A
Section A - A
Cellular arch
A
Longitudinal section
A
Section A - A
Techniques for reinforcement of the core-face
Core-face reinforcement by means of fibre glass structural elements
A
Longitudinal section
A
Section A - A

Fig. 4.2

Calibrating theoretical calculations using preconvergence calculation tables

Preconvergence calculation tables can also be used to effectively calibrate calculation approaches and to calibrate the characteristic line method in particular (see figure):

- if the strength of the *half core,* which may be reinforced, is known, the theoretical value for preconvergence at the face can be identified on the characteristic lines graph, which, by using the calculation tables in question, can then be compared with the real value obtained indirectly from extrusion measurements;

- if the value for measured extrusion is known and therefore, by using the calculation tables, also the value for preconvergence at the face, then the real point of equilibrium between the characteristic line for the half core and that for the cavity at the face can be found if the strength of the half core is known. Other points on the real characteristic line for the face can be identified in a similar manner as a function of the rigidity of the core.

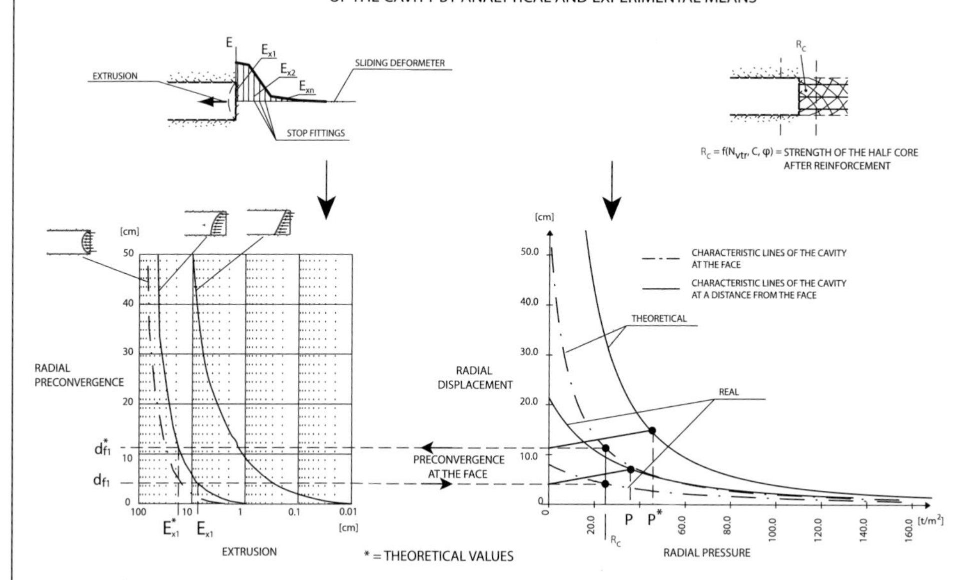

2. if the tunnel invert is cast closer to the face and the extrusion surface is progressively reduced, this produces an equally progressive decrease in extrusion (which tends to occur more symmetrically across the height of the face) and therefore in convergence also.

   They also show that:

- with the tunnel invert cast at the same distance from the face, the deformation calculated for half face advance is comparable to that obtained for full face advance (in other words, casting the tunnel invert a long way from the face is like top heading and bench type advance);

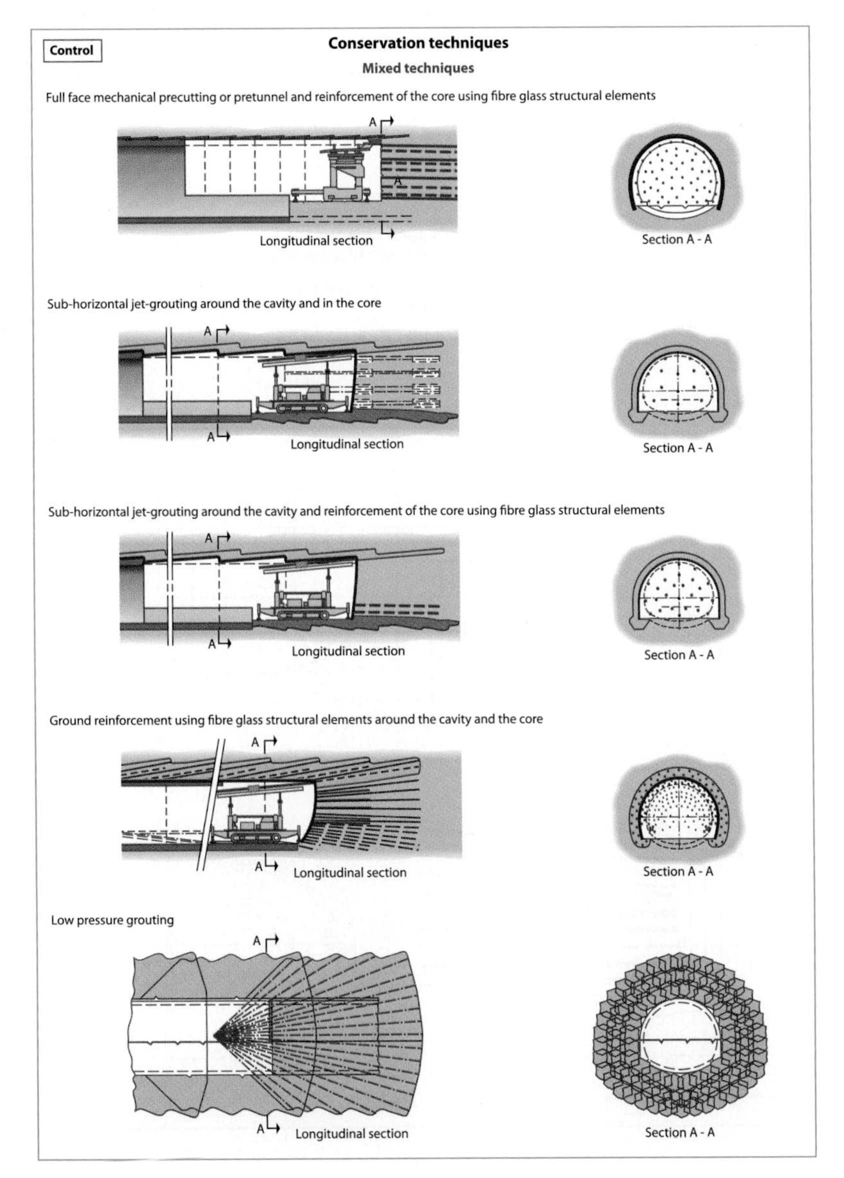

Fig. 4.3

Axial symmetric and three dimensional models

Three dimensional finite element or finite difference modelling conducted using appropriate calculation codes in the non linear field, appears, for now, to be the only path which produces results which match reality in terms, amongst other things, of stresses on tunnel linings in the study of tunnels to be driven after first reinforcing the advance core. The figure below shows some of the results obtained from three dimensional modelling conducted for the San Vitale Tunnel (Caserta-Foggia railway line).

The finite element model, consisting of 6,393 isoparametric elements with 8 nodes, simulates the excavation of the tunnel under a 100 m overburden. The face reinforcement consisted of 90 fibre glass structural elements cemented into the face and another 96, again in fibre glass, fitted with valves and injected around the theoretical profile of the tunnel [43].

Analyses were conducted by examining the curves for extrusion against the length of tunnel advance and extrusion against the number of fibre glass structural elements to identify the correct number and appropriate overlap in length of the fibre glass elements to be placed in order to prevent instability of the core-cavity system.

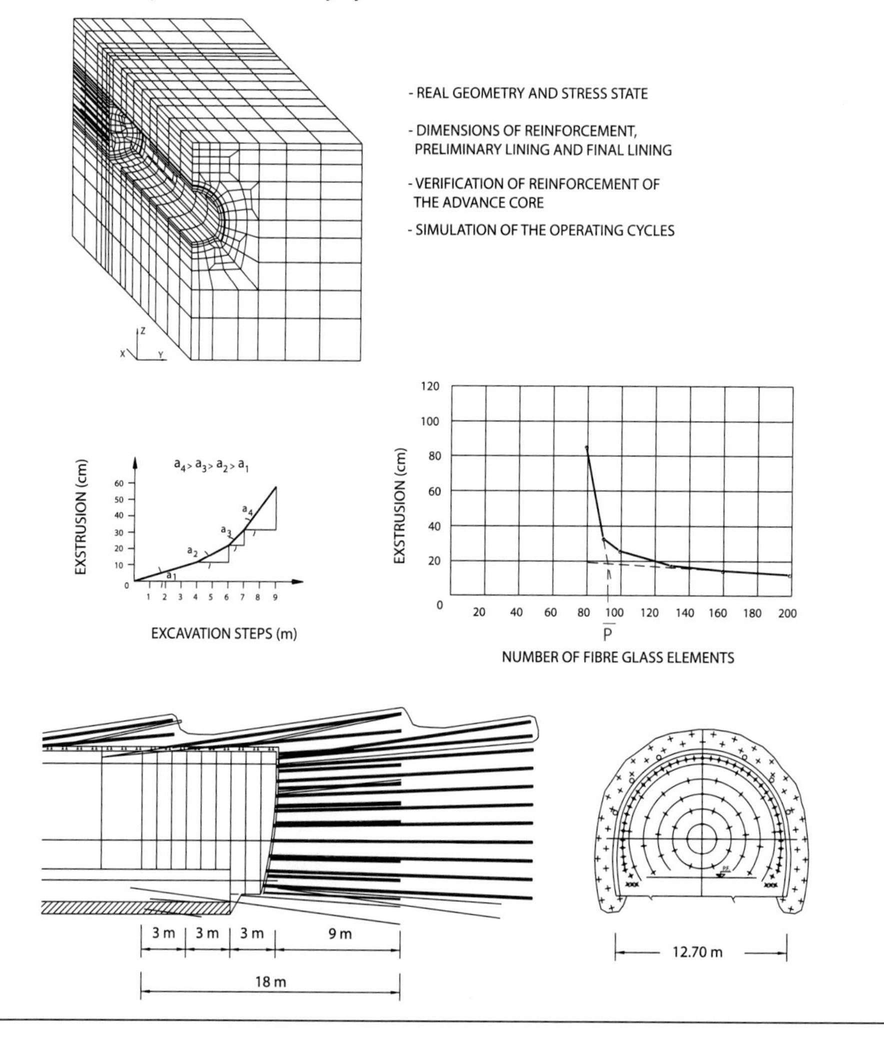

- half-face advance always produces greater deformation than full-face advance.

It follows that the tunnel design engineer, who has already started to control the deformation response with action to regulate the rigidity of the advance core, has the chance (which becomes of fundamental importance in extreme stress-strain conditions) to make that action a continuous process and to control extrusion from inside the tunnel by casting the kickers and the tunnel invert as close as possible to the face. To cast the latter far from the face results in a larger extrusion surface, asymmetrical extrusion and an advance core of greater

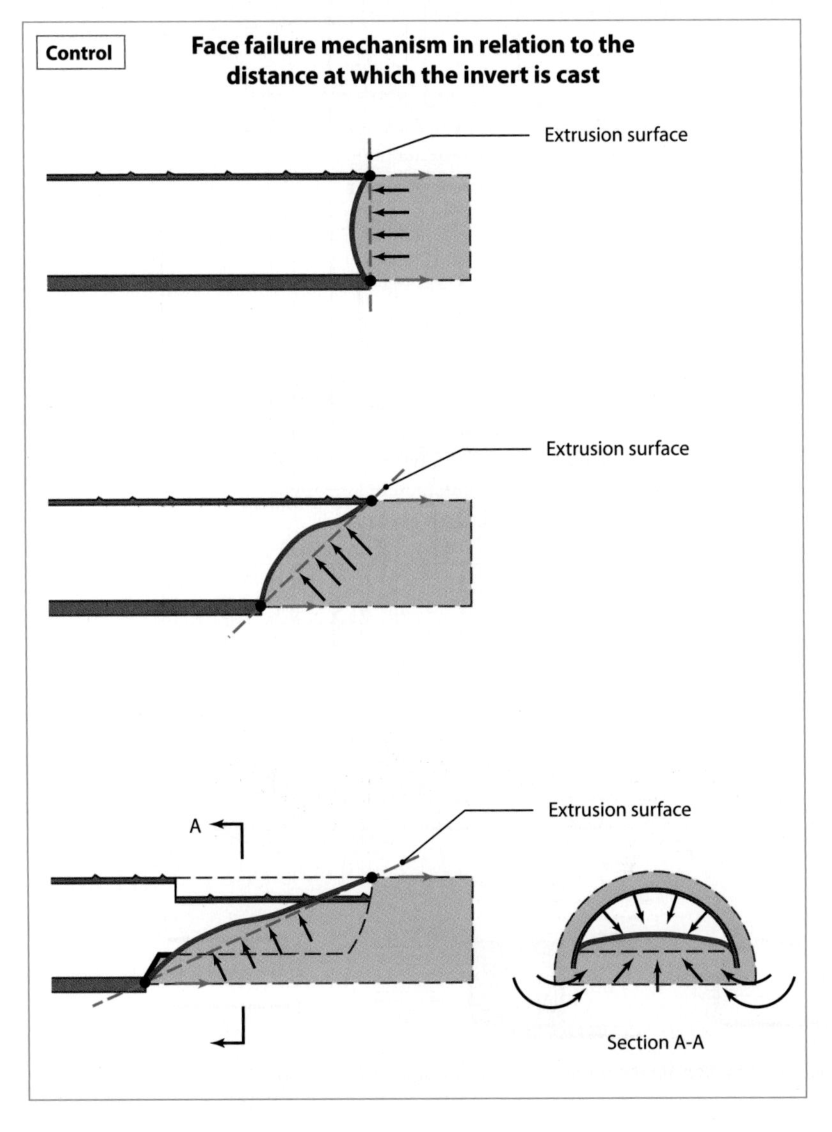

Fig. 4.4

The effectiveness of action to stiffen the core-face for controlling convergence

The results of the convergence and settlement measurements during the excavation of tunnels after first reinforcing the core-face confirm that deformation increases significantly as the length of the reinforcement elements diminishes (the thicker lines in the figures indicate the average displacements).

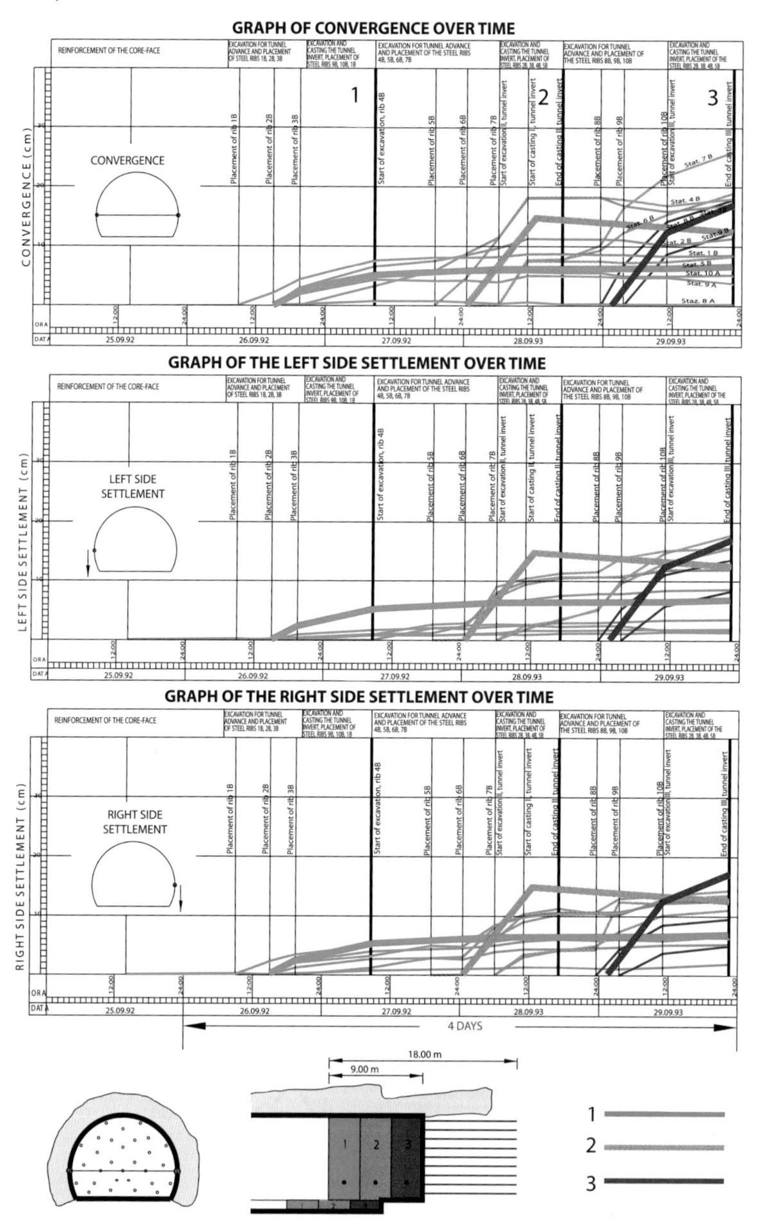

Measurements taken on the San Vitale Tunnel (Caserta-Foggia railway line).

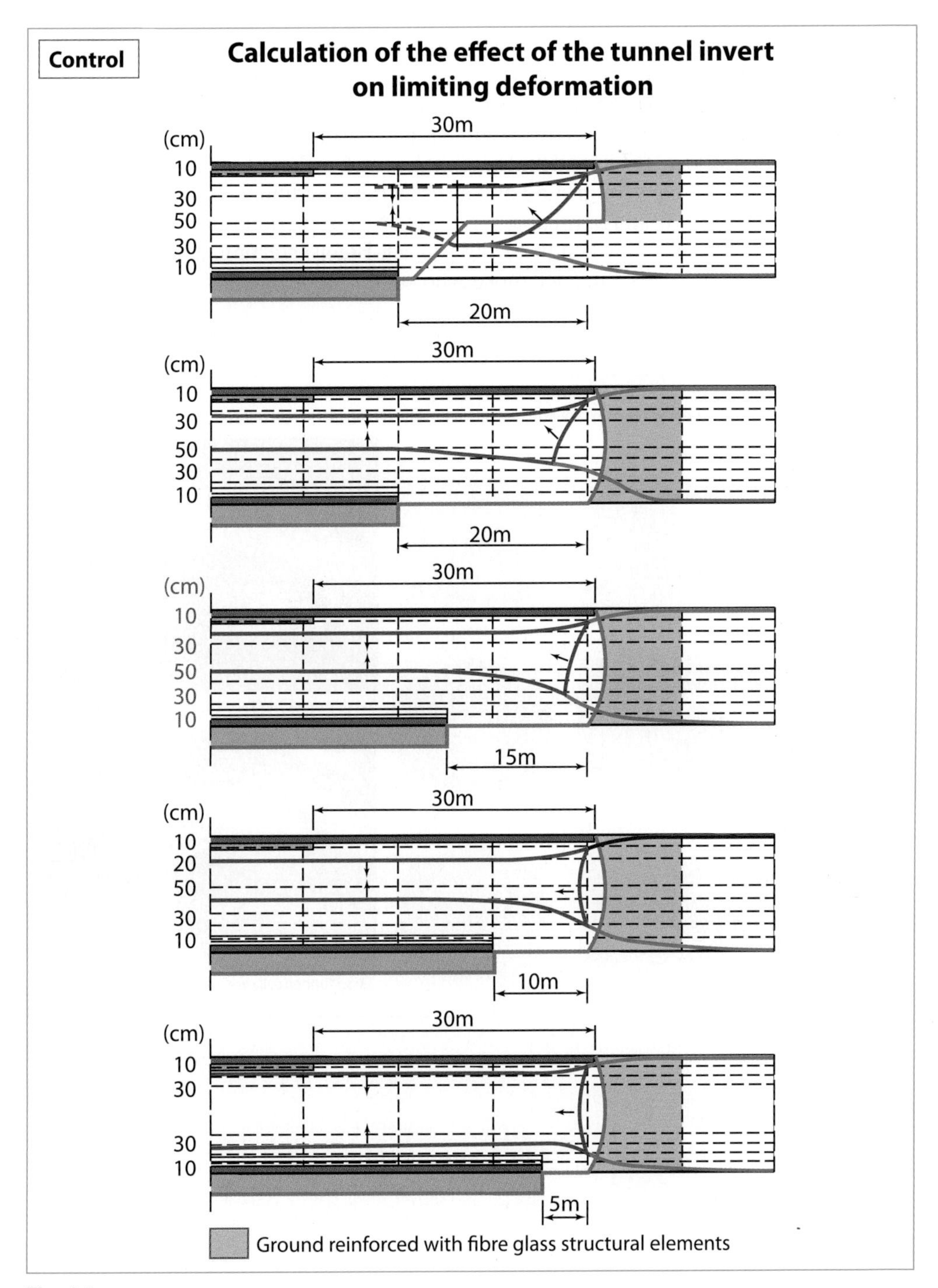
Control
Calculation of the effect of the tunnel invert on limiting deformation
(cm)
10
30
50
30
10
30m
20m
(cm)
10
30
50
30
10
30m
20m
(cm)
10
30
50
30
10
30m
15m
(cm)
10
20
50
30
10
30m
10m
(cm)
10
30
30
10
30m
5m
Ground reinforced with fibre glass structural elements

Fig. 4.5

dimensions that is more difficult to deal with and these are all conditions which lead to tunnel instability (Fig. 4.6).

At this point, the time was ripe to start to translate the theoretical principles of the ADECO-RS into a new approach to the design and construction of tunnels, which would overcome the limitations of conventional approaches and make it possible to design and construct tunnels in all types of ground and stress-strain conditions, to industrialise tunnel excavation and even to make reliable forecasts of construction times and costs as is normal for all other types of civil engineering project. Before starting, it was essential to set down guidelines as a reference for those who set out to design and construct underground works.

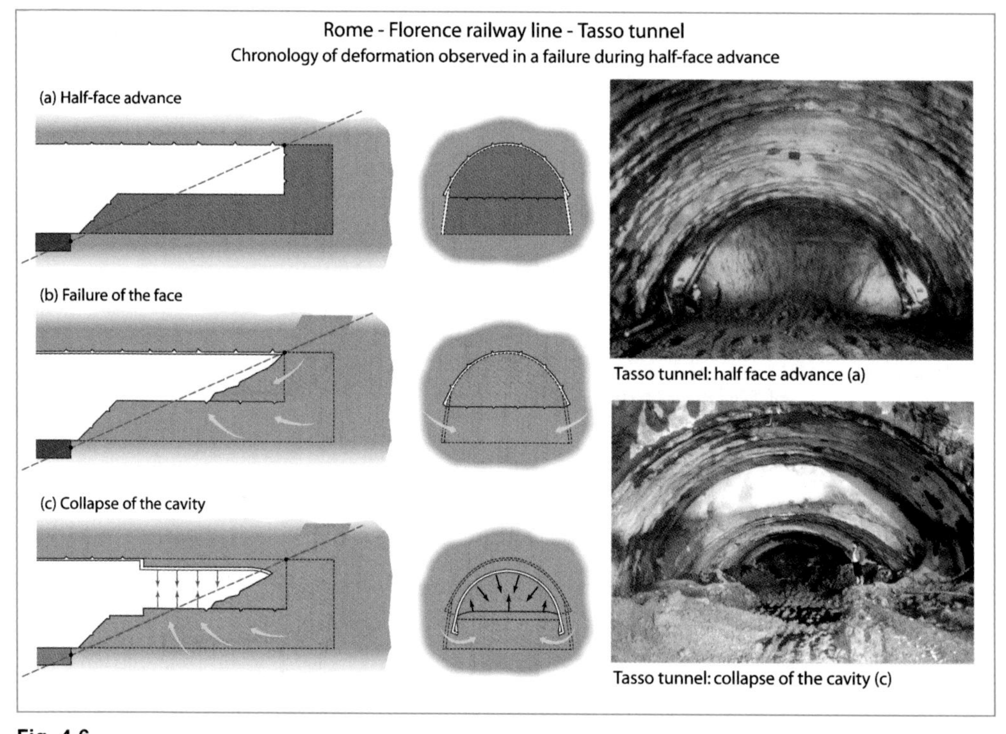

Fig. 4.6

CHAPTER 5

The analysis of controlled deformation in rocks and soils

5.1 Development of the new approach

It follows from what has been said in previous pages that today, a tunnel designer must no longer think immediately in terms of support structures and linings when designing a new tunnel, but must consider the entire resolution of statics problems in terms of the ground (medium) right from the outset. This is the true construction material, to be assisted where necessary, with action in advance to improve, protect and/or reinforce the core-face in order to guarantee the stable creation of an arch-effect (channelling of stresses) in a band of ground as close as possible to the profile of the tunnel that is able to resist the field of increased deviated stresses.

This is the only way, no matter what the geomechanical, environmental or architectural context of the project and no matter what the degree of initial uncertainty, in which the design and construction process can proceed if it is to be based on knowledgeable management of the complex mechanisms that govern the behaviour of an underground work right from the outset of excavation work.

From this viewpoint, we have seen that the analysis and control of the deformation response, understood as the reaction of the medium to the action of excavation, plays a fundamental role and constitutes indispensable steps to be taken in the correct design and construction of an underground work:

- analysis (designed to predict the deformation that will occur as a result of excavation) must be performed by theoretical means, using analytical or numerical calculation instruments at the *design stage*, during which the tunnel designer makes operational decisions, based on the predictions made, in terms of systems, stages, excavation cycles and ground reinforcement, improvement and stabilisation instruments;
- control of the deformation response, on the other hand, occurs at *the construction stage*, when design decisions are implemented and verified as excavation proceeds by assessing the deformation response of the medium to the action taken.

The following is therefore fundamental for the correct design and construction of an underground work:

- **at the design stage:**
  - to have a detailed knowledge of the medium in which one is to work, with particular regard to its strength and deformation properties;
  - to make a preliminary study of the stress-strain behaviour (deformation response) of the medium to be excavated in the absence of stabilisation measures;
  - to define the type of confinement or preconfinement action needed to regulate and control the deformation response of the medium to excavation;

Certainties

The development of the *Analysis of Controlled Deformation in Rocks and Soils,* the subject of this chapter, sprang from certainties, summarised in the figure below, ascertained and clearly formulated as the outcome of the theoretical and experimental research described in full in the preceding pages.

→ THE DEFORMATION RESPONSE (D.R.) CAN BE CONTROLLED

→ THE CORE IS AN EFFECTIVE AND PROPER INSTRUMENT TO USE

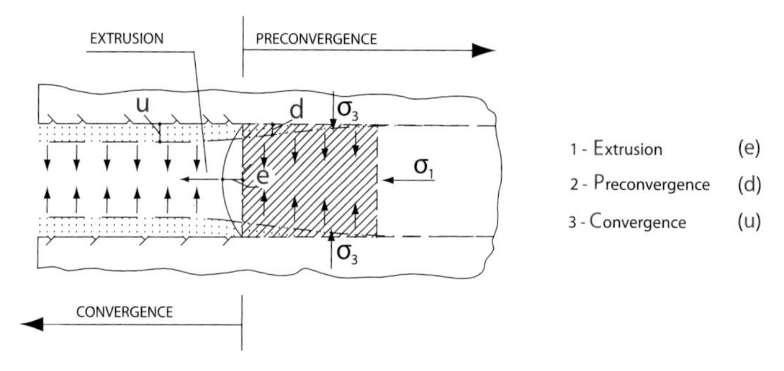

1) the D.R. has three components: extrusion (E), preconvergence (P) and convergence (C)

2) P and C are a consequence of E, which is the most significant component of the D.R.

3) if you can control E, then P and C are automatically also controlled

4) control of E is achieved by modifying the rigidity and therefore the strength and the deformability of the core-face

5) instruments (protective and reinforcement) exist to control the rigidity of the core-face

(E) Full scale measurement of E is possible and provides absolute values

(P) This can be calculated indirectly through E (volume balance)

(C) Full scale measurement of C is possible and provides relative values

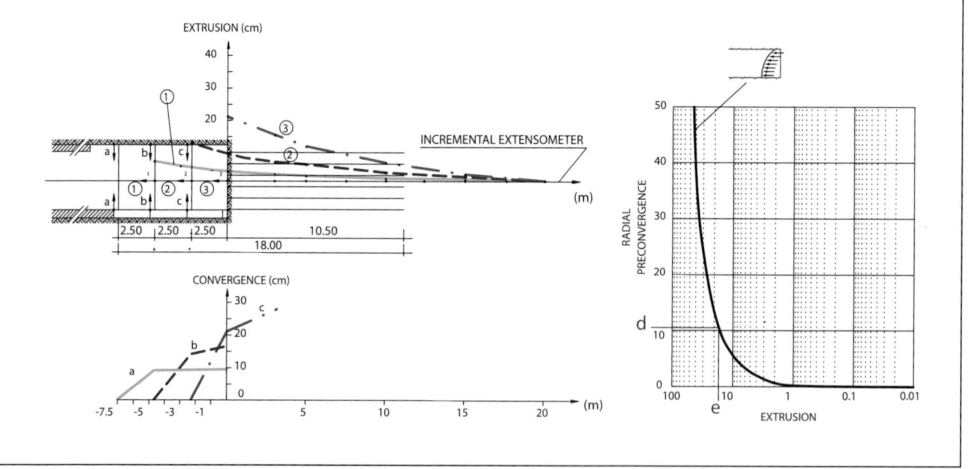

  - to select the type of stabilisation from those currently available among the existing technologies on the basis of the preconfinement or confinement action that they are able to provide;
  - to design tunnel section types on the basis of the predicted response of the medium to excavation, defining not only the most adequate stabilisation measures for the context in which they are expected to operate, but also the stages, cycles and timing with which they are to be implemented;
  - to use mathematical calculation to determine the dimensions of the intervention selected, in order to obtain the desired response of the medium to excavation and the necessary safety coefficient for the works and to test that intervention;

- **at the construction stage:**
  - to verify that the response of the medium to excavation during construction corresponds to the response that was calculated by analytical means at the design stage. To then proceed to fine tune the design, balancing the intensity of the stabilisation measures between the face and the perimeter of the cavity.

Consequently, if the design and construction of an underground work is to be located within the framework of a conceptually accurate and universally valid approach then it must necessarily be organised in the following chronological order:

1. a *survey phase*, for geological, geomechanical and hydrogeological knowledge of the medium;
2. a *diagnosis phase*, for predicting, by theoretical means, the behaviour of the medium in terms of the deformation response in the absence of stabilisation measures ;
3. a *therapy phase*, consisting firstly of defining the method of excavation and stabilisation of the medium employed to regulate the deformation response and then of using theoretical means to assess its effectiveness in achieving this;
4. a *monitoring phase*, for monitoring, by experimental means, the actual response of the medium to excavation in terms of the deformation response in order to fine-tune the excavation and stabilisation methods employed.

5.1.1 Conceptual framework according to the ADECO-RS approach

The approach based on the ADECO-RS differs in various important respects from other methods that have been commonly followed:

1. the *design* and *construction* of a tunnel are no longer seen as they were in the past but now represent two quite distinct stages with a clear and well defined physiognomy in terms of timing and practices;
2. the approach employs a new type of conceptual framework for underground works based on one single parameter common to all excavations: the stress-strain behaviour of the core-face;
3. the approach is based on the prediction, control and interpretation of the deformation response of the rock mass to excavation, which is the only reference parameter considered. It is considered firstly by theoretical means, to be predicted and regulated and then by experimental means, to be measured and interpreted as a means of fine-tuning the design during construction;

Development of a modern design and construction approach valid in all types of ground

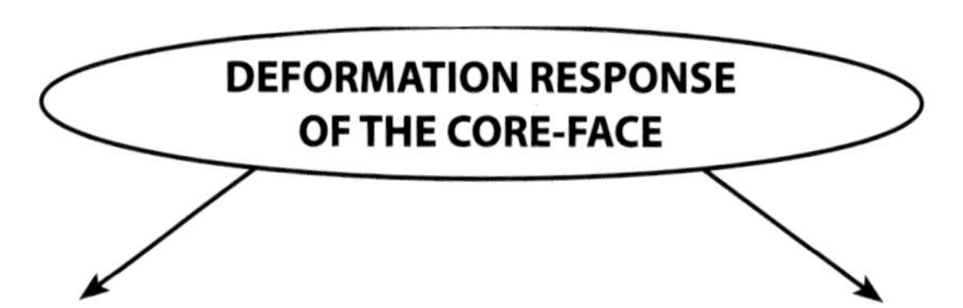

ANALYSIS	CONTROL
1 Knowledge of the medium	3 Definition of the excavation and stabilisation systems capable of controlling it
2 Forecasting the type and magnitude of the deformation response in the absence of intervention	4 Verification and calibration of intervention by monitoring

↓

	PHASE	CONTENT
Design stage	1 SURVEY	For geological, geomechanical and hydrogeological knowledge of the medium
	2 DIAGNOSIS	For prediction, by theoretical means, of the behaviour of the medium in terms of deformation response
Construction stage	3 THERAPY	For deciding the methods of excavation and improvement/reinforcement of the medium to control the deformation response
	4 MONITORING	For empirically monitoring the behaviour of the medium in terms of the deformation response for fine tuning the excavation and stabilisation systems

ADECO-RS
Analysis of COntrolled DEformation in Rocks and Soils

- valid and applicable in all types of ground and under all statics conditions
- capable of providing design and construction instruments that are able to solve the problems of different statics conditions in all types of ground
- capable of distinctly separating the design stage or 'moment' from the construction stage or 'moment'
- capable of planning projects reliably in terms of construction schedules and costs.

4. the concept of preconfinement is introduced as an addition to the already well known concept of confinement. This makes it possible to solve even the most difficult statics problems in a programmed manner without having to resort to improvisation during construction;
5. the approach involves the use of *conservation systems* designed to maintain the geotechnical and structural properties of the ground, seen as the "construction material", as unaltered as possible when these play a fundamental role in the speed and rate of underground construction.

The approach introduces a new conceptual framework for viewing underground works.

Since it has been observed that deformation of the medium during excavation and therefore the stability of the tunnel itself are dependent on the behaviour of the core of ground ahead of the face (advance core) the *stability of the core-face* has been taken as the basic element in the conceptual framework. Consequently, by making reference to one single parameter, valid for all types of ground (the deformation response of the core of ground ahead of the face), the method overcomes the limitations of systems used until now, especially in soils with poor cohesion.

As already illustrated, three fundamental behaviour categories can be identified (Fig. 5.1):

- *Category A*: stable core-face or rock type behaviour;
- *Category B:* stable core-face in the short term or cohesive type behaviour;
- *Category C:* unstable core-face or loose ground type behaviour.

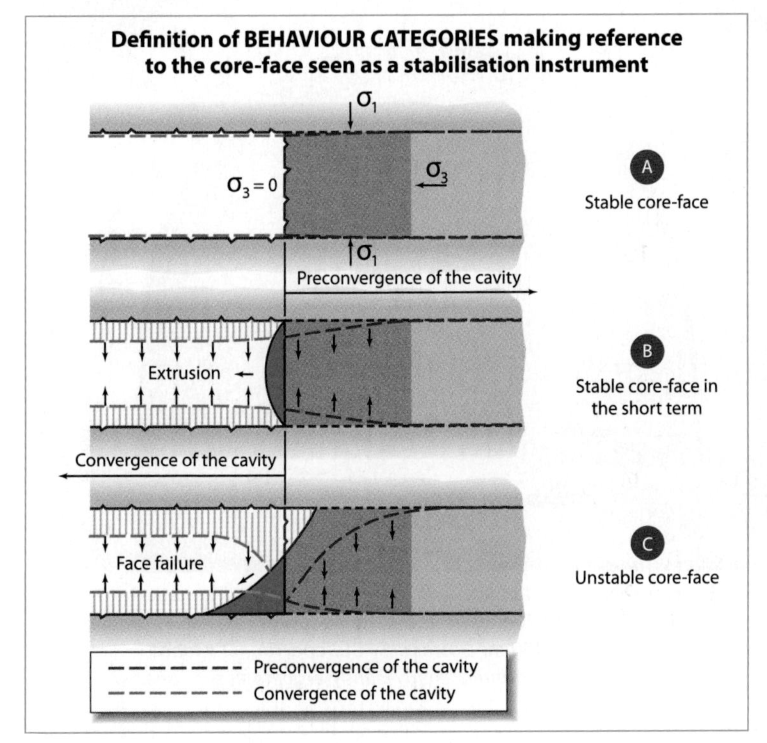

Fig. 5.1

Category A

Fig. 5.2 *Two examples of tunnels with stable core-face deformation behaviour. Top: S. Rocco Tunnel (Rome-L'Aquila motorway) 1967, ground: limestone, overburden ~ 400 m. Bottom: S. Stefano Tunnel (Genoa-Ventimiglia railway), 1984, ground: marly-sandy argillite flysch, overburden ~ 100 m*

5.1.1.1 Category A

Category A is identified when the state of stress in the ground at the face and around the excavation is not sufficient to overcome the strength properties of the medium. The closer the actual profile of the tunnel is to the theoretical profile of the tunnel, the closer the formation of an arch effect will be to that profile.

Deformation phenomena develop in the elastic range, occur immediately and are measurable in centimetres.

The face as a whole is stable (Fig. 5.2). Only local instability is found due to the fall of isolated blocks caused by unfavourable configurations of the rock mass. In this context a fundamental role is played by the anisotropic stress-strain state of the ground.

The stability of the tunnel is not normally affected by the presence of water even under hydrodynamic conditions, unless the ground is fractured or susceptible to weathering or the hydrodynamic gradient is so intense that washing away destroys the shear strength along the slip surfaces.

Stabilisation techniques are mainly employed to prevent deterioration of the rock and to maintain the profile of the excavation.

5.1.1.2 Category B

Category B is identified when the state of stress in the ground at the face and around the cavity during tunnel advance is sufficient to overcome the strength of the medium in the elastic range.

An arch effect is not formed immediately around the excavation, but at a distance from it and this depends on the size of the band of ground that is subject to plasticisation.

The deformation develops in the elasto-plastic range, is deferred and measurable in decimetres.

Time has a significant effect on the stability of excavations: at normal advance rates the tunnel is stable in the short term and stability improves or worsens as advance speeds increase or decrease (Fig. 5.3). Deformation of the advance core in the form of extrusion does not affect the stability of the tunnel, because the ground is still able to muster sufficient residual strength.

Instability manifesting in the form of widespread spalling at the face and around the cavity, allows sufficient time to employ traditional radial confinement measures after the passage of the face. In some circumstances it may be necessary to resort to preconfinement of the cavity, balancing stabilisation measures between the face and the cavity to contain deformation within acceptable limits.

The presence of water, especially under hydrodynamic conditions, reduces the shear strength of the ground, favours the extension of plasticisation and therefore increases the magnitude of instability. This must therefore be prevented, especially near the face, by channelling water away from the advance core, by launching drainage far ahead of the face.

Category B

Fig. 5.3 *Two examples of tunnels with stable core-face in the short term deformation behaviour. Top: S. Stefano Tunnel (Genoa-Ventimiglia railway), 1984, ground: marly-sandy argillite flysch intensely tectonised, overburden ~ 80 m. Bottom: Frejus motorway tunnel, 1975, ground: limey-schists, overburden 1,500 m*

5.1.1.3 Category C

Category C is identified when the state of stress in the ground is considerably greater than the strength properties of the material even in the zone around the face. An "arch effect" can be formed, neither at the face nor around the excavation because the ground does not possess sufficient residual strength. The deformation is unacceptable because it develops immediately into the failure range giving rise to serious manifestations of instability such as failure of the face (Fig. 5.5) and collapse of the cavity without allowing time for radial confinement intervention: ground reinforcement and improvement operations must be launched ahead of the face to develop preconfinement action capable of creating an artificial arch effect.

If due account is not taken of the presence of water under hydrostatic conditions, this favours the extension of plasticisation by further reducing the shear strength properties of the ground and basically increases deformation phenomena. Under hydrodynamic conditions, it translates into the transport of material and/or piping which is absolutely unacceptable. It must therefore be prevented, especially near the face, by channelling the water away from the advance core, with drainage launched far ahead of the face.

Stage	Phase	Description
Design	– Survey	– Analysis of existing natural equilibriums
	– Diagnosis	– Analysis and prediction of deformation phenomena (*) in the absence of stabilisation measures
	– Therapy	– Control of deformation phenomena (*) in term of stabilisation systems chosen
Construction	– Operational	– Application of the stabilisation instruments for controlling deformation phenomena (*)
	– Monitoring	– Control and measurement of deformation phenomena (*) as the response of the rock mass during tunnel advance (measurement of extrusion at the core-face and of convergence at the contour of the cavity and at varying distance from it, inside the mass of the ground)
	– Final design adjustments	– Interpretation of deformation phenomena (*) – Balancing of stabilisation systems between the core-face and the perimeter of the cavity

(*) Deformation phenomena in terms of extrusion at the core-face and of convergence at varying distance from it, inside the mass of the ground

Fig. 5.4

Category C

Fig. 5.5 *Two examples of tunnels with unstable core-face deformation behaviour. Top: Cassino Tunnel (aqueduct in western Campania), 1985, passing through a fault in dolomitic limestone, overburden ~ 120 m. Bottom: Pianoro Tunnel (new high speed Milan-Rome-Naples railway line between Bologna and Florence), 2004, ground: Complesso Caotico formation, overburden ~ 70 m*

5.1.2 The different stages of the ADECO-RS approach

The approach based on the analysis of controlled deformation in rocks and soils suggests that in the logical development of the design and construction of a tunnel one should proceed according to the stages summarised in Fig 5.4.

A **design stage** consisting of the following:

- a *survey phase*: during this phase the design engineer determines the characteristics of the medium through which the tunnel passes in terms of rock and soil mechanics. This is indispensable for an analysis of the pre-existing natural equilibriums and for the success of the subsequent diagnosis phase;
- a *diagnosis phase*: during this phase the design engineer uses the information collected during the survey phase to divide the tunnel into sections with uniform stress-strain behaviour according to the three behaviour categories, A, B, and C, as described above, defining the details of how deformation will develop and the types of loads mobilised by excavation for each section;
- a *therapy phase*: during this phase, the design engineer decides, on the basis of predictions made during the diagnosis phase, the type of action (preconfinement or simple confinement) to employ and the necessary means of implementing it to achieve complete stabilisation of the tunnel, with reference to the three behaviour categories, A, B and C. He designs, therefore, the composition of the longitudinal and cross section types and tests their effectiveness using mathematical instruments.

A **construction stage** consisting of the following:

- an *operational phase*: during this phase the works for the stabilisation of the tunnel are carried out on the basis of the design predictions. They are adapted in terms of confinement and preconfinement to match the real deformation response of the rock mass and are monitored on the basis of a quality control programme prepared in advance;
- a *monitoring phase*: during this phase deformation phenomena (the response of the medium to tunnel advance) are measured and interpreted during construction, firstly to verify the accuracy of the predictions made during the diagnosis and therapy phases and then to fine tune the design by adjusting the balance of stabilisation techniques between the face and the cavity. The monitoring phase does not end when the tunnel is completed, but continues during its whole life to constantly monitor safety when it is in service.

Correct design of underground works therefore means knowing, on the basis of the natural pre-existing equilibriums, how to *predict* the behaviour of the ground during excavation in terms of how deformation

Tunnels and behaviour categories. The Italian experience

It has been observed from experience acquired over more than twenty years of tunnel design and construction that all cases of underground works fall into one of these three behaviour categories, A, B, and C, hypothesised by the ADECO-RS approach. This is the basis of its validity for all types of ground.

TUNNEL	Km	CLIENT	TYPE OF BEHAVIOUR OF THE MATERIAL		
			ROCK $c \neq 0$ $\varphi \neq 0$	COHESIVE $c \neq 0$ $\varphi \cong 0$	LOOSE $c \cong 0$ $\varphi \neq 0$
GRAN SASSO	2x10,0	ANAS			
FREJUS	2x13,0	ANAS			
SIBARI	4,0	F.S.			
PRATO TIRES	13,0	F.S.			
S. STEFANO	3,5	F.S.			
FLERES	7,2	F.S.			
VALSESIA	2x0,6	ANAS			
MADONNA DEL CARMINE	4,0	F.S.			
S. ELIA	2x2,0	ANAS			
MONTE OLIMPINO	8,0	F.S.			
CARBONARA	2x2,0	ANAS			
DIRETTISSIMA 2	5,0	F.S.			
LANGENIA	2x1,2	ANAS			
ZUC DEL BOR	9,0	F.S.			
S. ROCCO	4,0	F.S.			
MASSINO VISCONTI	2x2,7	ANAS			
MOTTARONE 1	2x1,2	ANAS			
DIRETTISSIMA 1	3,5	F.S.			
QUINCINETTO	7,0	ENEL			
CARBOSULCIS	3,5	ENEL			
PASSANTE MI	0,27	F.S			
S. FRANCESCO	0,8	F.S.			
CAMPIOLO	2,0	F.S.			
LES CRETES	2x1,3	ANAS			
VILLENEUVE	2x3,1	ANAS			
ARVIER	2x2,4	ANAS			
LEVEROGNE	2x1,8	ANAS			
AVISE	2x3,1	ANAS			
VILLARET	2x2,7	ANAS			
CHABODEY	2x1,0	ANAS			
CARDANO	4,0	F.S.			
CERAINO	6,0	F.S.			
S. LEOPOLDO	6,0	F.S.			
MALBORGHETTO	8,0	F.S.			
TARVISIO	1,0	F.S.			
LEILA	3,0	F.S			
MOTTAVINEA	2x1,8	ANAS			
MOTTARONE 2	2x1,8	ANAS			
A.V. BO-FI: 9 GALLERIE	73,0	F.S.			
A.V. RM-NA: 22 GALLERIE	21,4	F.S.			
TARTAIGUILLE	0,9	S.N.C.F.			

CLASSIFICATION SYSTEM		ROCK	COHESIVE	LOOSE
	RABCEWICZ			
	BARTON			
	BIENIAWSKI			

is triggered and how it develops. The objective is to then define, within the framework of the three behaviour categories, the type of action to exert, (confinement or preconfinement) and the type of intervention to employ to maintain deformation within acceptable limits with the definition of the timing and work cycles to implement it as a function of tunnel advance and the position of the face.

Correct construction of an underground work, on the other hand, means working according to design decisions: firstly by *correctly interpreting* (using the advance core as the primary key to interpretation) the deformation response of the ground to the action of excavation and stabilisation intervention in terms of extrusion and convergence of the face and the walls of the tunnel at the surface of the cavity and in the ground at a distance from it; and secondly (once the results of the measurements have been *interpreted*), by deciding the length, speed and rate of tunnel advances, the intensity, location and timing of stabilisation operations and the balance of these between the face and the perimeter of the excavation.

The design stage

The first and most important task of a tunnel designer is to study if and how an arch effect can be triggered when a tunnel is excavated and then to make sure that it is formed and to calibrate it appropriately as a function of the different stress-strain conditions, methods of excavation and stabilisation works.

CHAPTER 6

The survey phase

6.1 Introduction

To perform excavation underground is to disturb natural pre-existing equilibriums. To design such an excavation with respect for environmental equilibriums and with minimum disturbance to stress states (and the deformation response that results) is only possible if one has the fullest possible advance knowledge of the natural state of equilibrium of the ground before excavation begins. Hence the design and therefore the construction of any underground work must be preceded by activity, termed the *survey phase*, to obtain that knowledge and this is performed by acquiring all possible information on the morphology, structure, tectonics, stratigraphy, hydrogeology, geotechnics, geomechanics and stress states, which characterise the geology of the body of ground in which excavation occurs.

6.1.1 The basic concepts of the survey phase

The **survey phase** is therefore intended as meaning that phase of acquiring, analysing and interpreting all the results of the investigations performed in the field and on the geological bodies affected by the construction of the underground work. Its purpose is to acquire knowledge in the greatest depth possible of the state of the natural equilibriums present in the rock mass to be excavated.

Since this knowledge is indispensable for reliable prediction of the response of the rock mass to excavation and then for tunnel design which observes those natural equilibriums in the subsequent *diagnosis and therapy phases*, the survey phase, which is when that knowledge is acquired, is of fundamental importance: in-depth and carefully co-ordinated implementation of this phase is a basic and necessary condition for the correct design and construction of an underground work.

A progressive and gradual approach to obtaining knowledge of geography and terrain must therefore involve the construction of a geological model by acquiring a proper understanding of the genesis and history of the stratigraphy and structure and also of the stress states in the geological bodies and the terrains in question, including the nature and geometry of the contacts. The objective is to identify both the currently persistent local and the regional morphological, tectonic and hydrogeological equilibriums and finally to arrive at the geotechnical and/or geomechanical characterisation of the ground to be excavated.

Three levels of design

Italian law on public works (Law No. 109 of 11th February 1994) states that public works must be designed at three levels of detail:

- preliminary;
- final;
- detailed.

As a consequence, the survey phase of the design of an underground work also requires different levels of detail according to the level of design it must provide. More specifically, substantial investment must be reserved to it, especially for the preliminary and then the final design stages, when the most important and decisive decisions are made which fully define the project.

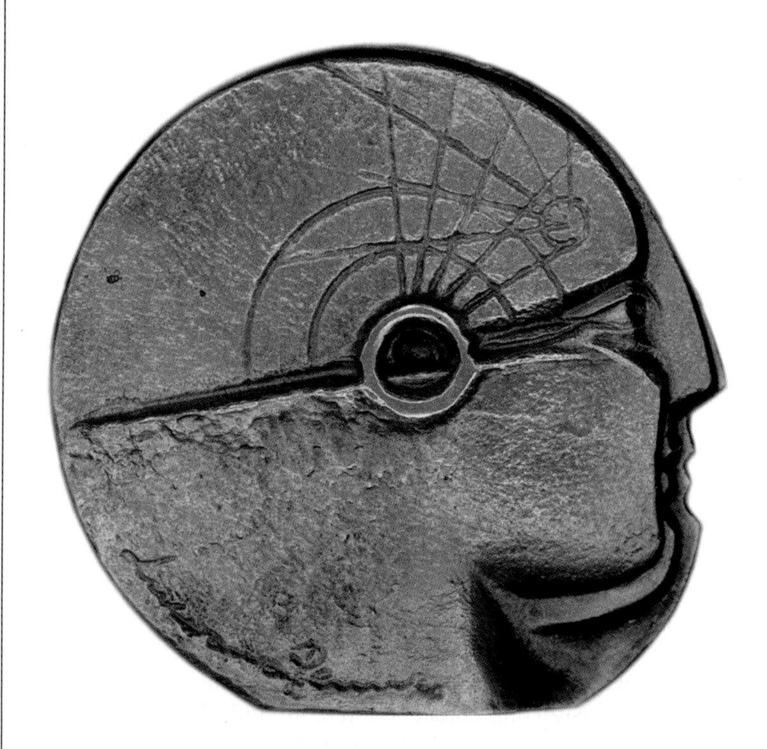

A bronze sculpture on the theme of tunnelling by the artist Luciana Penna

The level of detail of the survey phase will depend on the problem to be solved and will be proportional to its complexity and dimension. It is, however, fundamental that the study conducted has the intrinsic capacity to translate the main information acquired into quantitative or pseudo-quantitative form so that it can be employed appropriately in design activity. The survey phase must provide the following results:

- a clear assessment of the main geological, geomorphological and hydrogeological problems and the limits of the study performed;
- accurate identification and classification of the geological units involved with particular reference to spatial extent;
- an adequate geotechnical and/or geomechanical characterisation of the ground;
- careful identification of critical points on the alignment.

The design engineer's study of the stress-strain response of the medium to the action of excavation will be based on the results of the survey phase. It is therefore important for the planning of the geological survey and the results obtained to be developed on the basis of the preliminary geological model and to be studied and discussed in a spirit of active mutual co-operation between the applied geologist and the design engineer so that the latter can obtain an accurate picture of the real geological and geomechanical conditions and therefore tackle the subsequent diagnosis phase with success.

A few very brief **guidelines** are given below to orient the design engineer in planning survey phases for different types of project. Consideration is also given in various sections to particular problems concerning excavation using mechanised systems (TBM, shield) or where a pilot tunnel has been bored.

6.1.2 The survey phase for conventional excavation

The survey phase for conventional tunnel excavation involves techniques and methods and also interpretation and analysis which differs according to the geological and topographical characteristics of the tunnel alignment. The following deserve careful consideration:

- the geological-tectonic complexity of the area;
- the length of the alignment;
- the size of the overburdens.

Generally speaking the difficulty encountered in addressing the question increases as a function of the magnitude of the parameters listed above: a long tunnel, in a geologically complex context, which must pass through high and extensive mountains will always present

Preliminary design

The preliminary design defines the qualitative and functional characteristics of the works, the framework of the demands to be satisfied and the specific functions to be provided. It consists of a report illustrating the motivation behind the design decisions based on an assessment of potential solutions and it also makes reference to environmental aspects and the administrative and technical feasibility ascertained by means of the indispensable investigations and surveys with an initial estimate of the costs to be determined in relation to the forecast benefits. It also consists of drawings of the special characteristics, the types, functions and technologies involved in the works to be performed (*Law No. 109 of 11th February 1994; Technical regulations for constructions – Official Journal No. 222 of 23.09.05*).
The preliminary design therefore establishes the most significant aspects and characteristics of the output of the subsequent levels of design as a function of the economic dimensions and the type and category of the work. It will generally be composed of the following documents:

a) an illustrative report;
b) a technical report;
c) a preliminary feasibility study which will take account of the effects that the work will have on the surrounding environment;
d) preliminary geological, hydrogeological and archaeological surveys;
e) general plans and drawings;
f) initial indications and provisions for drawing up safety plans;
g) approximate calculation of costs.

Sample of marly limestone

great difficulties in analysing and interpreting geological data because of the problem of acquiring indispensable information at great depth.

Extreme cases in this respect are found in Alpine and Apennine tunnels where the characteristics listed above reach the maximum magnitude.

The survey phase is on the contrary, generally less difficult in cases of short and shallow tunnels typical of routes along valley bottoms, required to connect sections of the route running on viaducts or embankments.

Whatever the case, it must provide full information, with the maximum precision possible using the available techniques and technologies, on the following:

- geological, geomorphological and hydrogeological characteristics of the area;
- location and definition of the lithology through which the underground alignment passes and assessment of the volumes and reciprocal stratigraphic-tectonic contacts;
- tectonics, geological structure and stress state of the rock mass;
- hydrogeology of the ground;
- geotechnical and/or geomechanical characteristics of the materials.

It is important in acquiring the information in these points to follow a systematic method and more precisely to discover:

1. which elements are to be identified, and that is which **characteristics** to seek;
2. which methods of identification, and that is which are the most suitable **instruments** for that search;
3. when to seek those characteristics, and that is which **stage of the design** is the most appropriate for the search.
4. why seek those characteristics, and that is what are the **reasons** for the search.

6.1.2.1 The geomorphological and hydrogeological characteristics of the area

The salient **characteristics** to be analysed can be summarised as follows:

- *potential energy of the ground*: indicates the state of the development of the area and the nature of the outcroppings (rapidly rising massif, areas at a senile stage, etc.);
- *active forms and processes*: these suggest sectors that may be dangerous because appreciably active on a human time scale (landslides, terraces, cones, faults, etc.);

Final design

The final design fully identifies the works to be carried out in order to comply with the requirements, criteria, constraints, guidelines and indications contained in the preliminary design and it contains all the necessary elements for obtaining the authorisations and approvals required. It consists of: a report describing the criteria employed in making the design decisions and details of the materials specified and of how the project is set in its environment; an environmental impact study where required; general drawings, at an appropriate scale, describing the main characteristics of the works, the surface areas and volumes to be constructed, including those for the type of foundations; the preliminary studies and surveys required with regard to the nature and characteristics of the works; preliminary calculations of the structures and installations; a statement describing the technical and economical performance expected of the design as well as an estimated bill of quantities. The studies and surveys required, such as the geological, hydrological, seismic, agronomical, chemical and biological investigations and surveys are conducted to a degree of detail that will allow the preliminary structures, installations and estimated bill of quantities to be calculated (*Law No. 109 of 11th February 1994*).
The final design, prepared on the basis of the preliminary design already approved and of any "services conferences" there may have been, therefore contains all the necessary elements for the grant of planning permission and certification of compliance with general urban plans or other equivalent certification.
It will generally include:

a) a descriptive report;
b) geological, geotechnical, hydrological, sewage and water works and seismic reports;
c) specialist technical reports;
d) vertical and horizontal geometry drawings and an urban planning study;
e) graphs, charts and diagrams;
f) environmental impact study, where required by regulations in force or an environmental feasibility study;
g) preliminary calculations of structures and installations;
h) descriptive statement of the performance of the technical elements;
i) expropriation plan by parcel of land;
j) estimated bill of quantities;
k) financial and operating budget.

The drawings and reports and the preliminary calculations are prepared to such a degree of detail that there are no appreciable cost and technical differences between them and those of the subsequent executive design.

- *quiescent forms*: slope failures, landslide (slip) with a long life cycle of activity (decades) and so on, which is to say morphologies which indicate the need for further investigation;
- *morphological structures:* these generally represent morphological evidence of processes conditioned by structures; they allow large scale identification of the geological structure of the terrain (slope failures, structural plateaux, surface relicts, slope asymmetries etc.);
- *hydrological patterns*: observation of surface drainage gives important information on the nature and lithological-structural control of the terrain (twisted, angular, radial centrifugal, radial centripetal, etc.).

The **instruments** employed to identify the morphological and hydrological characteristics are as follows:

- analysis of the basic cartography and topography (detailed 1:25.000 scale and 1:5.000/10.000 ordinance scale maps);
- stereoscopic analysis of **aerial photograms** (generally on a scale of 1:13.000/1:8.000);
- analysis of simple photograms including the whole extent of the area to be tunnelled or important parts of it;
- analysis of the existing literature (universities, research institutes, central and local government agencies, etc.);
- geological and geomorphological surface survey.

The **stage** of the search for and acquisition of these characteristics must obviously precede all other survey phase activities; it will therefore correspond to the preliminary design stage and, in more detailed form, to the final design stage.

The **reasons** for the survey, as already mentioned, lie in the acquisition of information that is essential to an understanding of the nature and structure of the terrain, as well as its state of development on a human time scale.

6.1.2.2 Location and definition of the terrain through which the underground alignment passes

Immediately after the completion of the first part of the survey phase, the geological and petrographical **characteristics** of the tunnel are identified on the basis of the information acquired from it, together with the distribution or linear sequence in which it is predicted that they will be encountered. The relative geometrical relationships between the different lithologies and the different positional, stratigraphic or tectonic conditions are also identified.

Detailed design

The detailed design is drawn up on the basis of the final design and determines all the details of the works to be carried out and the relative budgeted cost. It must be prepared at a level of detail that allows the form, type, quality, dimension and price of each element to be identified. More specifically, this design consists of all the reports, the executive calculations of structures and installations and the drawings at an adequate scale, including any construction details, special contractors' performance or description specifications and the bill of quantities and unit price lists. It is prepared on the basis of the studies and surveys performed in the previous stages and of any further studies to verify and detail design hypotheses, which may be necessary, and on the basis of plan and profile surveys, measurements, staking out and surveys of underground utilities networks. The detailed design must also contain a maintenance plan for the project and its parts to be prepared on the basis of the terms and procedures specified in the regulations contained in article 3 *(Law No. 109 of 11th February 1994).*
Consequently the detailed design consists of the engineering for all the works and it therefore fully defines all the architectural, structural and engineering details of all the work to be done. Only construction site operating plans, supply plans and calculations and drawings for temporary works are excluded.
The design must comply fully with the final design and also with any specifications dictated when construction authorisations and certifications of conformity with urban plans were granted, or when "services conferences" were held, or certification of environmental compatibility was granted, or measures for the exclusion from procedures were taken, in the cases where these apply.
The detailed design will generally be composed of the following documents:

a) general report;
b) specialist reports;
c) drawings, which also include those of structures, installations and of environmental restoration and improvement operations;
d) executive calculations of structures and installations;
e) maintenance plans for the project and its parts;
f) safety and co-ordination plans;
g) final bill of quantities and financial and economic report;
h) time schedules
i) unit price list and analysis if necessary;
j) percentage allocation of labour to different parts of the project;
k) outline of contracts and special specifications for awarding contract.

These elements consist basically of the following:

- the geological nature of the formation (sedimentary, metamorphic or igneous);
- the lithological-petrographic characteristics of the terrain (limestones, sandstones, gneisses, schists, granites, basalts etc.);
- the spatial position and reciprocal contacts between different lithological units;
- the presence and frequency of the lithologies encountered along the alignment of the tunnel (prevalent, slight, negligible, etc.).

The **instruments** available today to perform these studies can be summarised as follows:

- analysis of the officially recognised geological maps available from certified sources (1:100.000);
- analysis of geological maps and literature available from various sources (universities, research institutes, central and local government and private sector agencies, journals, etc.) on varying scales (1:10.000/ 1:5.000);
- analysis of stereoscopic aerial photograms;
- geolithological-geostructural surface survey, conducted at a level of detail which will depend on the problem in question, but always at a greater level than the previous point (generally on the basis of 1:2.000/ 5.000 scale maps);
- creation of a database using suitable digital maps capable of enabling the construction of a 3D geological model;
- programming of direct surveys (e.g. continuous core bore sampling, exploration shafts, etc.) and indirect geophysical surveys (e.g. seismic, geoelectric, magnetometry, gravimetry, etc.).

The **stage** at which this study is carried out, after the cartographic study, constitutes the key and most important part of the preliminary design and as far as the performance of the direct surveys are concerned is fundamental to the final design.

The **reasons** consist basically of the need for knowledge of the ground to be tunnelled, the proportions in which different types of ground will be encountered as the tunnel is constructed and how and with what frequencies the types will be found along the tunnel alignment.

6.1.2.3 Tectonics, geological structure and the stress state of the rock mass

An understanding of the tectonic structure, the consequence of the processes that have affected the ground during the history of its de-

Five families of parameters

To complete the geomorphological, geological, lithological, tectonic, hydrological and hydrogeological studies of the rock mass to be excavated, the parameters to be identified in the survey phase required for the design for an underground work can be divided into five main families:

- parameters describing the original stress state;
- physical characteristics of the medium;
- mechanical characteristics;
- hydrogeological characteristics;
- "constructability" parameters.

Assessment of these is dependent on geological surveys and on *in situ* and laboratory tests to be selected in terms of the type, number, location, depth and so on, on the basis of information drawn from the preliminary model of the geology. The quantity of the surveys to be performed is often a very delicate decision and will depend not only on the level of detail of the design (preliminary, final or detailed), but also on the size of the project and the geological complexity, in terms of morphostructure and hydrogeology, the geological formations in question, the degree of heterogeneity and the possible presence of obstacles in them. These surveys are often very costly (they can amount to 5% of the total cost for some projects) and on some intensely urbanised or mountain sites can be difficult to perform. Nevertheless, they play an indispensable role in reducing areas of risk: to save on these is a grave mistake.
When account is taken of this and of the difficulties of *a posteriori* interpretation which surveys often present, it is essential for the survey phase to be performed under the supervision of specialists in geology, geotechnics, geomechanics and hydrogeology in possession of great experience in underground works.

Placing of a flat jack

formation during different geological periods and the process it is currently subject to, is fundamental to the study of grounds that are prevalently rocky.

The identification of recent tectonic activity or of substantial, even post-glacial, gravitational slope phenomena (deep slope gravitational deformation) can in some cases impose a constraint on the tunnel alignment or a change in the alignment both at a local or even a regional scale.

The main **characteristics** to be sought at this stage of study are as follows:

- structures that have been inherited in relation to past and present tectonic regimes (compressional, extensional, transform, etc.);
- nature and location of the main tectonic features on a macro-structural scale (thrust faults, folds, etc.);
- the nature, organisation and distribution of the mesostructural discontinuities present (diaclases, joints, strata, faults etc.).

The following **instruments** are employed to identify these characteristics:

- stereoscopic analysis of aerial photograms;
- analysis of detailed satellite images;
- macro-structural field survey;
- detailed geostructural (mesostructural) field surveys (using structural survey stations to survey the system of discontinuities on carefully selected and statistically significant measurement stations);
- direct surveys (core bore samples etc.);
- downhole measurements (TV camera downhole, sonic downhole, cross-hole, and log);
- indirect geophysical surveys (seismic refraction, seismic reflection, tomography etc.).

As in the previous cases, the **stage** at which these characteristics are surveyed coincides with the preliminary and final design stage since they are fundamental to geological knowledge of a site.

In addition to the **reasons** given previously concerning an understanding of the reciprocal relations between different formations, it is important to consider that a knowledge of the macro-structural-tectonic situation allows the intensity and number of the critical points in driving a tunnel to be defined, where opening a cavity will create particular and sometimes complex problems because of the presence of tectonic disturbance and particular stress states (high *in situ* stress, relaxed zones, etc.).

Detailed knowledge of the mesostructure will also allow numerical schematisation of the rock mass using continuous or discontinuous, isotropic or anisotropic models and the attribution of values to the geomechanical parameters.

Parameters describing the original stress state

The result of the geological and tectonic history of the rock mass, the stress state in the ground at the depth of the work before it is constructed (orientation and intensity of the stress tensor) conditions the behaviour of the work from moment excavation begins and for the whole of its useful life. Knowledge of the original stress state is therefore indispensable for all stability and deformation calculations, as well as being very important for perfecting construction systems.

It consists of two main components: the true natural stresses and stresses due to any overloading there may be (surface constructions, etc.).

The stress state in the rock mass can be defined by schematising the lithological bodies present and identifying the extension of these in space and the contact surfaces (discontinuities) resulting from deformation processes of a geological nature. If it is considered that these discontinuities separate lithological bodies that often have very different rigidities and that it is improbable that ordinary description by numerical models will identify these and if the geological complexity of the Italian peninsula is also considered (with particular reference to the Apennine and Alpine structure), then the problem of accurate measurement of the original stress state is very important in many cases (e.g. large cavities and tunnels at medium and great depth). In the less complex cases it is sufficient to estimate the following:

1. the weight of the overburden (and any overloading there may be);
2. the value of the λ coefficient (ratio of horizontal and vertical stresses in total terms) or of the K_0 coefficient (the same ratio in effective terms);
3. the hydraulic load.

In complex cases, the weight of the overburdens and the values of the λ coefficient and of the pressure coefficient K_0 must be verified with the assistance of numerical models capable of grasping the complexity of the geological structure that can substantially modify the action of gravity.

The existence of measurements, knowledge of the site, experience, studies of the literature and expert investigations are always of fundamental importance for accurate assessment of the original stress states. Assessment of the natural stress tensor cannot in fact be obtained by direct measurement. It is the fruit of interpreting the pressures or displacements that are induced in mediums characterised by complex structures. They are costly investigations difficult to perform, especially in very mountainous areas that have been subject to folding and great displacement due to slip. In these cases specialist interpretation is essential. What happens above all is that the distribution of the values measured is scattered, hence the need to employ iterative tests and interpretations.

The most established techniques in rocks are flat jack tests in the tunnel, down-hole hydraulic fracturing and doorstopper tests.

Calculation of the pressure coefficient K_0 at rest is a very delicate operation for soils and can be performed either in the laboratory using special tests or *in situ*. For fine grain soils, self-boring pressuremeters or equivalent systems can be used.

6.1.2.4 Hydrogeological regime of the rock mass

Fundamental **reasons** for a knowledge of the type (free, confined, etc.) and number of water tables present and of the hydrogeological regime within the rock mass to be tunnelled are given by the ability to predict:
- water entering the tunnel;
- hydrostatic and hydrodynamic regimes and consequent contribution to loads on final linings;
- hydrogeological risks attached to possible changes in the water table regime and in flow rates from springs and surface wells (consideration must be given to areas in which care must be taken to respect potable springs and wells for which precise regulations exist when plotting the tunnel alignment);
- large inflows captured during tunnel excavation that can cause surface subsidence and slow construction work, which can be particularly serious.

Essential **characteristics** of a definition of the hydrogeological regime consist of:
- identifying and describing the nature of the aquifers present in the geological mass;
- identifying and describing the nature of the water tables and attributing values to their geometrical parameters by measuring the hydrodynamic parameters;
- identifying the types and magnitude of underground flows both in the transition regime during construction and when excavation is complete and the final lining is in place;
- the nature and quality of the waters.

The survey **instruments** for this investigation are as follows:
- analysis of morphological, climatic and hydrological aspects;
- general census and measurement of the wells and springs present in the terrain;
- measurement of the level of the water table in bored holes (for each water table identified) using piezometers of various types;
- indirect measurements using geoelectric geophysical methods;
- measurement, where necessary, of underground water flows using special instruments (micro paddle wheel flow meters, etc.) or by using tracer techniques.

The **stage** at which hydrogeological surveys are performed is the same as for the previous surveys cited.

6.1.2.5 Geomechanical characteristics of the materials

A full summary of the phase for the geomechanical characterisation of the materials would require extensive discussion which is neither possible nor within the aims of this book.

Physical characteristics of the medium

Identification parameters

The physical characteristics of the medium add indispensable quantitative information to the detailed lithological description of the ground, which often even allow an initial estimate of behaviour.
For example, a knowledge of the volumetric weights will allow estimates to be made of natural vertical stresses, unit weight, water content, porosity, Atterberg limits, granulometry and so on. Appropriate correlations can then be used to give very useful indications of the mechanical and hydrogeological properties of porous mediums and in cases where the use of shields with face confinement is considered, they will also give indications for numerous essential parameters for selecting face confinement systems, for the nature of the fluid under pressure, the spoil removal process, debris separation etc.. Mineralogical analysis of grains of clayey particles (colloidal activity) then helps to improve knowledge of the ground by furnishing a fairly accurate idea of its weatherability, its swelling potential, its colloidal properties and of the probable wear of excavation tools. Geochemical analysis of the soil, on the other hand, allows the constituent components to be identified which can present varying degrees of potential harm to the life of a tunnel and/or to the health of construction workers (e.g. methane, hydrocarbons, pollutants etc.),

Discontinuities

It is well known that the stability of excavations in rock, at both shallow and great depths, is basically governed by the intersection of the discontinuities in the rock mass: schistosity, stratification planes, fissures, diaclases and faults. The larger the cross section of the excavation the more important the effect of these discontinuities becomes and how they are distributed spatially has a great effect on the performance of excavation machinery (advance speed, ease of excavation, cuttability).
A precise description of discontinuities observed at outcroppings and in core bore samples must be formulated right from the start of the survey phase, with specification of the family, orientation, spacing, persistence, aperture, quality and the material filling the joints.

Weatherability

The sensitivity of certain materials to water and to hydrometric variation (schists, foliated marls, chalks or anhydrites, certain limestones, etc.) can constitute important information when deciding the type of confinement to use. In addition to mineralogical petrographical and geochemical analyses on clayey fractions, essential for defining the weatherability of a soil or rock, it is also very useful to perform specific tests that simulate the new conditions that the material will be subject to following excavation.

Chemistry of the water

Analysis of the chemistry of the water contained in the ground and assessment of its aggressivity are indispensable for choosing the type of cement to use for linings. It is extremely advisable to seek to discover the pollutants in the water and to know the state of pollution under the water table before works begins.

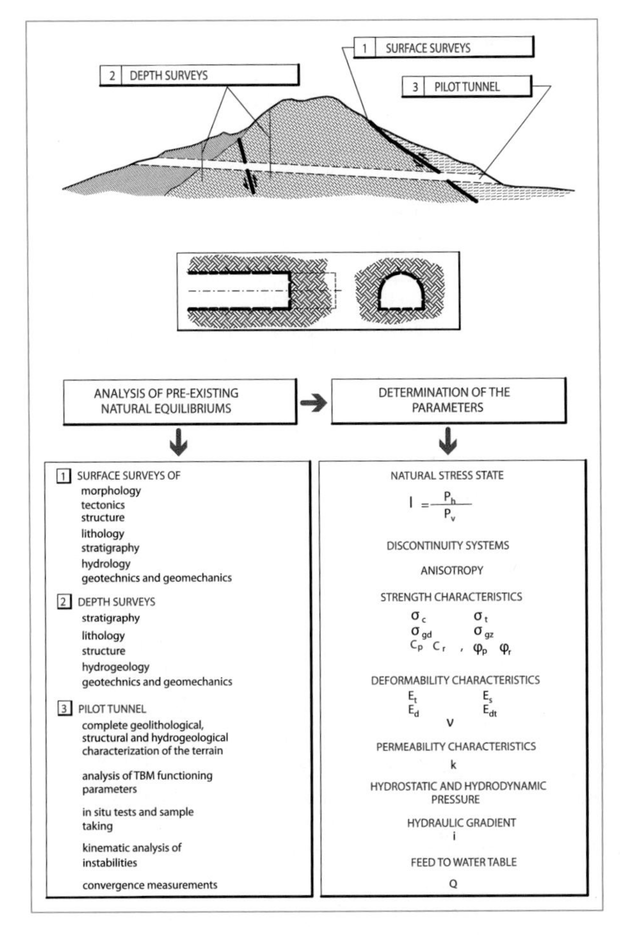

Fig. 6.1

Treatment of the subject is therefore limited here to listing the **key concepts connected with the attribution of values to the parameters**, understood as a procedure designed to identify the following in relation to the geological medium understood as intact samples and at the full scale of the rock mass:

- the structure of the material understood as the lithology and discontinuities at micro and mesostructural level;
- density, at the level of the whole significant volume of each single lithotype identified (unit weight);
- resistance to monoaxial, triaxial and shear stresses (σ_c, c', φ');
- deformability (E_t, E_s, E_d, E_{dt}, υ);
- primary permeability (k), for porous systems, secondary (K) in relation to the state of fracturation.

The **instruments** needed to identify the geotechnical and/or geomechanical characteristics of the materials are mainly (Figs. 6.1 and 6.2):

- laboratory tests to identify:
  - physical and chemical parameters;
  - geomechanical parameters (strength and deformability) at sample scale;

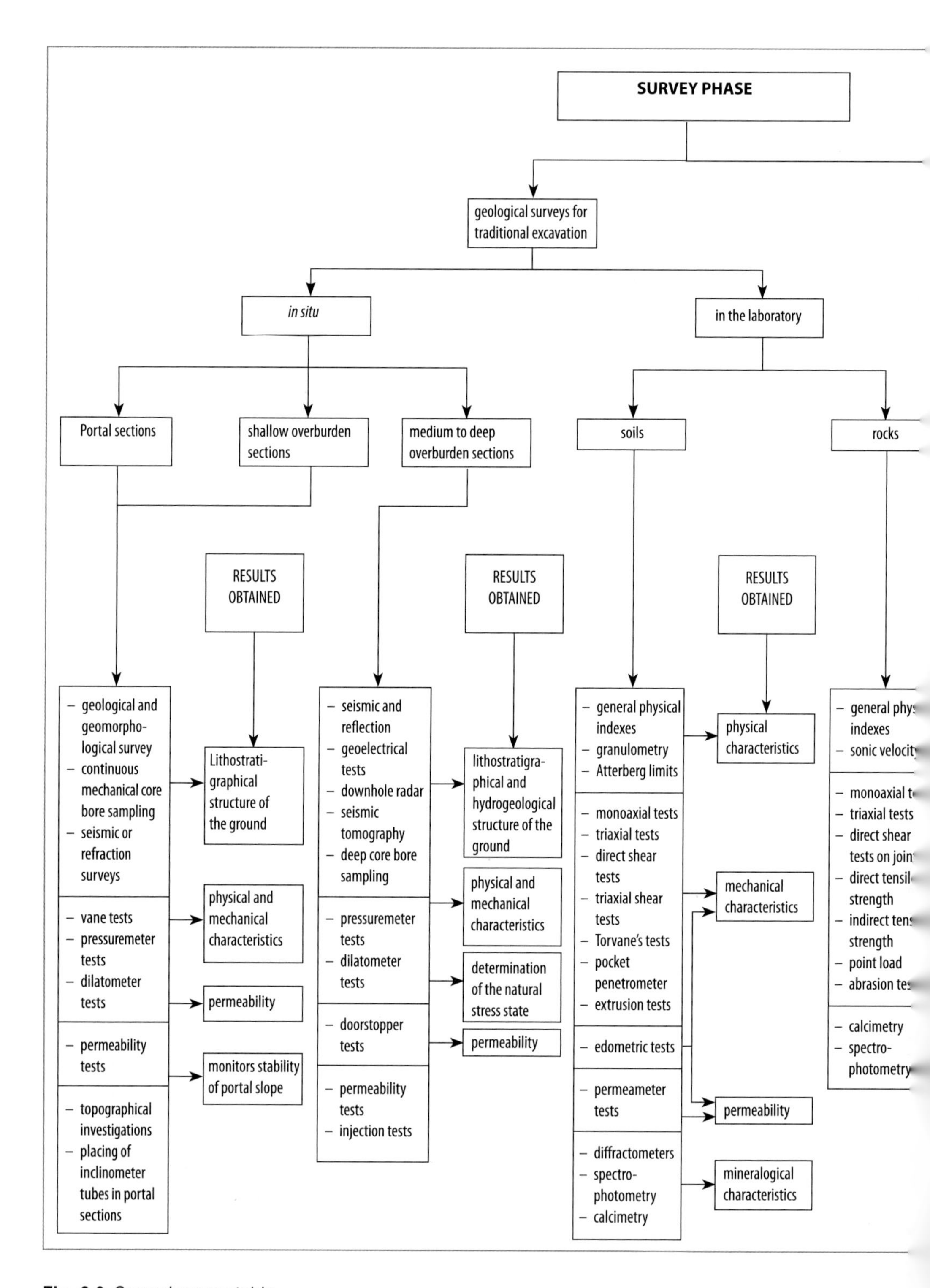

Fig. 6.2 *Ground survey table*

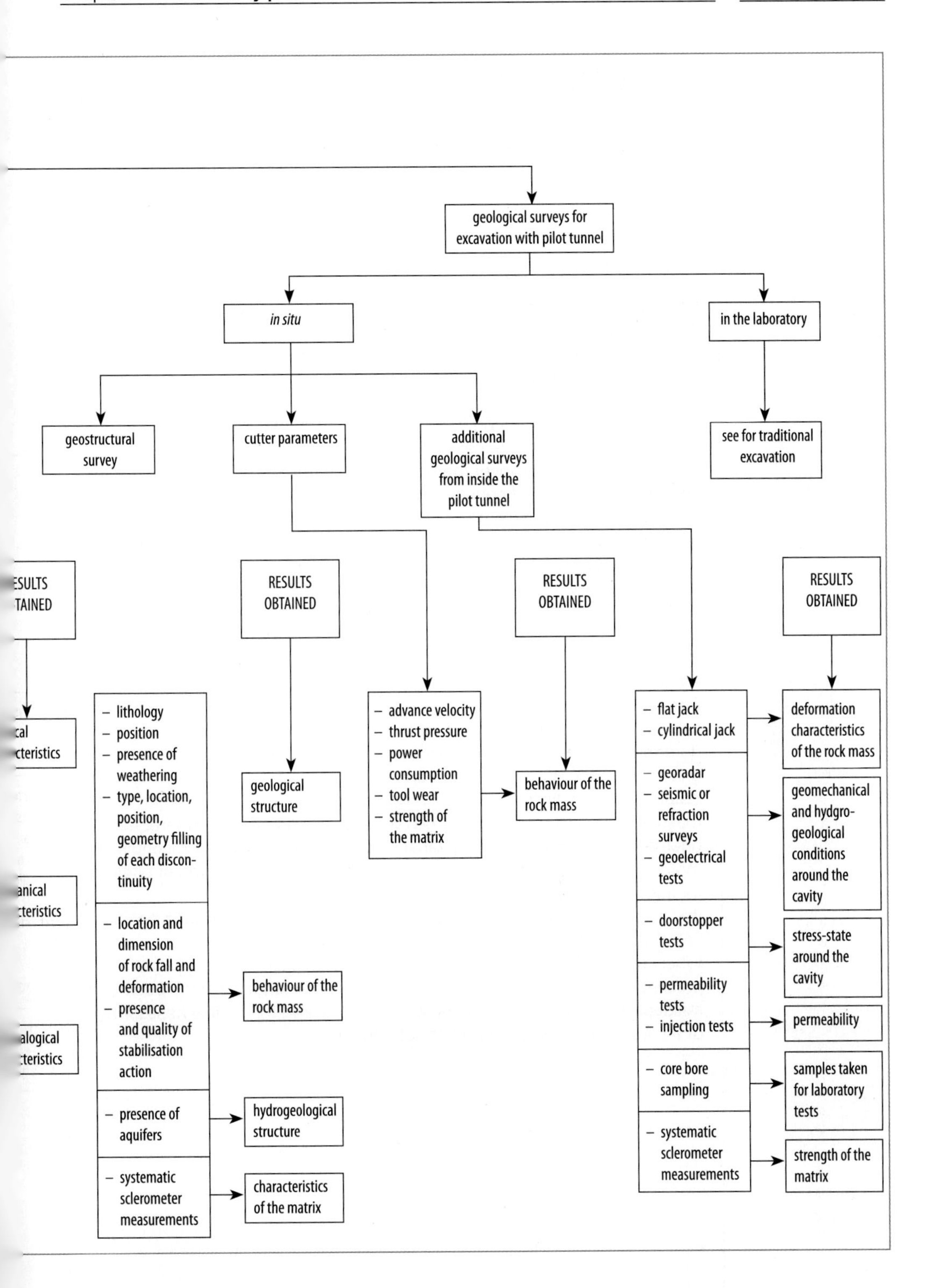
geological surveys for excavation with pilot tunnel
in situ
in the laboratory
geostructural survey
cutter parameters
additional geological surveys from inside the pilot tunnel
see for traditional excavation
ESULTS TAINED
RESULTS OBTAINED
RESULTS OBTAINED
RESULTS OBTAINED
cal cteristics
anical cteristics
alogical cteristics
– lithology
– position
– presence of weathering
– type, location, position, geometry filling of each discontinuity
– location and dimension of rock fall and deformation
– presence and quality of stabilisation action
– presence of aquifers
– systematic sclerometer measurements
geological structure
behaviour of the rock mass
hydrogeological structure
characteristics of the matrix
– advance velocity
– thrust pressure
– power consumption
– tool wear
– strength of the matrix
behaviour of the rock mass
– flat jack
– cylindrical jack
– georadar
– seismic or refraction surveys
– geoelectrical tests
– doorstopper tests
– permeability tests
– injection tests
– core bore sampling
– systematic sclerometer measurements
deformation characteristics of the rock mass
geomechanical and hydrogeological conditions around the cavity
stress-state around the cavity
permeability
samples taken for laboratory tests
strength of the matrix

Mechanical characteristics – 1

Mechanical characteristics include strength and deformability parameters. In most cases these are static parameters, corresponding to the static stresses generated during the different stages of construction. Where works that are subject to external dynamic action are considered (seismic activity, artificial explosions, etc.), then the dynamic parameters must also be investigated.

Strength parameters
It is indispensable to know the strength parameters if the stability of a tunnel at both the face and in a given section of a tunnel is to be assessed by means of calculation. To achieve this, different types of parameters are used depending on whether the tunnel passes through soil or rock.
For soils, the shear strength is characterised by the cohesion c' and the angle of friction φ'. In the case of non cemented sands, the cohesion c' is in theory nil; when however the sand is damp, it is possible to count on the short term existence of cohesion, which is apparently greater than zero and this is sometimes sufficient to guarantee the stability of the face for the time required, even in the absence of any substantial confinement action.
In the case of saturated clays, the slowness of the drainage makes it possible, more often than not, to analyse the short term stability of the cavity as a function of total stresses, since the strength of the ground is given by the undrained cohesion cu. However to study the long term behaviour of a tunnel, the stabilisation works and linings, the strength parameters in drained conditions must be known and used.
For rocks, a distinction must be made between the strength of the rock mass and the strength of the matrix. The simple compressive strength σ_c, determined by means of tests performed in the laboratory on rock samples gives the strength of the rock matrix rather than of the material *in situ*. To obtain a more realistic estimate of the strength of the latter (strength of the mass σ_{gd}), the value for σ_c must be reduced as a function of the frequency, orientation and strength characteristics of the discontinuities present in the material. For rocks too, the strength can be expressed in terms of angle of friction and cohesion in drained and undrained conditions.
It is also useful to know the tensile strength (σ_{gz}) of the material to be excavated when selecting the means of excavation.
When stressed beyond their elastic limit, both soils and rocks normally present appreciably lower strength values. Knowledge of both peak and residual strength values must therefore be acquired for reliable assessment of the long and short term stability of a tunnel.

- downhole tests to identify:
  - strength and deformability parameters at an intermediate scale;
  - permeability parameters;
  - *in situ* natural stress state;

- *in situ* tests to identify:
  - strength and deformability parameters at full rock mass scale;
  - hydrodynamic parameters (permeability, transmissivity etc.).

Laboratory tests (monoaxial and triaxial compression, direct shear, oedometric etc.) allow the intrinsic strength and deformability to be identified for samples and structures not affected by discontinuities of a mesostructural magnitude.

Downhole tests, on the other hand, are used mainly to investigate the characteristics of materials at an intermediate scale between that of intact samples and that of the rock mass itself depending on the properties of the mesostructural fracturing system (or systems).

Finally, *in situ* tests are used to assess the overall characteristics of the rock mass understood as the superimposition of the fracturing system over intact material and as therefore being generally discontinuous.

The **stage at which values are attributed to parameters** undoubtedly constitutes the final and conclusive part of the survey phase. It can be thought of as bringing together all the previously acquired information and as preparatory to true design testing.

The **reasons** for seeking these parameters lie in the need to know the geomechanical and stress-strain behaviour of the medium and the characteristics of how it interacts with excavation.

Precise knowledge of the structure of the medium and of its physical, mechanical and stress-strain characteristics allows calculation models to be selected and used effectively to forecast deformation that will occur when a cavity is opened. This therefore provides the necessary preparation for tackling the subsequent **diagnosis** and **therapy** stages.

6.1.3 The survey phase for TBM excavation

If the technical and financial feasibility of employing a full face cutter to drive a tunnel is considered, then the survey phase must necessarily involve greater caution, because there is less flexibility with this type of advance as compared with conventional methods if the operating conditions encountered are different from those forecast.

More specifically, in addition to the activities described for conventional excavation, a series of specific laboratory tests must be performed to investigate the 'cuttability' of the materials to be excavated.

Mechanical characteristics – 2

Deformability parameters

It is indispensable to know the value for the deformability modulus for all calculations of interaction between soil and structure used to assess the magnitude of deformation in a tunnel (extrusion, convergence) and any surface settlement that may be induced by excavation. Except for rare cases in which calculations can be made under conditions of isotropy and elasticity (for which the determination of the deformability of the ground can be reduced to an estimate of Young's modulus E and of the Poisson coefficient υ), assessing the value of the deformation modulus is a very delicate operation, because account must be taken of the magnitude of the stresses induced in the ground by excavation and of all the particular aspects dependent on the nature of the material (anisotropy, heterogeneity etc.).

Since the behaviour of the soils and rocks is neither linear nor reversible, account must be taken when deciding the value of the modulus of both the initial stress state and of the dynamics of the loading and unloading of the stresses induced by construction works. This makes it essential to estimate the modulus on the basis of a curve for cyclical loading and unloading tests.

In the case of tunnels excavated in fractured rock, a distinction must be made between the modulus of the rock matrix determined on a rock sample in the laboratory and that of the rock mass in the ground which depends on the conditions of the joints. Either an equivalent modulus must therefore be calculated which represents the deformability of the matrix and of the joints, or recourse must be made to more sophisticated models which take account of the geometry and the characteristics of the discontinuities present in the rock mass.

As is already the case for the study of stability, the drained deformability parameters must be known to calculate the long term deformability of underground works.

In the case of swelling materials the swelling potential must be known. This can be estimated from the swelling index C_{rig} and from the nil deformation swelling pressure σ_{rig}.

The dynamic characteristics

It is essential when verifying underground works subject to earthquake or explosions to know the dynamic parameters of the medium, such as the velocity of the propagation of the waves (the V_s above all), damping and dynamic moduli. It is also important for some soils to know the liquefaction potential.

Recourse to specialists in *rock and soil dynamics* and in *underground works* is needed to plan appropriate surveys and tests to determine this type of parameter.

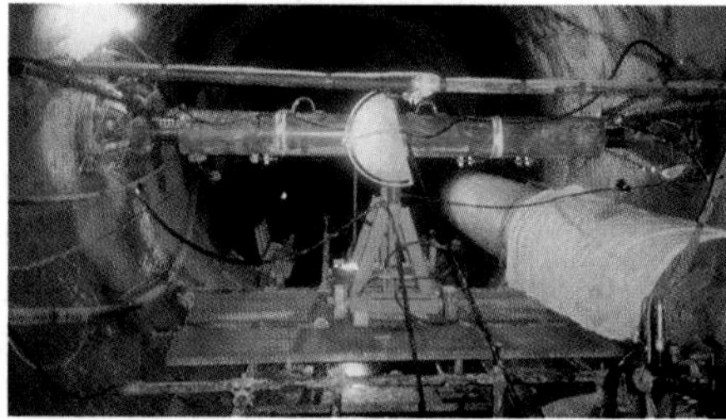

Loading and unloading test using a cylindrical jack

Maximum attention must be paid to identifying the main tectonic phenomena (faults, normal, reverse and so on, zones in which energy is stored) which, as experience has shown, constitute the main causes of interruptions and delays in construction work.

In addition to tectonic phenomena, material with poor geomechanical characteristics or with a tendency to swell also constitutes a risk for the efficiency of TBM's, which in some cases can get literally 'stuck' in the cavity. In this case, macrostructural and tectonic study of the terrain must identify the main discontinuities, the nature, location and frequency of which constitute without doubt the key elements in the design decision to employ tunnel boring machines. Where the presence of large overburdens or insurmountable morphological difficulties make continuous core bore sampling impossible, the basic geological model is used as a guide for detailed geophysical survey campaigns, such as for example (if the nature of the materials allows it) deep seismic reflection or gravimetry.

TBM's should always perform continuous recording of what is termed 'cutter-data' (power consumption, advance rate, tool wear, total thrust exerted on the face, etc.) during excavation. Knowledge of this data is often valuable for predicting the behaviour of the ground beyond the face with a fair degree of accuracy. Interpretation of this data is far from simple, because it must be done by defining index parameters that are characteristic of the interaction between the rock mass and the cutter (**RS method**) [44] [45].

These parameters do in fact have the particular quality of involving a number of different values relating to both the machine and the material at the same time. As a consequence, a certain degree of effort is required to obtain clean, and therefore fully reliable results on the strength and deformability of the medium tunnelled.

6.1.4 Geological surveys for excavation with preliminary pilot tunnel

The decision to drive a small diameter *pilot tunnel*, coaxial to the future tunnel using a TBM, may be taken for two fundamentally distinct reasons:

1. driving a pilot tunnel as a means of construction;
2. driving a pilot tunnel as a means of surveying.

One of the factors determining the decision may arise from a careful analysis of cost-benefits. In cases where studies and surveys have shown that the ground through which a tunnel must be driven is difficult in terms of stability and therefore of safety, because it is subject to high degrees of deformation, a pilot tunnel may be used as a *means of construction* which allows the future tunnel to be stabilised in advance from a small diameter drift. When difficult and complex geological and

Hydrogeological characteristics

Although it is of fundamental importance, assessment of the hydrogeological parameters of a rock mass is a very difficult operation, which requires great caution and experience.

The first and most important element to estimate is undoubtedly the hydrogeological system affecting the rock mass (number and type/s of fault/s). Next come the distribution of the piezometric load/s inside the rock mass, before excavation. The hydrodynamic parameters (permeability, gradients, transmissivity, etc.) constitute fundamental data which accompany these assessments. Assessment of the effects of excavation on a hydrogeological system and of the tunnel itself could even affect the feasibility of the project. Some measurements, such as that of *permeability* will be taken *in situ* using different methods according to the structural characteristics of the rock mass, its heterogeneity, its anisotropy and state of fracturation. Small scale permeability is considered for soils and the k coefficient can generally be assessed by means of *Lefranc* type tests and *variable* or *constant* load *pumping*. *Secondary permeability* is considered for fractured rocks and the K coefficient is assessed by means of *Lugeon* tests. Knowledge of the hydrodynamic characteristics of the aquifer in which operations take place is indispensable not just for general stability but also for the study of any drainage and impermeabilisation works.

The limit conditions must also be determined because they constitute an important parameter for estimating flow rates.

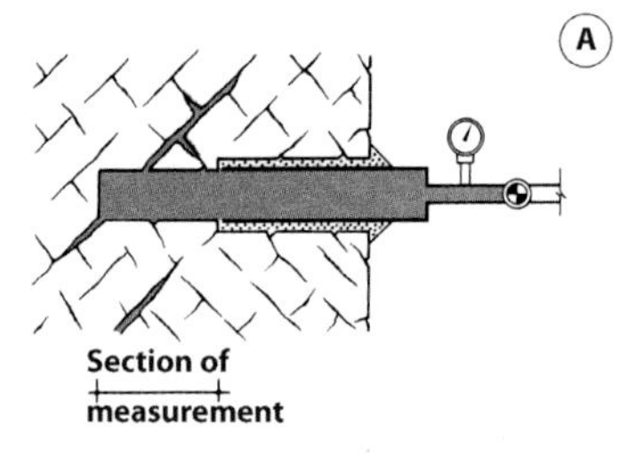

PRESSURE AND PERMEABILITY MEASUREMENT DIAGRAMS

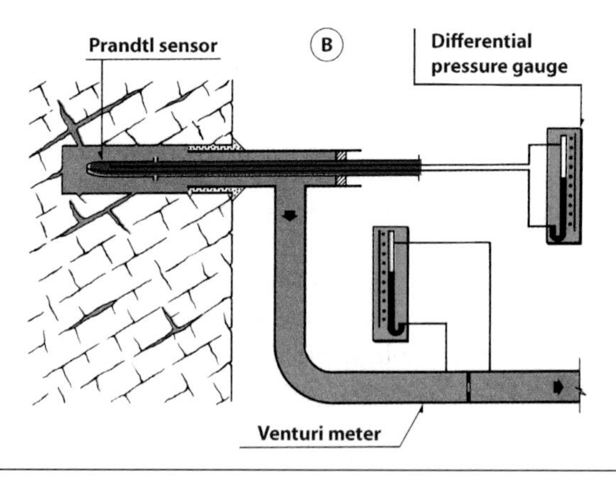

environmental conditions determine high survey costs to obtain sufficient information to guarantee adequate standards of design and functionality for the project, then a pilot tunnel may be chosen as a means of surveying. In this case a pilot tunnel allows geological surveying capable of providing continuous input to the subsequent design phase of a geostructural and geomechanical nature which is without interruption and matches the reality perfectly. It is a practice that is reserved to tunnels of a certain length, in a context of decidedly rocky ground, where it is not important to be able to count on the presence of a rigid advance core for the excavation of the final tunnel.

Although the small diameter of a pilot tunnel (generally between 3.50 m and 4.20 m) causes very limited disturbance to the rock mass, it is nevertheless a tunnel to be driven using a TBM. It is therefore indispensable to conduct a geological survey campaign, designed according to the criteria already described in previous sections, before any decision is taken concerning its benefits and feasibility.

Many advantages can be achieved by using a TBM to drive a pilot tunnel as a means of surveying. They include:

- the possibility of acquiring any information desired of a geostructural nature on the rock masses encountered, seamlessly in a direct and simple manner by taking systematic detailed observations and samples from the walls of the excavation;
- the possibility of knowing the distribution of the values for the strength of the rock mass along the tunnel alignment, by using the RS method to acquire and interpret the operating parameters of the TBM (thrust, advance rate, power consumed);
- the possibility of easily conducting numerous *in situ* tests to obtain extremely reliable results;
- the possibility of ascertaining the type of behaviour that the rock mass will exhibit when widened to the full diameter by observing the deformation behaviour of the pilot tunnel and taking account of the effect of the larger scale and the time factor;
- the possibility, if necessary, of performing radial ground improvement operations from the pilot tunnel before widening for the final tunnel.

All the information of a geological, hydrogeological, geostructural and geomechanical character obtained from the pilot tunnel must be integrated with a specific campaign of tests to be performed *in situ* and in the laboratory. In this respect the pilot tunnel constitutes a perfect geomechanics laboratory in which load plate, dilatometer and flat jack tests can be performed to measure the natural stress state of the rock mass (hydraulic fracturation, doorstopper etc.). They can be arranged along different lines of orientation according to the anisotropy of the medium and give results that are normally very reliable.

"Constructability" characteristics

Constructability characteristics are intended as meaning all those parameters which influence the choice of construction system. They are determined by means of specific tests and must normally be used in combination with other tests.

Excavation of material
The parameters that describe the properties of the ground in relation to excavation action must provide indispensable information on which to make decisions concerning the type of tools to use and the power of the machinery adopted. The parameters currently most commonly in use are: hardness, abrasiveness, velocity of seismic waves. They must be assessed in combination with the type and frequency of the discontinuities and the compressive and tensile strength of the material.

Excavation in clayey ground
The use of TBMs in prevalently clayey grounds requires advance knowledge of the tendency of the material to stick to tools and to the spoil removal system. This can be assessed by means of special mineralogical and geochemical analyses, the results of which are considered in combination with parameters such as for example the water content, the plasticity index and the activity index of the ground.

Sensitivity of the ground to vibrations
The sensitivity of the ground to vibrations transmitted by explosives or by TBMs is determined by their frequency, the damping characteristics of the material and the velocity of propagation. It must be assessed on the basis of the state of fracturation, the type and frequency of the discontinuities, the global fracture indices, the strength and the compressibility of the material.

Bolting operations
The choice of the type of bolts to use (end anchored, fully bonded etc.) depends on the general characteristics of the medium (nature, spacing and persistence of fracturation, mechanical strength etc.) and will also depend on the technology and conditions in which each type of bolt is placed. It is generally essential to investigate the following:

- the anchorage strength (for end anchored bolts);
- the value for friction or ground-bolt adherence (for fully bonded bolts).

The number will obviously depend on the quantity and type of the lithologies intercepted, on the complexity of the geostructural context and on the overburdens involved. On sites where the geological structure is significant or where deformability tests are performed, a variety of geological survey operations can be performed from the walls of the tunnel to obtain samples for the laboratory tests already discussed for geological survey campaigns performed prior to conventional tunnel excavation (section 6.1.2).

Linear surveys should be performed along the whole length of the pilot tunnel using seismic refraction or georadar techniques. These are useful both for detecting the possible presence of significant zones of material in the immediate vicinity of the tunnel that has been loosened by excavation and for obtaining information of a geostructural character which could be of great importance during widening operations. If it is necessary to obtain greater knowledge of particular hydrogeological aspects, then geolectrical surveys may also be performed.

The study of the stress states present in the rock mass where the future tunnel will run is considerably easier when a pilot tunnel is available. Bore holes can be drilled from it and fitted with instruments at points selected on the basis of the overburdens involved. Obviously when the geostructural context is complex, the number of these tests must be rather large to provide data that can be interpreted reliably.

Diffuse and concentrated deformation phenomena that is produced along the walls of the pilot-tunnel must always be carefully monitored by installing convergence measurement stations and also by taking a census of rock falls that occur during excavation with kinematic analysis of them. If stabilisation works have been performed when instability has occurred, then these must be carefully monitored and described in fine detail. As already mentioned, this information can be used to effectively predict the behaviour that will be exhibited by the rock mass during operations to widen the tunnel to full diameter, with account taken for the effect of the larger scale and the time factor.

6.1.5 Final considerations

The survey phase of the ADECO-RS system, which is preparatory to any type of excavation, must lead to the acquisition of the following information with the maximum possible degree of reliability, considering the wide range of technical resources currently available in the geological survey field:

- structural geological and hydrogeological conditions and the natural stress state of the ground to be tunnelled;
- the physical characteristics, the strength and deformability of the geological bodies affected by the excavation:
- the hydrogeological conditions in the rock mass.

2D and 3D tomography

One of the *in situ* types of investigation, tomography, deserves brief illustration because it is a recently introduced technology which has considerably improved the results of physical surveys. The method is based on the CAT (computerised axial tomography) technique used in clinical diagnosis and it can be seismic, sonic, electric, magnetic or electromagnetic. In all cases it gives the point distribution for magnitudes measured within plane sections inside the geological structure surveyed, which solves the problems encountered with traditional physical survey methods (sonic core boring, cross hole measurements etc.), with which only the average magnitudes measured can be determined for each measurement alignment.
The experimental part of the technology consists of acquiring the measurements for a dense network of alignments to cover the section or volume subject to survey as uniformly as possible and in the greatest possible number of directions.

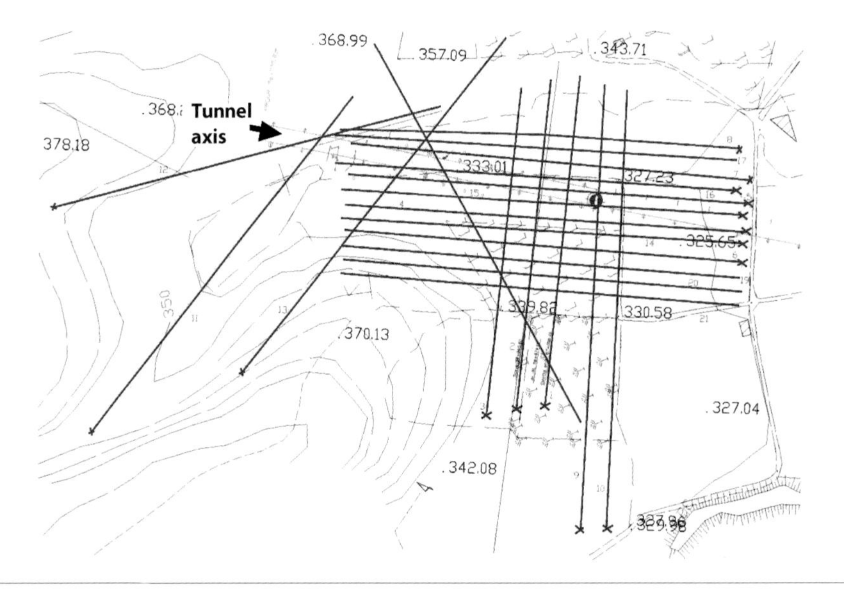

Profile 17
East longitudinal, distance from tunnel axis 26.5 m, angle of 6°

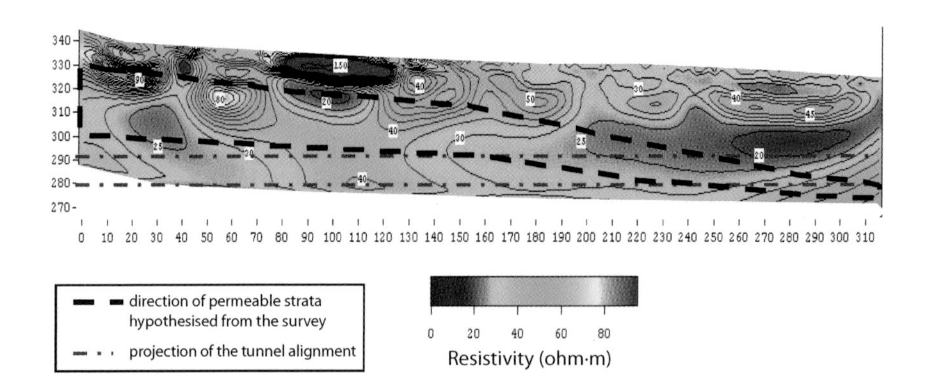

The data processing follows an inverse procedure which starts with the speed of propagation of the seismic signals to reconstruct the field of velocity in the medium examined. The section measured is discretised according to a rectangular grid for the calculation procedure.
The potential of tomography for characterisation of terrains is huge and still unexplored in part. The figure below shows the results in graphic form of a survey performed using electrical 3D tomography to study the hydrogeological characteristics of the rock mass for the Firenzuola tunnel, which formed part of the works for the construction of the new high speed Bologna-Florence railway line.

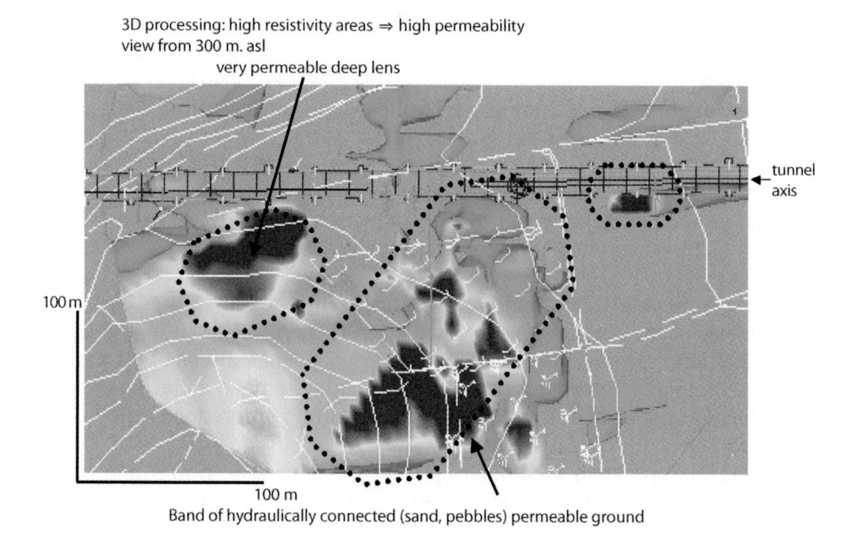

Profile 9
transverse, intercepts tunnel axis at chainage 57,000, angle of 98°

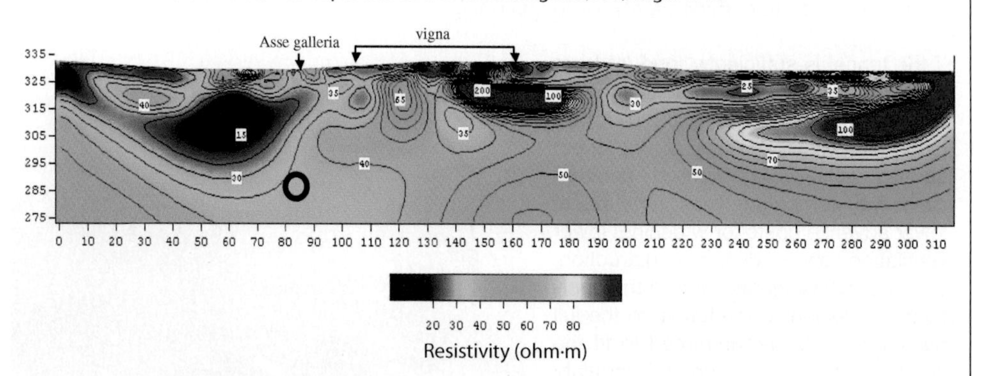

The RS method

The *RS* pilot tunnel *method*, taken together as a set of procedures, represents a true and genuine alternative to conventional methods of surveying for tunnel design. The pilot tunnel, driven more or less co-axially to the final tunnel to be excavated, using a full face TBM of small diameter (3 – 4 m), is the equivalent of horizontal core drilling, performed with destruction of the core, from which the rock mass concerned can be surveyed directly and then fully characterised from a lithostratigraphic, structural, geomechanical and hydrogeological viewpoint.
The necessary and indispensable conditions for correct application of the method are as follows:

- a preliminary geological survey which establishes the existence of geomechanical characteristics in relation to the overburdens which will guarantee a minimum level of stability for the pilot tunnel and a sufficiently fast advance velocity for the tunnelling machine;
- the use of a full face continuous excavation tunnel boring machine fitted with appropriate instrumentation;
- the implementation of a pilot tunnel along the whole of the route of the tunnel to be driven.

Opening a pilot tunnel, given the means and method of execution, makes it possible to use a series of characterisation methods which, taken together, provide a full and exhaustive picture of all the necessary elements for the final detailed design.

1. Because of the reduced disturbance produced by the action of the TBM, the surface of the perimeter of the tunnel is a full scale open book from which all the stratigraphic, lithological, structural, tectonic and hydro hydgrogeological features of the rock mass can be read. The method involves general collection and automatic archiving of all this data, which by means of special software can subsequently be reproduced graphically in any form required.
2. By monitoring the operating parameters (thrust, advance rate, power consumed) of the tunnel boring machine, which can be likened to a huge penetrometer, it is possible to calculate the precise energy required to excavate a given unit volume of material and hence the strength of the rock mass which correlates directly with it using a proportional coefficient determined *in situ* using flat jack tests.
3. A great variety of surveys, tests and measurements can be performed from within the pilot tunnel (core sampling, flat and cylindrical jack tests, geophysical investigations, a census and kinematic analysis of rock fall, convergence measurements, etc.), which give results that can be used immediately for tunnel design.

If the tunnel is sufficiently long (at least a few kilometres), then the initial greater expense incurred in driving the pilot tunnel is fully offset, not only by the direct savings achievable ("lighter" geological survey campaigns from the surface, lower advance costs for widening, lower ventilation costs during construction, etc.), but above all because all the operations for widening the tunnel to the full diameter can be programmed to fit the actual reality with consequent certainty of execution times and costs.

This information must be provided in reports, maps, charts and plans (plan views, plane and 3D sections), which define the technical geological model. The degree of uncertainty of the data provided must also be indicated as well as the studies and surveys performed to validate and support the technical-geological model furnished to support design activities.

These must include a geological and geomechanical profile of the route on which all the results of the studies and surveys performed are summarised (see the example in Fig. 6.3).

Proper interpretation of the data acquired and correct location of it in the existing geostructural situation constitute indispensable necessary conditions for adequate knowledge of the pre-existing natural equilibriums in the medium to be tunnelled in preparation for the diagnosis phase.

Naturally, the more extensive and detailed the survey phase is the lower the geological risk will be. The latter being understood as meaning the cost to be incurred for uncertainties at the construction stage over geological contexts which are generally identified as unforeseen at the design stage. As is known, these elements constitute one of the main causes of increased costs and delays in completing underground projects.

A distinction should be made between two levels of unpredicted geology, the result of insufficient knowledge of either the geological

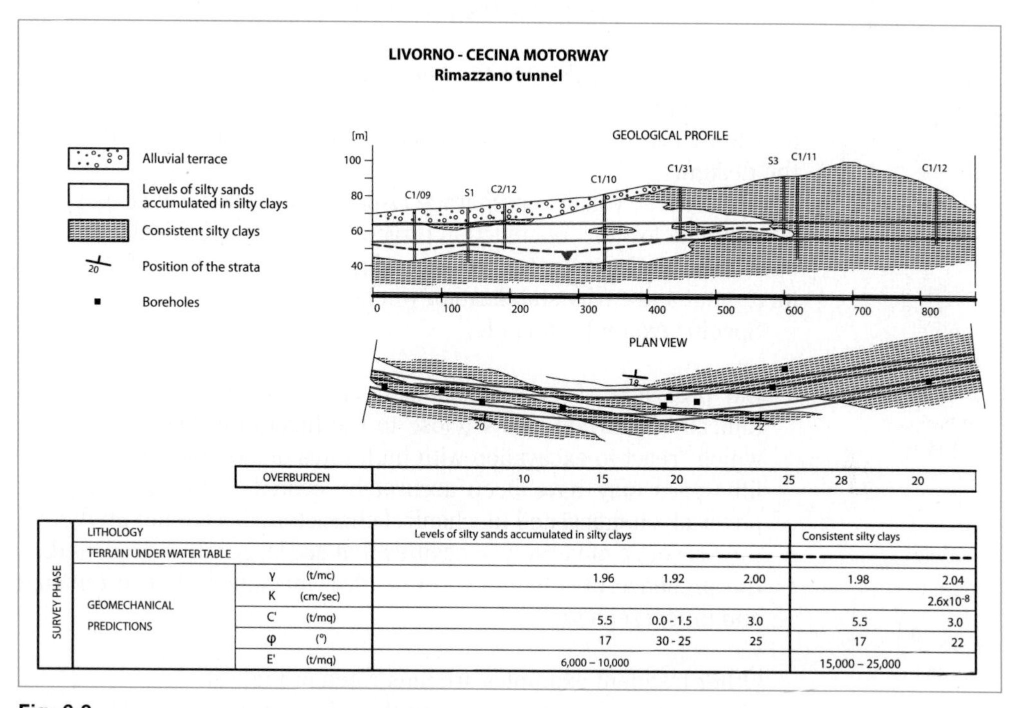

Fig. 6.3

nature of the terrain or of the geomechanical characteristics of the ground of which it consists.

In the *narrow sense* of the term unpredicted geology consists of an unfortunate geological feature, the presence and/or precise location of which was missed during the design stage and which as a consequence requires partial or total redesign of the project and significant changes to the methods of tunnelling and to the construction schedules. Generally speaking unpredicted geology generally occurs in one of the following forms:

- a tectonic element of primary importance such as a regional scale fault or fold or a sector with high residual stresses (compression zones, the hearts of folds) often associated with substantial flooding;
- inaccurate reconstruction of the geometrical position of highly chaotic and/or tectonised geological bodies, often associated with the presence of water and with poor geomechanical characteristics;
- deep slope gravitational deformation;
- a series of types of ground, not very extensive in surface area and poor, either because of the stratigraphical structure or of the vacuolar and/or porous nature, which are the site of large aquifers under pressure (e.g. confined levels of carbonatic terrains and Triassic or Miocene chalk, etc.). It is not infrequent for these terrains to display a tendency to swell or suffer chemical and physical weathering when decompressed in the presence of water.

Unpredicted geology in the *broad sense* of the term consists mainly of either a characteristic or a type of the geomechanical behaviour of a tectonic element or of a geological nature that was not predicted in the design.

The characteristic or characteristics in question cause negative phenomena for the construction project and more specifically both for excavation and for the functioning of ground improvement and reinforcement and cavity stabilisation works.

Specific examples include:

- the presence of overconsolidated clayey ground located in tectonically disturbed zones (close to the head of a thrust fault, etc.) which "react to excavation with high rates of swelling". While these lithotypes may have been accurately located in the design, their physical, chemical and mechanical characteristics are not predicted;
- the presence of karst type cavities that are larger than predicted;
- the presence of well-known geological formations but in unusual and unforeseen *facies*.

Other frequent examples are illustrated in Fig. 6.4.

Modern survey techniques and the considerable knowledge accumulated by modern geology mean that the geological risk attached

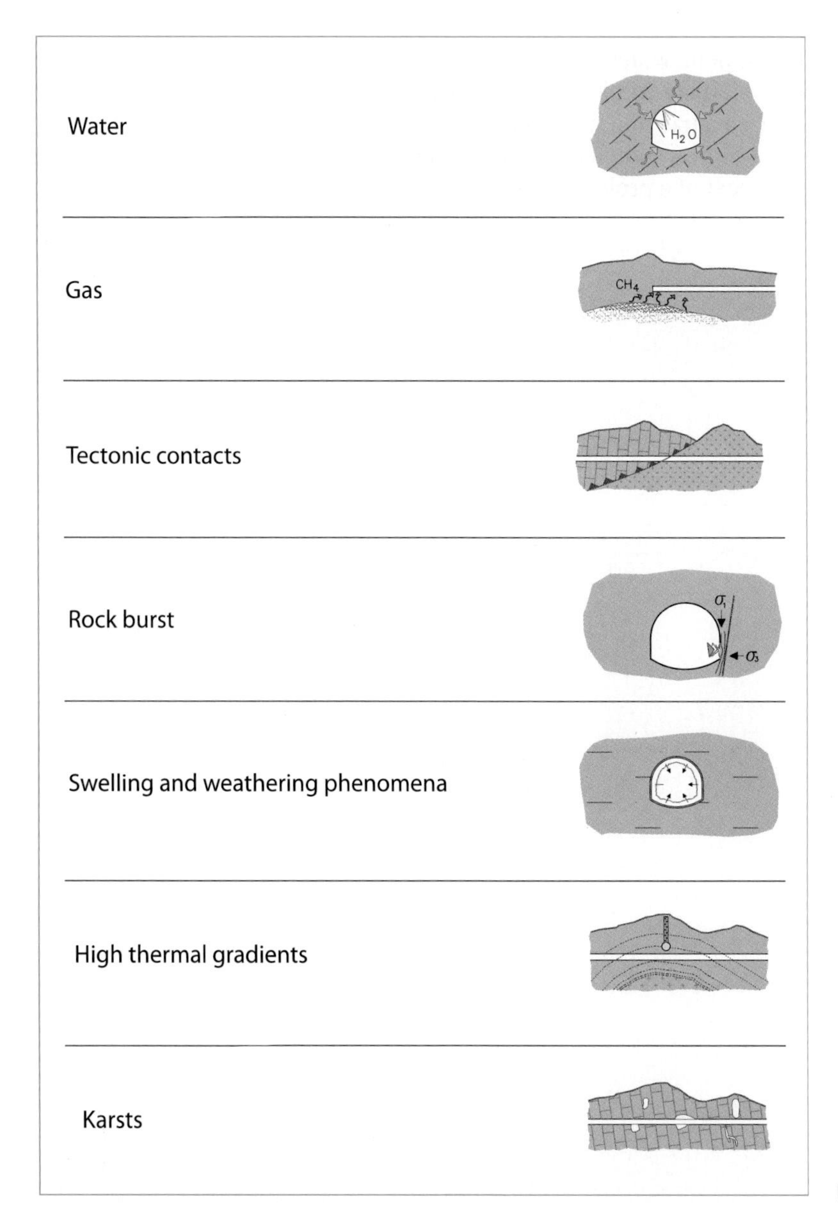

Fig. 6.4

to unpredicted geology can be greatly reduced and that underground construction can be performed with the risk of only unpredictable events, which is to say those unfortunate events of exceptional gravity and size which it is virtually impossible to predict or which are in any case very unlikely given reasonable technical and financial investment.

A careful and detailed survey campaign is therefore indispensable to substantial reduction of the risks intrinsic to underground construction. In this respect, resources committed to the survey phase, preparatory to construction, must be considered as a true and genuine investment, which will generate certain and beneficial returns in terms of the ability to plan and industrialise tunnel advance, with certainty over construction times and costs.

It seems completely reasonable to allocate not less than 2% of the total cost of a project to an investment of such importance.

CHAPTER 7

The diagnosis phase

7.1 Background

The diagnosis phase is that in which the design engineer uses the information collected during the survey phase to formulate predictions of the deformation behaviour of the tunnel (extrusion, preconvergence and convergence) in the absence of stabilisation intervention. The result is the division of the underground alignment into sections with uniform deformation behaviour, identified in terms of three basic behaviour categories:

- *category A*: stable core-face or rock type behaviour;
- *category B*: stable core-face in the short term or cohesive type behaviour;
- *category C*: unstable core-face or loose type behaviour

It is important for the study of excavation methods and stabilisation techniques designed to guarantee the feasibility and long life of an underground project to be preceded by a phase in which a clear idea is acquired of the type of deformation response to be expected in the core-face zone as a result of excavation. This emerges clearly when the evidence produced by the theoretical and experimental research illustrated in previous chapters is considered: the rigidity of the core of ground ahead of the face, and therefore the conditions of stability in that ground, have a decisive effect on that deformation response and determine how an arch effect is triggered and consequently the tone of the stress-strain response in the whole tunnel.

It is therefore fundamental for the design engineer to know how to predict it accurately since he will be able to formulate essential guidelines based on it for the design of stabilisation intervention to guarantee the success of the enterprise.

7.2 The basic concepts of the diagnosis stage

Before starting on the actual diagnosis study itself, it is always useful to divide the tunnel alignment into sections along the alignment with similar geomechanical characteristics in relation to the lithology, structure, morphology hydrogeology and so on. The design engineer can then identify sections with a uniform stress-strain behaviour for which the type and magnitude of the expected deformation can be evaluated, on the basis of the magnitude of the lithostatic loads and

Sections with uniform stress-strain behaviour

The purpose of the diagnosis phase is to divide the underground route into sections with uniform stress-strain behaviour, identified in terms of three fundamental behaviour categories:

- category A: stable core-face or rock type behaviour;
- category B: stable core-face in the short term or cohesive type behaviour;
- category C: unstable core-face or loose type behaviour.

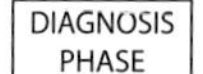

DIVISION OF THE UNDERGROUND ROUTE
INTO SECTIONS WITH UNIFORM STRESS-STRAIN BEHAVIOUR

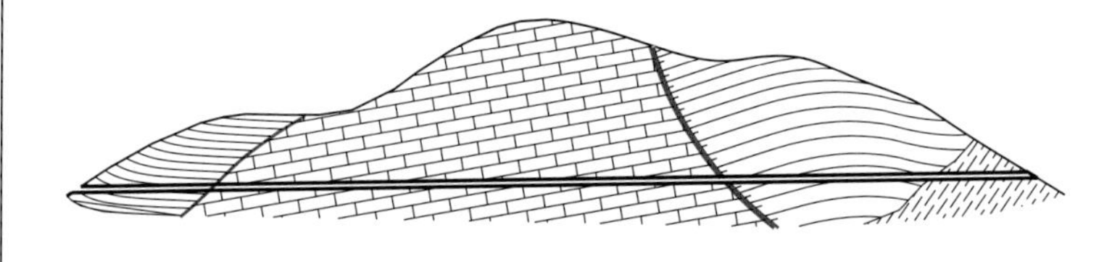

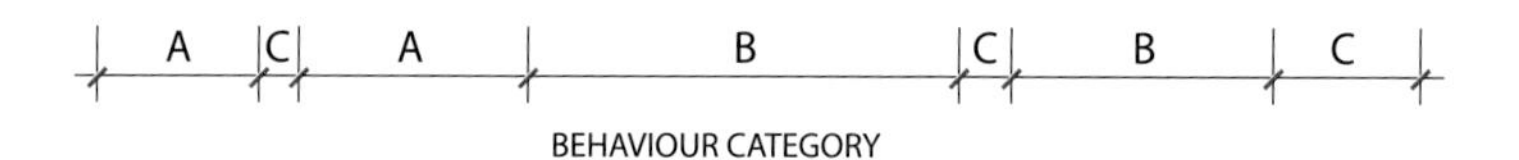

BEHAVIOUR CATEGORY

Definition of behaviour categories

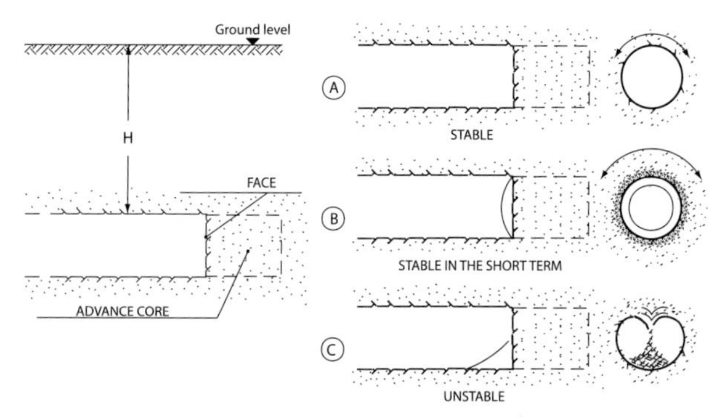

To achieve it, the design engineer uses the information collected during the survey phase to formulate predictions of the deformation behaviour of the tunnel, in terms of extrusion, preconvergence and convergence in the absence of stabilisation intervention, by using the experimental and mathematical instruments available to him.
With regard to the former, triaxial cell extrusion tests are very useful for tunnels to be driven in prevalently clayey terrains.
With regard to the latter, characteristic line theory [12] can be used very successfully in diagnosis assessment for most tunnels.

any other determining factors (tectonic stresses, hydraulic gradients, pre-existing buildings and structures, etc.).

The prediction (diagnosis) of these deformation phenomena is then performed on the basis of the parameters acquired during the survey phase, taking account of environmental, morphological, hydrogeological, structural and stress aspects and using the analytical or numerical systems of calculation currently available, in both the elastic and the elasto-plastic range. It is important for this purpose for the design engineer to also have a precise knowledge of regulatory and environmental constraints which must be observed during construction and service of the infrastructure (e.g. the need to conserve areas of particular interest for wildlife, causing a minimum of disturbance to inhabitants, maintaining the levels of water tables which supply aqueducts and not disfiguring the landscape, etc.).

Without dwelling on the criteria that should be employed to divide the tunnel alignment into sections with similar geomechanical characteristics, and it must be considered there may not be only one set of criteria, in this chapter we will illustrate those which should be followed to be able, on the basis of the results of the preceding survey phase:

- to divide the underground alignment into sections with uniform stress-strain behaviour;
- to study the genesis and development of deformation phenomena fully and in detail for each individual section, assessing the magnitude in order to identify the corresponding behaviour category;
- to take account of how the presence of ground water, or the need to limit deformation because of adjacent buildings and structures, can affect the behaviour category;
- to recognise the type and magnitude of the prevalent instability phenomena which will manifest in each section as a possible development of deformation phenomena;
- accurate assessment of the stability of natural and artificial slopes above and around tunnel portals.

7.3 Identification of sections with uniform stress-strain behaviour

A section with uniform stress-strain behaviour is defined as any portion of the underground alignment within which it is reasonable to hypothesise that the deformation response will develop in a qualitatively constant manner for the whole of its length as a reaction of the rock mass to excavation.

Triaxial cell extrusion tests for the study of the deformation response

In the case of underground works to be driven in prevalently clayey ground, triaxial cell extrusion tests, already discussed in section 3.1.2, constitute an excellent instrument for studying and predicting the deformation response of the ground to excavation. These tests simulate tunnel advance on a small scale by reconstructing the situation that is created in the ground for a cross section of the ground ahead of the face when the horizontal confinement pressure in a direction parallel to the tunnel alignment decreases until it is completely zero when the face arrives.
The apparatus needed to perform them is similar to that used for standard triaxial tests. The main difference consists of a special "extrusion chamber" applied to the top of the sample. This consists of a hollow cylinder in stainless steel which penetrates the inside of the test sample along its axis for a given length. A thin lattice membrane separates the inside of the chamber from the ground. The latter is connected to an external burette filled with water. The extrusion of material in the chamber as the pressure inside it is reduced, causes the water to come out and the level in the burette to change.
The test is performed on samples of ground prepared by cutting them with a die. A cavity with the same dimensions of the extrusion chamber is cut in the end of the test piece. It is then positioned, saturated and consolidated, inside the triaxial cell at a pressure equivalent to the overburden of the tunnel. At this stage, the pressure inside the extrusion chamber is always kept at the same level as the total vertical pressure acting on the test piece. Once the pre-established consolidation condition has been reached, the extrusion phase is started: the pressure inside the chamber is reduced in steps, while maintaining the test piece under drained conditions. At each step the extrusion is protracted until it finishes, or at least until it has been accurately determined, and the change in volume resulting from the extrusion of the material and the change in the height of the test piece is measured. Finally the results of the test are summarised by an *extrusion-confinement pressure* graph (see figure). The stress-strain response predicted for the tunnel can be easily deduced from the extrusion curve as a function of the decrease in pressure:

- if the extrusion curve remains in the elastic range when the confinement pressure reduces to zero and is more or less straight, the behaviour of the future tunnel will be of the stable core-face type (category A);
- if the extrusion curve deforms in the elasto-plastic range without reaching failure when the confinement pressure reaches zero the behaviour conditions will be stable core-face in the short term (category B);
- if, finally, the extrusion curve deforms into the failure range when the confinement pressure reaches zero the behaviour conditions will be of the unstable core-face type (category C);

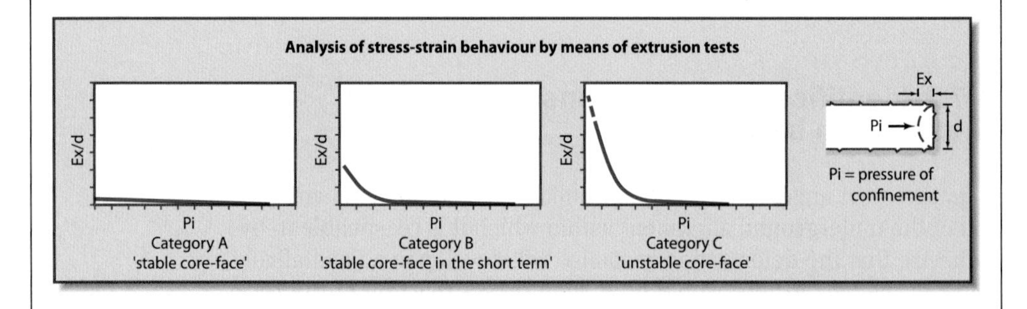

The division of the underground alignment into sections with uniform stress-strain behaviour constitutes the main purpose of the diagnosis phase.

The starting point is the complete characterisation of the terrain on the basis of a detailed geological and geomechanical profile of the tunnel to be driven, produced during the survey phase, and the identification of zones along the alignment with similar geological and geomechanical characteristics, in order to be able to treat them in a uniform manner in the subsequent prediction phase.

To achieve this the design engineer must give particular consideration to the following:

- the geological and geomechanical characteristics of the rock mass, with particular reference to the tectonic-structural situation and to the strength and deformability parameters;
- the general hydrogeological conditions.

He will therefore be able to make some preliminary considerations on the probable genesis of the deformation response and on how it may propagate when an underground cavity is opened.

To do this, in addition to the factors already mentioned, he must also take into consideration:

- the geomorphology;
- any pre-existing buildings and structures and environmental constraints there may be;
- the natural stress state in the ground at the depth of the tunnel;

each of which have a greater or lesser direct effect on the type, magnitude and limits of acceptability of the deformation response.

More specifically:

1. the geomorphological and tectonic-structural situation in the rock mass has a direct effect on how stresses are channelled within the radius of influence of the face and around the cavity. For example: a walled tunnel may have its side walls loaded in an appreciably asymmetrical manner; particular geometrical configurations of rock discontinuity planes in relation to the position of the tunnel walls can give rise to the so-called "rock bursts";
2. passing through an aquifer in which it is indispensable to conserve the hydrogeological equilibrium unaltered by preventing or limiting drainage during tunnel advance and when it is in service, will definitely require costly preventative works to be performed ahead of the face (water proofing injections etc.). This, as we will see later, will automatically place the section considered in the behaviour category C, regardless of the structural stability of the core-face;

Calculation methods for study of the deformation

Today the design engineer who specialises in underground construction has numerous calculation methods available for diagnosis of the deformation response of the rock mass to the action of excavation.
Each method differs from the others in terms above all of the hypotheses on which they are based making each more appropriate for reliable prediction of determined situations rather than for others.
It is, however, possible to divide the different methods into two large families: solid load methods and plasticised ring methods.
Solid load methods are those which assess the deformation response of the rock mass to excavation as a consequence of the formation of volumes of material at the face and around the cavity that are released under their own weight. They are therefore methods suitable for studying this deformation response in soils lacking in cohesion and in fractured rock masses, the behaviour of which is determined by the combination of the discontinuities and their geomechanical characteristics rather than by the strength of the matrix, which remains basically good.
Plasticised ring methods, on the other hand, assess the deformation response of the rock mass as a consequence of the formation of a continuous band of disturbed ground around the face and around the cavity. This tends to progressively extend into the rock mass with the geometry of a ring coaxial to the tunnel and with a radius that increases as a function of the natural stress state and the strength and deformability of the material as tunnel advance proceeds.

Rockfall

3. the same is true for sections of tunnel in terrain with poor geomechanical characteristics which pass close to pre-existing structures such as buildings, the foundations of bridges or viaducts and so on, which impose severe limits on admissible deformation;
4. the magnitude and orientation of the tensor for the pre-existing natural stresses in the rock mass at tunnel depth have a determining effect on the stress-strain response of a tunnel. Knowledge of these is therefore of fundamental importance for correct design. It is known for example that residual tectonic stresses can orient the natural stress tensor in a different direction to the position it would assume as a result of lithostatic loads alone. This can give rise to markedly dissymmetrical deformation.

A general picture can therefore be drawn from detailed analysis of the factors mentioned above of how the behaviour of the tunnel will develop during excavation to arrive at an initial basic division of the underground alignment into sections with uniform stress-strain behaviour. A behaviour category can be assigned to each section by comparing the stress states induced with the strength of the rock mass.

It should be pointed out that this first division of the underground alignment into sections with uniform stress-strain behaviour is based on very simple considerations, many of which are of a qualitative nature. Naturally the general picture that the design engineer draws from them can be modified to a greater or lesser degree when more detailed analysis is performed using the instruments that mathematics provides.

7.4 Calculation methods for predicting the behaviour category

From a general viewpoint, the study of the genesis and development of deformation phenomena in the diagnosis phase must be conducted using elasto-plastic calculation methods valid for all types of ground that are capable of modelling the statics in three dimensions to a sufficient degree of approximation. Each new scientific contribution which makes it possible to predict deformation phenomena to obtain a closer match with reality should be given due consideration.

In ordinary cases, when the situation can be interpreted within the basic hypotheses assumed by it, the design engineer can make a sufficiently reliable assessment of the behaviour category of a tunnel by using the characteristic line method, which as is known, allows:

- account to be taken of the contribution made to the stability of a tunnel by the core of ground ahead of the face. This therefore shows how the stability of the core-face influences deformation phenomena which occurs following tunnel advance;

Solid load calculation methods. Block theory

Block theory [46] is a three dimensional calculation technique which uses the methods of analytical geometry for in-depth study of discontinuity systems and the reciprocal interaction between them. It identifies *key blocks* on which the stability of the whole zone surrounding an excavation depends. "Key blocks" are therefore intended as meaning those wedges of rock, which, when they move, create a space which other previously confined wedges can move towards to trigger a progressive and uncontrollable series of collapses. It is therefore sufficient to appropriately secure the key blocks, with anchors for example, to guarantee the safety of the entire tunnel or to place confinement works able to resist the gravitational thrust of the unstable masses.

Basically block theory is useful for kinematic study of the deformation response in rock masses stressed mainly in the elastic range and characterised by fairly regular discontinuities.

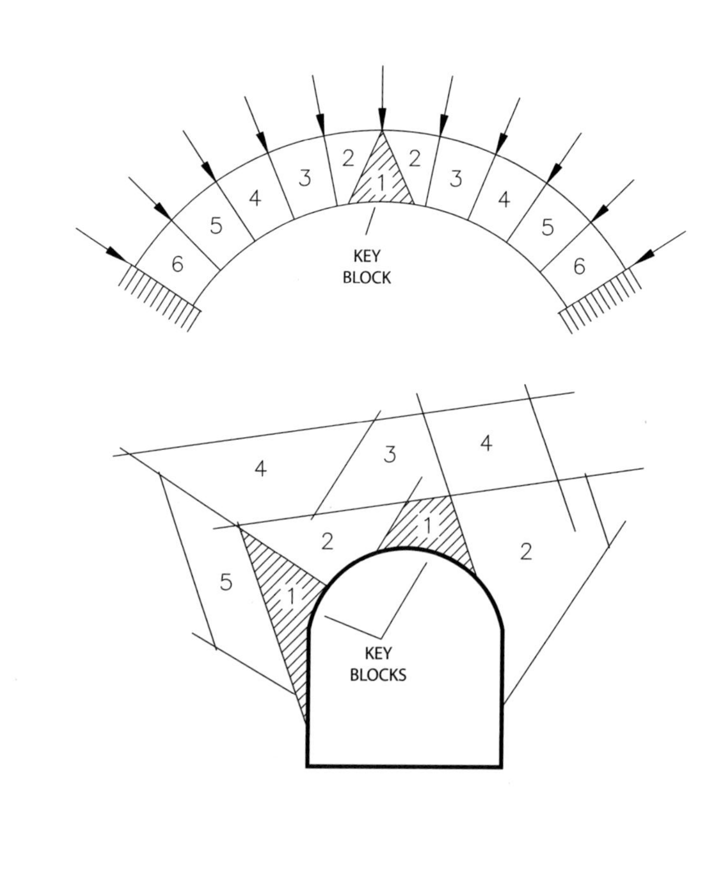

- the detailed construction, based on the geotechnical and hydrogeological parameters of the rock mass, of the course taken by deformation of the advance core and of the cavity of the future tunnel at the face as a function of the loosening produced in the rock mass by excavation.

More specifically, by using these well-known analytical formulas in the elasto-plastic range, the design engineer will plot:

- the characteristic line of the core-face;
- the characteristic line of the cavity, valid at the face;

and, if necessary, also:

- the characteristic line of the cavity, valid for all sections outside the zone of influence of the core-face.

To achieve this, it is important to adopt the following criteria:

- the strength and deformability parameters to be employed in the calculation must be those for the rock mass at full tunnel scale;
- in the case of tunnels to be driven by TBM's and shields, the pressure exerted on the face by the machine must be treated as confinement of the core and as such must be taken into consideration in diagnosis calculations;
- in the case of twin bore tunnels, the calculation must consider the increased stresses that each tunnel exerts on the other (see the Index Area method on the focus box of the same name).
- in the case of tunnels under the water table, the effect of pore pressure on each characteristic line must be considered appropriately by employing special formulas.

The behaviour category of the future tunnel in the section considered can be easily predicted from the characteristic lines and more specifically from the curve of the core-face characteristic line in relation to that of the characteristic line for the cavity at the face (Fig. 7.1):

- if the characteristic line for the core-face crosses that for the cavity at the face while remaining in the elastic range, then the future tunnel will display stable core-face behaviour (category A);
- if the characteristic line for the core-face crosses that for the cavity at the face under elasto-plastic conditions, then the conditions will be those of stable core-face in the short term (category B).
- finally, if the characteristic line for the core-face does not cross that for the cavity at the face because it has deformed in the elasto-plastic range to the point of failure, then the conditions will be those of unstable core-face (category C).

Plasticised ring calculation methods. Characteristic line theory

The *theory of characteristic lines* [12], which has already been mentioned in chapter three on the analysis of the deformation response, is a more economical alternative to computer methods for the study, without any large errors, of situations which meet the basic hypotheses of the method (circular tunnel in a homogeneous and isotropic medium, stressed by a field of natural hydrostatic type stresses), characterised by markedly elasto-plastic phenomena, which is either ductile or fragile.
It is one of the few elasto-plastic systems of calculation which allows the static situation of the core-face of a tunnel to be predicted during tunnel advance by using simple formulas without having to solve complex systems of equations.
This makes it useful for predicting the behaviour category of a tunnel during the diagnosis phase, without the need to resort to numerical methods, which, while they are definitely more precise, are also more difficult to manage and in any case much more powerful than what is required for the purpose.
The static and dynamic effects of water, if present, on the stress-strain state of the rock mass can also be properly accounted for using special formulas [47] [48].

Convergence of the walls of a tunnel in plasticised rings or bands

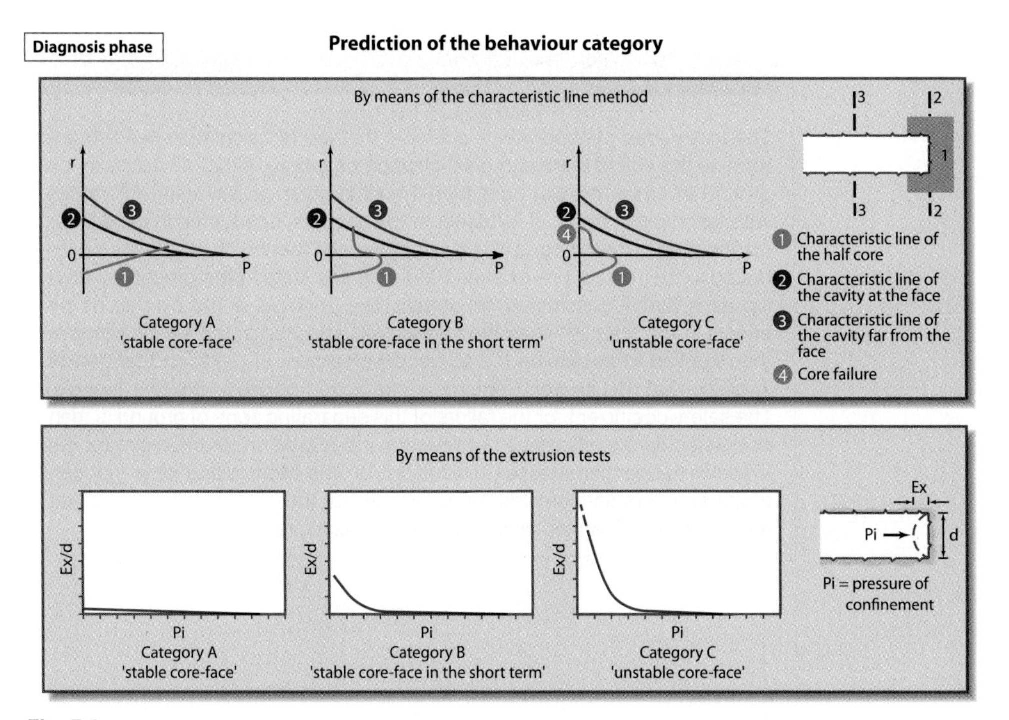

Fig. 7.1

For sections of the underground alignment that pass through prevalently clayey terrains, the design engineer must formulate similar and very reliable predictions for the stability of the core-face and the behaviour category, based on triaxial cell extrusion tests performed on undisturbed samples taken *in situ*. Here too the stress-strain behaviour category can be easily identified from the extrusion-confinement pressure curve, as shown in the focus box *Triaxial cell extrusion tests for the study of the deformation response.*

In particularly complex situations, where the basic hypotheses for the characteristic line method do not apply (tunnels at shallow depths, tunnels close to the sides of slopes, tunnels adjacent to pre-existing buildings and structures, etc.), the determination of the behaviour category must be performed using more powerful, three dimensional means of calculation, such as, for example, finite element or finite difference numerical methods in the elasto-plastic range.

It is worth repeating that the assignment of a behaviour category during the diagnosis phase is not necessarily always connected with the stability of the core-face. As has already been said, if, for example, a tunnel is to be driven through an aquifer for which the design engineer considers it indispensable to conserve the hydrogeological equilibrium, by preventing or containing drainage during tunnel advance or when it is in service, this means that ground improvement and waterproofing must be performed ahead of the face. The section in question

The "Index Area" method

The *Index Area* method [49] is a simple method of calculation used to determine the stress state and plasticisation phenomena that develops in the ground in cases of twin bore tunnel construction, widely used for routes with fast moving traffic. It is based on the analysis, conducted in the elasto-plastic range according to the Kastner-Fenner theory, of modifications induced in the natural pre-existing natural stress state in the ground by driving each tunnel considered separately. The principle of the overlap of the effects in the area between the two tunnels stressed in the elastic range is then applied to determine the actual development of radial and tangential stresses that results from reciprocal interaction between the two tunnels. The safety coefficient for the failure of this separating zone of ground is then calculated as the difference between the index area under the curve for the available tangential stresses (calculated on the Mohr plane as a function of the actual radial confinement stress and of the intrinsic curve) and that under the actual tangential stresses in the elastic range.

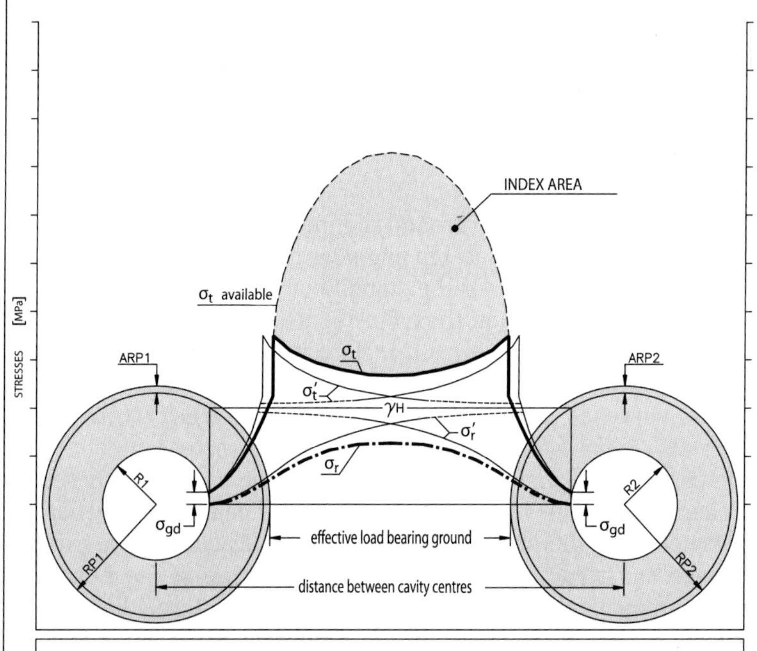

LEGEND

σ_r = actual radial stresses

σ'_r = radial stresses for each individual cavity

σ_t = actual tangential stresses

σ'_t = tangential stresses for each individual cavity

σ_{gd} = monoaxial breaking strength of the rock mass

will therefore be assigned to behaviour category C, regardless of the structural stability of the core-face.

Similarly, when the alignment of a tunnel passes close to pre-existing infrastructures, the design engineer must also consider the acceptable limits to the deformation induced on those infrastructures by the new tunnel, when a behaviour category is assigned. For example, if a section under examination has a shallow overburden and passes under a building, then it is probable that preconfinement intervention will be necessary to limit surface subsidence, unless the ground consists of competent rock; this will automatically place the section considered in category C, regardless of the stability of the core-face.

Obviously the passage through a seismic zone will not in itself determine assignment to one category rather than to another. Nevertheless, the design engineer must identify sensitive sections (such as, for example, rock-detritus sections), which will need to be studied with particular attention in the therapy phase.

7.5 Assessing the development of the deformation response

Once each section of the alignment has been classified as belonging to one of the three behaviour categories, the following must also be identified in the diagnosis phase for each category:

- the types of deformation that will develop at the face and around the cavity;
- consequent and expected manifestations of instability, such as:
  - rockfalls and spalling at the face, produced by extrusion of the core and by preconvergence;
  - failure of the core-face, produced by extrusion and by preconvergence;
  - rockfalls and spalling around the cavity, produced by convergence of the cavity;
  - collapse of the cavity, produced by failure of the face;
- loads mobilised by excavation according to solid load and plasticised ring models.

In consideration of what has already been stated in chapter one on the question of the deformation response, it is important to predict whether the formation of solid loads or of plasticised rings will be prevalent (see also the focus box, *The type and the development of the deformation response (reaction)* in chapter one). There will be:

- predominant formation of **solid loads**, if the deformation predicted is of the elastic type or develops towards failure;

Manifestations of instability. The effect of the geological structure of the rock mass

The kinematic analysis of rock falls and expected deformation in the cavity following tunnel excavation provides extremely useful information for subsequent decisions made in the therapy phase [50].

The deformation response of the rock mass to excavation depends basically, not only on the size of the overburdens (stress state), but also on the geological structure. The table below identifies those found to occur most frequently on the basis of experience acquired over twenty years of research into the development of the deformation response and the consequent manifestations of instability.

Longitudinal view	Cross section	Description
	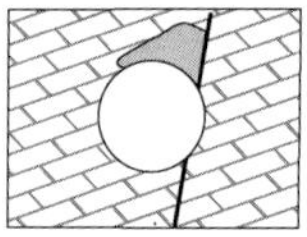	Collapse caused by the geometry of discontinuities
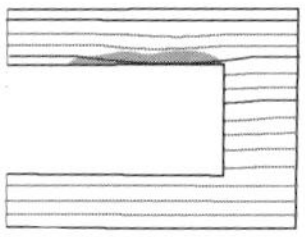	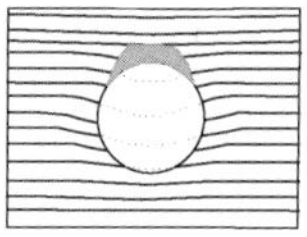	Flexure, failure and collapse of rock slabs in the crown
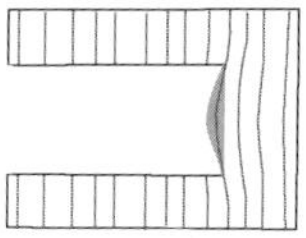	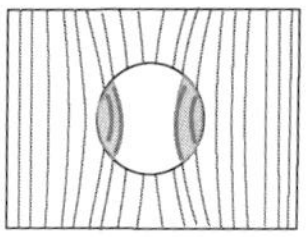	Flexure and failure of rock slabs in the wall
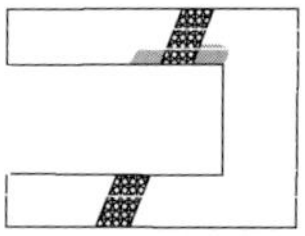	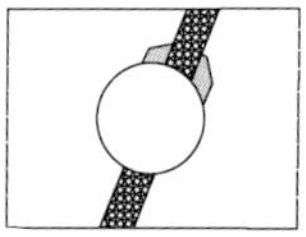	Collapse caused by the presence of a single joint
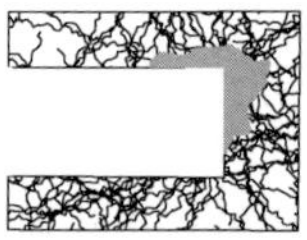	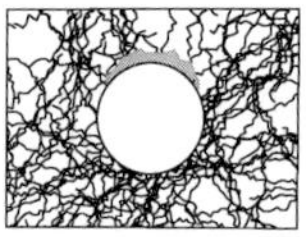	Collapse in densely fractured rock mass
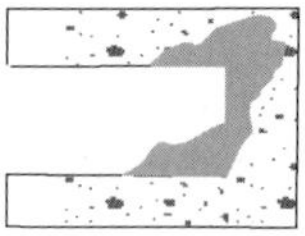	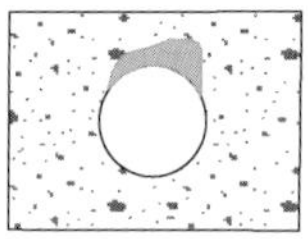	Collapse of cataclastic material with reduced or no cementation
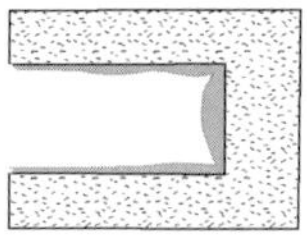	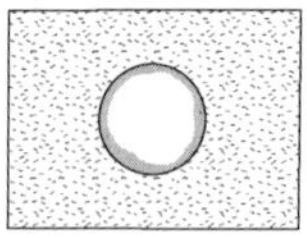	Narrowing of the cavity due to the presence of plastic, squeezing and/or swelling material

- predominant formation of **plasticised rings**, if the deformation predicted is of the elasto-plastic type.

In the first case, study of the deformation response is conducted using **solid load calculation methods.** For basically good quality rock masses, characterised by fairly regular discontinuities, which govern the deformation behaviour of the cavity while the rock matrix remains stressed in the elastic range (rock type behaviour), study is aimed essentially at assessing the magnitude and location of gravitational phenomena triggered by the structural anisotropy of the rock mass.

Since the influence of the discontinuities increases as the stress state around the cavity (arch effect) decreases, it is important in this case to conduct a preliminary study to identify the formation of zones that may be subject to tensile stress in the crown and around the springline of the tunnel.

In soils with no cohesion or rock masses that are intensely fractured at the scale of the diameter of the tunnel (loose type behaviour), then the magnitude of the solid load that is formed at the face and in the crown of the tunnel must be assessed along with the vertical and horizontal pressures that may manifest as a consequence.

In the second case, study of the development of the deformation response is conducted using **plasticised ring methods of calculation** and will be aimed at assessing the probable extension of the radius of influence of the face and the magnitude of the ring of plasticised ground around cavity on which decisions to intervene and stabilise the ground to be made in the therapy phase will depend. All those factors which could worsen the stability of the tunnel, such as chemical and physical weatherability, *fluage* and creep, the effects of possible hydrodynamic pressures and so on are identified and quantified in terms of deformation (extrusion and convergence). In both situations account must be taken, when studying deformation phenomena, of the effects of underground water under hydrostatic or hydrodynamic regimes, if it is present. In the case of twin tunnels, the design engineer must also pay particular attention to stress and deformation effects resulting from the presence of two tunnels close to each other which, as is known, induce a field of increased stress on each other.

7.6 Portals

It is indispensable to spend a few words on the question of portals before concluding the diagnosis phase, because there are often delicate aspects to these.

In this case too, the objective of the diagnosis phase is to predict the stress-strain response of the slope that is produced, in the absence of stabilisation works, when excavation work commences to start the tunnel.

Predicting the stability of portal excavation – 1

The stability of a portal excavation in rocky ground characterised by joints and fractures is heavily influenced by the reciprocal orientation of the discontinuity families with respect to the free wall of the portal. Careful study of the geological structure of a rock mass will allow a fair initial estimate to be made of the problems for stability that will be encountered. This interaction is clearly shown in the figure below which shows how, with other factors remaining constant, a change in the position of a portal face affects the situation to be considered.

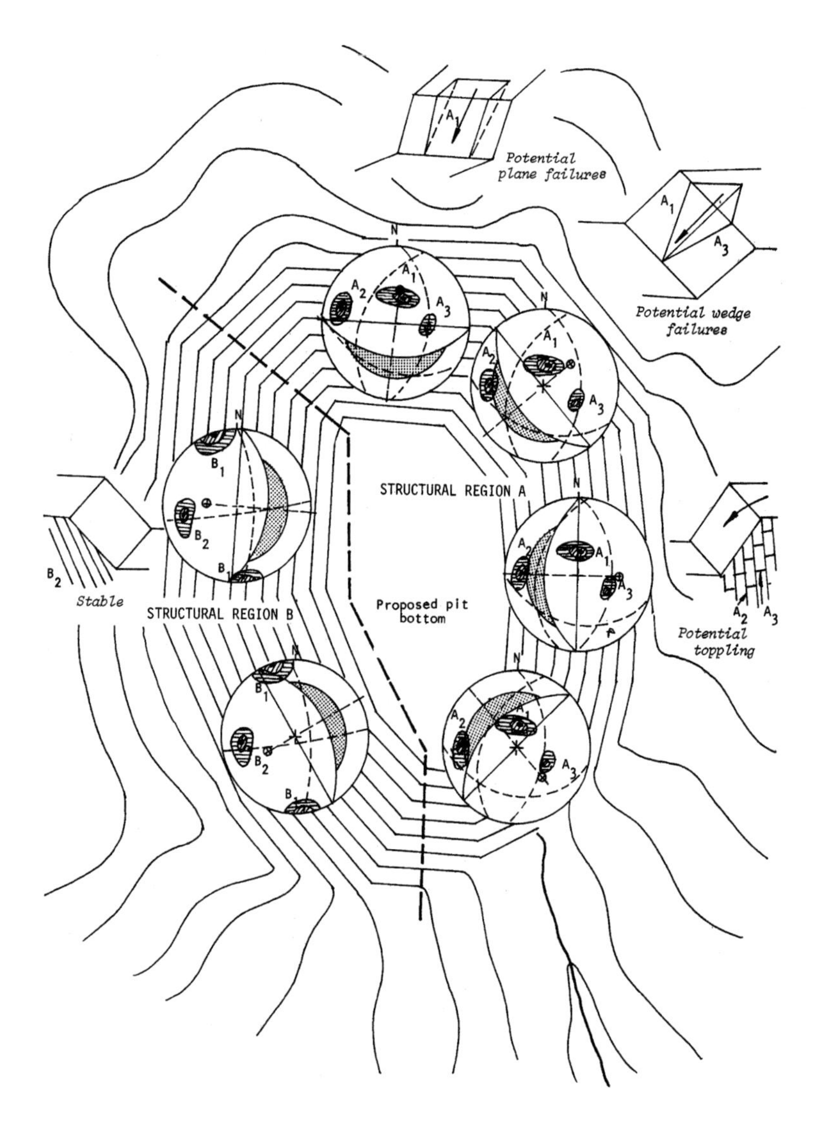

Source: Hoek E., Bray J.W., Rock slope engineering. Institution of Mining and Metallurgy.

To achieve this, thorough study of the following results of the survey phase must be conducted:

- the lithology, morphology, tectonics and structure of the slope to be entered;
- the hydrology, the pre-existing buildings and structures and the environmental constraints affecting the slope;
- the geomechanical characteristics of the ground of which it is formed.

It is basically a question of predicting whether the excavation performed will be generally stable or unstable in the absence of specific confinement intervention.

As we will see, with the aid of a few examples, in-depth analysis of all the data acquired in the survey phase can direct the design engineer to the formulation of reliable predictions in this sense.

7.6.1 Lithology, morphology, tectonics and structure of the slope to be entered

Knowledge of the lithology, morphology, tectonics and structure of the slope to be excavated is indispensable to reliable prediction of its deformation behaviour. It is in fact quite clear that sheer cliff consisting of high quality rock material will be very unlikely to pose problems for overall stability, while it is certain that a mass of pseudo-cohesive material cannot be excavated under any circumstances without being first confined.

7.6.2 Hydrology, pre-existing buildings and structures and environmental constraints

The presence of water is always a factor to be given serious consideration because, not only does it reduce the shear strength of the ground due to the effect of neutral pressures, but it also acts as a lubricant in the material that fills structural discontinuities in ground consisting of rock. Any pre-existing buildings or structures and environmental or archaeological constraints must be considered carefully, because they may require particular construction methods (e.g. to eliminate all deformation) not actually needed to guarantee the stability of the excavation itself.

Manifestations of instability that occur on a portal excavation

STRUCTURE	TYPICAL PROBLEMS	CRITICAL PARAMETERS
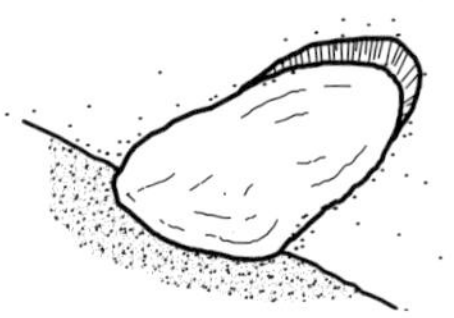 Portals in ground subject to landslide	Complex failure with sub circular slip surfaces along faults or other structural features which may also affect the matrix	• Presence of regional faults • Shear strength of the material along the slip surface • Distribution of ground water in the slope, as a consequence of rain or submersion of the foot of the slope in particular • Potential seismic load
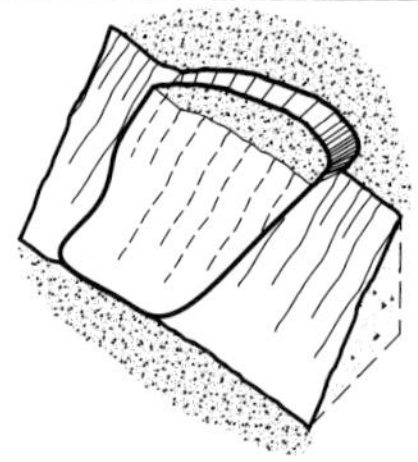 Portals in soils or heavily fractured rock masses	Failure along a spoon shaped surface, in soils or heavily fractured rock masses	• Height and gradient of the face • Shear strength of the material along the slip surface • Distribution of groundwater in the slope
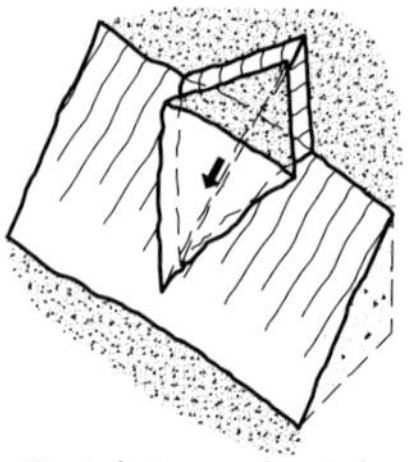 Portals in moderately fractured rock masses	Plane or wedge shaped slip on a structural feature or along a line on intersection between two structural features	• Height, gradient and orientation of the slope • Immersion and direction of the structural features • Distribution of groundwater in the slope • Potential seismic load • Sequence of excavation and stabilisation operations
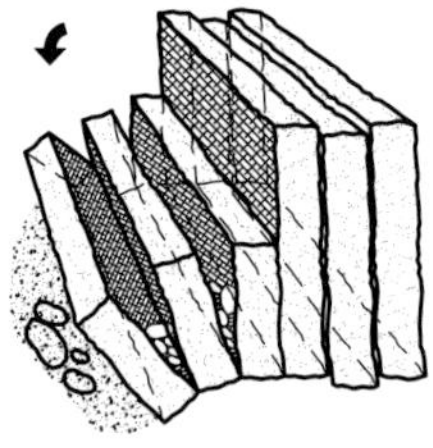 Portals in sub-vertically fractured rock masses	Toppling of columns separated from the rock mass by sub vertical, parallel or sub parallel features at the face of the slope	• Height, gradient and orientation of the slope • Immersion and direction of the structural features • Distribution of groundwater in the slope • Potential seismic load • Sequence of excavation and stabilisation operations
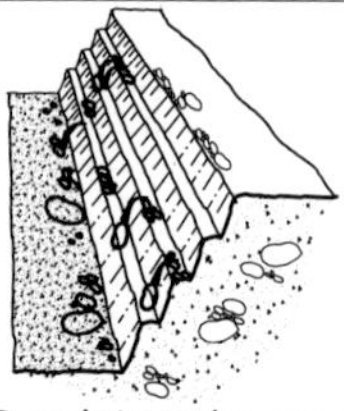 Portals in rock masses with loose blocks of rock	Slip, roll, fall and toppling of blocks of rock or erratic boulders on the slope	• Geometry of the slope • Presence of loose boulders. • Restitution coefficient of the material of which the slope consists • Presence of structures to halt the fall and bounce of rock

7.6.3 Geomechanical characteristics of the ground

Knowledge of the geomechanical characteristics of the ground of which a slope is formed will provide valuable indications. Various tables can be used on the basis of the strength of the ground to assess the maximum height of entry that is possible without the need for confinement. By analysing the data from the survey phase, a skilled and experienced design engineer will therefore be able to form a realistic picture of the natural equilibrium of a slope which a tunnel must enter and of how excavation will affect the statics of the situation.

Naturally this picture, formed from an initial analysis, can be modified to a greater or lesser degree following in-depth verification performed in doubtful cases by using mathematical instruments.

7.6.4 Forecasting the deformation behaviour of the slope

Except for those cases of portals to be constructed in slopes consisting of material with no cohesion, and therefore unstable if not adequately confined, the prediction of the stress-strain response of a slope must be conducted on the basis of calculation methods capable of modelling the real morphological, hydrogeological, stratigraphical, geomechanical and environmental situation to a sufficient degree of accuracy. It is worth mentioning here the finite element, distinct element, block theory and limit equilibrium methods. The choice of the calculation method most suitable for modelling the situation under study must be made by the design engineer on the basis of experience and must take account of all those factors which can have a negative effect on the safety of excavation such as high rainfall, seismic activity, the vibrations of possible rock blasting, etc.

On conclusion of the analysis the design engineer must be able to establish:

- whether excavation without confinement will be stable or unstable:
  - if it is stable, whether the deformation suffered by the slope will be compatible with any pre-existing buildings and structures there may be;
  - if it is unstable, the type of instability. More specifically, it is necessary on the one hand to predict whether instability will occur as a result of slip along discontinuities or whether it will also involve the matrix of the material, and on the other, whether failure will occur in a fragile or ductile manner. Fragile behaviour normally occurs each time the material is unable to develop sufficient residual strength (e.g. sand, stratified rock mass susceptible to landslide and characterised by discontinuities with poor shear strength). Ductile behaviour, on the other hand, normally occurs in material with a good reserve of residual strength (e.g. argillites, plastic clays, etc.);

Predicting the stability of portal excavation – 2

Predicting the general stability of a portal excavation can be performed using the same methods of calculation normally used to assess the safety conditions of natural and artificial slopes. The design engineer will select, from the many available, those able to schematise the type of slip (circular, plane, wedge shaped, etc.) with the best fit to the reality of the situation according to each individual case. Numerous calculation charts and tables can be found in the literature for rapid assessment of the stability of a portal without the need to resort to computer models.
The figure below shows a typical chart taken from the book "Rock Slope Engineering" by Hoek and Bray [51].

CIRCULAR FAILURE CHART NUMBER 1

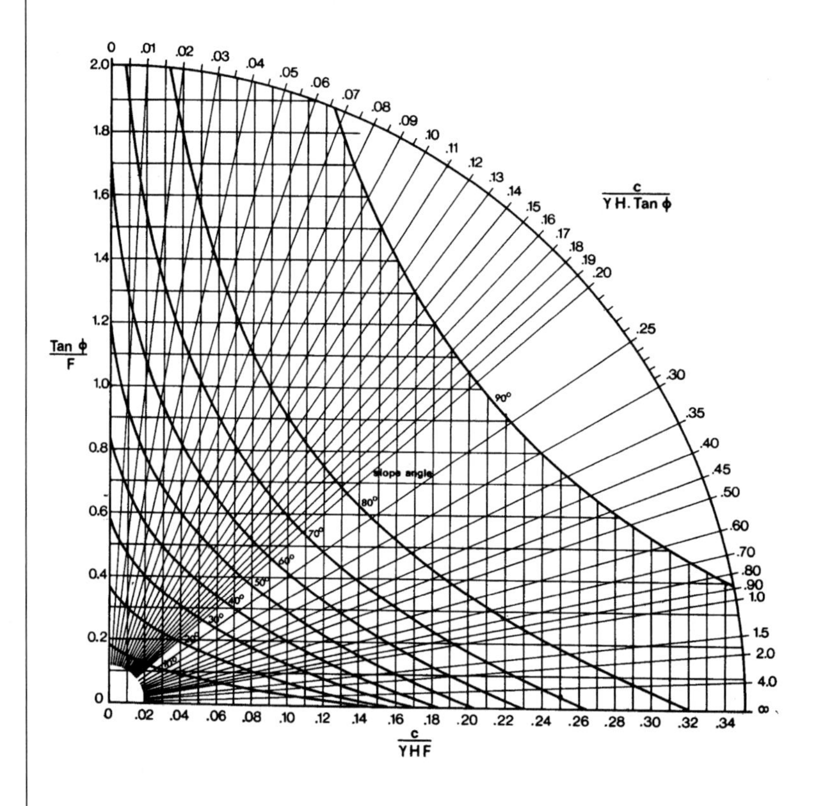

- the confinement action needed to contain potential instability.

The design decisions to be taken in the therapy phase will be based on these answers.

7.7 Final considerations

To conclude, by studying the deformation behaviour of the cavity, the diagnosis phase enables the design engineer to acquire the indispensable information required to make the design decisions necessary in the therapy phase to perform excavation and stabilise the tunnel. It consists of the type and magnitude of the prevalent instability that will manifest as a consequence of deformation along the different sections of the tunnel alignment.

The results of the diagnosis study are summarised on a geological and geomechanical profile of the tunnel (see the example in Fig.7.2), which will then constitute the "ID card" of the future tunnel to which reference is made in the subsequent design and construction phases.

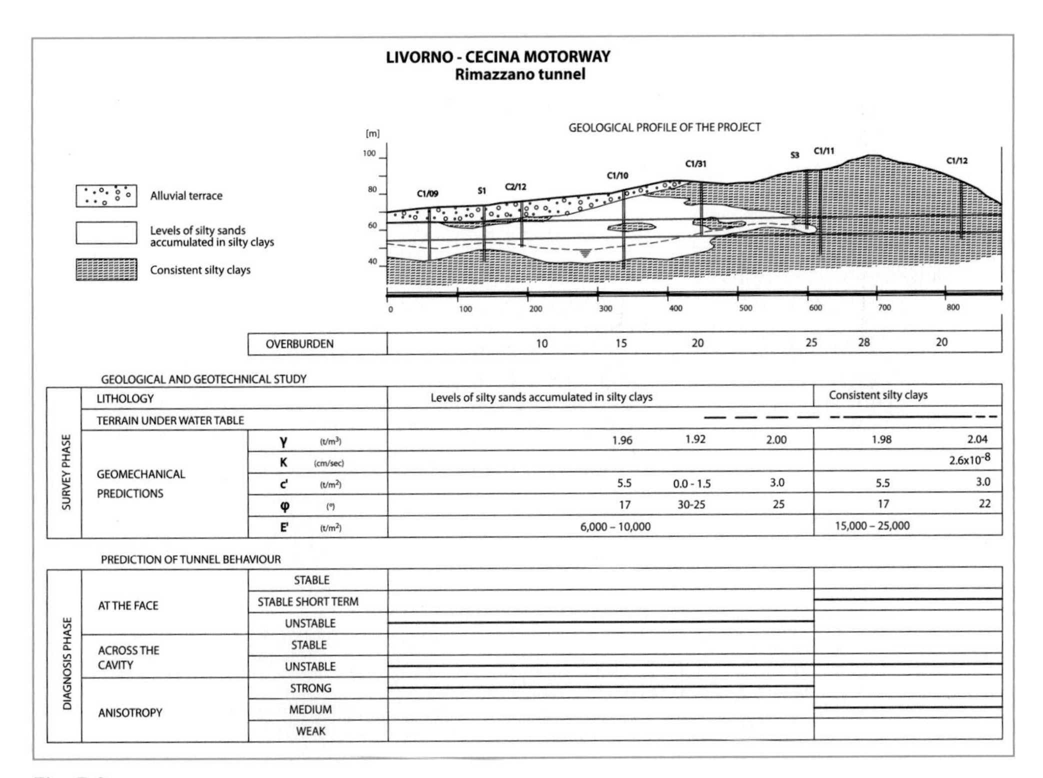

Fig. 7.2

CHAPTER 8

The therapy phase

8.1 Background

The therapy phase is that in which the design engineer makes operational **decisions** on the basis of the predictions of the quality, location and magnitude of the deformation phenomena that might be generated during tunnel advance, made in the diagnosis phase. They concern systems, advance rates and excavation stages, but above all **stabilisation instruments** and how these are balanced between the core-face and the perimeter of the cavity, in order to achieve the following:

- the stability and safety of the excavations and of any nearby pre-existing buildings or structures there may be through the formation of an arch effect, close to the walls of the tunnel;
- the industrialisation of construction work, compatible with the need to work with full respect for the environment, by careful design of the methods that will be used for tunnel advance and the placement of any stabilisation structures if necessary.

It is in this decision-making context that it must be decided whether and how to regulate deformation, whether to allow it or to counter it with pre-confinement and/or confinement operations to limit it or eliminate it completely.

In this chapter we will therefore illustrate the criteria that a design engineer must follow on the basis of the previous "survey" and "diagnosis" phases to be able to unequivocally decide:

- the most appropriate system of excavation (mechanised or conventional) in relation to the project and the geohydrological and geomechanical context;
- the stabilisation instruments to employ to ensure the stability of the excavation and to conserve the natural equilibriums;
- how these instruments act on the geomechanical characteristics of the rock mass and on how it relaxes (in terms of cohesion, friction, principal minor stress σ_3) to guarantee the control of deformation phenomena;
- where these instruments must be placed with respect to the position of the face for them to have a structural function;
- the accessory operating procedures for the construction of the tunnel.

Consideration is therefore given firstly to the different types and characteristics of modern tunnel boring machines, followed by a discussion of the criteria for tunnel design with an examination of the most commonly used stabilisation instruments in tunnel design and construction in the case of conventional tunnel advance. The following aspects are considered for each of these:

Method of excavation and stability

While surface construction consists of assembling materials (bricks, cement, aggregates, steel, etc.), which will constitute its structure in elevation, underground construction consists of removing earth, an operation which creates a cavity produced by face advance. It occurs during the excavation stage, when the earth is demolished, and it inevitably disturbs the natural pre-existing equilibrium in the rock mass, producing changes of a stress, geomechanical and hydrogeological character within it (impact of excavation on the medium).

From the viewpoint of stresses, the original stress field, which is uniformly distributed in the ground concerned, is deviated around the advancing cavity (channelling of stresses) with the consequent formation of zones of increased stress on the walls. **From a geomechanical viewpoint**, if the stress in the material is increased in the elastic range, then, depending on its strength and deformability characteristics, its fractures and discontinuities *tighten*, which translates into an increase in shear strength. The channelling of the stresses can then take place close to the profile of the cavity, which will ensure excellent long and short term conditions of stability. If, however, the material is over-stressed in the elasto-plastic range, then the ground around the profile of the tunnel will plasticise, which, as is known, gives rise to a significant loss of strength. This phenomenon will cause the stresses to be channelled (the arch effect) away from the profile of the excavation as it spreads progressively deeper into the rock mass. The stability of the tunnel will change as a consequence, in the sense that the stability of the tunnel will become more precarious the further the plasticisation extends.

From a hydrogeological viewpoint, the piezometric pressure will fall drastically when under the water table in the absence intervention and as a consequence this will trigger the infiltration of water towards the tunnel under construction, which if it is not properly controlled can give rise to the entrainment of material or piping which can be very dangerous for the existence of the cavity itself.

The magnitude of the changes caused by tunnel advance on pre-existing natural equilibriums constitutes the disturbance (impact) caused in the ground by excavation and determines the stability over time of the arch effect that is formed in it. The magnitude of the changes and therefore the impact of excavation on the medium will obviously depend on the methods used to remove or excavate the ground and, more specifically, on the lesser or greater degree of gradualness with which the passage from the old to the new equilibrium can be achieved.

The method employed in the excavation of a tunnel will therefore determine the long and short term stability of the tunnel itself and also of any adjacent pre-existing buildings or structures.

Since the stability of a tunnel and of the pre-existing structures is the main objective of an underground design engineer, it follows that his task is to minimise the impact of excavation both in terms of stress and the geomechanics and hydrogeology.

- the statics function of the intervention;
- the effects on the conservation line and on the intrinsic curve of the ground (conservative or improvement action) and action taken to stabilise the tunnel (preconfinement and confinement);
- the zone within which intervention must be implemented for it to perform its action adequately (zone of intervention);
- field of application.

8.2 Basic concepts of the therapy phase

A clear understanding of the mechanisms that govern the genesis, development and consequences of the deformation response of the ground to excavation is indispensable to a correct approach in the therapy phase.

A good or a poor understanding of these mechanisms will inevitably translate into good or poor design decisions. It is worth recalling in this respect what has been stated in previous chapters and can be summarised as follows: "the deformation response of the medium to excavation is an integral response, the result of the design and construction decisions adopted to regulate its relaxation **within the radius of influence of the face**".

As already occurred in the diagnosis phase, the therapy phase must also start by taking the conditions of stability in the core-face of the tunnel as its main point of reference.

While **predictions** of deformation in the absence of stabilisation within the radius of influence of the face were made in the diagnosis phase, with characteristic line theory as the main point of reference, in the therapy phase the design decisions required to control the predicted deformation are made with the intrinsic curve of the ground and the conservation line as the main points of reference.

Until just a few years ago, poor knowledge of the phenomena that govern the genesis and development of the deformation response of the medium to the action of tunnel advance resulted in a very reductive approach to the problem of the stability of the core-face. Attention was paid to this for the sole purpose of ensuring safety for construction workers, while the placement of the real stabilisation structures was left for later inside the tunnel.

Resort was made to *header and bench* type excavation as a remedy for the potential or actual instability of the core-face, by leaving a central core of ground in place to perform the function of temporary confinement. By operating in this manner, however, the generation of a rapid reduction in the confinement pressure σ_3 inducing the stress state to exceed the intrinsic strength of the rock mass even ahead of the face could not be prevented.

Back from the face inside the tunnel, the ground which had already collapsed required costly and substantial stabilisation structures

Methods of excavation and production

The more the design engineer is able to ensure the following, then the greater production, which is to say the speed of face advance, will be

- excellent construction site safety conditions, above all in the proximity of the face;
- ample space for personnel and machinery to manoeuvre;
- industrialisation of tunnel advance operations.

Clearly, these conditions can only be guaranteed by careful design of the tunnel advance cycle beforehand, and that is of the methods with which face advance will be produced.
It is worth underlining here that there is a theoretical maximum achievable production rate for any given tunnel, which will depend on its geometry, the geological and geohydrological conditions, the stress state and so on, in which the technologies available at the time of excavation are employed. If this question is approached properly then production rates close to that theoretical maximum possible will be achieved. If it is not approached properly, perhaps with the idea of making the conditions "fit" the means possessed by the contractor rather than the contrary, this will almost always translate into diseconomies and in the worst of cases into tunnel collapse. Great progress has been made in the area of safety and industrialisation of excavation in Italy over the last fifteen years by applying the criteria illustrated in this book.
The figure on the left gives the graphs for the production rates achieved during the excavation of tunnels on the new high speed/capacity railway line between Bologna and Florence under a great variety of geological and stress state conditions using the ADECO-RS approach. Not only does it show the high average advance rates achieved in relation to the different types of ground tunnelled, but above all it shows the excellent linearity of the production, an indicator of industrial type construction performed at constant rates with no unforeseen events.
If it is true that the method employed to excavate a tunnel determines not only the long and short term stability of the tunnel itself and of any adjacent pre-existing buildings and structures there may be, but also the rate and linearity of the production that can be achieved, both important objectives for an underground design engineer, then it follows that *excavation and how it is conducted constitute primary elements of the design and must therefore form an integral part of the design of every tunnel.*

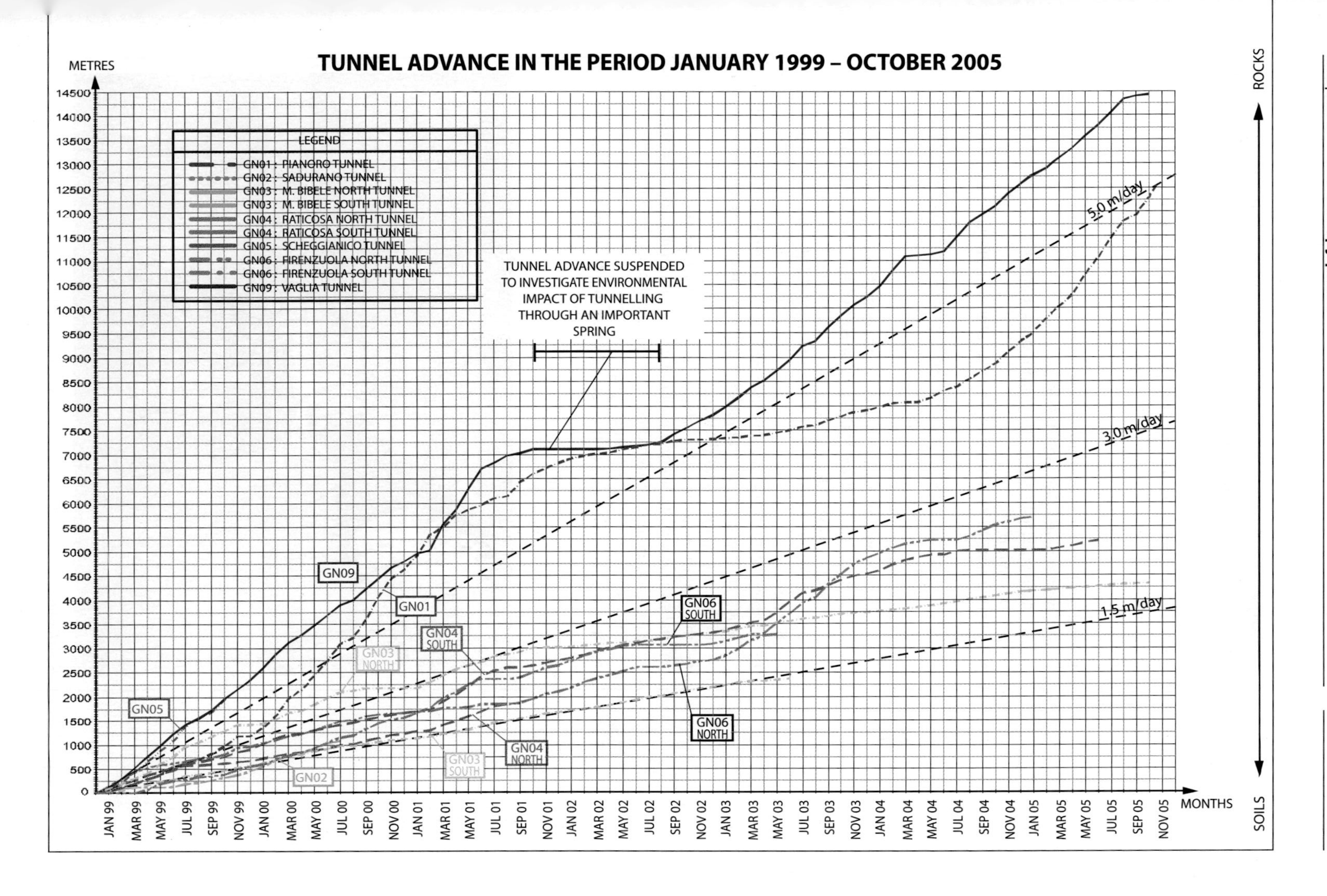
TUNNEL ADVANCE IN THE PERIOD JANUARY 1999 – OCTOBER 2005
METRES
LEGEND
GN01 : PIANORO TUNNEL
GN02 : SADURANO TUNNEL
GN03 : M. BIBELE NORTH TUNNEL
GN03 : M. BIBELE SOUTH TUNNEL
GN04 : RATICOSA NORTH TUNNEL
GN04 : RATICOSA SOUTH TUNNEL
GN05 : SCHEGGIANICO TUNNEL
GN06 : FIRENZUOLA NORTH TUNNEL
GN06 : FIRENZUOLA SOUTH TUNNEL
GN09 : VAGLIA TUNNEL
TUNNEL ADVANCE SUSPENDED TO INVESTIGATE ENVIRONMENTAL IMPACT OF TUNNELLING THROUGH AN IMPORTANT SPRING
5.0 m/day
3.0 m/day
1.5 m/day
GN09
GN01
GN04 SOUTH
GN03 NORTH
GN05
GN02
GN03 SOUTH
GN04 NORTH
GN06 SOUTH
GN06 NORTH
ROCKS
SOILS
MONTHS
14500
14000
13500
13000
12500
12000
11500
11000
10500
10000
9500
9000
8500
8000
7500
7000
6500
6000
5500
5000
4500
4000
3500
3000
2500
2000
1500
1000
500
0
JAN 99
MAR 99
MAY 99
JUL 99
SEP 99
NOV 99
JAN 00
MAR 00
MAY 00
JUL 00
SEP 00
NOV 00
JAN 01
MAR 01
MAY 01
JUL 01
SEP 01
NOV 01
JAN 02
MAR 02
MAY 02
JUL 02
SEP 02
NOV 02
JAN 03
MAR 03
MAY 03
JUL 03
SEP 03
NOV 03
JAN 04
MAR 04
MAY 04
JUL 04
SEP 04
NOV 04
JAN 05
MAR 05
MAY 05
JUL 05
SEP 05
NOV 05

Factors that influence excavation

It is essential, before considering the criteria for the design of excavation, to have a clear idea of the factors that influence it and over which the design engineer can exercise effective control.
These are:

- the system of excavation, which may be:
  - mechanised (full face TBMs);
  - conventional (blasting, roadheaders, hammer, ripper, mechanical bucket, etc.);
- intervention to reinforce/improve or stabilise the ground, which may be:
  - conservative (reinforcement of the core-face, sub horizontal jet-grouting, mechanical precutting, end anchored radial rock bolting, etc.);
  - improvement (conventional injections, freezing, truncated cone 'umbrellas' of drainage pipes ahead of the face, rings of ground reinforced with fully bonded rock bolts, etc.):
- the excavation stages (full face, half face, face divided into many sections);
- shaping of the face in a manner which, depending on the characteristics of the medium tunnelled, facilitates the formation of an arch effect in a longitudinal direction;
- the length of excavation steps: as a function of the type of material excavated, its homogeneity, the intensity of the stress states, the presence or not of reinforcement/ground improvement intervention, the excavation system employed;
- the rate of advance, i.e. the velocity and regularity of face advance.

Roadheader

around the cavity because it could only react with its own residual strength.

In other cases, when it is important to prevent ground deformation because of possible interference of underground excavation with pre-existing buildings and structures, resort has been made to face confinement using **mechanised shields** or when under the water table using **pressure balanced shields** (air, slurry pressure, etc.) capable of providing confinement to balances the thrust of the ground against the face. They are, however, rigid systems with limited fields of application because they cannot adapt to geological variation and most importantly they cannot be used in squeezing of swelling ground.

Today:

- having understood the importance of the stability of the core-face even for the long term stability of a tunnel;
- being able to predict the possible instability of the core-face on the basis of considerations made in the diagnosis phase;
- and being in possession of instruments able to adequately deal with all types of geotechnical or geomechanical conditions;

it is possible to intervene ahead of the face with action capable of anticipating deformation phenomena and halting it at the initial onset.

The main concern of an underground design engineer must therefore be to avoid tunnel advance in a medium that has already collapsed. To achieve this he must first of all:

- conserve the integrity of the advance core resorting, when necessary, to appropriate preconfinement of the cavity.

Subsequently, and compatibly with logistic problems, he must:

- increase the advance rate as much as possible;
- maintain excavation rates constant;
- shape the face wall appropriately to give it a concave form.

It follows that on the basis of the results of the diagnosis study:

- he may limit himself to simple confinement action in the case of tunnels with stable core-face stress-strain behaviour (category A);
- he will have to think in terms of producing vigorous preconfinement action – in addition of course to confinement action – in the case of tunnels with unstable core-face stress-strain behaviour (category C);
- he may choose between preconfinement or simple confinement of the cavity as a function of the rate of advance he estimates he can achieve in the case of tunnels with the stable core-face in the short term (category B).

Excavation systems

Before the advent of explosives, tunnels were excavated using thermal shock (first heating the rock with fire and then cooling it with water), or by elementary manually powered mechanical means, with very great use of manpower.
Gunpowder was not used until the end of the 16th Century to replace fire as a means of reducing the mechanical strength of the rock to be demolished when the first "round" was detonated in Connecticut. The first industrial manufacture and the relative marketing was by Du Pont, in Delaware (1802).
The modern industry of high explosives was born in the middle of the 19th Century: the preparation of fulminate of mercury (an essential ingredient for detonators) was described by Howard in 1800; Schonbein nitrated cellulose in 1846 and nitro-glycerine was first prepared a little later. Finally Nobel started to manufacture dynamite in 1867.
Blasting technology has made considerable progress since the second half of the Nineteenth Century in terms of efficiency and the variety of safety products and also as a result of the developments in blast hole drilling. Developments in mechanical means of excavation, on the other hand, have also made considerable steps forward in terms of power, automation and control systems to the extent that selecting which to use is far from easy and requires careful analysis.
Typical tunnel excavation processes are generally divided into two categories. Energy is transferred to the medium extremely rapidly with those belonging to the first category (excavation by blasting), while it is transferred very much more slowly with those belonging to the second category (mechanical excavation). The latter category is further divided into a few sub categories to distinguish between excavation using hydraulic demolition machinery (hammers), excavation using excavators, rippers, dozers and so on and demolition using cutters (roadheaders or full face).

Mechanical excavation using a hydraulic demolition hammer

Once the type of action to be exerted has been decided, the action must be perfected in terms of systems, rates, excavation stages and above all stabilisation methods and tools. For the latter, how and where they are to be employed with respect to the position of the face and the behaviour category, A, B or C, involved, must be established so that the desired action is produced.

8.3 Excavation systems

The tunnelling or excavation system is a tunnel characterisation parameter which a design engineer must select with great caution, skill and intelligence. It can be **mechanised or conventional**.

There are really a great many possibilities. Even if we wanted to narrow the concept of mechanised excavation to include only those systems for which all stages of underground construction from excavation of the ground to the placement of the final lining are performed in sequence by a machine, as in a large assembly chain, the design engineer can choose from:

- full face rock cutters (TBM), open or shielded, with single or double shield;
- simple shields, blade shields, compressed air shields, hydroshields, EPB's, mixshields.

The choice for conventional excavation methods is hardly narrower. It ranges from blasting to mechanical excavation, which may be performed with roadheaders, demolition hammers, rippers and mechanical buckets of various types depending on the type of ground and in some limited cases with rocksplitters or non explosive demolition agents (Bristar, Cardox etc.).

The choice of excavation system will depend mainly on the type of ground involved, its morphology, geology, hydrology and hydrogeology, its physical and mechanical characteristics, its homogeneity, stress states and the geometry of the section. Often, however, other, financial, technical and organisational factors come into play.

Although the complex and generally poor geology in Italy has favoured the development of conventional excavation systems as opposed to mechanised systems, the availability of much more flexible and reliable tunnel boring machines on the market in recent years, able to guarantee good performance even in soft grounds, means that this type of excavation is increasing in Italy too. It has been found particularly useful, above all:

- as a means of surveying with the construction of a small diameter pilot tunnel along the alignment of the main tunnel to be driven

Mechanised or conventional excavation?

Conditions \ System of excavation		Conventional	Mechanised with full face TBM	
			Without confinement of the core-face	With confinement of the core-face
Characteristics of the tunnel	Length	Cost of equipment relatively low. Cost of excavation not greatly influenced by the length of the tunnel	Cost of the machine: generally high. Suitable for longer tunnels	Cost of the machine: generally high. Suitable for longer tunnels
	Shape of the cross section	Various cross section shapes are possible. The shape of the cross section can be changed during construction	Shape of excavation circular. After excavation, the shape can be modified using conventional systems	Shape of excavation circular. Other shapes are possible using special machines
	Surface area of the cross section.	There are no limits. Even cross sections larger than 800 m² have been excavated	Tunnels have been driven with a maximum diameter of 12 m approx	Tunnels have been driven with a maximum diameter of 14 m approx
Geological conditions/ behaviour category	Sound rock Behaviour category A	Suitable	Suitable, except for extremely hard rock (≥ 200 MPa)	Unsuitable
	Sound rock. or with little fracturing. Behaviour category B	Suitable	Suitable	Unsuitable
	Rock with poor stratification or very fractured. Aquifer zones. Behaviour category B	Suitable with adequate counter measures	Not very suitable in zones with soft ground or where water inflow under pressure is frequent	Unsuitable
	Soils. Behaviour category C	Suitable with adequate counter measures	Unsuitable	Suitable provided deformation phenomena are contained
Environmental conditions	Noise and vibrations	Blasting and demolition hammers are not very suitable near residential areas and important infrastructures	Causes little disturbance to the surrounding rock mass	Almost no disturbance to the surrounding rock mass

(analysis of the operating parameters of the TBM using the RS method);
- as an alternative to conventional technologies in the construction of tunnels with shallow overburdens, especially in urban environments where the requirement to limit and control subsidence, to construct rapidly and to pass safely under the foundations of buildings, utilities and pre-existing structures in general provides the conditions for mechanised excavation to display its full potential;
- in driving tunnels under the water table in loose material.

The strong point of mechanised excavation definitely lies in its speed and its continuous operation, characteristics which enable substantial savings to be made in terms of time and costs compared to conventional excavation. It is also, however, important to underline that it is consequently destined to fail when there are prolonged and repeated machine down times. That is why the quality of the design of a tunnel to be driven by TBM must be higher than that of a tunnel to be driven using conventional techniques, which should, it is hoped, be already very high.

The possibility of employing TBM technology should be considered at the preliminary design stage and the decision should be made on the basis of detailed assessment of the risk and a comparison of the cost/benefit analyses for mechanised and conventional excavation.

To achieve this, it is essential to have extremely thorough geological-stratigraphical, geostructural, hydrogeological and geotechnical-geomechanical surveys in sufficient quantity to minimise geological risk on the one hand and to allow informed selection of the machine on the other, because, as is well known, the machine is the element which determines the success of the work, especially when it is considered that changes to it cannot be made during construction (except for some details in its construction planned during the design stage).

The decision will be heavily conditioned, not only by the dimensions of the tunnel cross section, but also by the physical, chemical, geotechnical and geomechanical characteristics of the ground, as well as the intrinsic stability of the core-face and of the cavity and finally by how this relates to the possible presence of water.

8.4 Mechanised or conventional excavation?

To furnish a few basic criteria on which to decide between conventional and mechanised excavation it is perhaps worthwhile briefly considering the aspects listed below. Analysis of these is essential for a reliable assessment of the risks involved in employing mechanised excavation and for in depth cost/benefit comparison of the two methods of excavation.

Mechanised excavation. Risks and choice of machine.

The choice must fall on a machine with the greatest probability of success in terms of advance rates and advancing through critical points on the alignment identified in the survey and diagnosis phases. The possible risks during tunnel advance are summarised in the list below with an indication of the possible remedies in terms of the characteristics of the machine.

Problems of core-face stability	• Adequate design of the cutter head and tools • Possibility of reversing and rotating the head in both directions • Possibility of intervening to reinforce and improve the ground in the core-face and around the cavity
Passing through faults with cataclastic zones	• Possibility of reversing and rotating the head in both directions • Possibility of drilling boreholes ahead (probing ahead) • Possibility of intervening to reinforce and improve the ground in the core-face and around the cavity
High convergence	• Possibility of intervening to reinforce and improve the ground in the core-face and around the cavity • Possibility of overbreak • High thrust capacity on jacks • Bentonite on the extrados of the shield • Decreasing diameter (cone shaped shield) • Collapsible prefabricated lining
Very cohesive soils alternating with rock	• Openings for loading buckets • Possibility for rapid replacement of cutters • Fluidising Polymers in the pressure chamber (EPB) • Full possibility to control the speed of rotation of the cutter head
Erratic boulders in a sandy-silty matrix	• Jaw for breaking rocks up to 1 m^3 • Use of disks • Dual pressure chamber • Polymer additives • Possibility of blasting at the face • Possibility of homogenising the ground with ground improvement intervention
Large inflows of water in soils with permeability >10^{-4} m/s (impossible to use compressed air at the face)	• Possibility of drilling boreholes ahead (*probing ahead*) • Possibility for drainage and injections • Dual pressure chamber • Use of special slurries
Cemented interstrata between loose soils	• Possibility for rapid replacement of cutters • Limiting loss of compressed air with polymers or other additives • Sufficiently high torque even for rock • Systems for automatic control of the direction of the excavation • Possibility to absorb strong asymmetrical longitudinal thrusts • Adequate design of the cutter head (number of spokes, shape, retractable overbreak cutters, etc.)

1. Cross section and length of the tunnel

Circular cross section geometry must be employed for mechanised excavation using a full face cutter or shield and this normally requires a larger excavation surface and lining than is necessary for a similar polycentric cross section typical of conventional tunnelling methods. Furthermore, although it is efficient from a statics viewpoint in the majority of situations, a circular cross section is not the best for conditions with markedly asymmetric natural stress states.

The current dimensions for road and rail tunnels require excavation diameters of between 10 and 14 m and modern technology provides TBM's of those dimensions. They are, however, complex machines with a high cost that require a long time to construct and assemble and dismantle *in situ.* The supply and assembly of machines of this type and size requires average times of around 15 months, a period that can only sometimes be arranged to partially or totally coincide with the time taken for setting up the construction site and preparing the portal. It follows that the construction times and costs can only be reduced for tunnels of a certain length.

Stress due to the nature of the ground and hydrogeological phenomena are amplified considerably for large diameter tunnels and the risk of delays and machine downtime must be must be carefully considered. A very large face increases the probability of having a mixed and inhomogeneous surface to excavate with terrains of different nature and geomechanical behaviour.

Conventional excavation systems are undoubtedly more flexible. Obviously construction times are longer, but this difficulty can be reduced or even eliminated completely by increasing the number of faces and opening up intermediate adits, inclines and shafts for access if necessary. When tunnels are very long, the time and investment required for conventional excavation becomes very great. One need only consider the removal of spoil (a 3 m tunnel advance results in approximately 650 m^3 of ground to be removed).

Over long distances, a well organised TBM system allows effective rationalisation of excavation and spoil removal and a finished tunnel, already lined, is obtained with great reductions in construction times and costs. Tunnel advance can in fact proceed from two to three times faster with a TBM than it can with conventional methods.

2. The vertical and horizontal geometry of the tunnel alignment

The vertical and horizontal geometry of the tunnel alignment can be important in deciding between mechanised and conventional excavation. A very small radius of curvature would in fact be beyond the capacity of large size TBM's or shields.

3. Logistic constraints

Careful attention must be paid to possible obstacles to the transport, assembly and supply of the machine in the location where it is used.

Full face cutters (TBMs)

Full face cutters are machines that allow good industrialisation of tunnel advance operations and lining placement and give the fastest advance rates if properly designed and used. Their energy performance is in fact very good, being second only to that achievable with excavation by blasting.

One common denominator of this type of machine is that it has a large rotating front part, termed the cutter head, fitted with picks or disks able to perform full face tunnel excavation. This part must consist either of a series of arms in a spoke configuration or a full cross section disk fitted with only those apertures needed for the passage of spoil and for the passage of a man for maintenance purposes. The cutterhead is supported by a series of roller thrust bearings able to withstand the high stresses, potentially highly asymmetric, which may manifest during excavation, and is driven by electrical or hydraulic motors.

Full face cutters are commonly referred to as TBM's (*Tunnel Boring Machines*) and can be classified in various ways according to the type of equipment and functioning.

For the purposes of the therapy phase, they are best classified on the basis of the type of action they are able to exert on the core-face of the tunnel during excavation:

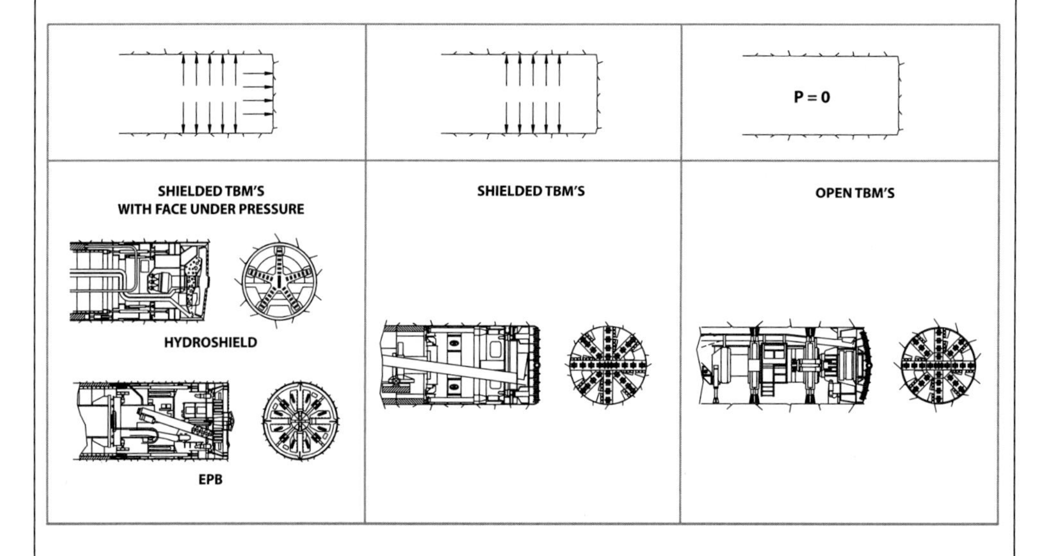

Some models of TBM's, termed *mechanical pressure* TBM's, are able to exert significant pressure on the face thanks to their mass and the geometry of the cutterhead. However, the more modern, versatile and effective *hydroshields* or EPB's are normally preferred to these.

4. The nature of the ground, stress states, hydrogeology and the presence of gas

The main characteristic to focus on is not so much the quality of the geomechanics of the ground as its homogeneity and the possible presence of inconvenient geological features (faults, karst phenomena, etc.) of a size greater than the diameter of the tunnel.

Clearly there is much less risk involved in driving a tunnel using conventional methods that are more flexible than mechanised excavation in locations where formations change frequently and present very inhomogeneous mechanical characteristics or the tunnel alignment runs through numerous problematic geological features.

If, however, the tunnel alignment is fairly homogeneous from a geomechanical viewpoint, or there are in any case sufficiently long homogeneous sections of tunnel with a relatively low number of problematic geological features, then TBM tunnel advance can be employed to advantage.

The cuttability of the material to be excavated is also of importance in selecting the method of excavation. For example, high abrasive properties might make the tool wear so great as to make the use of a cutter uneconomical.

If the overburden and therefore the stress state is great in relation to the strength and deformability characteristics of the medium, then the resulting deformation will also be great. In this case, and also in swelling ground, the problems of full face cutter excavation should be analysed with particular care and attention, especially if there is a shield. If the diagnosis phase predicts substantial and rapid convergence (greater than 15 cm), the cutter head and the shield could actually become stuck.

Even if machines are available today that are able to operate under complex stress-strain conditions by exercising significant confinement pressure on the face (EPB shields, hydroshields, etc.), conventional means of excavation remains indispensable where it is necessary to control the deformation response of the rock mass to excavation by using preconfinement of the cavity. It is in fact wrong to think that it is possible to operate in all types of material with the protection of a shield and a prefabricated lining. It could be excessive for most of the length of a tunnel, but insufficient in zones of poor ground which cannot be identified in the presence of a shield and a lining. In some tunnels built some time ago using this technology, deformation and failure of prefabricated linings is occurring because of the condition of the surrounding rock mass, which, in the absence of the necessary ground reinforcement and improvement (which should have been performed during excavation), has destabilised with the passage of time (creep).

The *presence of water* under pressure is a difficult problem to deal with both with conventional and TBM tunnel advance. In the case of

Shielded TBM's

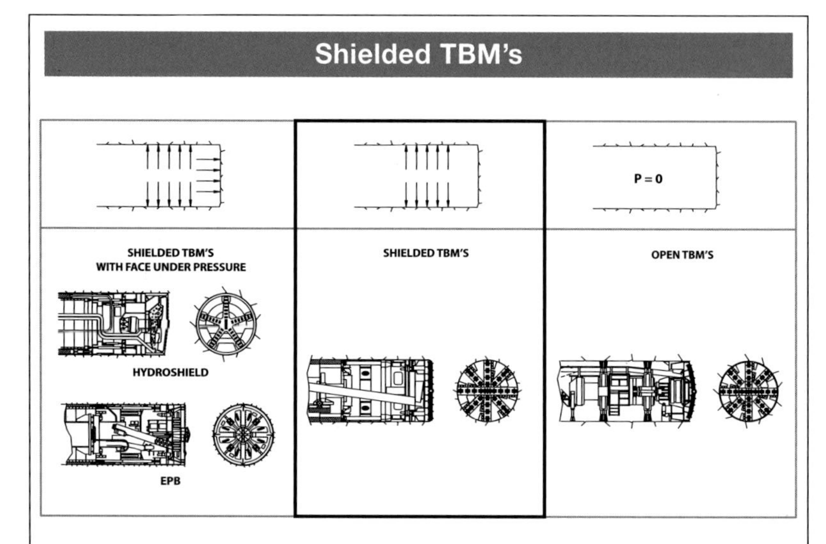

The cutter on shielded machines is fitted inside a cylindrical steel shell (the shield), which has two main functions:

- to gradually counter thrusts from the ground, maintaining the excavation intact and protecting personnel;
- to prevent any water that is encountered from entering by means of a system of seals.

The shield may be in one piece, or articulated with two segments connected by hydraulic jacks in order to improve the steerability of the machine.
The tunnel lining is erected immediately behind the shield or inside the rear of it and often consists of a series of prefabricated reinforced concrete segments placed by a rotating erector. Water-proofing between the segments is by means of elastic seals. Mortar or bentonite is injected under pressure behind the lining on the extrados (backfill) to seal it against the surrounding rock mass.
The force required to drive the machine forward is produced by a series of hydraulic jacks which generally push against the last ring of the lining placed. Some models (double shields) have both 'grippers' and jacks which push on the lining already placed. The cutterhead is able to advance independently of the shield by pushing on its grippers and consequently excavation and lining placement operations can be performed simultaneously. This possibility and their greater flexibility in steering allow these double shield machines to reach advance rates of 1.5 to 1.7 times higher than single shield machines under the same conditions. Furthermore, the dual thrust system and the possibility of driving the machine backwards and forwards as if they consisted of two independent shields, without having to remove rings of the lining already placed, allows these machines to retreat more easily from difficult situations. The length of the machine also plays an important role in this respect. It is more difficult for machines with a length of around one tunnel diameter ($L/\varnothing \cong 1$) to become trapped in the rock mass when the stress-strain conditions are difficult.

conventional excavation it means working in the presence of water and mud, with great inconvenience for personnel and machinery in addition to the considerable use of pumps and piping. In the case of TBM tunnel advance everything is easier to manage and programme, especially if the machine is large, but there is a higher risk of enormous damage, especially for some machines with shields which can become stuck in the tunnel at any moment as a result of transported material becoming clogged in the gap between the outside of the shield and the perimeter of the excavation. Currently available machines are able to withstand maximum hydraulic pressures of around 7 – 8 bar.

Finally the *presence of gas* is easier to manage with TBM advance provided the machine is adequately designed and equipped (depressurisation of the excavation zone in order to keep the presence of gas close to the face, monitoring of the air in the confined zone, protection of engines and electrical equipment, etc.).

5. Pre-existing buildings and structures

The problem of pre-existing structures arises primarily in urban environments where careful analysis of subsidence and vibrations generated by excavation systems is essential.

6. Safety and the working environment

A TBM is virtually a factory in which all the processes are extremely rigorously defined and repetitive with respect to conventional tunnel excavation. Machinery, plant and equipment and personnel are less exposed to mud, water, knocks and dust. Electrical and hydraulic motors can be sound-proofed and operated more easily. Furthermore, personnel are never directly exposed to rockfall and tunnel collapse in machines with shields, especially if a prefabricated lining is placed immediately behind it.

8.5 Tunnel boring machines in relation to the confinement action they exert

If it is decided on the basis of the characteristics of the project at the preliminary design stage to opt for mechanised excavation, the design engineer must first of all select the type of machine to use. This task, which until quite recently was performed by the contractor and the TBM manufacturer, has now become the responsibility of the design engineer for the project who works with the designer of the machine providing precise input to optimise it on the basis of the geological, geomechanical and stress-strain context in which it must operate.

Generally speaking, the choice must fall on the machine with the greatest probability of success in terms of advance rates and advancing

Shielded TBM'S with the face under pressure

P = 0

SHIELDED TBM'S WITH FACE UNDER PRESSURE

HYDROSHIELD

EPB

SHIELDED TBM'S

OPEN TBM'S

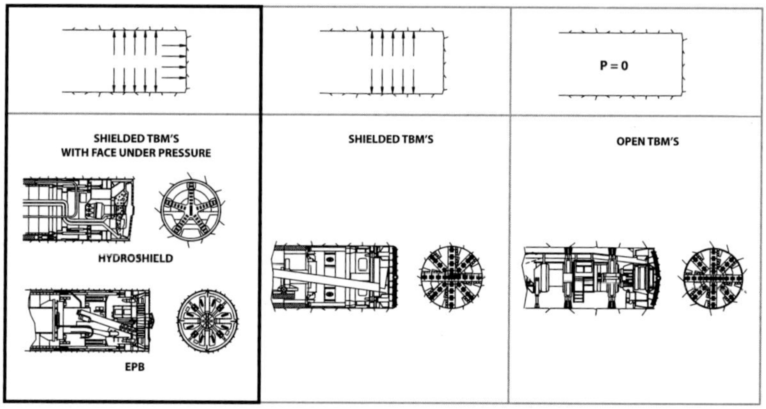

Under difficult stress-strain conditions, the mechanical counter pressure furnished by the cutterhead alone is normally not sufficient to prevent the ground in the core-face from relaxing. Shielded cutters which maintain the face under pressure were therefore designed to solve this problem. They are known as EPB (*Earth Pressure Balance*) or *slurry shield* machines depending on the technology used to achieve the purpose.
The use of one or the other generally depends on the type of ground tunnelled. Traditionally EPB TBM's were chosen for cohesive soils and slurry shields for non cohesive soils, but with the advent of the latest generations of TBM's, it can be said that the two systems are two extremes of the same type of machine, which control the face-core with a medium which differs only in its specific weight.
The slurry TBM's, in which the pressure of the bentonite is not controlled directly but by means of a bubble of compressed air maintained in communication with the slurry, are known as *hydroshields*. This more modern system of control has the advantage of always keeping the fluid pressure on the core-face constant, because whenever there are sudden falls in the pressure of the slurry, the compressed air is immediately able to balance the change.

through critical points on the alignment identified in the survey and diagnosis phases.

To do this, the design engineer must know the types and characteristics of the TBM's available today. They are normally considered in terms of the equipment with which they are fitted and the type of functioning. We will consider them above all in terms of the type of action they are able to exert on the core-face of a tunnel during excavation. It must be observed in this respect that at the current state of the art, no machine is able to systematically produce action for preconfinement of a cavity. It is possible, on some recent machines, to open a certain number of apertures in the bulkhead and in the shield for survey bore drilling and limited ground improvement work. It is, however, an operation reserved for exceptional circumstances along the alignment, which reduces the advance rate of the machine appreciably and therefore also the advantages of using it.

Machines in existence today which exert confinement action on the core-face and around cavity include the following:

- *slurry shields* or *hydroshields* in which direct confinement of the core-face is obtained by employing bentonite in suspension pumped under pressure into the excavation chamber;
- *earth pressure balanced (EPB) TBM's* in which it is the muck itself, fluidified using special foam, which is maintained under pressure in the excavation chamber where control of the advance rate and of the spoil removal flow rate is synchronised;
- *mixshield* or *polyshield TBM's* are multi-purpose machines capable of operating in the two modes already mentioned and also dry, with no fluid pressure on the core-face.

The following, on the other hand, only exert radial confinement action on the cavity:

- shield TBM's with no equipment to generate pressure on the core-face.

Finally, open cutters are unable, by themselves, to exert any confinement action at all on the surrounding rock mass; however they do allow the rock mass immediately behind the cutterhead to be easily confined with appropriate action (rock bolts, steel ribs, shotcrete, etc.).

Once the type of machine has been selected on the basis of the input from the survey phase on the type of ground to be tunnelled and from the diagnosis phase on the stress-strain behaviour expected along the tunnel alignment, the design engineer will then inform the machine manufacturer whether the following is necessary:

Shielded TBM'S with the face under pressure. The EPB system

An EPB system is one in which the material excavated by the cutter is mixed and accumulated under pressure in the spoil chamber and then extracted in a controlled manner by a screw conveyor. The result is that the core-face is confined by the pressure exerted by the grains of soil that have just been excavated on those of the ground still to be excavated. Once the internal pressure reaches equilibrium, tunnel advance proceeds at constant volume (the volume of material extracted from the spoil excavation chamber is equal to that of the material excavated from the face). The confinement pressure exerted on the core-face to balance that coming from the ground is controlled by co-ordinating the forward thrust of the advancing machine and the speed with which the conveyor extracts the spoil.

EPB SYSTEM

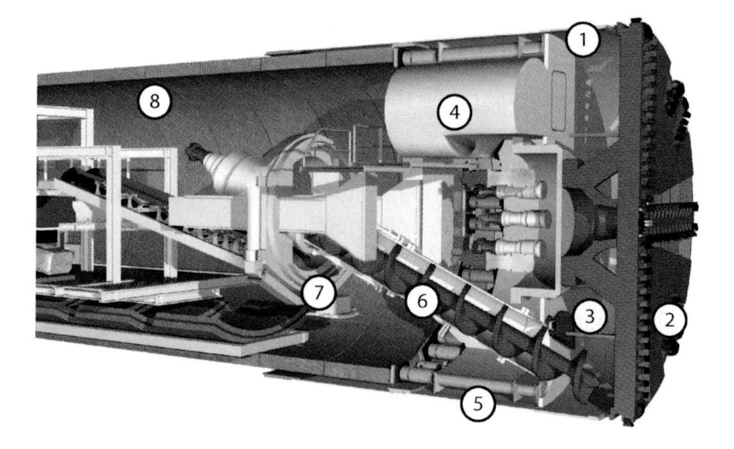

- 1. Shield
- 2. Cutter picks and disks foam nozzles
- 3 Excavation chamber-fluidised earth under pressure
- 4. Pressure lock
- 5. Thrust pistons
- 6. Spoil extraction screw conveyor
- 7. Erector
- 8. Rings

To drive a tunnel with success using an EPB system depends on the degree of control there is over the core-face. To achieve this, a determining role is played by the characteristics and properties of the fluid used as a support medium which define its ability to transfer uniform and constant pressure on the core-face. These properties are only found naturally in soils in exceptional cases. That is why additives, able to improve the properties as desired, are generally mixed with the materials. The additives used depend on the characteristics requiring improvement. The most recent developments in this field involve the use of chemical foam and polymer injections (water-proofing agents, coagulants, dispersants, fluidisers, etc.). Modern EPB machines are therefore easily able to drive tunnels even through sections of tunnel that are properly speaking not ideally suited to this type of system.

- to monitor the real conditions of the rock mass about to be excavated by drilling boreholes ahead of the face to a depth of at least twice the diameter of the excavation (performing this bore drilling is clearly inconvenient from an operational viewpoint in certain situations, but is fully justified by the advantages obtained in terms of safety and the reduction in the risk of damage to machinery and personnel which would have unfortunate repercussions on final construction times and costs);
- to perform ground reinforcement/improvement of the core-face and of the cavity from inside the machine (specifying the type and quantity in order pass through critical points on the alignment relatively rapidly).

8.6 Design using conventional excavation

Those factors that affect conventional tunnel excavation and as a consequence the stability of the tunnel during operations and the advance rate achieved are basically as follows:

- action to reinforce/improve or stabilise the ground;
- the excavation stages;
- the actual excavation system;
- the length of excavation steps;
- the shaping of the face;
- the advance rate.

Intervention to reinforce/improve or stabilise the ground occupies first position and not just for clear reasons concerning the long and short term stability of the tunnel. Whether excavation takes place in the absence or presence of this type of intervention will not only be determining for the effectiveness of the arch effect that will be created around the cavity, but it will also have repercussions in design terms on other factors that affect excavation. This intervention, termed either preconfinement or simple confinement of the cavity, which, as has already been mentioned in section 8.2 (*The basic concepts of the therapy phase*), must be selected on the basis of the stress-strain behaviour of the rock mass predicted in the diagnosis phase, is examined in detail later in the book.

The decision concerning **excavation stages** has an enormous influence both on the stability of the tunnel (especially during construction, but also in the long term) and on the tunnel advance rates that may be reached.

If it is recalled that tunnel advance can be full face, half face or by dividing the face into a number of sections, it is interesting to observe that the results of research always favour full face excavation whenever

Shielded TBM'S with the face under pressure The bentonite slurry or hydroshield system.

The bentonite slurry system is based on techniques used to construct sheet diaphragms and piles in the presence of bentonite. The general principle is that of generating core-face confinement by using a liquid with thixotropic behaviour like bentonite slurry. It provides support to the core-face and at the same time acts as a vehicle for removing spoil. The pressure on the core-face is controlled by automatic valves that regulate the outflow of the slurry. Normally the slurries containing material that has just been excavated are first sieved to remove large size pebbles and are then unloaded through screw valves from which they pass into pipes and are sent to the surface. Outside the tunnel, a special plant separates the excavated material from the bentonite and sends the latter back to the face appropriately purified and integrated with new slurry. In some cases, in large diameter tunnels with ample space, part of the separation plant is installed directly on the *back-up* of the TBM, in order to prevent the excavated material from dissolving completely in the slurries, which then makes it more difficult and costly to separate.
With slurry systems, the mucking process is contained entirely within piping systems and the tunnel and the working environment remains clean. The system has advantages for situations where harmful gases are present, since they are piped directly to the surface and prevented from spreading in the tunnel. However, if the machine passes through a contaminated zone, the whole system and the slurry also become contaminated.
One disadvantage to a slurry system is the need for costly separation plant which requires a large amount of space on the construction site. The separation plant is normally very noisy and sound proofing barriers are needed in urban environments; if the construction site is located in a residential area, then tunnel production may have to be limited to a day shift only.

HYDROSHIELD SYSTEM

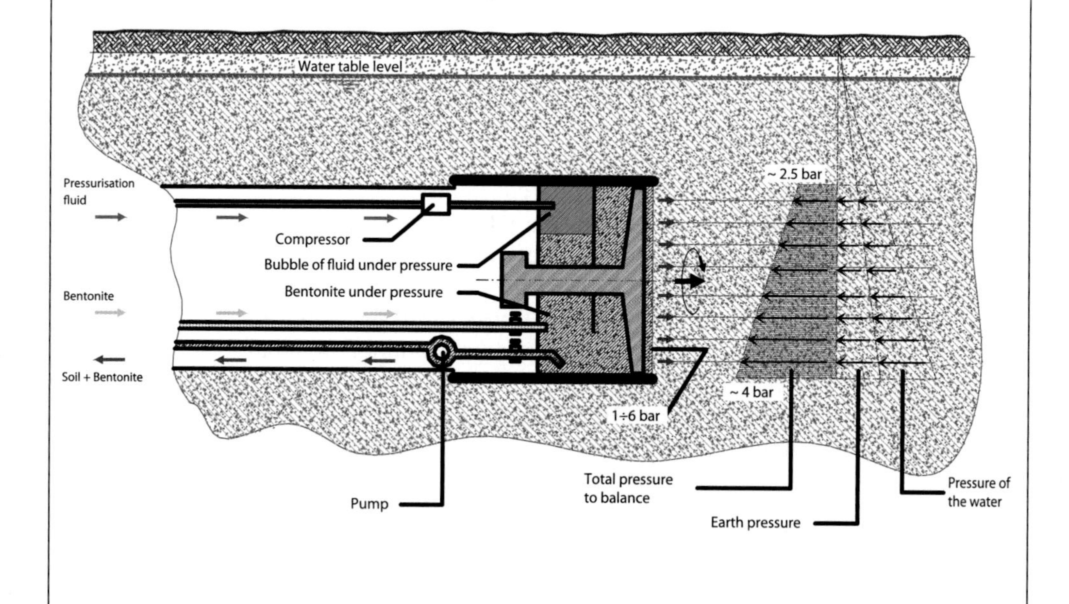

it is possible (and therefore almost always if tunnels of unusual shape and size are excluded), even, and in fact above all, in grounds with the most difficult stress-strain conditions. In the past, construction by driving side drifts and casting the side walls in advance occupied an important position in complex situations. The method is not very common today because of the narrow working conditions in the side drifts, which translate into slow advance rates and in the best of cases these are never greater than 1 m/day. Although it allows a greater degree of mechanisation since the height of the top heading in the crown is 4.5 m or higher, even half face advance does not enjoy the same popularity as it once did, because not only does it give no guarantee of face stability without adequate intervention, but it also generally creates serious statics problems when it comes to excavating the bench.

While the use of full face excavation under the most difficult conditions seems paradoxical at first sight because of the potential instability of the face, which may even reach a height of 9–12 m, it is in fact often found to be very reliable giving good performance, provided the stability of the face is ensured (the only critical zone with this type of advance) with appropriate ground reinforcement and improvement.

The statics advantages afforded by this method and by the intense use of large and powerful machinery, which is possible in the large spaces available, often means that full face advance can be achieved under excellent safety conditions with advance rates that are beyond the capacity of other methods.

The choice of the **excavation system** is governed by various factors. Those concerning the medium are as follows:

- the strength of the matrix (σ_f);
- discontinuities and inhomogeneities in the structure of the rock mass (velocity of seismic waves);
- capacity to transmit energy (acoustic impedance);
- size of acceptable overbreak (repercussions on the arch effect).

Other important factors which might exclude the use of explosives or very powerful demolition hammers are the presence of:

- nearby buildings and structures;
- inhabited areas or areas of wildlife interest.

There are numerous studies in the literature, with the results summarised in easy to use calculation charts and tables, for assessing the compatibility of a material with the different methods of excavation available (blasting, hammer, ripper, mechanical bucket, boom header) and they also provide forecasts of the production rates that can be achieved with each of them.

The table below provides general indications.

Open TBM'S

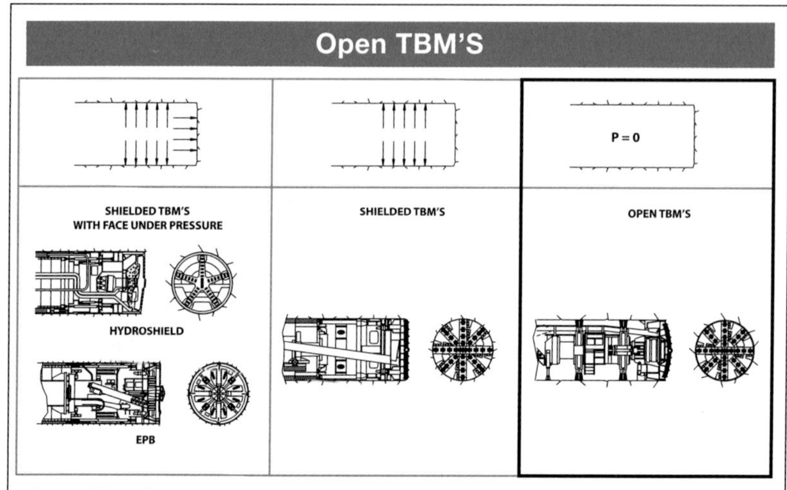

Open TBM'S are very versatile machines, capable of competing successfully with shielded machines in most situations. They can in fact guarantee high rates of production operating in both formations of consistent and soft rocks.

It can be said that, except for cases where mainly loose or very poor quality ground must be tunnelled, which would require substantial core-face confinement, serious consideration should always be given to their use. The advantage of having direct contact with the condition of the materials excavated, which is precluded with shielded cutters, is important for distributing stabilisation intervention intelligently along the tunnel, avoiding excessive stabilisation in some zones and insufficient stabilisation in others.

Open TBM's are irreplaceable above all for tunnelling material that does not give rise to marked instability in the core-face, but which produces immediate convergence at a distance of less that one tunnel diameter from the face.

By improving or reinforcing the ground in advance and stabilising it immediately behind the cutterhead, these machines can be usefully employed with good results even in difficult situations, where, for example, a shielded machine could become trapped. A prefabricated lining can also be placed immediately with open TBM's, because it can be erected without problems by machinery installed on them.

Generally speaking the presence of large quantities of water causes less disturbance to an open TBM, especially when solids are transported by it, because the machine facilitates the drainage of the water, while, on the contrary, a shielded machine presents an obstacle to the water and runs the risk of becoming blocked in the tunnel by the transported material, which can clog the space between the shield and the wall or the tunnel.

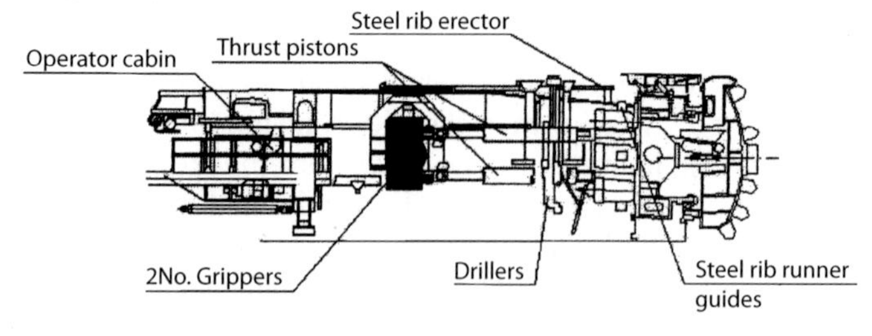

The best system of excavation for each formation is therefore identified for each case on the basis of careful cost-benefit analysis.

Parameter	Strength of the matrix [MPa]		
	< 3	3–20	> 20
Excavation system	ripper	hammer, roadheaders[1]	blasting

[1]useable only on small to medium cross section tunnels

The **shaping of the face** is another factor over which a design engineer can exercise control. To do this he must decide whether the type of ground and the stress-strain conditions in question will allow excavation to be performed in safety without adopting particular measures concerning the shape of the face or whether, on the contrary, it is indispensable to give it a concave shape to assist the longitudinal formation of the arch effect needed to minimise extrusion and to ensure the safety of work performed in proximity to it.

The **length of excavation steps** can vary from half a metre to several metres depending on the type of ground, its homogeneity, the intensity of the stress states, the presence or not of intervention to stabilise the ground and the system of excavation employed.

The **rate of advance,** which is to say the velocity and regularity with which advance proceeds is clearly of economic importance, but it also has far from negligible importance for the long and short term statics of the cavity. It has in fact been demonstrated that maintaining adequate advance rates under difficult stress-strain conditions has the effect of causing less disturbance to the surrounding rock mass and this reduces the extent of the plasticisation around the cavity and therefore also the magnitude of the consequent deformation. In ground with marked rheological behaviour (swelling, creep etc.), this behaviour is contained considerably. The consequence is a tunnel that is more stable in the construction stage with less pressure on the preliminary and final linings.

The table below gives some criteria that may be useful in guiding the choice of the parameters cited above, if conventional excavation has been decided [the length of excavation steps and the advance rates are for tunnels of present day diameter (12–13 m)].

8.7 Stabilisation intervention

There are a series of instruments available to the design engineer to perform all types of stabilisation required to obtain the appropriate action for a given stress-strain situation (preconfinement or simple

Mechanised excavation. Adaptability of machines to soft soils in relation to the stability of the core-face

From "Recommendations and Guidelines for TBM's, ITA Working Group" No. 14 (Mechanised Tunnelling), Sept. 2000.

Type of machine / Type of ground	N-value	Water content or permeability	Open type		Closed type					
			Mechanical excavation type		EPB type				Slurry type	
					Earth pressure type		High-density slurry type			
Alluvium clay	0–5	300%–50%	□	– Core-face stability – Settlement	o	– Difficulty in extremely soft clay – Control of the ground loss	–	– Earth pressure type is more suitable	□	– Difficulty in extremely weak clay – Slurry spouting on surface – Need for plant to treat slurry
Diluvium clay	7–20	< 50%	o	– Existence of water bearing sand – Blockage in slit chamber	o	– Fluidity of the soil – Control of the ground loss	–	– Earth pressure type is more suitable	o	– Need for plant to treat slurry solution
Soft rock (mudstone)	>50	< 20%	o	– Existence of water bearing sand – Wear of cutter bits	–	– The earth pressure type with slurries is more suitable when there is water bearing sand	–	– Suitable when there is water bearing sand	–	– Suitable when there is water bearing sand
Loose sand	5–30	10^{-2}–10^{-3} cm/s	X	– Unstable core-face	□	– Fine particle contents	□	– Highly advanced excavation control	o	– Highly advanced excavation control – Quality control of slurry solution
Dense sand	>30	10^{-3}–10^{-4} cm/s	□	– Core-face stability – Ground water level, permeability	□	– Fine particle contents	□	– Cutter bits wear – Dosage of additives	o	– Quality control of slurry solution
Sand gravel	>30	10^{0}–10^{-2} cm/s	□	– Core-face stability – Ground water level, permeability	□	– Fine particle contents	□	– Cutter bits wear – Dosage of additives	o	– Running away of slurry – Gravel crusher – System for fluid transport
Sand and gravel with boulders	>50	10^{0}–10^{-1} cm/s	X	– Core-face stability – Boulder crush – Cutter bits and head wear	□	– Fine particle contents – Cutter bits and head wear – Boulder crusher – Diameter of boulders in relation to screw conveyor	□	– Cutter bits wear – Boulder crusher – Diameter of boulders in relation to screw conveyor	□	– Running away of slurry – Boulder crusher – System for fluid transport
Applicability for ground condition changes			Impossible to change excavation system		Fair. Need for additional injection equipment.		Good. Widely applicable in a variety of ground conditions		Good. Widely applicable in a variety of ground conditions	

o Applicable □ Consideration required X Not applicable

Factors to consider when applicable or, when not applicable the reasons why it is not applicable.

Parameter	Behaviour category		
	A	B	C
Systematic improvement intervention	never	often	always
Concavity of the face	recommended	imperative	imperative
Length of excavation steps (m);	2.5–5	1–2.0	≤ 1.25
Rate of advance (m/day)	> 5	1.5–5	1.0–2.5

confinement) when tunnel advance is performed using conventional means of excavation.

They can be grouped into the following categories:

- **preconfinement intervention**, when they act ahead of the face and form or facilitate the formation of an artificial arch effect in advance with a structural and protective function;
- **confinement intervention** when they act inside the tunnel back from the face with the function of countering deformation phenomena which develops after the passage of the face;
- **presupport intervention** when they act ahead of the face without, however, having any influence on the formation of an arch effect because they are able neither to contain the decay of the minor principal stress to any appreciable degree, nor to improve the shear strength of the ground.

The intervention can be:

- **conservative** when the main effect is that of containing the decay of the minor principal stress σ_3, and of conserving the shear strength characteristics of the ground;
- **improvement** when the main action is that of increasing the shear strength of the medium.

On the Mohr plane where the **intrinsic curve** is usually plotted to describe the strength characteristics of rocks and soils in graph form, the effect of the first type of intervention can be schematised by plotting a "conservation line" which sets a limit on the decay of the natural confinement stress σ_3; the effect of improvement intervention on the other hand translates into a general raising of the intrinsic curve which increases the domain of stability (Fig. 8.1).

In effect, stabilisation intervention very often results not so much in an improvement in the mechanical characteristics of the ground as

Shielded TBM's. Blockage under difficult stress-strain conditions

Examination and analysis of the problem

A shielded TBM advancing through a rock mass can be compared to a steel cylinder with a given radius that is moving at a certain velocity in a medium, which, for simplicity, we will assume is homogeneous and isotropic.

Because of a certain normal overbreak of a few centimetres designed to prevent the machine from becoming trapped, the point at which the contact between the ground and the shield begins will be at a certain distance from the cutterhead. From that point onwards the ground will exert increasing radial pressure on the shield as a function of [56]:

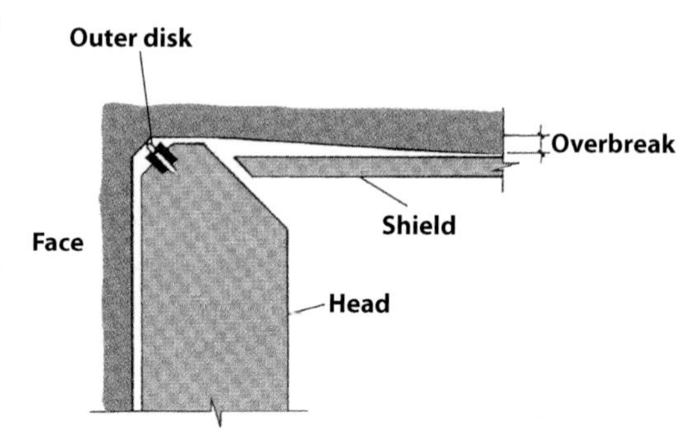

- the progressive weakening of the three dimensional effect of the face;
- the progressive deterioration of the geomechanical characteristics of the material around the cavity from peak to residual values;
- the viscous characteristics of the medium (the magnitude of which is inversely proportional to the advance rate of the machine).

Analysis of the problem requires the use of 3D calculation software codes able to take realistic account of the rheological behaviour of the medium. These can be used to calculate the distribution of the radial pressure exercised by the ground on the shield (see the figure) as a function of the daily advance rate of the machine and then to obtain the magnitude of the total radial pressure by integration on the length and circumference on which it is applied. Consequently the installed thrust capacity required to resume machine advance is given by:

Distance from the face
0 2 4 5 8 10 11 12 (m)
δ_{rad} 0 2 3 4 5 6 (cm)
Overbreak = 5 cm
Shield length: 11 m
a

Daily advance rate: 10 m
0 2 4 5 8 10 12 (m)
Pressure on the shield 25 50 75 100 (t/m²)
b

Machine halted
0 2 4 5 8 10 12 (m)
Pressure on the shield 25 50 75 100 (t/m²)
c

$$S_i = [(S_{rt} + P_m) \cdot \mu + S_t] \cdot F_s$$

where:

S_{rt} = Total radial pressure acting on the shield

P_m = Weight of the machine

μ = Coefficient of shield-ground friction (~ 0,25)

S_t = Thrust on the head

F_s = Safety coefficient

When the safety coefficient falls below 1.4 it is advisable to fit the machine with shield lubrication systems using bentonite or synthetic foams to help it to slide through the material to be excavated.

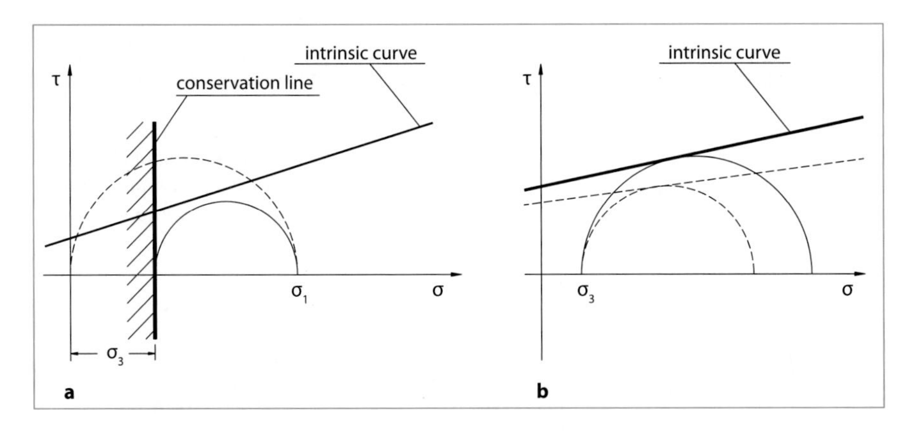

Fig. 8.1

in a degree of conservation of the natural stress state of the ground; the division into improvement and conservative intervention is designed more than anything else to highlight which aspect of each stabilisation instrument is prevalent.

Figure 8.2 groups together the most commonly used stabilisation instruments in design and construction practice showing the type of action exerted for each of them (preconfinement, confinement or presupport) and indicating the effect on the intrinsic curve and on the conservation line.

Figure 8.3, on the other hand, shows, schematically, the zones in relation to the position of the face where stabilisation instruments must be placed in order to perform preconfinement, confinement or simple presupport of the cavity during tunnel advance.

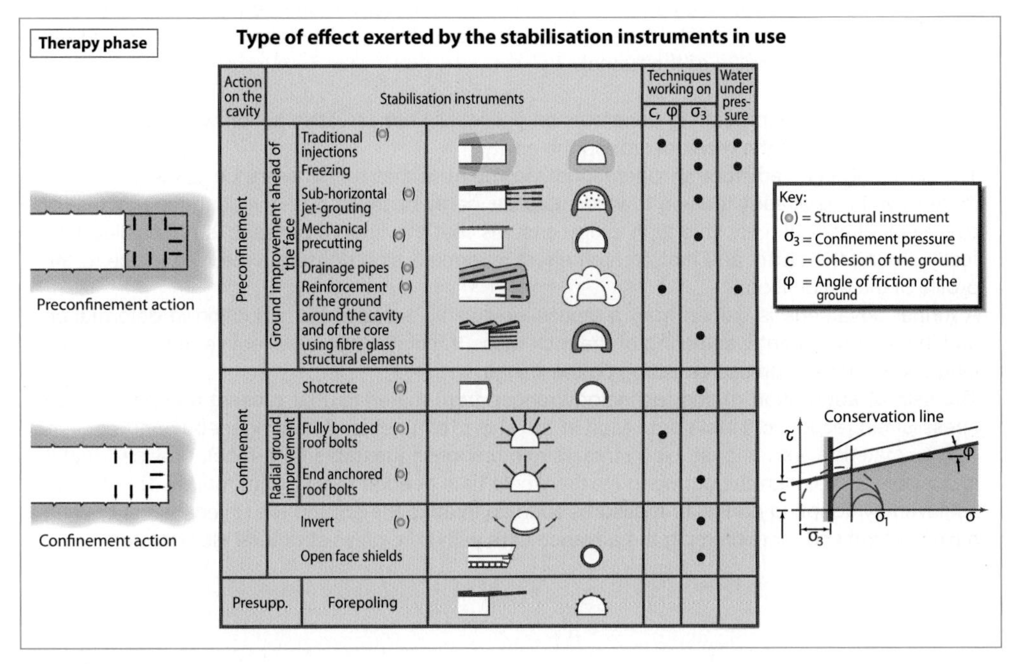

Fig. 8.2

Shotcrete tiles created by mechanical precutting technology

Tunnel advance following mechanical precutting (see also the focus box *The tunnels on the Sibari – Cosenza railway line (1985)* in chapter 2) consists of creating a continuous preliminary lining in the ground before excavation actually starts, which is able to exert radial preconfinement action to prevent the confinement pressure σ_3 from falling to zero and to appreciably improve the response of the ground to excavation.

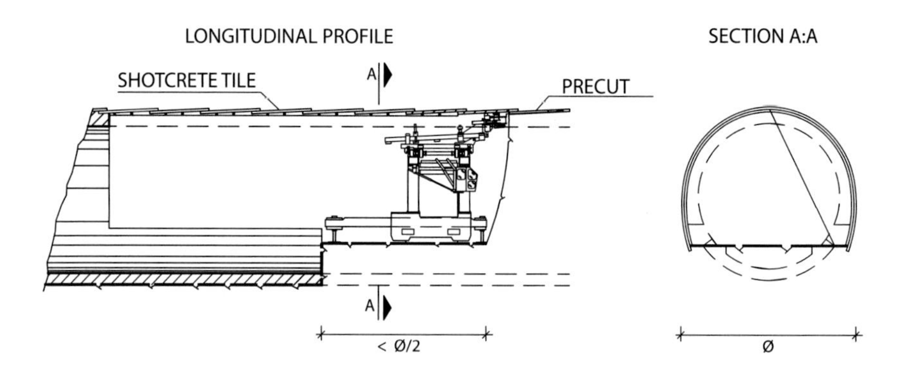

The mechanism by which this result is achieved is twofold:

- by extending beyond the face, the shotcrete tile shifts the zone in which potential extrusion phenomena is triggered deeper into the rock mass postponing its appearance and reducing its effects;
- by preventing the principal minor stress σ_3' from disappearing, even only temporarily, the radial preconfinement action reduces the size of the zone of plastic behaviour around the cavity or even eliminates it and the consequent deformation phenomena. This places the ground in a condition to develop an arch effect ahead of the face which will guarantee the transverse stability of the tunnel.

Basically mechanical precutting generates an essentially conservative action with regard to the stress-strain state of the medium around the cavity.
The advantage gained from an operational viewpoint is that tunnel advance takes place in a material which is already tending towards stability, partly because by eliminating the problem of overbreak almost entirely, this technology ensures that the flow of stresses is channelled perfectly around the cavity and that an arch effect is formed, as a consequence, very close to the profile of the excavation.
A further advantage is gained from a statics viewpoint, because the reduction in deformation and in the consequent weathering of the rock mass translates into an effective reduction in the long and short term pressures acting on the linings.
The field of application of this technology ranges from materials that present a cohesive type stress-strain response (stable core-face in the short term), even if heterogeneous and aquiferous, to materials with a loose type stress-strain response (unstable core-face), provided that it is possible to maintain the cut open for the whole time needed to fill it with mortar. This can be achieved by adopting particular measures such as treating the ground with injections of binding mixes beforehand, or reinforcing the advance core with fibre glass structural elements etc.

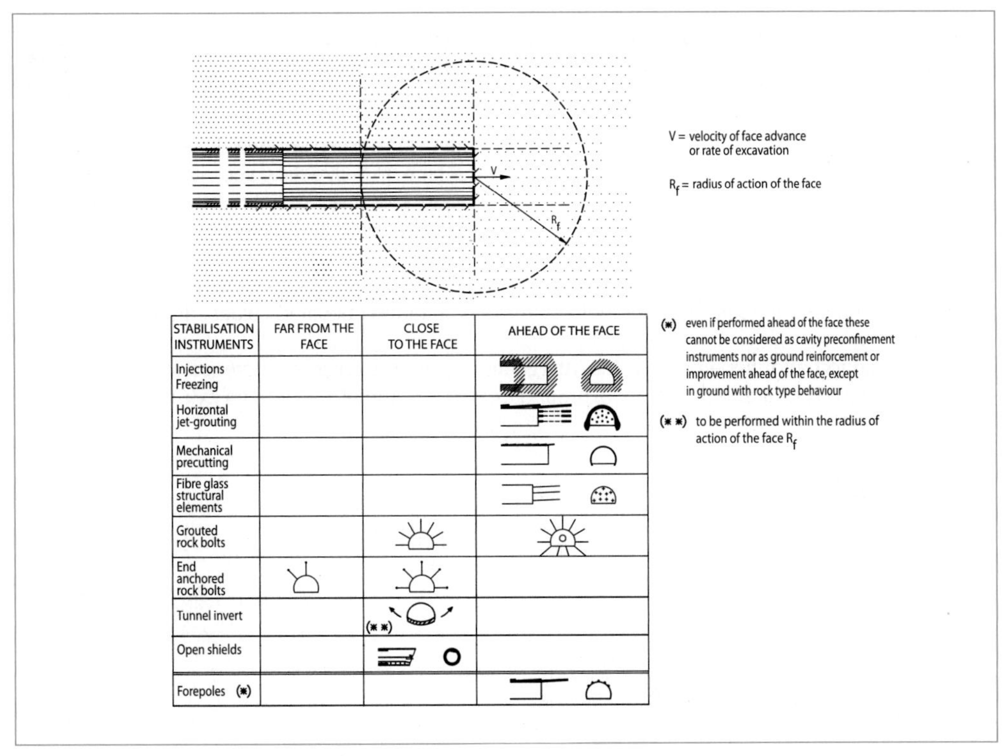

STABILISATION INSTRUMENTS	FAR FROM THE FACE	CLOSE TO THE FACE	AHEAD OF THE FACE
Injections Freezing			
Horizontal jet-grouting			
Mechanical precutting			
Fibre glass structural elements			
Grouted rock bolts			
End anchored rock bolts			
Tunnel invert		(* *)	
Open shields			
Forepoles (*)			

Fig. 8.3

8.7.1 Preconfinement intervention

Intervention for preconfinement of the cavity acts inside the rock mass ahead of the face, when it is still in a triaxial stress state. Its function is to prevent the principal minor stress σ_3 from decaying to zero by acting when deformation phenomena can still be controlled.

This characteristic is particularly important because, as we know, any loosening of the ground in the advance core resulting from extrusion causes a fall in the principal minor stress σ_3 around the future cavity before the face arrives and makes the ground susceptible to radial deformation (convergence), which once triggered is extremely difficult to control using traditional confinement techniques (steel ribs, rock bolts, shotcrete etc.). It is therefore essential to conserve adequate confinement pressure ahead of the face in order to avoid having to advance constantly through loosened and collapsed ground and then to be faced with very difficult situations and costly intervention to stabilise the tunnel. This can only be done by maintaining the advance core intact and solid.

It is consequently fundamental for intervention to protect and/or reinforce the advance core, whether to conserve or to improve it, to become operational when the core itself is still subject to triaxial stresses and when deformation is limited.

Appropriate intervention to stabilise the core must be performed during the therapy phase in order to transform the core of ground ahead of the face into a **structural element** with characteristics that are able to solve stability problems. This intervention must act either to conserve the shear strength of the core by means of an "arch effect", with a protective function, or to actually improve it by direct reinforcement of the core itself.

Reinforcement of the core-face by means of fibre glass structural elements

Tunnel advance after first reinforcing the core-face using fibre glass structural elements consists of drilling a series of evenly distributed holes into the face, in a direction sub parallel to the axis of the tunnel to a depth of not less than its diameter, in which special fibre glass elements are inserted and then immediately injected with low pressure, controlled shrinkage, cement mixes. This reinforcement stage is then followed by that of excavation, during which the section of reinforced tunnel core is demolished. When the length of the fibre glass elements after tunnel advance is no longer sufficient to contain extrusion within the specified values, another series is placed.
If well designed and implemented this intervention gives a significant increase in the strength and deformability characteristics of the ground treated and therefore reduces extrusion phenomena drastically.

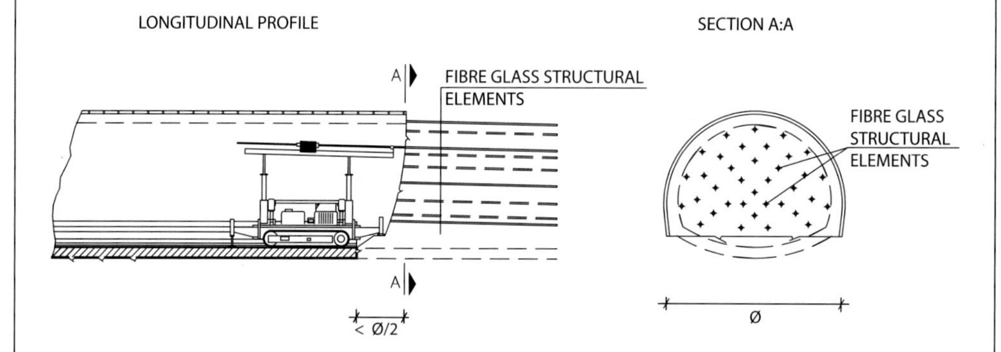

The substantial improvement in the stability of the face that is obtained translates immediately from an operational viewpoint into greater safety on the construction site. From a statics viewpoint, to limit extrusion by minimising the decompression of the medium, reduces the magnitude of weathering in the ground, which lies at the origin of the pressure on linings that increases over time and is observed in underground works even at a distance of years after construction.
The results achieved in the many applications performed to date confirm that this type of intervention exerts an improvement action on the ground in the advance core which, with regard to the stress-strain state of the rock mass (and therefore the long and short term stability of the cavity), translates into a marked conservative effect.
The intervention is normally applied to semi-cohesive and cohesive soils with mainly elasto-plastic behaviour, susceptible to chemical and physical weathering and subject to extrusion of the core. It is nevertheless possible to extend its application to include grounds with loose type behaviour, providing the core-face reinforcement is performed in combination with other ground improvement and reinforcement intervention capable of producing an artificial arch effect (horizontal jet-grouting, mechanical precutting, drainage etc.).

A number of instruments for structural stabilisation of the advance core are available to the design engineer, which are capable of exerting effective cavity preconfinement action because they produce an artificial arch effect ahead of the face. Those which have a mainly **conservative** effect are as follows:

- fibre reinforced shotcrete tiles created by means of mechanical precutting along the profile of the tunnel using the precut itself as formwork [19], [57];
- reinforcement of the advance core to a depth not less than the diameter of the tunnel by means of fibre glass structural elements set into the ground with cement mortar; the intensity of this operation will depend on how much the shear strength of the core is to be increased [18], [19], [20], [21], [22], [57];
- truncated cone umbrellas, consisting of sub horizontal columns of jet-grout improved ground, side by side, and partially overlapping [57], [58].
- cellular arch [59]
- artificial ground overburdens [60].

A mainly **improvement** action is exerted by:

- truncated cone umbrellas of ground improved by traditional injections or freezing;
- truncated cone drainage umbrellas, (under the water table).

A brief description of each of these types of intervention is given in the special focus boxes from pages 204 to 216 with details of the statics function, the effect produced (conservative or improvement), the zone in which it must be implemented and the field of application.

8.7.2 Confinement intervention

Confinement intervention acts on the ground down from the face inside the tunnel and its function is to control deformation around the cavity by exerting the confinement pressure needed to guarantee the stability of the tunnel, even after the face has moved a distance ahead. Clearly in difficult stress-strain conditions the task of this intervention is made much easier if appropriate preconfinement intervention has been performed beforehand to prevent the rock mass from loosening and relaxing.

This intervention usually consists of two types:

- preliminary intervention designed to fully stabilise short term deformation by channelling of stresses around the cavity;
- final intervention designed to guarantee the long term stability of the cavity, by absorbing any deferred pressure (swelling, *fluage*, etc.) which might manifest and also by providing the tunnel with a reasonable safety coefficient when in service.

Confinement intervention, like preconfinement intervention before it, produces a stabilising effect by acting on the ground either to conserve or to improve it.

Truncated cone 'umbrellas' consisting of partially overlapping sub horizontal columns of ground improved by means of jet-grouting

Truncated cone 'umbrellas' consisting of partially overlapping sub-horizontal columns of ground improved by means of jet-grouting are created ahead of the face to form a resistant 'pre-arch' prior to excavation, able to produce preconfinement action around the cavity (artificial arch effect).
The deviation of the stresses produced by the presence of a resistant artificial arch beyond face (shell of improved ground), makes it possible to take the load off the core-face and protect it,

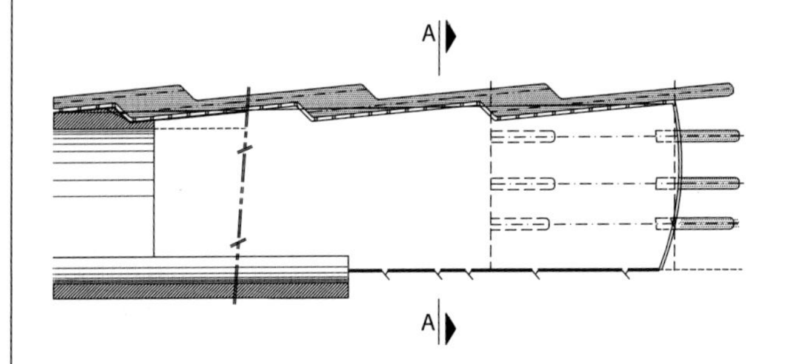

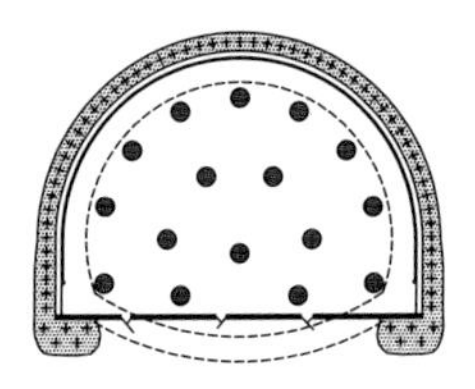

preventing or at least limiting decompression of the rock mass following excavation. Furthermore, since the problem of overbreak is practically eliminated with this technology, an arch effect is always artificially produced close to the profile of the excavation.
All this has considerable advantages from an operational viewpoint because tunnel advance occurs through a material which has practically been already stabilised, while from a statics viewpoint, it translates into an appreciable reduction in action that will put stress on the preliminary and final linings.
It is therefore conservative intervention because it prevents the principal minor stress σ_3 from decaying to zero.
The field of application of this technology ranges from non cohesive to cohesive soils with shear strength values which will allow them to be improved by the very high pressure of the jet-grouting treatment. It is therefore possible to apply it even in heterogeneous formations and obtain sufficiently uniform results.

The principal cavity confinement instruments which have a **conservation** effect are as follows:

- a primary lining shotcrete shell that produces confinement pressure around the cavity as a function of its thickness;
- radial rock bolting performed by means of end anchored bolts (see the special focus box on page 220), which apply "active" confinement pressure on the walls of the tunnel, to an extent that is predetermined by the pre-stress with which the bolts are tightened;
- the tunnel invert, which closes the shells of the preliminary and final linings and multiplies their capacity to produce very high levels of confinement pressure around the cavity.
- open shields.

A mainly **improvement** action is exerted by:

- a ring of reinforced ground around the cavity, created by means of fully bonded rock bolts capable of increasing the shear strength of the ground treated (raising its intrinsic curve) (see the relative special focus box on page 226).

8.7.3 Presupport and support intervention

Those tools, which fall into neither of these two categories because they produce neither preconfinement nor confinement action, are termed presupport or support methods according to whether they act ahead of the face or not. They are to be considered as passive intervention and have no effect on the formation of an "arch effect" since they are unable to either contain in any appreciable manner the decay of the minor principal stress σ_3, or to improve the shear strength of the ground.

An example of one of these presupport methods is forepoles which, although constituting structural elements arranged around the profile of the tunnel resting on ribs already placed, are incapable of producing any appreciable arch effect ahead of the face because there is no reciprocal transverse action between them (see the special focus box *Presupport intervention*),

8.8 Composition of typical longitudinal and cross sections

In the preceding paragraphs we have seen that the stability of the core-face plays a determining role for the quality of the deformation response of a medium to the opening of an underground cavity and consequently for the long and short term stability of the tunnel itself.

We have also seen that the stability of the core-face can be described in terms of three fundamental behaviour categories, which characterise and classify the type of tunnel to be excavated for each section and it is completely logical to refer to these categories when deciding the stabilisation methods to use to guarantee the stability and safety of the works.

The cellular arch

The *cellular arch* construction method can be used for the construction of large underground cavities in any type of ground by placing the lining before excavation begins.
It consists of a composite structure consisting of a lattice-like framework with a semi-circular cross section where the longitudinal elements (cells), consisting of tubes filled with reinforced concrete, are connected together to form a single structure by means of a series of large transverse ribs, again in reinforced concrete (arches).
There are five main stages in the practical implementation of the cellular arch method:

1. excavation of side tunnels and casting of the sidewalls of the final tunnel lining;
2. pipes in reinforced concrete are jacked into the ground side by side (diameter 2 m approx.) around the profile of the future tunnel (minitunnels) from a completely independent upper construction site;
3. transverse tunnels are excavated at appropriate intervals from the minitunnels to be used as the formwork for casting the connecting arches in reinforced concrete; the reinforcement for the arches and cells is then placed and they are cast;
4. excavation of the ground inside the section of the tunnel for the station under the protection of the cellular arch, which is already practically active;
5. casting of the tunnel invert.

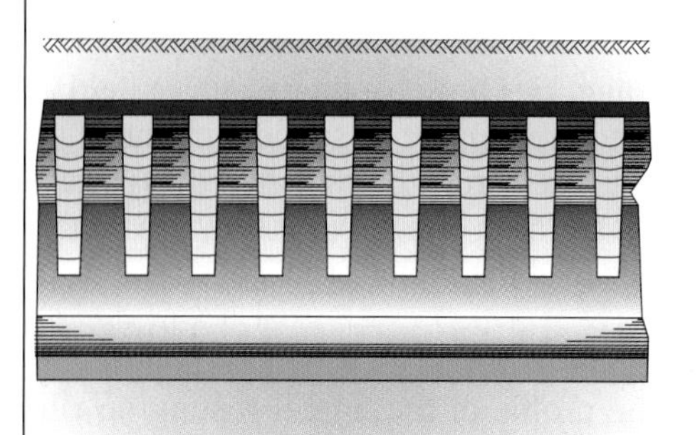

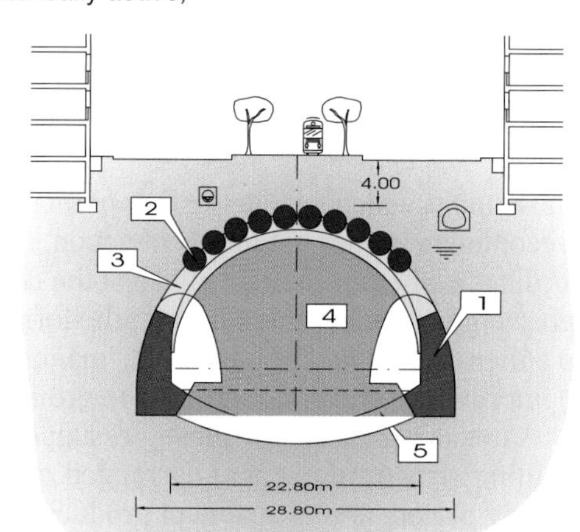

OPERATIONAL STAGES	
1	EXCAVATION AND CASTING OF THE SIDEWALLS
2	DRIVING THE PREFABBRICATED PIPES INTO THE GROUND (MINITUNNEL)
3	CASTING OF THE CELLS AND ARCHES (CELLULAR ARCH)
4	EXCAVATION OF THE TUNNEL FOR THE STATION
5	CASTING OF THE TUNNEL INVERT

A semi-circular framework in reinforced concrete is therefore formed before the tunnel is actually excavated. It consists of cells and arches which ensure that stresses are channelled perfectly around the cavity to artificially generate an arch effect indispensable to the long and short term stability of the tunnel to be constructed.
The theoretical limits of this system allow shallow cavities up to 60 m in diameter, regardless of the type of ground, without causing any appreciable surface settlement.
This technique, for which the author received the "man of the year award in the construction field" from the United States journal *Engineering News-Record*, was used to construct Venezia Station on the Milan Urban Railway Link Line (excavation diameter 30 m approx.), despite the difficult conditions (non cohesive soil under the water and very shallow overburdens, historical buildings on the surface, etc.) on schedule and with no additional costs with respect to conventional methods (see appendix D).

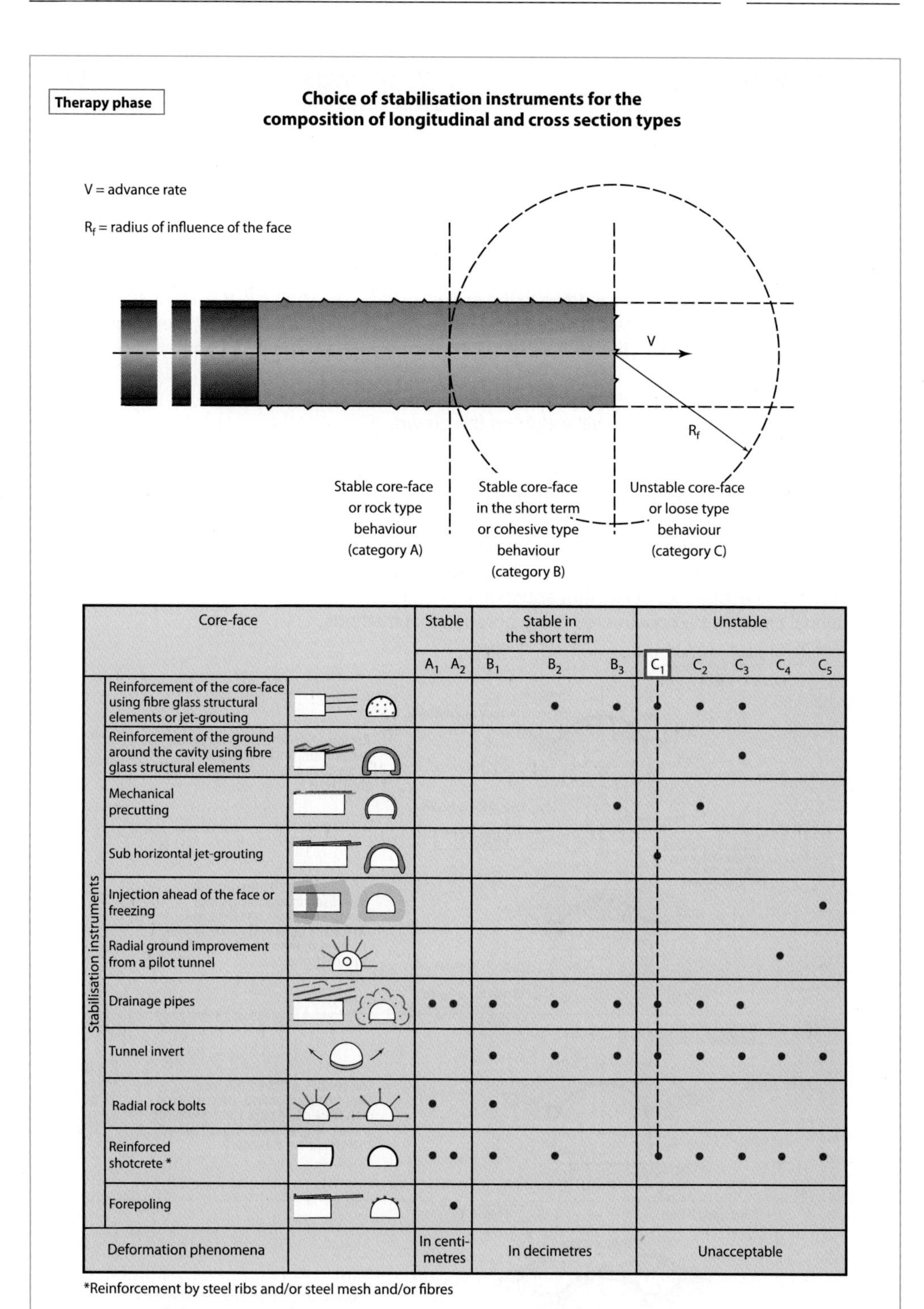

Core-face	Stable		Stable in the short term			Unstable				
	A_1	A_2	B_1	B_2	B_3	C_1	C_2	C_3	C_4	C_5
Stabilisation instruments										
Reinforcement of the core-face using fibre glass structural elements or jet-grouting				•	•	•	•	•		
Reinforcement of the ground around the cavity using fibre glass structural elements								•		
Mechanical precutting					•		•			
Sub horizontal jet-grouting						•				
Injection ahead of the face or freezing										•
Radial ground improvement from a pilot tunnel									•	
Drainage pipes	•	•	•	•	•	•	•	•		
Tunnel invert			•	•	•	•	•	•	•	•
Radial rock bolts	•		•							
Reinforced shotcrete *	•	•	•	•		•	•	•	•	•
Forepoling		•								
Deformation phenomena	In centimetres		In decimetres			Unacceptable				

*Reinforcement by steel ribs and/or steel mesh and/or fibres

Fig. 8.4

Artificial Ground Overburdens (AGO)

The term "AGO" is used for a special construction technology used to solve the problem of insufficient overburden over the crown of a future tunnel, by creating the overburden with *in situ* material after first treating it appropriately. This technology allows tunnels which would otherwise lack the necessary overburden to be driven using bored tunnel methods with very real statics, operational, environmental and economic advantages (see also Appendix E).
After first having removed the surface layer of material present over the tunnel to be constructed (following the profile of the crown down to the springline level, as shown in the figure) the ground on which the future AGO will rest is improved if necessary. A 10 cm layer of steel mesh reinforced shotcrete is then sprayed on the surface of the excavation. It has a dual function:

- to shape the future tunnel;
- to distribute the loads that will weigh on the crown.

If there is no overburden at all at the beginning, then the shape of the crown is formed by adding treated ground where necessary and then, as illustrated below, a layer of steel mesh reinforced shotcrete is placed.
At this point it is embanked in 30 cm layers with lime stabilised ground (or other suitable system to increase the strength characteristics *in situ*), until a layer of adequate thickness is formed over the crown of the future tunnel (3–4 m approx.).
The tunnel can now be constructed using bored tunnel methods.

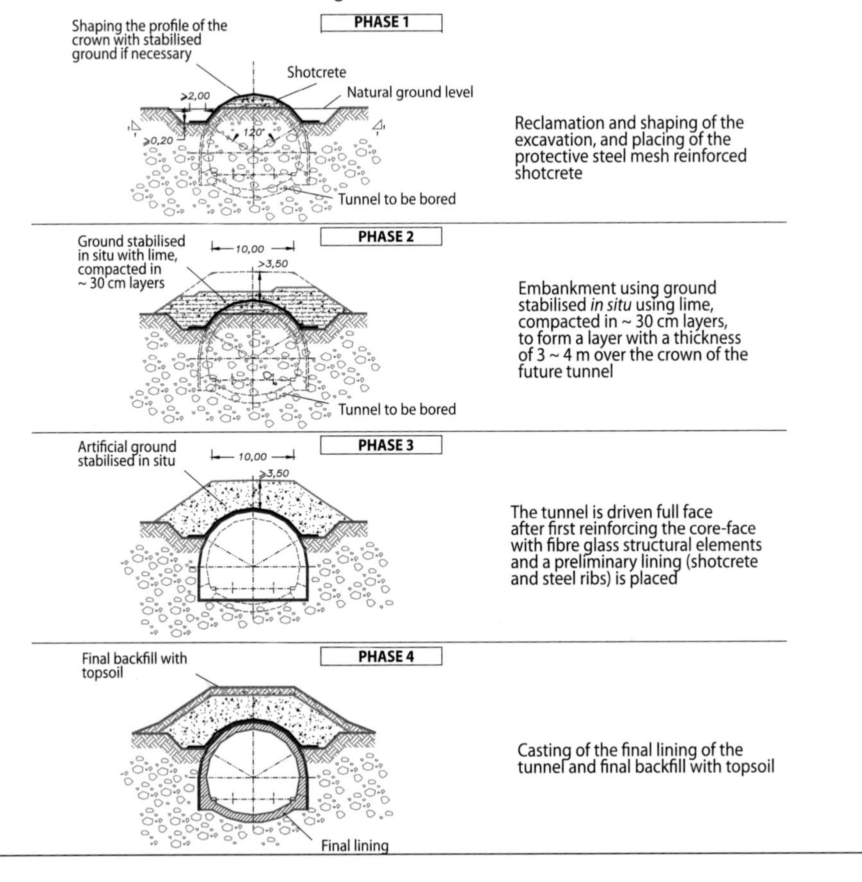

In this respect, Fig 8.4 shows the range of application of the different stabilisation instruments available to a design engineer in this framework. Taken together, these will determine the longitudinal and cross section tunnel types which will guarantee the feasibility of the excavation and the long and short term stability of the tunnel. For example:

- in sections of **stable core-face** tunnel (behaviour category: A, with stresses in the elastic range; typical manifestations of instability: rockfall), the function of the stabilisation intervention is above all protective and is determined by the structure of the rock mass and by the possible presence of water (drainage). If no rockfall of any significant dimension is expected then it will be sufficient to place a shotcrete lining reinforced with steel ribs (section type A1 in Fig. 8.4) and to secure any unstable blocks of rock with end anchored rock bolts.
  If, however, the rock mass presents fairly widespread fracturing, then forepoles should be placed ahead of the face (section type A2) in addition to the reinforced shotcrete.
  Although advisable, a tunnel invert has no structural function and it can therefore be placed to close the final lining at a distance from the face that is most convenient for construction site operations;
- in sections of tunnel with **stable core-face in the short term** (behaviour category: B, with stresses in the elastic-plastic range; typical manifestations of instability: spalling due to extrusion of the core, preconvergence and convergence of the cavity), the stabilisation intervention must guarantee the formation of an arch effect as close as possible to the profile of the excavation. Instruments are therefore employed that are able to prevent the ground from losing its strength and deformation properties, particularly in the core-face, by developing confinement or preconfinement action that is sufficient to counter the onset of plasticisation of the ground or at least to limit its extent.
  If the dimensions of the band of ground with plastic behaviour at the face and around cavity remains within certain limits (because of the high residual strength of the material in relation to the stress field present), the preliminary lining may consist of reinforced shotcrete, with possible fibre reinforcement, and fully bonded rock bolts and it must be completed with a tunnel invert which, because it has a structural function, has to be cast at a maximum distance from the face of not more than three times the diameter of the tunnel (section type B1).
  When, however, the plasticisation is so great as to require the integrity of the advance core to be preserved, then resort must be made to ground improvement and reinforcement techniques such as treating the core-face with fibre glass structural elements (section type B2) or using mechanical precutting (section type B3).
  While in the case of stable core-face behaviour (behaviour category A) the reinforcement for the shotcrete could consist of steel ribs placed with a lesser or greater frequency, in the case of stable core-face in the short term behaviour (behaviour category B), it is preferable to use a steel fibre additive to the shotcrete in order to obtain a lining which not only has high strength properties but is also ductile. In fact a preliminary lining which is too rigid would be unable to absorb the plastic deformation of the rock mass by deforming sufficiently itself and would end up by fracturing and be unable to fulfil its statics function;
- in sections of tunnel with an **unstable core-face** (behaviour category: C with stresses in the failure range; typical manifestations of instability: failure of the core-face, col-

Truncated cone "umbrellas" of ground improved by conventional grout injections

Tunnel advance after first creating truncated cone "umbrellas" of improved ground by means of conventional grout injections consists of alternating tunnel advance with phases of ground improvement in which a layer of ground of varying thickness around the cavity, and also the ground in the core-face if necessary, is injected with chemical or cement mixtures under low to medium pressure. A load bearing arch of improved ground with low permeability and better geomechanical characteristics than the original ground is thereby created, which limits the decay of the confinement stress σ_3 and at the same time facilitates the channelling of stresses around the cavity and the formation of an artificial arch effect close to the profile of the cavity.

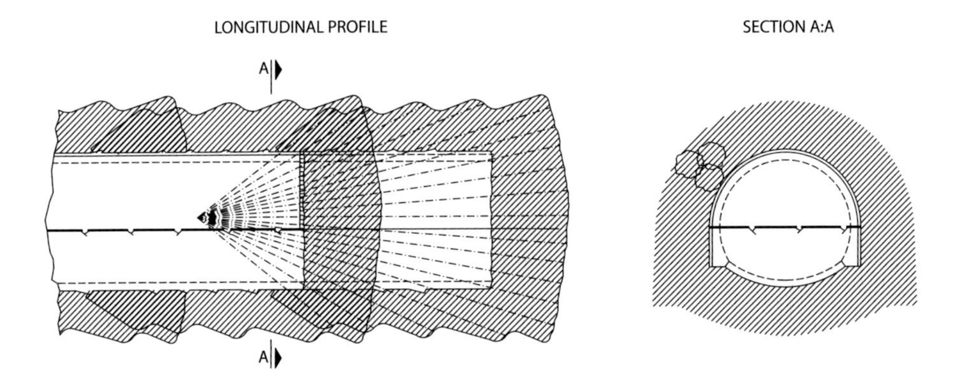

It is therefore an operation which produces both improvement and conservative action with considerable advantages from an operational (impermeabilisation of the cavity, construction site safety) and statics viewpoint (reduced pressure on linings). Conventional grout injections are employed above all in non cohesive or semi cohesive soils even under the water table, which are sufficiently homogeneous and permeable to ensure uniform diffusion of the bonding mixture.

lapse of the cavity), the stabilisation intervention must guarantee the formation of an artificial arch effect in advance ahead of the face. Instruments for preconfinement of the cavity are therefore employed which, by ensuring the stability of the core-face, prevent the minor principal stress σ_3 from falling to zero when deformation phenomena can still be controlled. The indispensable confinement pressure needed to prevent the decay of the stresses in the core-face can be produced by combining different types of intervention. For example, reinforcement of the core-face with fibre glass structural elements or with jet-grouting can be combined, depending on the type of ground, with improvement of the ground around the future cavity using truncated cone umbrellas consisting of columns of improved ground placed side by side and created by means of sub horizontal jet-grouting (section type C1, see the example in Fig. 8.5) or, alternatively, using mechanical precutting (section type C2) or, as a further alternative, reinforcement of the future cavity using fibre glass structural elements (section type C3).

If the behaviour of the rock mass to be excavated does not allow acceptable tunnel advance even with the intervention mentioned above, then it will be necessary to drive a pilot tunnel from which a series of radial ground improvement operations can be performed before widening it to the full cross section of the tunnel (section type C4). Finally, if the permeability and porosity of the ground will allow it, a large zone of the ground around the cavity and ahead of the face can be improved using conventional grout injections ahead of the face, or performed from ground level if the overburden

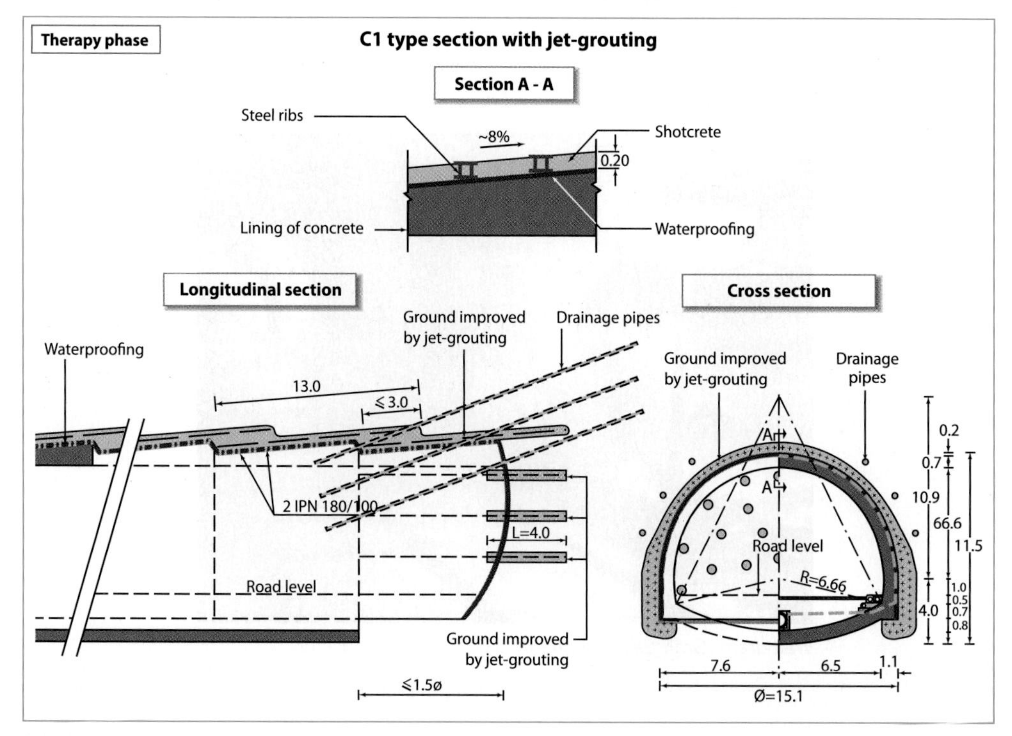

Fig. 8.5

Truncated cone “umbrellas” of drainage pipes (under the water table)

The implementation of truncated cone “umbrellas” of drainage pipes under the water table, when the design engineer considers it is best to advance under hydrodynamic conditions, creates a zone of high permeability around the cavity which will partially or fully bring down the level of the water table and also, as a consequence, that of the hydraulic gradient.
The hydraulic pressure on linings is also reduced appreciably.
Since from a geomechanical viewpoint, the shear strength of the material is negatively affected by the presence of neutral pressures inside the pores and discontinuities of the ground, lowering these neutral pressures will improve the shear strength considerably.
In oder for this improvement to be effective for the band of ground around the tunnel and for all the ground within the advance core, these drainage pipes must bc launched ahead of the face in a truncated cone ‘umbrella’ configuration and they must always be positioned outside the core-face. This is the only way to prevent entrainment effects from being produced in the core resulting from the seepage flow which would damage its stability.

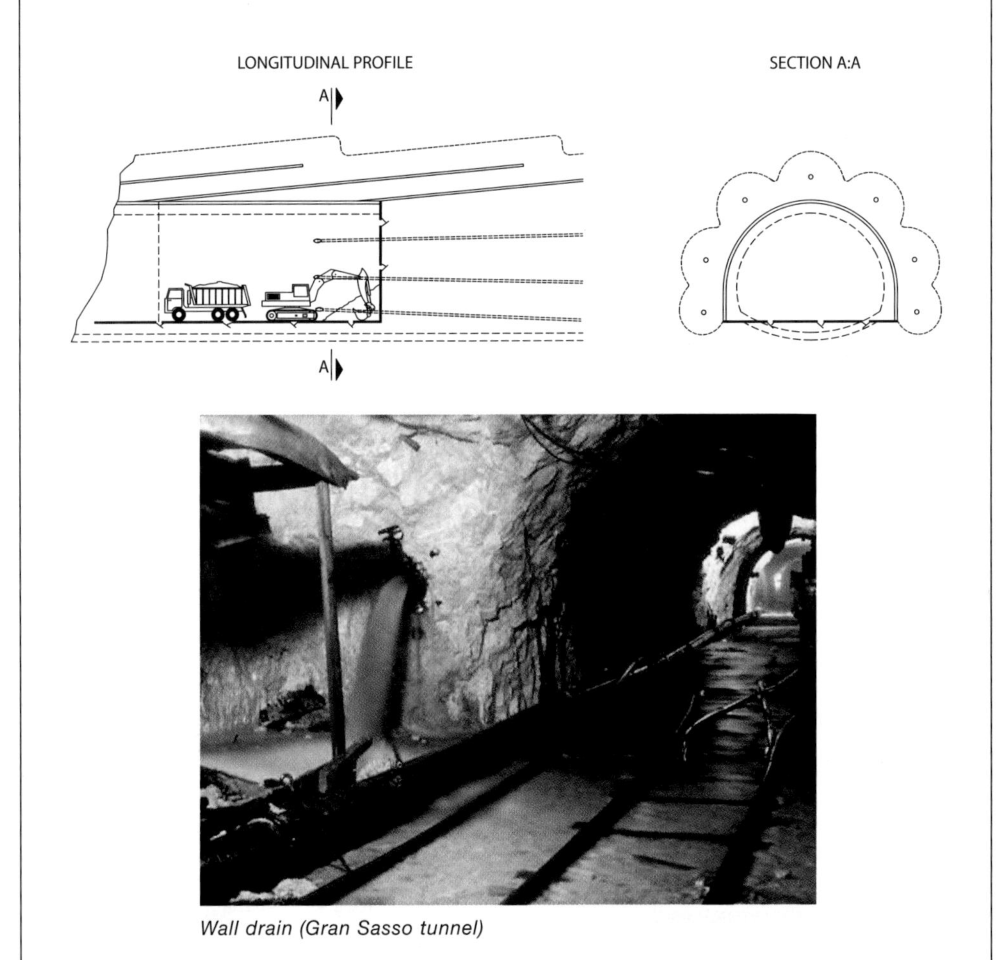

Wall drain (Gran Sasso tunnel)

is slight, in order to form an arch of load bearing ground with appreciably better geomechanical characteristics than the natural ground (section type C5).
It is always indispensable to systematically line not only the tunnel but also the face, appropriately shaped, with a sheet of fibre reinforced shotcrete to counter instability typical of behaviour category C (unstable core-face). The entire resistant structure must then be completed with the tunnel invert, which must be cast at a distance from the face of not more than 1–1.5 times the diameter of the excavation, where the beneficial arch effect is still appreciable because of the nearness of the face (radius of influence of the face R_f).

Regardless of the behaviour category of the tunnel for which a section type must be composed, it is wise to observe the following general guidelines:

- unless particular local environmental requirements prevent it, an adequate system of drainage should be implemented ahead of the face and external to the core whenever a tunnel passes through aquifers. The design engineer must carefully assess the effects of entrainment forces resulting from drainage on the stress-strain state of the core-face and the cavity;
- even in grounds which are not strictly speaking aquifers, it is important to place a drainage system along the foot of sidewalls and a layer or water-proofing between the preliminary lining and the final lining: this is to prevent small infiltrations of water, which are always possible, from damaging the concrete lining in the long term (e.g. in cases of aggressive water or potential frost damage, especially in shallow tunnels).

When this design approach is employed, in many situations the final lining can be placed even at a considerable distance from the face since it has no short or long term structural function.

8.9 Construction variabilities

The design engineer should also design the construction variabilities for each longitudinal and cross section tunnel type to be applied when statistically probable situations arise, which, however, cannot be precisely located on the tunnel alignment on the basis of the available data.

For each tunnel section type, the variabilities in the type, intensity, stages and rates of implementing intervention must be unequivocally defined and the range of geological-geomechanical and stress-strain conditions (extrusion and convergence) within which the section must be employed must be clearly identified.

Figure 8.6 gives an example of the application of a section type and of the relative variability in given geological-geotechnical and stress-strain ground conditions. It is essential to identify the variabilities for each tunnel section type that are admissible in relation to the actual response of the ground to excavation, which will in any case always be within the range of deformation predicted. This is because it allows a high level of definition to be achieved in the design and it also gives the flexibility needed to be able to avoid situations where each variation in conditions requiring even minor changes to the design

Shotcrete shell preliminary linings

Whether it is reinforced with steel ribs and/or steel mesh or fibre reinforced with special steel needles in a *concrete* grout mix, a shotcrete shell preliminary lining will develop good confinement pressure within a few hours from being sprayed (see figure). It performs a basically conservative action on the medium and is very effective from a statics viewpoint in ensuring an adequate safety coefficient to the tunnel during construction.
Its field of application comprises practically all possible situations:

- in rock type behaviour grounds (stable core-face), it fills cracks and protects against possible rockfalls;
- in ground with cohesive behaviour (stable core-face in the short term), it forms a very strong structure, which at the same time is fairly deformable and capable of absorbing deformation of the ground around the cavity without collapsing;
- in grounds with loose behaviour (unstable core-face), it complements intervention to preconfine the cavity (mechanical precutting, reinforcement of the core-face by means of fibre glass structural elements etc.).

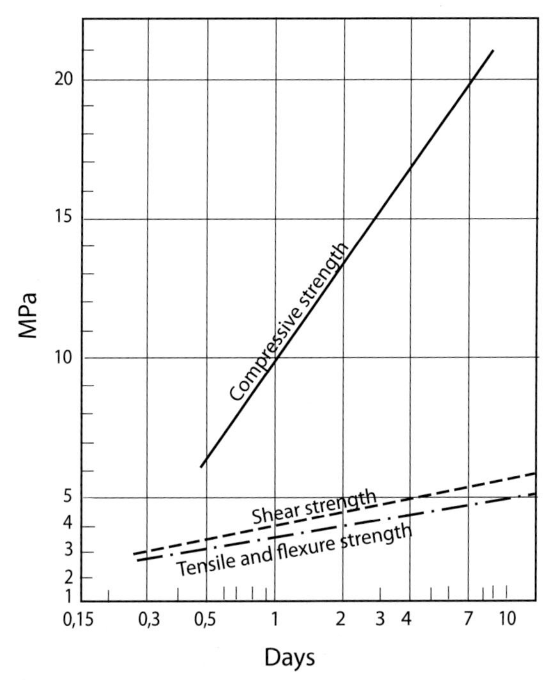

Change in the strength of shotcrete over time

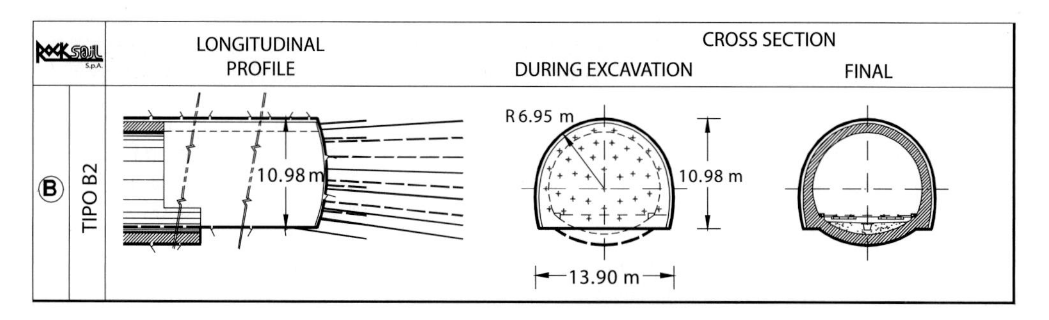

GEOLOGY	GEOMECHANICS	OVERBURDEN		DESIGN SECTION TYPE			EXPECTED DEFORMATION RESPONSE		VARIABILITY	
				INTERVENTION					INTERVENTION	
	intrinsic range	from	to	Type	PRELIMINARY	FINAL	(face)	(cavity)	PRELIMINARY	FINAL
Sandy-silty complexes in massive or poorly stratified facies from slightly to moderately cemented. Low permeability of the ground	φ =14–32 c=0.25–0.5 MPa E=0.5–3.0 GPa	50	100	B2	45 fibre glass structural elements in the core-face, overlap > 5.00 m Lattice girders at intervals of 1.20 m Shotcrete, thickness 30 cm	Concrete lining, thickness 90 cm Tunnel invert, thickness 100 cm	<10cm	5–18cm	35 – 55 fibre glass structural elements in the core-face, overlap > 5 – 7 m Lattice girders at intervals of 1.20 – 1.40 m Shotcrete, thickness 25 – 30 cm	Concrete lining, thickness 90 cm Tunnel invert, thickness 100 cm

Fig. 8.6

result in **non conformities** (differences between the design specification and what is constructed) arising, which will require official, partial redesign.

It is possible in this way to employ quality assurance systems in compliance with ISO 9000 during the construction stage without debasing the basic principles of the standard.

■ 8.10 The dimensions and verification of tunnel section types

Having decided the type of action to exert, designed the intervention to achieve it and composed the longitudinal and cross section types, the design engineer still has the task of deciding the intensity of the stabilisation action to be exerted in order to then specify the dimensions of the section types and verify them.

To achieve this the calculation methods already illustrated in the diagnosis phase may be used (solid loads or plasticised rings) depending on the predicted stress-strain response of the cavity. Particular attention must be paid to the appropriate distribution and the correct balance of intervention between the face and the perimeter of the excavation and to assessing the effectiveness of the intervention in relation to the magnitude of the deformation phenomena that it is considered correct to accept following the intervention.

The calculations must be carried out on the basis of the stress-strain behaviour predicted in the diagnosis phase by employing simple "convergence-confinement" models or, on the other hand, more complex "extrusion-confinement" or "extrusion-preconfinement" models (Fig. 8.7).

Radial bolting using end anchored bolts

Radial bolting using end anchored bolts apply an *active* confinement pressure (i.e. not the result of deformation of the rock mass around the cavity) on the wall of a tunnel, the magnitude of which is determined by the prestress set by the tightening of the bolts (conservative action).
From a statics viewpoint this type of intervention performs the function of preventing the fall of potentially loose rocks.
These are therefore normally employed in ground with rock type behaviour with little fracturing and good geomechanical characteristics.

LONGITUDINAL PROFILE

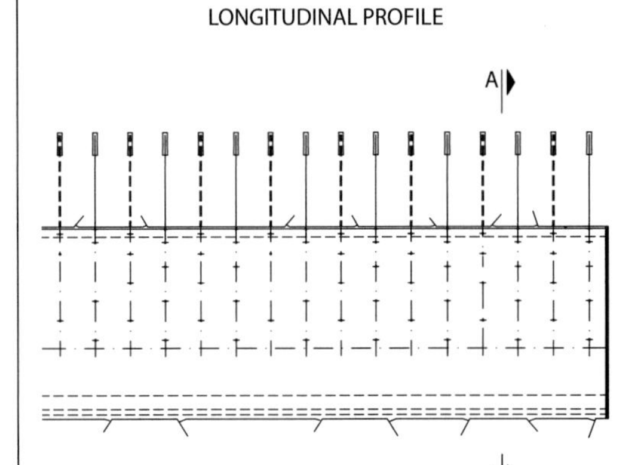

SECTION A:A

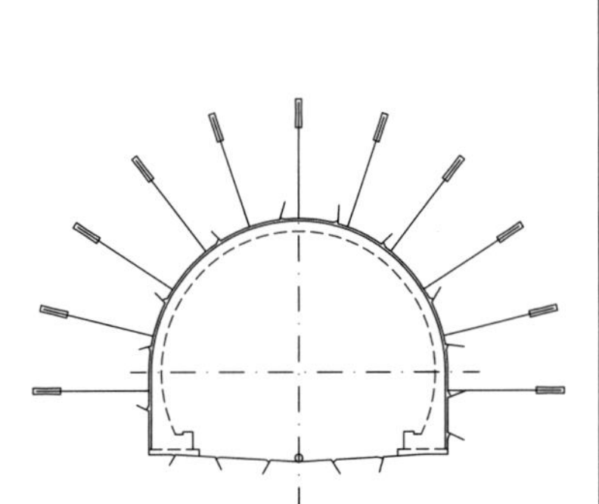

Radial bolting (Frejus motorway tunnel)

Whichever the case, more or less simple and approximate calculation methods may be used depending on the complexity of the situation to be analysed, with the more costly numerical analysis on computers reserved to the more complex (e.g. geometry of the excavation, characteristics of the rock mass, excavation system to be used, etc.) or more delicate cases (e.g., where precise measurement of surface settlement must be performed), when, that is, the less precise predictions typical of simpler calculations will not allow the problem to be studied satisfactorily.

8.10.1 Solid load calculation methods

In cases of a basically healthy rock mass in which the stress-strain response is determined by a network of fairly regular discontinuities, with the rock matrix stressed in the elastic range (stable core-face, rock type behaviour), deciding the preconfinement and/or confinement action to be exerted and therefore the dimensions and verification of the stabilisation structures can be conducted, for example, by using block theory or the *"distinct element" numerical method.*

However, in cases of soils with no cohesion or rock masses that are intensely fractured at the scale of the diameter of the tunnel (unstable core-face, loose type behaviour), the magnitude of the solid load that will be formed on the crown of the tunnel must be determined (see Fig. 8.8) along with the vertical and horizontal pressures that will act on the stabilisation structures, with account also taken for any asymmetries there may be as a result of particular morphological and structural conditions. In the simplest of cases this can be done using, for example, the methods suggested by Cacquot-Kerisel or Terzaghi [62], [63], with a special iterative calculation of a framework on elastic

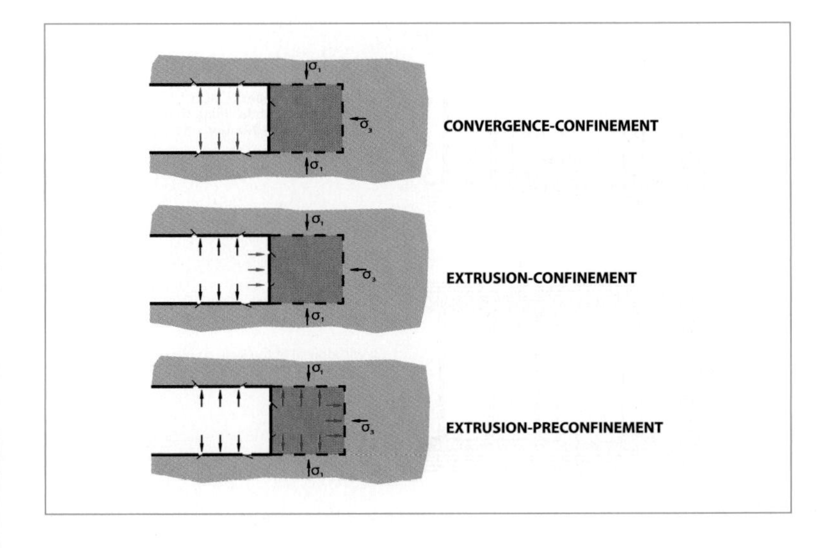

Fig. 8.7

Open shields

Open shields consist of three main parts:

- the cutterhead, consisting of the front part of the shield, is a particularly sturdy structure because it must withstand the immense stresses caused by tunnel advance. It is appropriately lined with anti-wear plates;
- an open cylindrical body in steel fitted with a hydraulic jacking system, which normally is used to drive it forward by pushing against the last ring of the lining placed (the thrust required is of the order of 70 tonnes per square metre of the cross section to be excavated and varies as a function of the geometry of the tunnel and the type of ground);
- the tail, which is the last part of the shield, within which construction workers use an erector to place the lining of the tunnel in safety, consisting generally of prefabricated concrete or steel segments.

Excavation is performed from inside the shield and can be manual or mechanised using cutter booms or excavators depending on the type of ground.

The characteristics of open shields are such that they generate effective radial confinement during excavation but, although they are always in close contact with the face, they are unable to help control core extrusion because they exert no preconfinement or longitudinal confinement action.

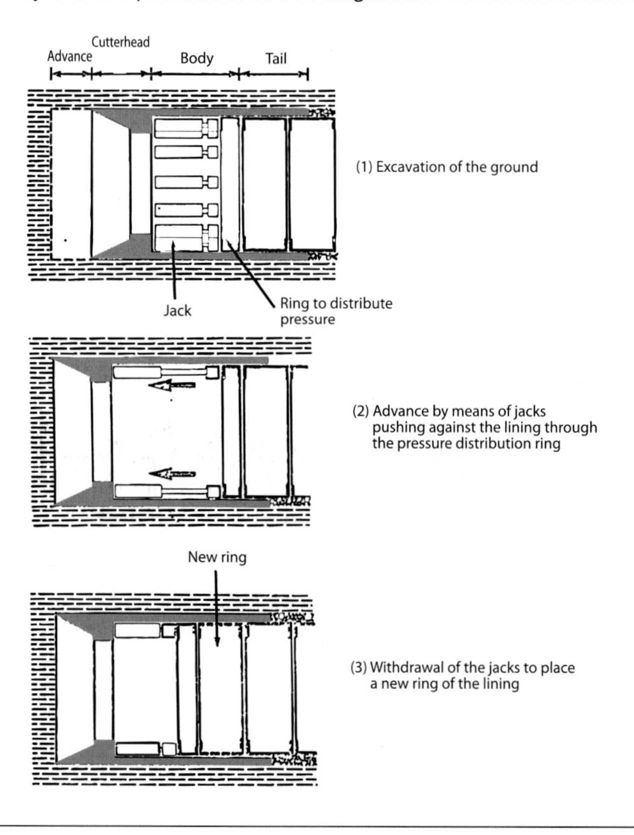

supports employed subsequently for the problem left unsolved of the ground-structure interaction. Numerical calculation using the two or three dimensional finite element method may be used in the more complex cases.

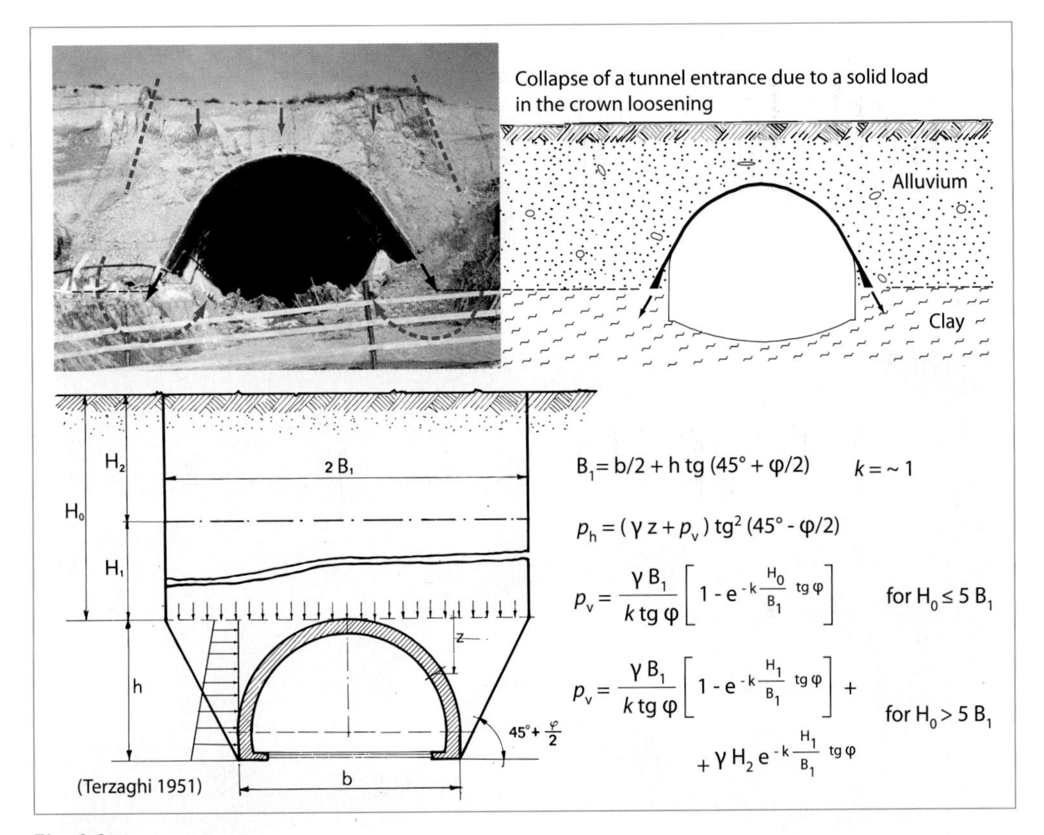

Fig. 8.8

8.10.2 Plasticised ring calculation methods

In cohesive grounds in which there is deformation in the elasto-plastic range in the core-face and around the tunnel (stable core-face in the short term or unstable core-face), the determination of the pre-confinement or confinement action to be exerted and therefore of the dimensions and the verification of the stabilisation structures should be conducted by using numerical methods such as, for example, three dimensional finite element methods in the non linear field.

Characteristic line theory may also be used, but only for section types with no preconfinement intervention [12]. Its use is not recommended, however, when the core-face is reinforced because it would furnish excessively conservative results in terms of final stresses on linings (see also the special focus boxes *Some reflections on the use of characteristic lines–1/2*).

The tunnel invert

The casting of the tunnel invert (in ordinary or reinforced concrete) completes the preliminary and the final lining to create a closed structure capable of developing considerable confinement pressure.

It therefore performs an important structural function and is used whenever stabilisation pressure of large magnitude is required. Generally this occurs in ground with cohesive or loose behaviour. In these cases it is important for it to be cast at a very short distance from the core-face so that by exploiting the arch effect produced by it, the tunnel invert becomes statically active before the principal minor stress in the ground decays completely.

In the case of grounds with rock type behaviour, the tunnel invert has no structural function and it can therefore be placed even at a distance from the face without any problems.

The tunnel invert must be adequately reinforced in zones close to portals, in sections of artificial tunnel and in fault and seismic zones.

Reinforcement of the tunnel invert (Appia Antica tunnel, Rome motorway ring road)

8.11 Particular aspects of the therapy phase

8.11.1 Tunnels under the water table

The importance of the presence of water in a tunnel has often been underestimated in the past. It has often been only considered in terms of how to remove it so that it does not interfere with construction work. However, it has been generally agreed for some years now that control and regulation of the infiltration of water plays a primary role in the stability of an underground excavation and also in mitigating its impact on the surrounding environment.

A tunnel that advances inside an aquifer with no intervention to water-proof it has the effect of a huge drain: a natural infiltration movement is created in the direction of the excavation which if not adequately dealt can lead rapidly to the collapse of the core-face and of the cavity in some situations.

It is in fact well-known that water has a harmful effect on the strength and deformability characteristics of rocks and soils. This effect, which is the result of the reduction of the effective cohesion and friction values, can become particularly negative in hydrodynamic conditions when the fine fractions that cement the fissures in rock are washed away at the same time.

As a consequence, whenever a tunnel intersects an aquifer, it is extremely important in the therapy phase for the design engineer to assess the effects of the presence of water on the stress-strain state of the rock mass around the future tunnel and to take the action required to ensure the long and short term stability of the cavity.

More specifically, he must decide the following on the basis of the water table recharge flow rate, the hydraulic gradient, the geomechanical characteristics and the primary and secondary permeability of the rock mass and the radius of the tunnel (Fig. 8.9):

- whether to reduce permeability artificially in order to perform excavation under hydrostatic conditions;
- whether, on the contrary, to decide to advance under hydrodynamic conditions by channelling the entrainment forces caused by the movement of the fluid away from the perimeter of the excavation by means of drainage;
- finally, whether to employ combined water-proofing and drainage intervention.

In making that decision the design engineer must consider that a hydrostatic regime is unacceptable when the hydraulic load is very high; on the other hand to advance under hydrodynamic conditions may not be tolerated for environmental impact reasons if the rate at which the water table recharges is not sufficiently fast.

A reinforced band of ground around the cavity created using fully bonded radial rock bolts

The creation of a band of reinforced ground around the cavity using fully bonded radial rock bolts (see the figure) constitutes an arch of rock which contributes to the statics of the tunnel on a par with any concrete lining. It also has the advantages of possessing deformability characteristics extremely similar to those of the surrounding ground and that it can be placed very close to the face. The function of the rock bolts is to improve the behaviour of the material by holding discontinuities together and preventing new cracks from opening as a result of the increased stresses induced by excavation operations.

A **direct improvement** in the geomechanical characteristics of the rock mass is obtained within the band of ground treated as a result of the **increased shear strength** and an **indirect improvement** is obtained as a result of the **expansion confinement** which is to say that phenomenon which precedes the decay of the intrinsic strength from peak to residual values in many rocks.

The rock bolts must be placed as close as possible to the face, because that is the only way to prevent the majority of the deformation and to limit decompression. It is intervention for application in cohesive grounds.

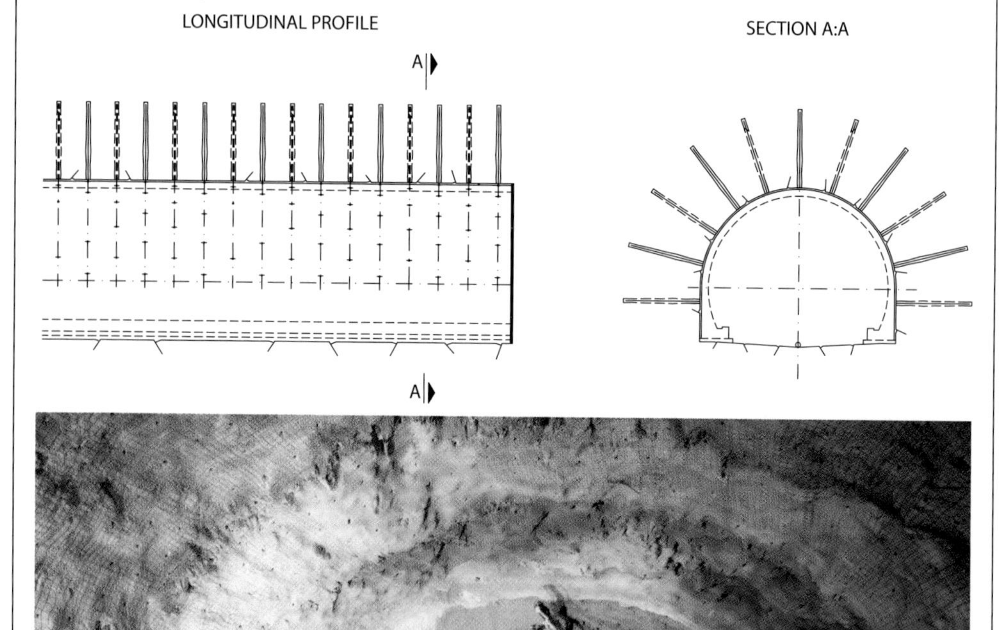

Band of reinforced ground around the cavity (cavern for the Brunico sewage plant)

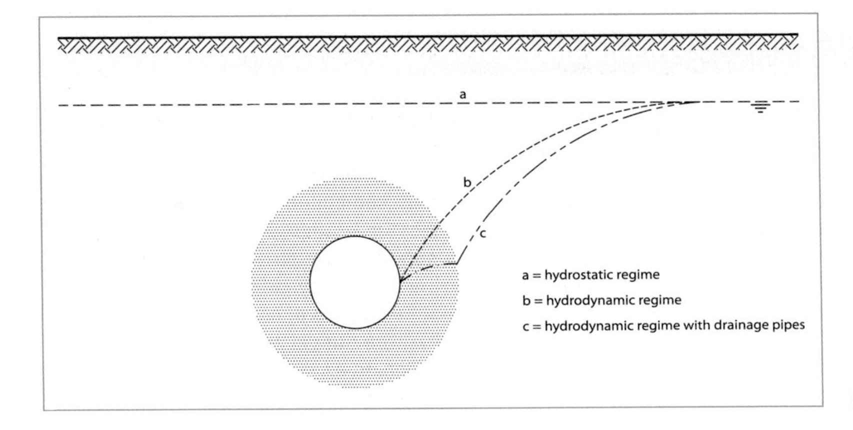

Fig. 8.9

The table in Fig. 8.10 summarises the criteria for selecting the hydraulic regime (and the consequent intervention) on the basis of the hydraulic pressure (reduced, medium or high), the importance of the hydraulic gradient J in relation to the value for it at which entrainment phenomena start to manifest (the critical hydraulic gradient J_c) and the feed to the water table.

The last factor is of great importance from the viewpoint of conserving the hydrogeological equilibrium of the aquifer in question. More specifically, when the supply to the water table is low and the hydraulic gradient is below the critical level, although it is possible and often more advantageous from a construction cost viewpoint to advance under hydrodynamic conditions (draining the ground), the design engineer should carefully consider the possible need to advance under a hydrostatic regime, in order to avoid altering the hydrogeology of the aquifer, even temporarily (exhausting water supplies, possible environmental damage, etc.).

<table>
<tr><td colspan="2"></td><td colspan="3">HYDRAULIC PRESSURE</td></tr>
<tr><td colspan="2"></td><td>LOW
U < 5 bar</td><td>MEDIUM
5 bar < U < 25 bar</td><td>HIGH
U > 25 bar</td></tr>
<tr><td rowspan="2">J < Jc</td><td>POORLY FED WATER TABLE</td><td>Regime:
hydrostatic or hydrodynamic*
Intervention:
water-proofing** or drainage*</td><td>Regime:
hydrostatic or hydrodynamic*
Intervention:
water-proofing** or drainage*</td><td rowspan="2">Regime:
hydrostatic or hydrodynamic*
Intervention:
water-proofing** or drainage*</td></tr>
<tr><td>WELL FED WATER TABLE</td><td>Regime: hydrostatic
Intervention: water-proofing**</td><td>Regime: hydrostatic
Intervention: water-proofing**</td></tr>
<tr><td rowspan="2">J > Jc</td><td>POORLY FED WATER TABLE</td><td rowspan="2">Regime: hydrostatic
Intervention: water-proofing**</td><td rowspan="2">Regime: hydrostatic
Intervention: water-proofing**</td><td>Regime: hydrostatic
Intervention: water-proofing**</td></tr>
<tr><td>WELL FED WATER TABLE</td><td>Regime:
hydrodynamic controlled*
Intervention:
water-proofing** + drainage*</td></tr>
</table>

*Only if altering the hydrogeological equilibrium does not lead to damaging consequences for the environment
**Achieved by means of injections or freezing, depending on the permeability of the ground

Fig. 8.10

Presupport intervention

This term is applied to all those types of intervention which produce neither preconfinement nor confinement of the cavity (passive intervention).
One of these is *forepoles*. They are load bearing pipes placed at the face and ahead of it on the extrados of the crown of a tunnel injected with cement grout (see figure). Reinforcement with a truncated cone configuration is created in this way to protect the excavation as it advances.
The pipes function individually as continuous beams on supports (the steel ribs placed during tunnel advance). No artificial arch effect ahead of the face is created by placing them nor is the decay of the principal minor stress σ_3 contained effectively. Furthermore, since the load bearing pipes are obliged to function as unbending beams, when they are not yet supported by the steel ribs inside the tunnel, they weigh on the advance core and worsen its stability.
They can be employed in fractured rock masses to hold discontinuities together.

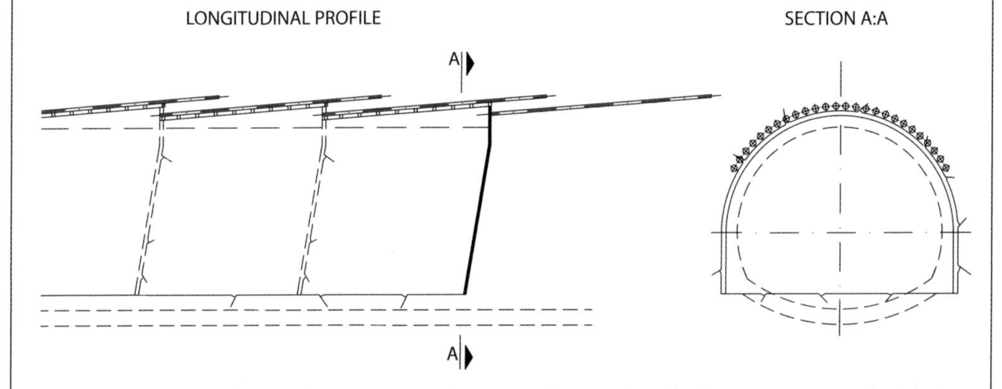

Presupport intervention with forepoles (Solignano tunnel)

If, however, he has decided to allow drainage by the cavity, the hydraulic regime will pass through a transitory period during which the hydraulic pressure will fall sharply as a result of water filtering through towards the walls of the tunnel.

In the absence of adequate counter measures, these phenomena will affect the advance core first, with an appreciable reduction in its geomechanical strength and therefore also in its rigidity. Since, as we have seen, the rigidity of the advance core plays a determining role in the long and short term stability of a tunnel, it is important for water to be prevented from circulating inside it. This can be achieved by intercepting the water three or four tunnel diameters ahead of face with special drainage pipes arranged in an umbrella configuration around the future tunnel (protection of the advance core). It is very important for the design engineer to give very precise specifications to ensure that the drainage pipes are constructed exactly as required in order to prevent drainage ahead of the face from being ineffective because implemented wrongly with effects that may be the opposite of those desired. More specifically it must be absolutely forbidden to insert them in the ground from the surface of the face. They must be arranged in a truncated cone configuration and must start from the side walls of the tunnel or at the most from the perimeter of the face so that the core is never intersected. If this is not done, then the water that is drawn in by them will soak the ground that forms the core with disastrous effects for its stability and also therefore for the stability of the cavity. In order to prevent this danger, it is also important to specify that the drainage pipes must also have no perforations in them for a length of a few metres from the end closest to the tunnel.

Similarly, and for the same reason, the design engineer must give clear instructions for the correct execution of all reinforcement treatment which involves drilling and then inserting reinforcement structures. It is important to drill one hole at a time and for it to be filled immediately and sealed perfectly with mortar.

This is the only way to prevent it from rapidly becoming a channel for water to flow through with devastating consequences for the advance core, which once soaked and softened, would no longer be able to perform its stabilising action effectively.

The design of the final lining also requires some extra attention. It should be protected from the potentially aggressive action of water by placing a water-proof sheet and a protective geotextile layer.

If the water table is eliminated permanently, then the design engineer must also specify the placement of channels at the foot of the sidewalls or under the tunnel invert to run drainage water away. However, if it has been decided to restore the natural piezometric level when construction is complete, then the water-proofing must be sufficient to make the tunnel inactive hydraulically. There will then be a second transition regime which will end when the original hydraulic head is restored.

Distinct elements

The distinct elements method [65] is a numerical approach used to simulate the discontinuous behaviour of a jointed medium. It is useful for studying tunnels where the deformation behaviour is dominated by large deformation phenomena which is the direct result of dislocations between blocks of rock. Three characteristics make it particularly suitable for this purpose:

1. the medium is simulated as a set of blocks which interact along the edges and at the corners;
2. the interactions between the blocks are governed by the friction and cohesion characteristics of the discontinuities. Maximum consideration is therefore given to the influence of the state of fracturation of the rock mass on its deformation behaviour rather than to the strength of the matrix;
3. the calculation algorithm is very versatile: the calculation can be performed for hypotheses of large displacements and rotations; the non linear behaviour of the material can be approximated for both the matrix and the discontinuities with account taken for the possible presence of interstitial fluid; different phases of tunnel excavation and lining placement can be simulated with solutions which proceed step by step.

The main input parameters are: the mesh network of elements (one for each wedge of rock) which determines the domain to which the analysis applies; the geomechanical properties of each element and of the discontinuity surfaces; the surrounding conditions; the loads acting. The simulation is performed by means of a series of distinct calculation steps so that the stress-strain development corresponding to any construction situation can be observed, including the intermediate situations. The results for each step consist of the numerical values for the displacements and the stresses in the barycentre of each element of the mesh network.
Quantitative assessment of the confinement needed for the stability of a cavity, is therefore derived from a knowledge of the stress state around the tunnel.
The greatest difficulty with this method is that of reconstructing the network of distinct elements so that it matches reality as closely as possible.

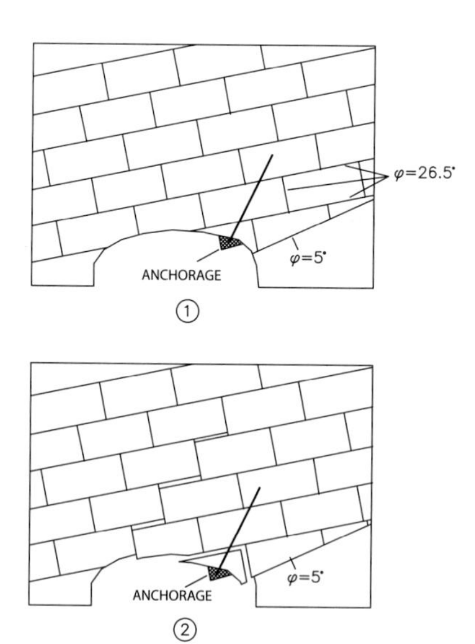

8.11.2 Adjacent tunnels

In the case of twin bore tunnels the problem faced by the design engineer in the therapy phase is that of the size of the distance between centres and, that is, of the ground that is to separate the two bores.

Whenever the morphological conditions of the rock mass and the characteristics of the alignment allow it, he will prefer to put a distance between the two cavities and stagger tunnel advance in such a way as to make the reciprocal interference between them negligible. If that is not possible, then the overall stability of the two adjacent cavities (safety coefficient of the separating ground) and the adequacy of the stabilisation intervention planned must be verified for the construction and service phases with account taken for the increased stresses which each cavity induces on the other. These effects can be lessened appreciably by staggering the advance of the two tunnel faces during the construction stage.

The index area theory [49], already discussed for the diagnosis phase, can be used to calculate the dimension of the zone of ground separating the two tunnels and to determine the safety coefficient.

8.11.3 Tunnels with two faces approaching each other

When tunnels of great length are driven, they are very often driven simultaneously on more than one face to cut construction times. The problem then arises of how to effect the meeting of two opposing faces under conditions of stability and safety. Although there are a few exceptions [64], the problem is unexplainably neglected by the technical and scientific literature and by the regulations and guidelines currently in force, but it is an extremely delicate operation and deserves attention.

As already mentioned, a face which is advancing deviates the flow of existing stresses in the medium towards the outside of the tunnel, channelling them in front of the face and laterally around the cavity (see Fig. 8.11).

This causes the appearance of zones of material with increased stress near the walls of the excavation and, if the strength of the medium is exceeded, plasticisation is triggered to varying degrees with consequent problems for stability.

In the case of two faces approaching each other, while the distance between them remains greater than a critical value (generally between 6 and 4 times the diameter of the tunnel), the stress states affecting the respective advance cores is no different from that which would affect them if each bore were isolated. When, however, that distance falls below a certain critical value, as the two advance cores penetrate each other until they become one single core for both bores of the tunnel, the channelled stress flows tend to superimpose on each other. It follows that the stress on the core or dividing ground, the strength

Finite elements

The three dimensional form of the finite element calculation method is the most complete and general for studying situations characterised by cohesive type (category B) or by loose type behaviour (category C). It can in fact be used to take simultaneous account of anisotropic stresses and materials, of tunnel advance stages and of any cavity preconfinement and confinement intervention there may be.
The main input parameters are the mesh network of elements which determines the domain to which the analysis applies, the geomechanical properties of each element, the surrounding conditions and the loads acting. As with the distinct element method the simulation is performed by means of a series of distinct calculation steps, one for each of the construction stages.
The stress-strain development corresponding to any construction situation including the intermediate situations can therefore be observed using this calculation method. As opposed to the distinct element method, however, the algorithm for the finite element method requires displacements to be congruent with the nodes of contiguous elements. The results for each step consist of the numerical values for the displacement and the stresses for each node of the mesh network.
If detailed knowledge of the mechanical characteristics of the ground is possessed and ground reinforcement and improvement intervention ahead of the face and in the tunnel is modelled correctly, the effects of the intervention on the long and short term stability can be calculated with great precision.
The finite element method seems to be indispensable for constructions that are difficult and complex either because of interaction with pre-existing buildings and structures or because of very shallow overburdens. This is the case, for example, of tunnels in urban areas which require accurate prediction of the deformation response of the ground and of surface settlement.

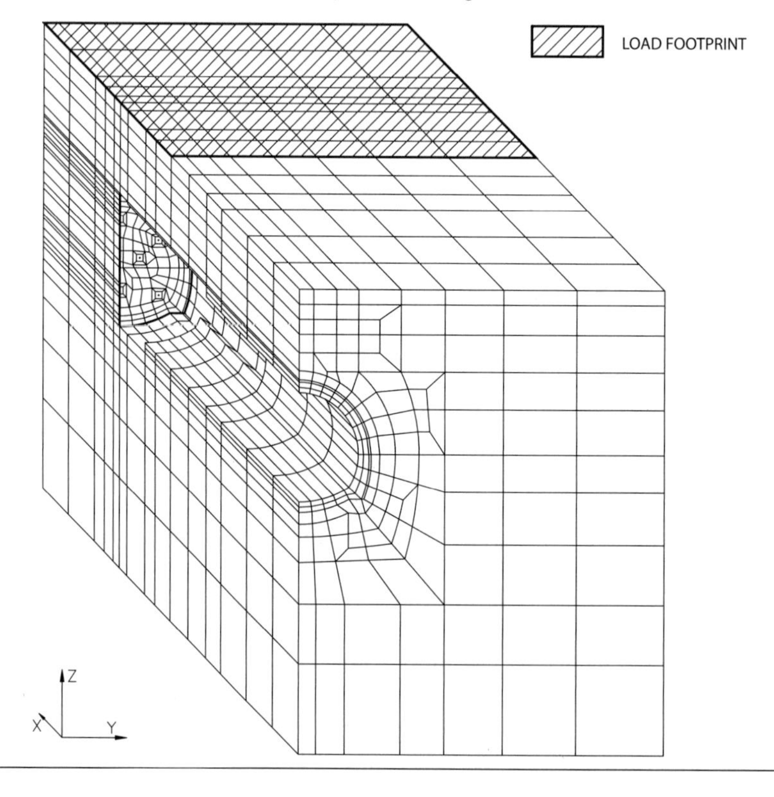

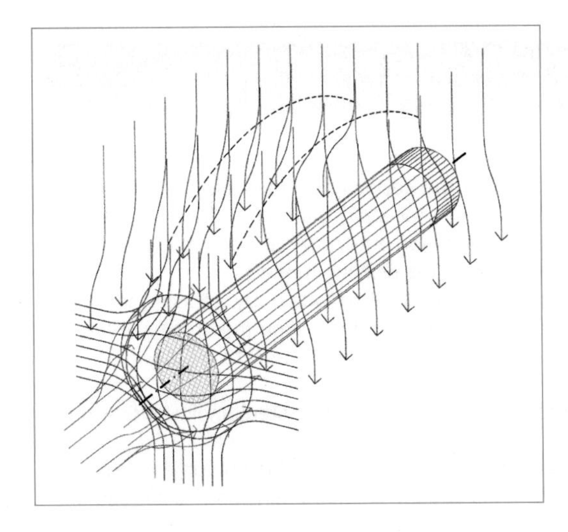

Fig. 8.11

of which in any case progressively diminishes as the bores advance, increases progressively and is much higher than in the case of a single isolated bore.

In difficult stress-strain conditions, specially when tunnel advance is performed with reinforcement of the advance core, it is easy for the magnitude of the increased stresses induced to become great enough to lead rapidly to the collapse of the advance core between the two bores and as a consequence to the collapse of the cavities, if fully adequate counter action is not taken (Fig. 8.12).

It is therefore imperative for the design engineer to give careful consideration, whenever a tunnel is to be driven on two faces, to the stress-strain phenomena that results when the distance between them falls below the critical value and to give precise specifications on how to conduct advance operations from that point onwards. To achieve this he must act effectively:

- to reduce the stresses in the advance core or dividing ground, by channelling stresses laterally around the future tunnel rather than longitudinally ahead of the two faces;
- to achieve the formation of a longitudinal arch effect as close as possible to the walls of the excavation capable of guaranteeing the stability of those sections of the tunnel not yet lined.

Both these objectives can be reached if:

1. the advance cores are adequately reinforced, but above all if they are protected with appropriate ground improvement in advance around the future tunnel;

Widening of road and mainline and metropolitan rail tunnels without interrupting traffic

There is an increasing requirement today to increase the capacity of road and rail infrastructures currently in service without interrupting traffic even for short periods. If there are tunnels along the infrastructure routes, then the only feasible solution until now has been to make costly changes to the routes to drive new tunnels in addition to the existing tunnels.
An ingenious method has recently been developed in this context which, by applying the tunnel advance principles of the ADECO-RS approach, can be used to widen the cross section of a road or rail tunnel under safe conditions without interrupting service and to reliably plan the construction times and costs of the project.
All the work is performed while traffic is protected by a self-propelled steel shell under which vehicles can continue to pass in safety.
A tunnel is widened by using specially designed machinery fitted with all the equipment required to perform all the necessary work ranging from the ground improvement and reinforcement ahead of the face and in the tunnel to the demolition of the existing tunnel, widening operations and placement of the new tunnel lining.
The operation is achieved in four main phases in which all the necessary work for the preconfinement and confinement of the tunnel is performed in a co-ordinated fashion:

1. cavity preconfinement intervention if necessary, performed on the face for wideing the tunnel using the most appropriate technology in relation to the geology and geomechanics of the ground;
2. excavation of the ground in steps under the protection of the intervention already performed until the theoretical profile of the widened tunnel is reached and demolition of the old lining;
3. the construction and immediate activation of the final lining immediately behind the face by placing one or more arches of prefabricated concrete segments in succession according to the "active arch" principle;
4. the construction of the foundation (tunnel invert).

There is no use of any type of preliminary lining, which is entirely replaced by the "active arch" type final lining that interacts by itself with the surrounding ground as a result of the precompressed action developed by small jacks fitted in the key segment.
The method was experimented for the first time in the world to widen the A1 motorway tunnel at Nazzano in the province of Rome from two to four lanes in both directions (see Appendix G).

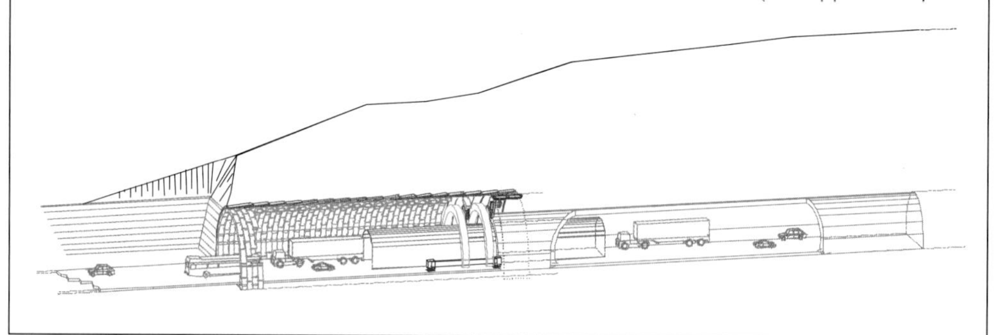

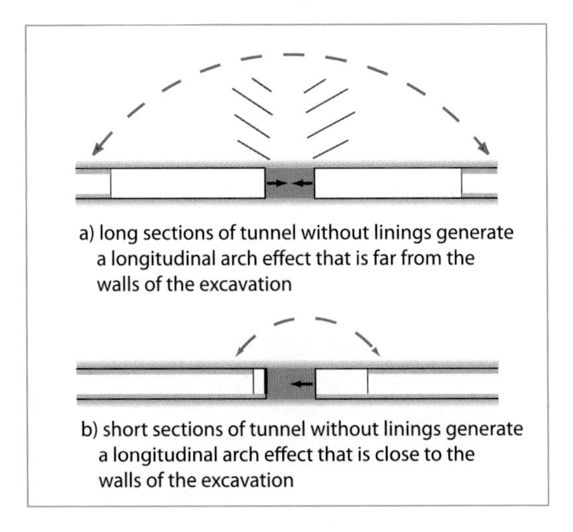

Fig. 8.12

2. advance is on one face only with the other halted and protected by a layer of fibre reinforced shotcrete, while the final lining is placed as close as possible to it and closed with the placement of the tunnel invert;
3. the final lining, closed with the tunnel invert, is placed as close as possible to the face that advances.

8.11.4 Portals

Before concluding this chapter on the therapy phase, we should look briefly again at the question of portals in order to complete the discussion on the diagnosis aspects contained in section 7.6 with those on therapy.

There are a number of techniques available to a design engineer for the construction of the portal of a tunnel in safety without destabilising the slope concerned, which once again can be divided into:

- **preconfinement intervention or ground improvement/reinforcement in advance**, when it is performed before the slope is actually excavated (e.g. shells of ground improved by jet-grouting, reinforcement of the rock using fibre glass structural elements, injections, shafts, drainage pipes etc.);
- **confinement intervention or ground improvement/reinforcement**, when it is performed as earth is excavated (e.g. Berlin type pile walls, active or passive steel anchors, support walls, etc.);
- **protective intervention**, when performed after cutting into the slope, basically to prevent the band of exposed rock from deteriorating and from possible local instability, resulting from the detachment and fall of rock and boulders (e.g. local rock bolting, steel mesh, a layer of shotcrete, etc.).

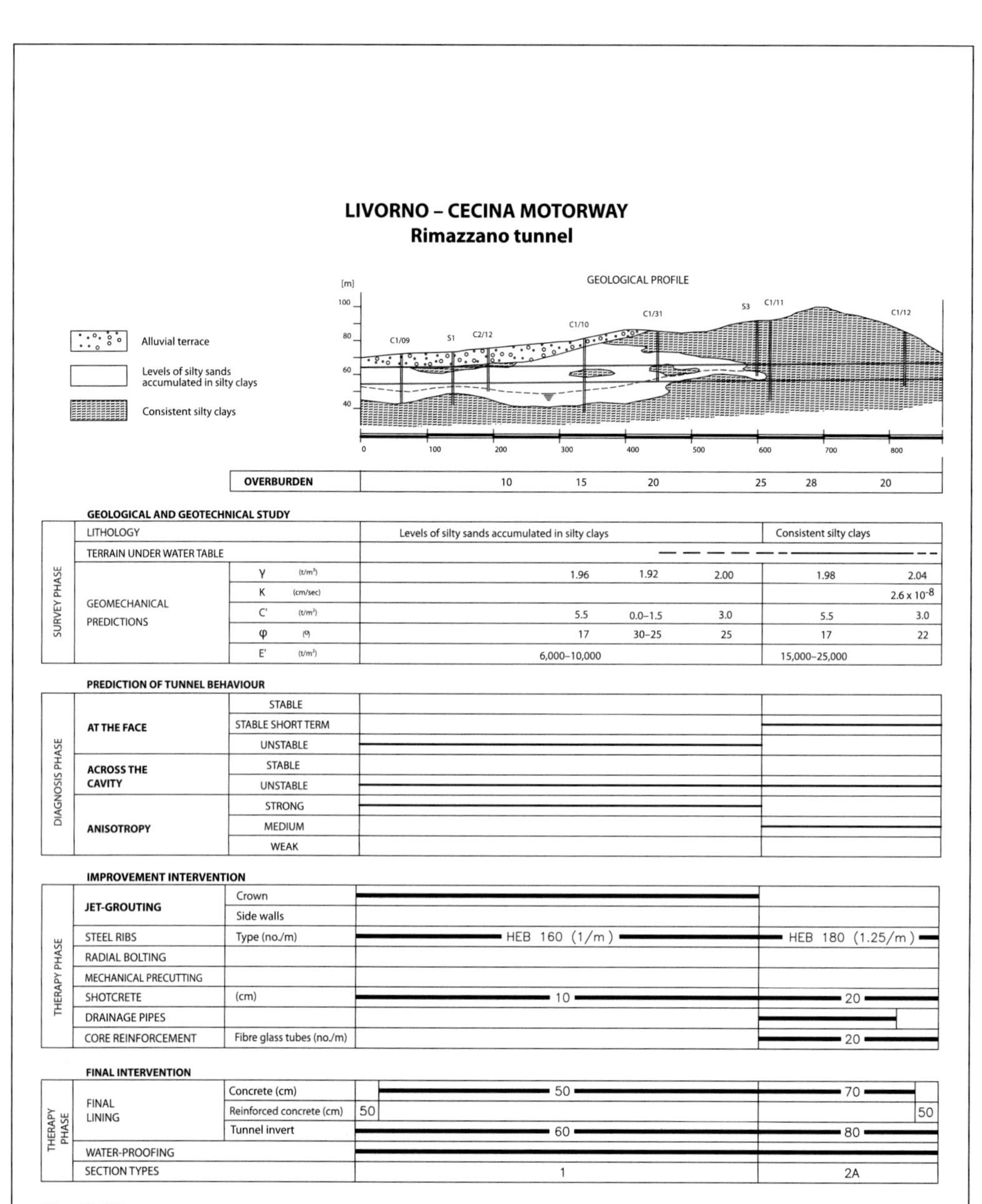

LIVORNO – CECINA MOTORWAY
Rimazzano tunnel
[m]
GEOLOGICAL PROFILE
100
80
60
40
C1/09
S1
C2/12
C1/10
C1/31
S3
C1/11
C1/12
Alluvial terrace
Levels of silty sands accumulated in silty clays
Consistent silty clays
0
100
200
300
400
500
600
700
800
OVERBURDEN
10
15
20
25
28
20
GEOLOGICAL AND GEOTECHNICAL STUDY
SURVEY PHASE
LITHOLOGY
Levels of silty sands accumulated in silty clays
Consistent silty clays
TERRAIN UNDER WATER TABLE
GEOMECHANICAL PREDICTIONS
γ (t/m³)
1.96
1.92
2.00
1.98
2.04
K (cm/sec)
2.6 x 10-8
C' (t/m²)
5.5
0.0–1.5
3.0
5.5
3.0
φ (°)
17
30–25
25
17
22
E' (t/m²)
6,000–10,000
15,000–25,000
PREDICTION OF TUNNEL BEHAVIOUR
DIAGNOSIS PHASE
AT THE FACE
STABLE
STABLE SHORT TERM
UNSTABLE
ACROSS THE CAVITY
STABLE
UNSTABLE
ANISOTROPY
STRONG
MEDIUM
WEAK
IMPROVEMENT INTERVENTION
THERAPY PHASE
JET-GROUTING
Crown
Side walls
STEEL RIBS
Type (no./m)
HEB 160 (1/m)
HEB 180 (1.25/m)
RADIAL BOLTING
MECHANICAL PRECUTTING
SHOTCRETE
(cm)
10
20
DRAINAGE PIPES
CORE REINFORCEMENT
Fibre glass tubes (no./m)
20
FINAL INTERVENTION
THERAPY PHASE
FINAL LINING
Concrete (cm)
50
70
Reinforced concrete (cm)
50
50
Tunnel invert
60
80
WATER-PROOFING
SECTION TYPES
1
2A

Fig. 8.13

Which intervention is selected will obviously depend on the results of the diagnosis stage. While protective intervention is reserved for slopes that are generally stable, it is imperative to employ preconfinement intervention or ground improvement and reinforcement whenever material subject to brittle failure is involved, whether it is of a rock or granular nature (sand, rubble-slopes etc.) or when deformation of a slope must be prevented completely in order to protect pre-existing buildings and structures.

Otherwise if the material of which the slope is formed has good cohesion and above all if this is able to develop appreciable residual strength, then more normal confinement or ground improvement and reinforcement intervention can be employed.

8.12 Final considerations

Once the type of action to be exerted has been **decided** on the basis of the deformation response of the rock mass to excavation predicted in the diagnosis phase and the tunnel advance operations have been **designed** in terms of systems, stages, excavation rates and stabilisation intervention and once the tunnel longitudinal and cross section types have been **composed and the dimensions determined** and **analysed** by calculating the effect of the design intervention on the expected deformation response and once the variabilities have been **specified** for each tunnel section type on the basis of the probable range of deformation phenomena that will manifest during construction, the ADECO-RS approach requires one last task to be performed by the design engineer in order to consider the results of his work (the design of the tunnel) as really complete and reliable: *the preparation in advance of an appropriate monitoring programme.*

This approach bases the entire procedure for the design and construction of an underground work on one single parameter, the deformation response of the rock mass to excavation. Once it has been predicted and analysed theoretically at the design stage, it is controlled with appropriate stabilisation intervention and measured in reality as work proceeds during the construction stage. It is therefore important to be able to make a systematic comparison between the predicted deformation response and the actual deformation response:

- during construction (operational phase), because this is the only way to seriously test the validity of the decisions taken in the therapy phase in terms of whether the deformation response is controlled as predicted, and to acquire information indispensable for optimising the design in terms of distributing and balancing stabilisation instruments between the face and the cavity;
- when construction is complete and the project is tested and during service to assess its suitability for use and its safety.

It is, however, only possible to acquire thorough and accurate knowledge of how the deformation response of the rock mass to excavation develops in reality, if the necessary measurement instrumentation has been installed and activated at the appropriate time during construction. This will then furnish information which is very useful and easy to interpret when it can be immediately compared with the predictions calculated in the design stage.

That is why it is essential for the design engineer to consider the best type of measurement instruments and the location for them beforehand at the design stage in order to then be able to acquire the data needed in the monitoring phase and subsequently for testing the tunnel.

The ADECO-RS approach furnishes basic criteria that should be followed to achieve this purpose satisfactorily. For the sake of clarity they are examined together in the chapter on the monitoring phase, which the interested reader may consult.

On conclusion of the therapy phase, as was done previously for the survey and diagnosis phases, the decisions taken concerning excavation systems, rates, stages and stabilisation instruments are summarised on the geomechanical profile of the tunnel (see the example in Fig. 8.13).

Since the average overall costs (per metre of finished tunnel) of each tunnel section type specified can be easily estimated, it is not difficult at this point for the design engineer to make a reliable forecast of the construction times and costs for the project.

The construction stage

The greatest art in digging a tunnel is to keep the pressure of the mountain at a distance. It is a much greater art than that required later to withstand it. The former is solved with shrewdness, the latter with heavy physical work. (Rziha, 1872)

CHAPTER 9

The operational phase

9.1 Background

The operational phase is that phase during construction in which the flow of the stresses in the medium through which the tunnel advances is deviated. It proceeds hand in hand with the monitoring phase, in which the design and all the operating decisions are monitored and perfected. This last phase is illustrated in the next chapter.

The tunnel section types specified by the design engineer guide the contractor and the project manager in the operational phase in the type of intervention that must be performed to channel stresses around the excavation with the least possible disturbance to the surrounding ground and to guarantee the long and short term stability of the cavity as a consequence.

Consideration is given above all in this chapter to that intervention consisting of the preconfinement and confinement of the cavity recently developed by the author, with the definition from a construction viewpoint, where necessary, of the criteria to follow in order to make the intervention function structurally so that the various parts work together as an indispensable component of the structure of a tunnel lining and are resistant even in the long term. After first briefly describing the operational phases for the implementation of each type of intervention, information is given, where appropriate, on the characteristics that must be possessed by the materials employed. On site experience does in fact show that these are of great importance to the effectiveness of construction and to its life.

Finally a description is given of the types of monitoring designed to:

- verify, during construction, that intervention has been performed correctly and that each operation complies with the design specifications;
- optimise the intensity and the balance of the intervention between the face and the cavity and to fine tune the rate and stages of tunnel advance.

9.2 The basic concepts of the operational phase

The success of the operational phase obviously depends on the accuracy of the predictions made in the diagnosis phase and on the design decisions made as a consequence. However, a considerable part is also played by perfect implementation during tunnel advance, particularly with regard to stabilisation work, which not only has to comply with the design specifications but must also integrate well with the surrounding rock mass.

To achieve this it is important that the operations are performed correctly and promptly, both with regard to placing the individual stabilisation instruments of which it is composed and with regard to the intervention as a whole, which must be continuous and even in order to avoid single points of concentration and stress.

Excavation: the core-face with the charges set before detonation (new high speed/high capacity railway line between Bologna and Florence on the Pianoro tunnel, 1999, ground: weakly cemented sandstone, overburden: ~ 70 m)

Excavation: the rock face after detonation of the charges (Genoa- Ventimiglia railway, S. Stefano tunnel,1984, ground: clayey sandy-schists, overburden: 120 m)

It is also important for each operation to be suited to the geomechanical context in which it is set, with particular regard to those aspects which, although apparently secondary, can affect functionality (particular geostructural situations, the presence of aggressive water, substantial inflows of water, etc.). Particular attention must be paid to the following:

- excavation and advance methods;
- timing, stages and methods of placing stabilisation instruments;
- the geometry of treatment, in terms of individual elements and the treatment as a whole;
- the appropriateness of the materials used, which must rigorously comply with design specifications and, when assembled on site, this must be performed using proper means and according to precise operating instructions (quality control).

It is the duty of the direction of the works to carry out inspections for which he has full discretion. Their purpose is to see that the actual reality matches the design specifications.

Safety is of particular importance for personnel working on site during the operational phase. All plant and equipment, especially lighting and ventilation equipment must satisfy operating requirements perfectly in order to minimise the risk to personnel. Precise safety rules and regulations should be drawn up and scrupulously observed for road and rail transport access, while criteria of maximum caution should be followed for excavation and stabilisation work, which should be performed under the direct supervision of the technicians in charge and account should also be taken of the possible toxicity and aggressivity of the substances employed. To achieve this the work should be planned, from the viewpoint of safety, in the finest detail and be subject to a high degree of surveillance.

9.3 Excavation

The method of tunnel advance adopted can have a considerable influence on controlling the deformation response of the rock mass to excavation and therefore on the formation of an arch effect close to the walls of a tunnel. Consequently, although some of the decisions concerning excavation methods will be made by the contractor on the basis of his organisation and equipment, it is nonetheless essential that these do not conflict with design specifications.

The criterion to follow when making operational decisions concerning excavation methods is always that of causing the minimum disturbance to the environment and the surrounding rock mass. This criterion can be fully complied with:

- by adopting excavation systems appropriate to the materials in question and to the context;
- by employing full face advance whenever possible, giving the face a concave shape and avoiding *overbreak;*
- by performing all the stabilisation work rapidly in full compliance with the tunnel section types specified in the design before and after each excavation step;

Excavation: mechanical excavation of the core-face using a ripper (Rome motorway ring road, Appia Antica tunnel, 1999, ground: granular pyroclastites, overburden: ~ 4 m)

Excavation: mechanical excavation of the core-face using a hammer (Caserta-Foggia railway line, S. Vitale tunnel, 1993, ground: scaly clay, overburden: ~ 130 m)

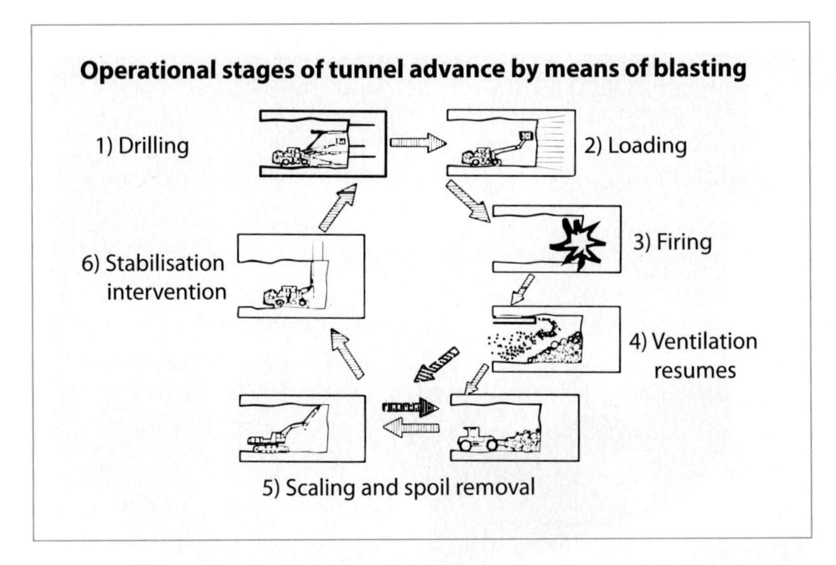

Fig. 9.1

- by preventing, when under the water table, water from standing freely at the face or from running along the bottom of the cavity by using drainage systems to convey it to the destination specified in the design.

When advancing under *stable core-face* or *stable core-face in the short term* conditions, excavation can be performed with the use of explosives (providing no ground improvement or reinforcement ahead of the face is specified) (Fig. 9.1). It is therefore important to adopt controlled blast systems which maintain tunnel profiles. Disturbance to the surrounding material can be minimised by performing careful drilling around the profile of the future tunnel and adopting appropriate priming to obtain even excavation of the cross section and a minimum of overbreak [66]. To achieve this it is necessary:

- take maximum care to ensure that holes are parallel;
- reduce the distance between them;

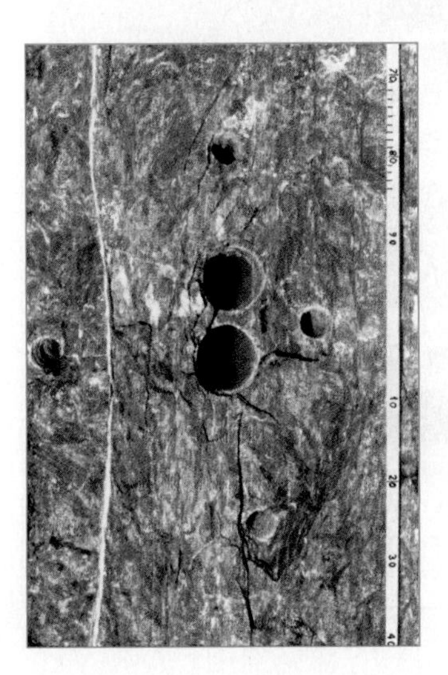

Fig. 9.2 *Parallel cut holes*

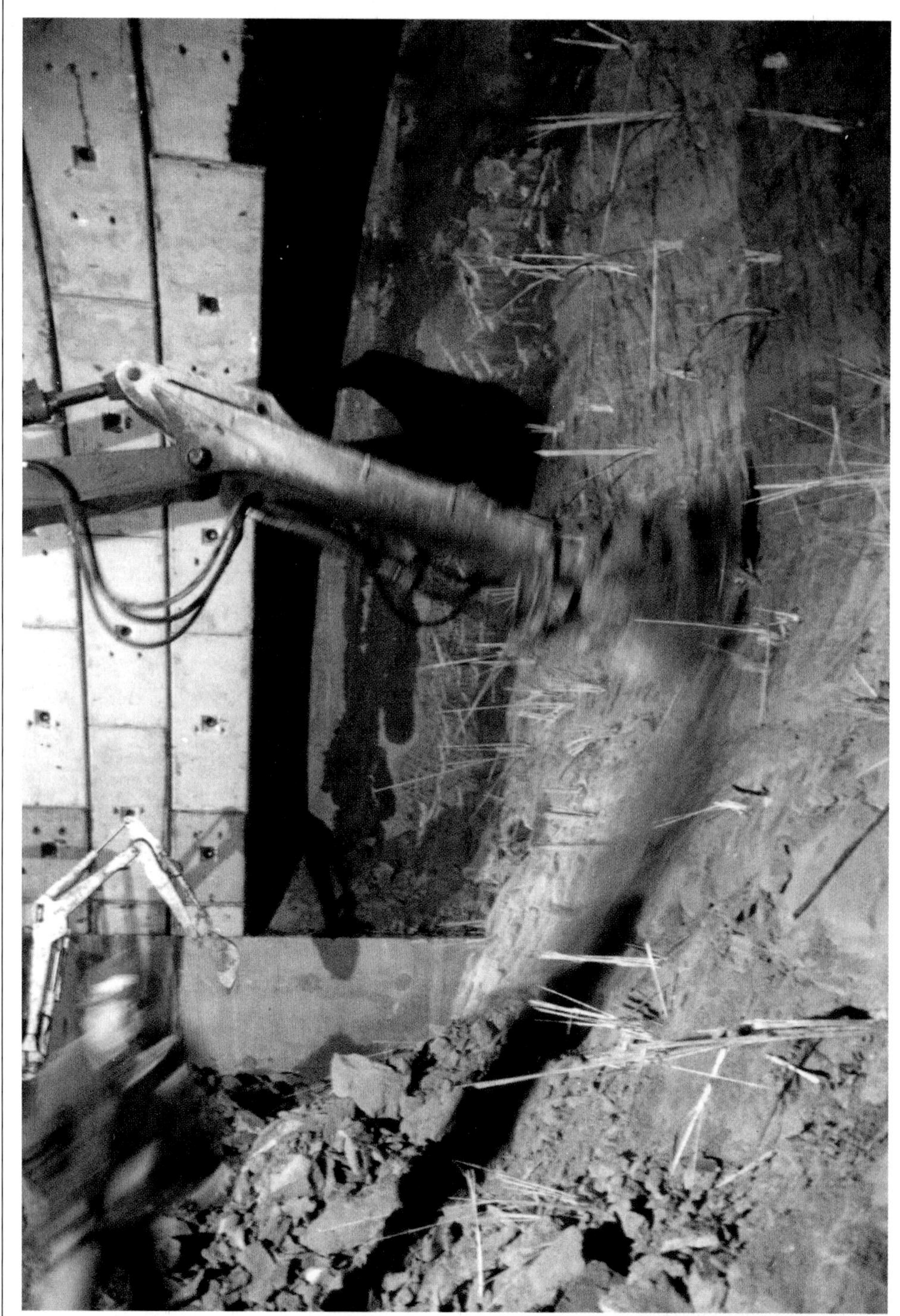

Excavation: giving the face a concave shape (line A of the Rome metro, Baldo degli Ubaldi Station, 1996, ground: clay with sand strata, overburden: ~ 18 m)

- use explosive that is precisely tailored to the situation, for example a detonating cord with a high gram weight (60–80 grams of pentrite per metre).

As concerns other aspects, the design of the rounds to give an optimum profile do not differ from other situations: the type of cut (Fig. 9.2), the distribution and depth of the blast holes, the power of the charges and the time delays must be decided on the basis of the circumstances and characteristics of the materials tunnelled. The definition of an efficient blast hole pattern is a highly specialised task [67].

The walls of the excavation must be scaled with the greatest care immediately after the each round of shots.

In sections of *unstable core-face* tunnel and also in some cases of *core-face unstable in the short term* tunnel, excavation is normally performed in the present of cavity preconfinement intervention such as treating the perimeter using sub horizontal jet-grouting, mechanical precutting etc.. It is best in these cases to excavate material using mechanical means only (excavators, demolition hammers, roadheaders, etc.), without ever resorting to explosives which would damage the intervention placed ahead of the face. It is essential to shape the latter so that is concave in order to encourage the formation of a longitudinal arch effect ahead of the face and to line it with shotcrete, fibre reinforced if necessary, after each tunnel advance.

It may be necessary when tunnel behaviour is characterised by extrusion and convergence of a certain magnitude, to complete the load bearing ring structure by casting the tunnel invert at a short distance from the face; excavation must therefore be shaped appropriately for it. Often an acceleration in deformation phenomena is measured during excavation of the tunnel invert. It is therefore best to cast it immediately, especially if it performs a structural function, or in cases of strongly squeezing ground a strut can be placed at the foot of the steel ribs to stiffen the ring and withstand the pressure before the concrete sets.

9.4 Cavity preconfinement intervention

9.4.1 Cavity preconfinement by means of full face mechanical precutting

Full face mechanical precutting consists of making an incision of a predetermined thickness and length around the line of the extrados of a future tunnel. The incision is made by using a special machine equipped with a chain cutter which moves on a rack and pinion portal that reproduces the shape of the tunnel outline (see photo on next page) and is immediately filled with fibre reinforced sprayed concrete with appropriate additives to give it rapid strength.

Cavity preconfinement by means of full face mechanical precutting: precutting machine constructed for the excavation of tunnels on the Sibari-Cosenza railway line (1985)

Cavity preconfinement by means of full face mechanical precutting (Sibari-Cosenza railway line, tunnel 2, 1985, ground: clay and silt, max. overburden: ~ 110 m)

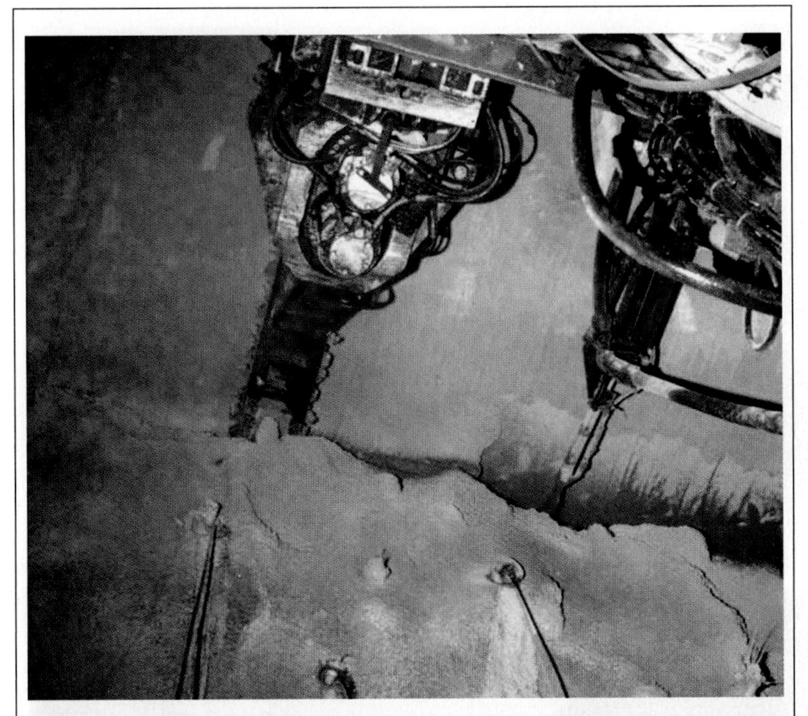

Cavity preconfinement by means of full face mechanical precutting: execution and filling of the precut (tunnels on the high speed/high capacity Milan-Rome-Naples railway line between South Arezzo and Figline Valdarno, 1988, ground: clayey silt, max. overburden: ~ 90 m)

Cavity preconfinement by means of full face mechanical precutting: close-ups of operations (tunnels on the high speed/high capacity Milan-Rome-Naples railway line between South Arezzo and Figline Valdarno, 1988, ground: clayey silt, max. overburden: ~ 90 m)

A pre-lining *tile* is thereby created with a truncated cone shape and good mechanical characteristics, which projects well ahead of the face to provide radial preconfinement of the surrounding ground sufficient to prevent the rock mass around it from loosening.

Its truncated cone shape allows a succession of partially overlapping tiles to be cast (see Fig. 9.4), alternating the placing of each tile with appropriate tunnel advance. A practically continuous arch of tunnel lining is obtained in this manner, which is immediately held rigid by casting the side walls and the tunnel invert to close the ring.

The scope of application ranges from soft rocks to clayey soils and silty-sandy soils, including heterogeneous grounds and aquifers, provided that they allow the incision to remain open, perhaps with a little artificial assistance, for the whole time needed to fill it.

Important characteristics of the method are as follows:

- the almost total elimination of overbreak with a consequent appreciable reduction in the need for backfill injections between the preliminary lining and the ground;
- a reduction in the amount of temporary confinement placed because it is practically all replaced by the precut shell;
- the very high degree of mechanisation and regular advance rates with advantageous repercussions on construction costs and production rates (industrialisation of tunnel advance);
- the construction of a preliminary lining which works as closely as is possible with the statics of the final lining, so that the thickness of the latter can be conveniently reduced, if the specifications allow the effect of that combined action to be taken into account.

The good stability of the core-face is determining for the success of the system. If the protection provided by the precut shell is not sufficient to guarantee this, then it is essential to reinforce the advance core by placing, for example, fibre glass structural elements (Fig. 9.4). It is important when under the water table to conserve the integrity of the core-face by the systematic creation of a fan of drainage pipes ahead of the face, always launched from outside the core-face.

The potential of mechanical precutting has increased considerably since it was first used. Modern technology for cutting the ground and filling the incision will cut tiles to a depth of more than 4.5 m with a thickness of 24 cm, while in terms of tunnel diameter, the system has already been successfully applied on spans of around 21.5 m.

From a construction viewpoint, it is important to limit the size of the incision so that it will support itself for the time needed for the shotcrete to set. Each tile must therefore be cast in separate segments, the number of which will depend on the diameter of the tunnel and the characteristics of the ground to be cut. The cut for each segment must be immediately filled with grout. It is important to never start a new cut until the previous cut has been completely filled.

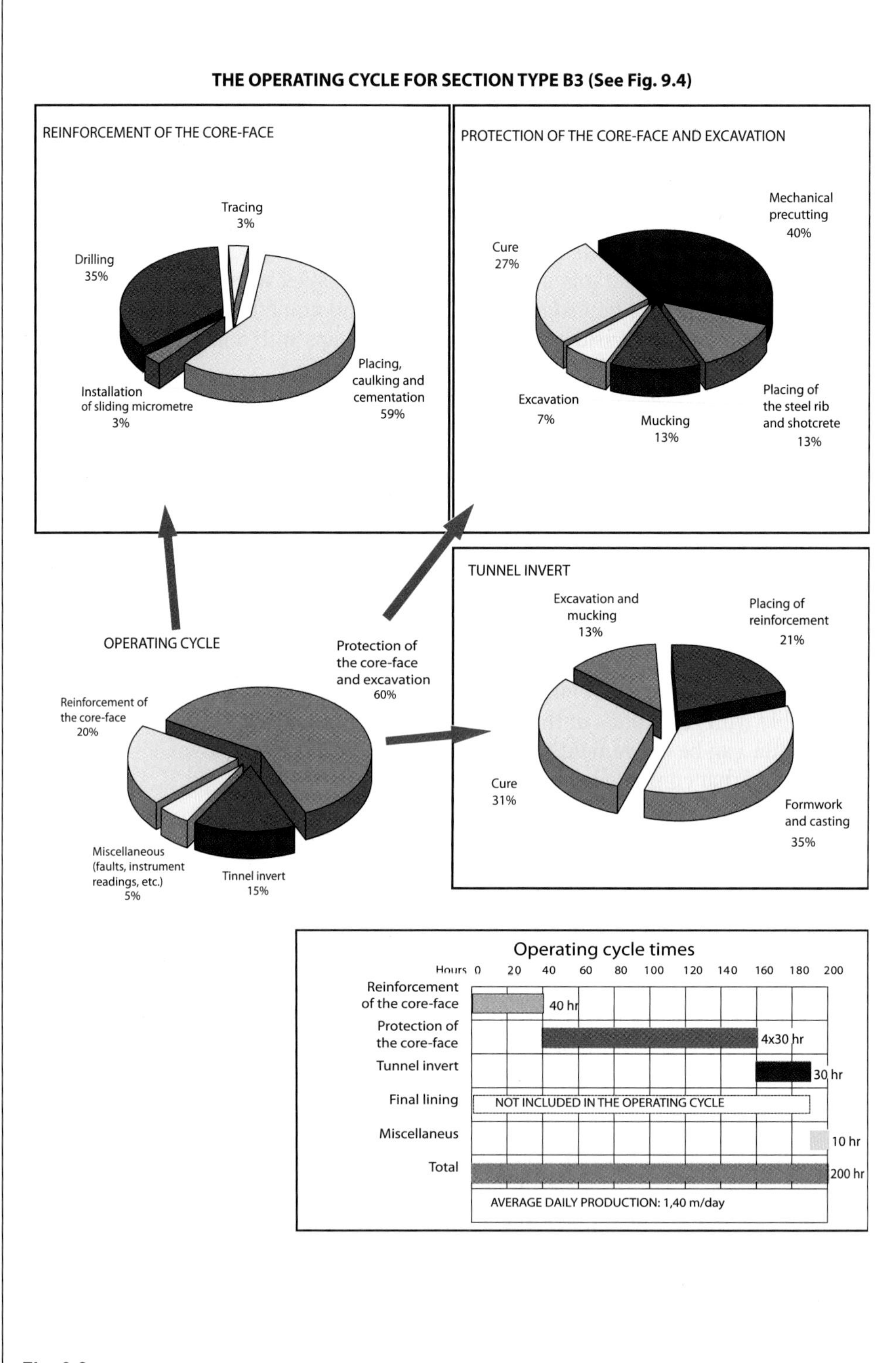
THE OPERATING CYCLE FOR SECTION TYPE B3 (See Fig. 9.4)
REINFORCEMENT OF THE CORE-FACE
Tracing 3%
Drilling 35%
Installation of sliding micrometre 3%
Placing, caulking and cementation 59%
PROTECTION OF THE CORE-FACE AND EXCAVATION
Mechanical precutting 40%
Cure 27%
Excavation 7%
Mucking 13%
Placing of the steel rib and shotcrete 13%
OPERATING CYCLE
Reinforcement of the core-face 20%
Protection of the core-face and excavation 60%
Miscellaneous (faults, instrument readings, etc.) 5%
Tinnel invert 15%
TUNNEL INVERT
Excavation and mucking 13%
Placing of reinforcement 21%
Cure 31%
Formwork and casting 35%
Operating cycle times
Hours 0 20 40 60 80 100 120 140 160 180 200
Reinforcement of the core-face
40 hr
Protection of the core-face
4x30 hr
Tunnel invert
30 hr
Final lining
NOT INCLUDED IN THE OPERATING CYCLE
Miscellaneus
10 hr
Total
200 hr
AVERAGE DAILY PRODUCTION: 1,40 m/day

Fig. 9.3

If difficulties are encountered in making cuts because of the sporadic presence of non cohesive materials, it is possible to give the ground the cohesion required to accept the cut by injecting cement mortar into the non cohesive zones close to the profile of the excavation.

Care must be taken to guarantee the structural continuity of the precut arch by ensuring that no discontinuities are generated between adjacent segments when a new segment is started, which might prejudice the strength of the arch. To achieve this it is important to design the division of the arch into segments so that no joint is actually located in the key.

The creation of each tile is followed by the excavation of a corresponding portion of the core-face, for which the statics function is assumed by the pre-arch of concrete. Excavation is performed using roadheaders or excavators under the direct control of an operator to avoid the risk of damaging the tile of the prelining that has just been placed and it must proceed to a depth that is less than that of the prelining in order to ensure that the core-face is always protected. A new tile is placed after each tunnel advance which must overlap with the previous tile by at least half a metre.

The stability of the cavity is guaranteed immediately behind the face, by the precut arch of shotcrete. Immediately after the core has been removed, the precut arch is normally reinforced with steel ribs and shotcrete and the ring is closed by casting the kickers and the tunnel invert at not more than 1.5 ∅ from the face. The final lining should then be completed at a distance of no more than 4–5 ∅ from the face.

The operational stages for full face, mechanical precut, construction are illustrated schematically in Fig. 9.5, which illustrates the very common case where this technology is applied in combination with reinforcement of the core-face.

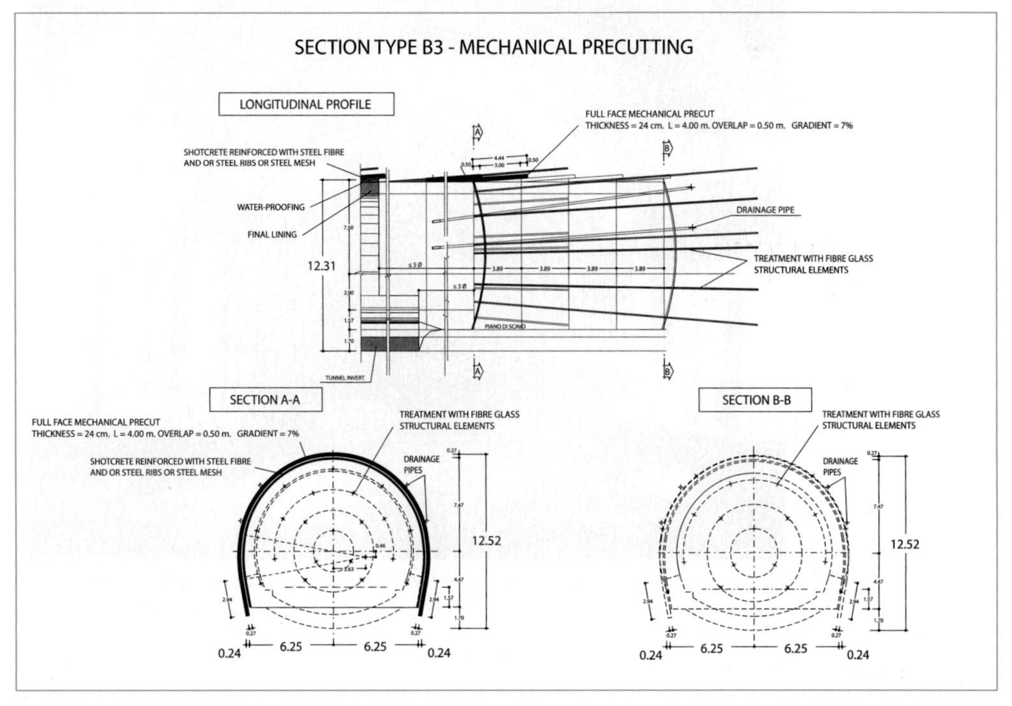

Fig. 9.4

Cavity preconfinement by means of full face mechanical precutting: view of the core-face protected by a mechanically precut shell (Sibari-Cosenza railway line, tunnel 2, 1985, ground: clay and silt, max. overburden: ~ 110 m)

Cavity preconfinement by means of full face mechanical precutting: view of preliminary lining consisting or mechanically precut tiles in the same tunnel as above

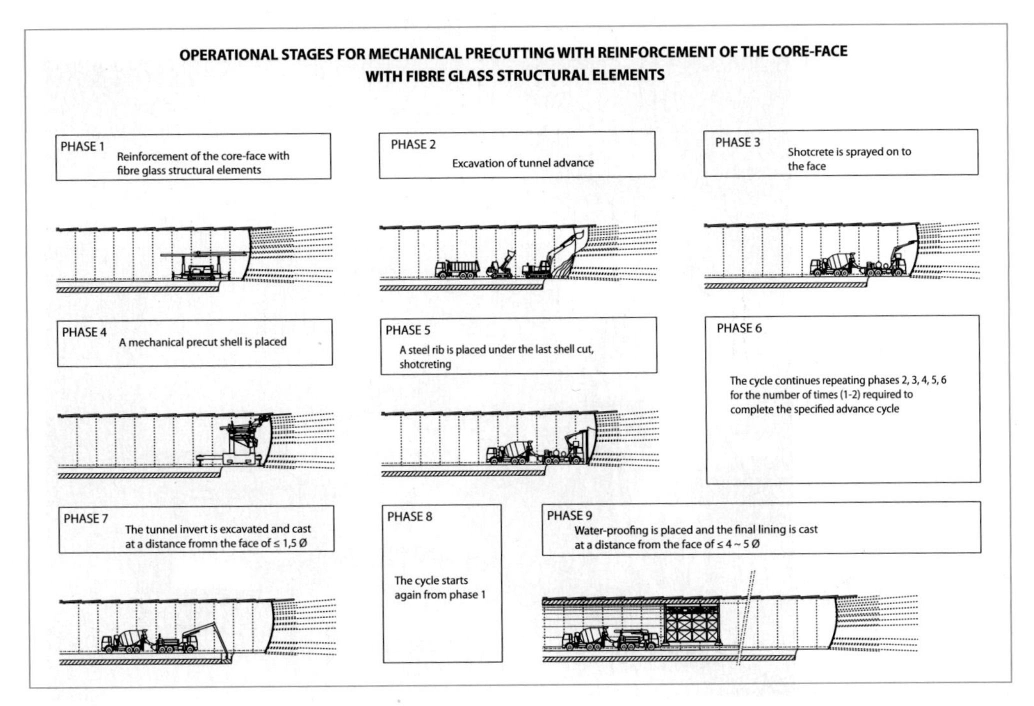

Fig. 9.5

Theoretical studies and on site experience have shown that the cross section of the treatment must be as circular as possible if stresses are to be channelled evenly around the precut arch with no local accumulations of stresses.

As with all construction work it is wise to perform quality controls to ensure that the actual work done complies with the design specifications.

The following tests should normally be performed with mechanical precutting:

- the compressive and flexure strength of the shotcrete grout should be monitored. More specifically quality control tests should be performed at regular intervals so that a graph can be plotted of the strength values over time, which starts as soon as it has set and continues until it is fully cured, in order to be able to make comparisons with the design specifications in terms of the increased strength the material must develop;
- monitoring the quality of the precut arches by taking core bore samples from precut arches with particular attention paid to bonding between castings;
- monitoring the ductility of the shotcrete, if it is fibre reinforced.

Cavity preconfinement by means of full face mechanical precutting: the equipment (tunnels on the high speed/high capacity Milan-Rome-Naples railway line between South Arezzo and Figline Valdarno, 1988, ground: sandy silt, max. overburden: ~ 90 m)

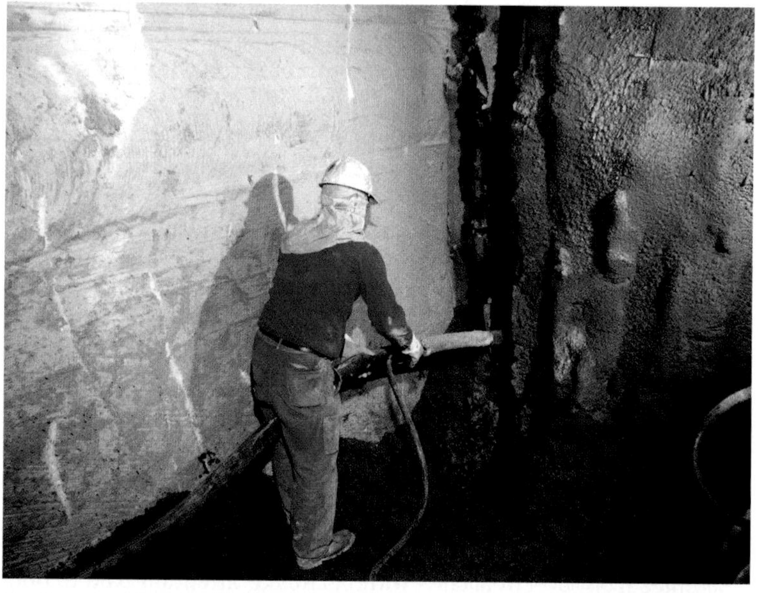

Cavity preconfinement by means of full face mechanical precutting: manual filling of the precut with shotcrete (tunnels on the high speed/high capacity Milan-Rome-Naples railway line between South Arezzo and Figline Valdarno, 1988, ground: sandy silt, max. overburden: ~ 90 m)

9.4.2 Cavity preconfinement using pretunnel technology

The innovative *pretunnel* technology can be considered to all effects and purposes an extension of precutting technology and a compromise between TBM mechanised advance and technology involving excavation after first performing cavity preconfinement. It combines the benefits of the latter with the advantages of the former.

The innovation lies in the possibility of constructing the tunnel lining before it is excavated without the need for preliminary ground improvement and reinforcement around the cross section to be excavated, nor for the placement of a preliminary lining in successive stages consisting of steel rib and fibre reinforced shotcrete

Like the mechanical precutting technology already described, pretunnel technology involves the creation of a truncated cone of concrete (see Fig. 9.6) in advance ahead of the face, which can become an integral structural part of the final tunnel lining or even replace it altogether, depending on the thickness adopted and the design decisions [68].

Pretunnel technology basically differs from mechanical precutting in the following ways:

- tile thickness (40–80 cm instead of 18–24 cm) and length (8–10 m instead of 3–4 m);
- use of concrete instead of shotcrete or mortar as the filler;
- possibility of excavation while the lining is being cast;
- the ring of the lining tile is closed immediately with a temporary tunnel invert in concrete or steel.

The advantages that can be gained from this new technology are clear:

- excellent protection of the core-face, thanks to the presence of the concrete lining placed in advance and therefore better stability of the excavation;

Fig. 9.6 *Pretunnel operations (high speed/high capacity Rome-Naples railway line, Castello 1 tunnel)*

Cavity preconfinement by means of full face mechanical precutting: 21.5 m span mechanical precutting machine (line A of the Rome metro, Baldo degli Ubaldi Station, 1996, ground: clay with sand strata, overburden: ~ 18 m)

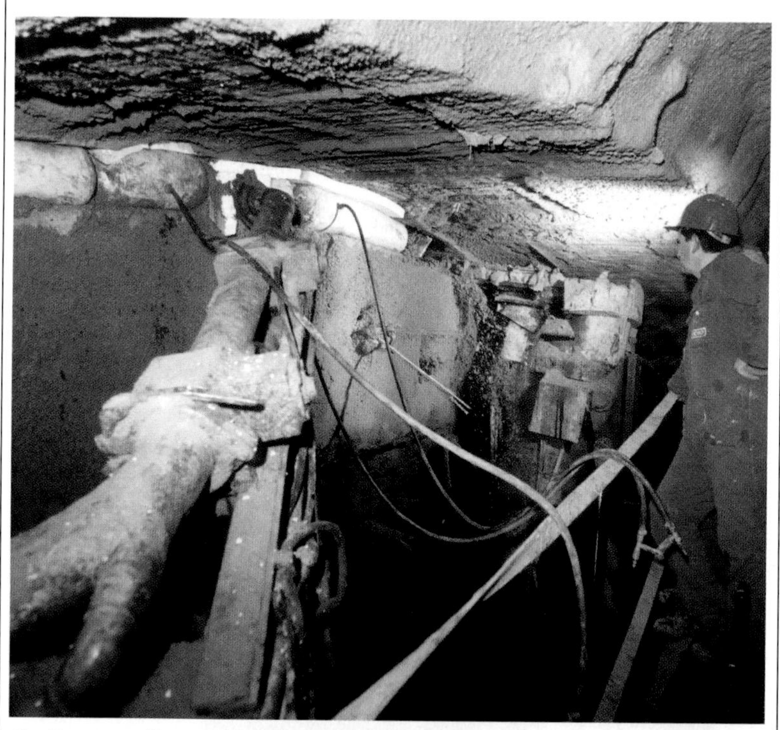

Cavity preconfinement by means of full face mechanical precutting: execution and filling of the 21.5 m span precut with pumped concrete confined using pneumatic formwork (line A of the Rome metro, Baldo degli Ubaldi Station, 1996, ground: clay with sand strata, overburden: ~ 18 m)

- maximum safety for site personnel who work under the protection of a lining already in place;
- considerable reduction in extrusion and convergence phenomena;
- drastic reduction in surface settlement, in the case of tunnels with reduced overburdens, thanks to the bearing capacity and structural rigidity of the concrete arch that is placed.

It can also be more competitive than conventional methods in terms of construction times and therefore also in terms of costs. This is especially true when conventional tunnel advance would require long and costly ground improvement operations.

When pretunnel technology is employed, tunnel construction is achieved in a sequence of not more than four operational phases:

- ground reinforcement (if necessary) of the core-face using fibre glass structural elements;
- the construction of a concrete tile in advance;
- excavation of the core of ground enclosed within the concrete tile that has been built;
- the creation of a temporary tunnel invert at the foot of the tile to close the ring of the lining.

Tunnel advance therefore occurs by simply removing the ground enclosed inside the concrete tile that has been created which performs the function of a lining. It proceeds for a length that is less than the depth of the tile in order to allow overlap between contiguous tiles and to give a margin of safety with regard to the stability of the core-face

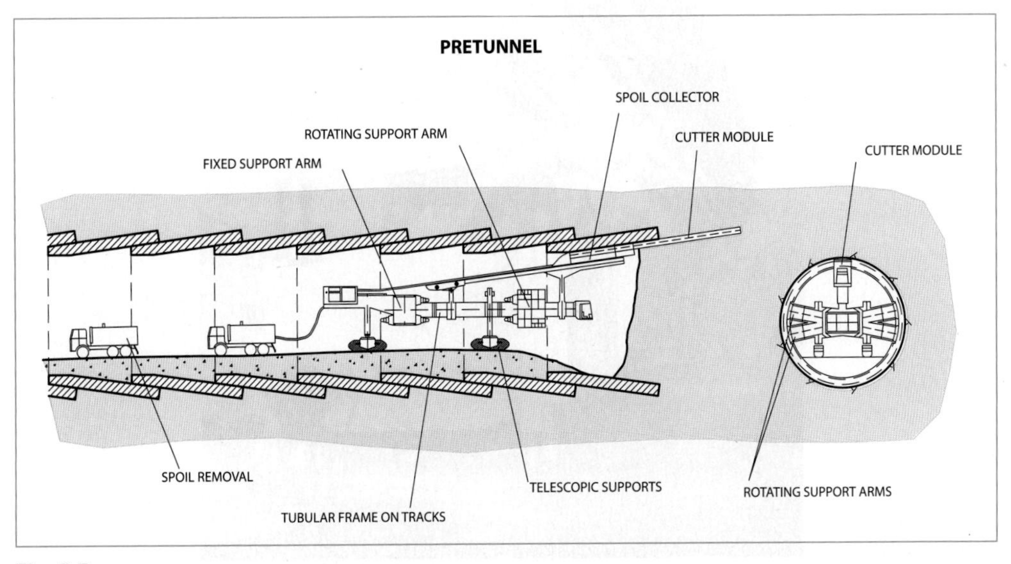

Fig. 9.7

Pretunnel:
the machine

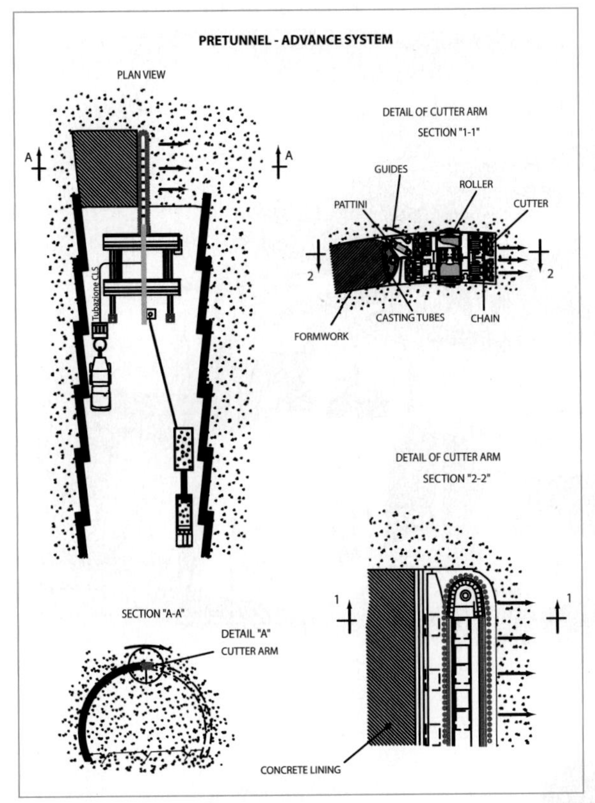

Fig. 9.8

which, as in the case of mechanical precutting, is also an indispensable condition for success in the execution of pretunnel technology.

Pretunnel technology is closely related to the equipment designed to perform this work. It consists of a self-propelled tubular frame to which two telescopic rotating arms are fitted, capable of rotating through 270° (Fig. 9.7). The arms act as the supports for a boom on which the cutter module runs longitudinally.

The machine moves on two tracks connected to the frame by special supports which allow the machine to be centred on the longitudinal axis of the tunnel, once it is in front of the face. The machine is locked in place during the various operational phases by special telescopic stabilisers.

The cutter module consists of two chains, assembled on a single box shaped rigid arm to which disks and picks are fitted in proportion to the excavation conditions expected with adjustable geometry and angle of inclination (Fig. 9.8). The chains, which can operate at different velocities will rotate in opposite directions for drill type cutting and synchronously for continuous excavation. This allows the machine to be adapted for the excavation of all types of material including heterogeneous grounds (soils with blocks and/or erratic boulders, grounds with rocky interbedding, alternating layers of rocks and detritus, etc.).

Pretunnel: experimentation on a test site (1996)

Pretunnel: details of the cutter head

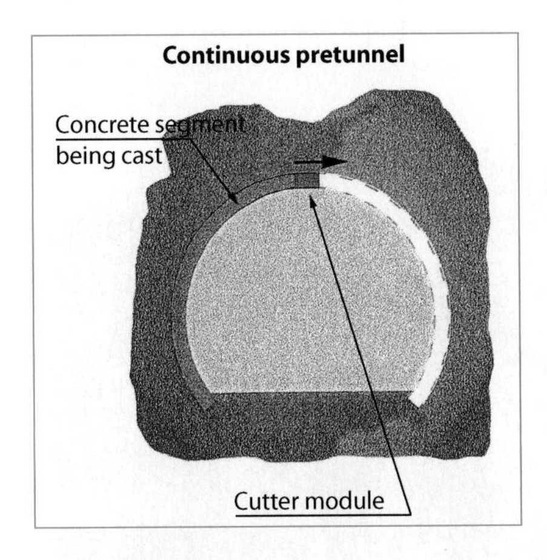

Fig. 9.9

Finally a formwork box is fixed to the back of the cutter module and runs up and down with the module itself and concrete is pumped from inside it through special jet tubes to fill the section that has been excavated.

Drills positioned on the front section of the machine are used to perform reinforcement of the advance core with the necessary fibre glass structural elements.

From an operational viewpoint the accurate positioning of the machine is a necessary condition for it to function correctly because once the geometry of the rotation of the cutter module is set, as determined by the rotating telescopic arms, the work is performed by the rotation of these on their axis so that the trajectory followed by the cutter module is that of a truncated cone.

The first operation consists therefore of positioning the machine in front of the face, using the telescopic supports so that the axis of rotation of the equipment corresponds to the axis of the truncated cone to be excavated. The boom that carries the cutter module is then positioned by setting the extension and longitudinal inclination so that at the end of each stage the extrados of each excavation coincides with the intrados of the previous one.

Two different operating methods can be used to create the concrete tile. With the **continuous pretunnel** technique (Fig. 9.9), it is performed after opening an appropriate slot in the ground using the cutter arm. To perform this the cutter tool first penetrates the ground longitudinally and then once the penetration is complete it moves sideways to follow the contour of the cross section to be excavated.

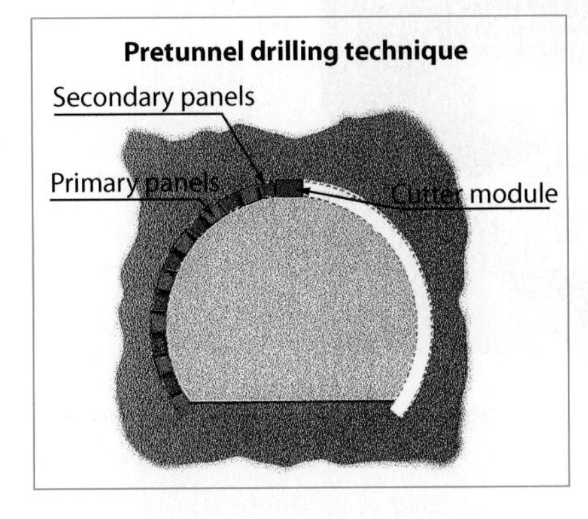

Fig. 9.10

Pretunnel: the start of excavation (Castello 1 tunnel, high speed/high capacity Rome-Naples railway, 1996, ground: cineritic tuff, max. overburden: 9 m)

The slot can be filled with concrete immediately behind the cutter module while cutting is being performed. The concrete is pumped from inside the formwork box fixed to the cutter module on the opposite side to that where cutting takes place and becomes the "engine" which drives the cutter module with the pressure that it exerts.

With the **pretunnel drill** technique (Fig. 9.10), the tile is constructed by creating contiguous panels which overlap transversally.

The ground is cut exclusively by boring. The cutter module is pushed longitudinally so that only the ground ahead of its front end is excavated.

When the excavation of each panel is complete, it is filled with concrete from a jet tube fitted inside the box frame of the cutter module. The pressure of the jet pushes the cutter module backwards, which guarantees that the cavity excavated is evenly filled.

The operating sequence consists of creating primary opening panels and then secondary closing panels which are set between the primary panels and partially overlap with them to guarantee the structural continuity of the arch.

The maximum depth of a cut is between 8.50 and 10.00 m, depending on the thickness of the cutter module. Two cutter modules have been used so far, one with a thickness of 80 cm the other of 45 cm; the minimum internal radius can be varied continuously from 5.75 to 7.40 m, depending on the cross section of the tunnel to be bored. Once the time necessary for concrete to harden has elapsed, a new tunnel advance starts. This must be of between 6.50–8.00 m to allow adequate an overlap between contiguous arches of around 1.50–2.00 m.

The structural continuity of the pretunnel must be ensured by taking great care with bonding between segments and especially with the closing segments. It is important for the concrete used to have the right characteristics to ensure that it will meet design specifications for continuity, impermeability and strength.

Standard quality controls for fibre reinforced concrete are performed (which include not only compressive strength tests but also those for flexural and punching shear strength).

Samples are taken while concrete is being cast and they are tested for compliance with the design specifications for the granulometry curve and the quantity of fibres.

It may be useful to plot the ultimate strength of the concrete as a function of time, because, as has been said, tunnel advance cannot take place until the pretunnel arch has hardened sufficiently to guarantee absolute safety for excavation.

Once the lining is complete, it is best to take some radial core bore samples to check the quality and continuity and that it adheres perfectly to the ground behind it. Particular attention should be paid to inspecting the quality and strength of the lining in the joints between separate castings of individual segments.

The electronic instrumentation fitted on the pretunnelling machine enables both the geometrical parameters of the machine and the main construction process parameters to be monitored in real time to provide the information required to take corrective action if necessary when constructing the arch.

All the data transmitted by the control and measurement equipment is collected and memorised in a centralised data acquisition system so that it can be subsequently reprocessed to reconstruct the history of the construction phases of the pretunnel arch as a guarantee of the quality of the work performed.

Experience acquired in pretunnel construction with 80 cm arch thickness has revealed not only the technical and structural validity of the method, but also the great complexity of the work with repercussions on construction times and costs.

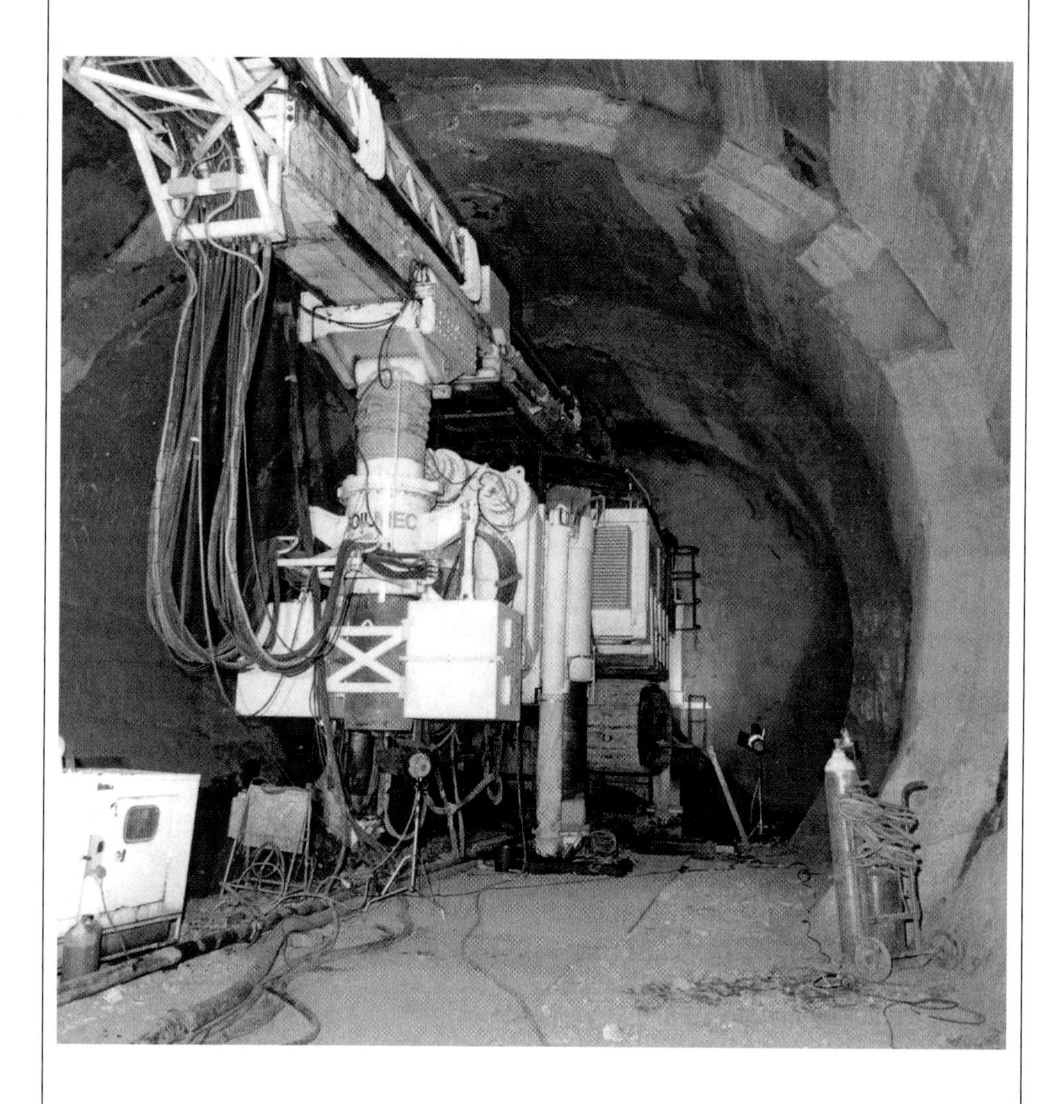

Pretunnel: the machine in operation (high speed/high capacity Rome-Naples railway line, Castello 1 tunnel, 1996, ground: cineritic tuff, max. overburden: 9 m).

The construction of arches with a thickness of up to 45-50 cm is very much less demanding and probably also responds more closely to real operational needs, constituting a competitive alternative in many situations. In this case the cutter module consists of one chain only. Given the reduced thickness of the excavation, the equipment is able to cut the ground with the power from the hydraulic motors alone and it does not therefore require the pressure from the concrete required for incisions of greater thickness. In many cases this allows excavation and casting operations to be performed separately, thereby reducing the operational and logistical complexity of performing them simultaneously.

If this system is employed it is therefore indispensable for the ground excavated to have sufficient cohesion to ensure that the walls of the incision will support themselves at least until they are filled with concrete. In order to reduce the time during which excavations are left open and also the probability of failure at the same time, it is good practice to cut the truncated cone in stages, dividing it into between five and seven sections, varying in length depending on the radius of the tunnel. This method of operation also simplifies the systems adopted to contain the concrete and confine it in the face considerably.

9.4.3 Cavity preconfinement after strengthening the core-face with fibre glass reinforcement

The technology in question consists of *dry drilling* a series of holes into the face, sub parallel to the axis of the tunnel, evenly distributed over the face and normally longer than the diameter of the tunnel. Special fibre glass reinforcement is then inserted into the holes and this is immediately injected with cement mortar. When the remaining length of the reinforcement inserted in the core-face after tunnel advance is no longer sufficient to guarantee preconfinement of the cavity (a circumstance that can be recognised immediately by careful reading of extrusion measurements) another series is placed.

The use of fibre glass for the reinforcement plays a determining role in the success of the operation, because this material combines properties of high strength with great fragility, which makes it easy to break it during tunnel advance using the same mechanical tools as those employed to excavate the ground. Stranded metal cable on the other hand could not be used to reinforce the ground because good shear strength is essential for the system to function properly.

The parameters that characterise this intervention are the length, frequency, overlap, cross section and geometrical distribution of the reinforcement. The geometry, as defined at the design stage, (see the example in Fig. 9.11) must be scrupulously complied with at the construction stage in order to ensure maximum effectiveness. Some corrections may be made during construction to adapt the treatment to fit local structural features of the rock mass not detected in the survey phase.

The technology can be used in cohesive and semi cohesive soils and, with a few measures taken to ensure the integrity of the drill holes, even in soils with very poor cohesion. If it is well designed and performed, it produces an appreciable improvement in the stress-strain characteristics of the core-face which makes it a structural element with a predictable and controllable stress-strain response, capable of developing extremely effective cavity preconfinement action.

If needed, it can be combined to advantage with other advance reinforcement techniques such as mechanical precutting, pretunnelling, etc.

Cavity preconfinement after reinforcement of the core-face with fibre glass reinforcement: drilling for insertion of the reinforcement (Caserta-Foggia railway line, S. Vitale tunnel, 1993, ground: scaly clay, overburden: ~ 130 m)

Cavity preconfinement after reinforcement of the core-face with fibre glass reinforcement: caulking with rapid setting cement before injection (Caserta-Foggia railway line, S. Vitale tunnel, 1993, ground: scaly clay, overburden: ~130 m)

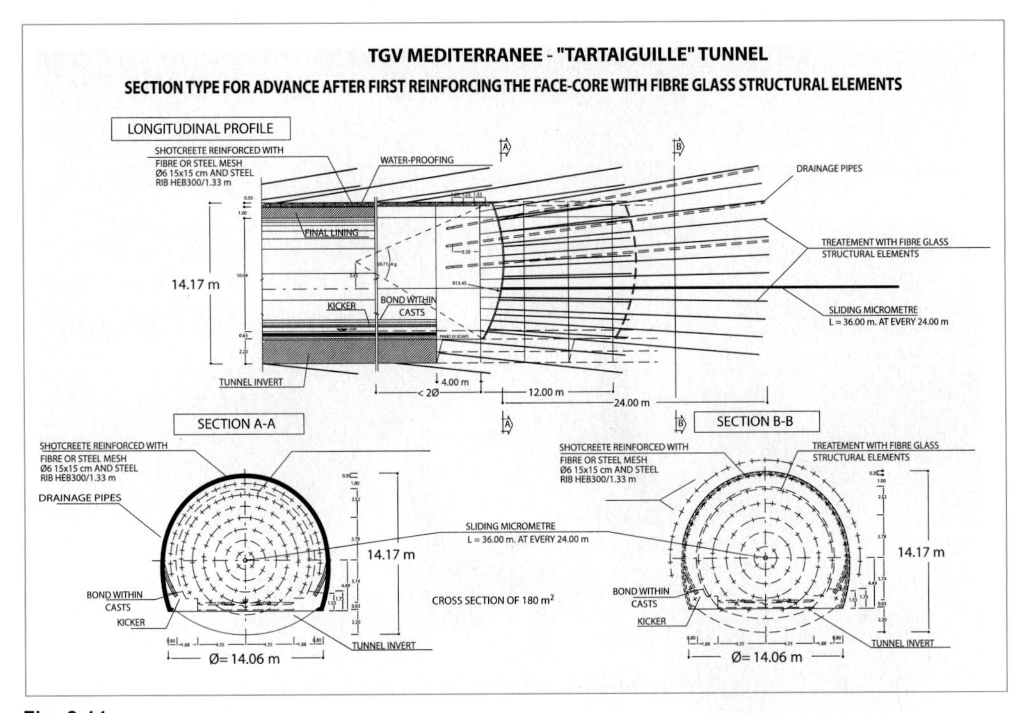

Fig. 9.11

The materials used to assemble the fibre glass reinforcement (*pool extrusion* process) have a thermoset polyester resin base, with up to 50% fibre glass content by weight.

They can consist of tubes (Fig. 9.12) or bars of varying length and can be joined when necessary if care is taken. The surface of tubular reinforcement is normally milled with a spiral groove to improve its resistance to slipping out and it can have a maximum length of from 15 to 18 m in a single piece, while the bars can be much longer even extending even for dozens of metres. If they must be joined, then it is important to use sleeve collars and glue (epoxy resin) which will ensure that the joint has the same tensile and shear strength as the rest of the reinforcement (it should be left to dry for a few days to be sure that the joint will not leak).

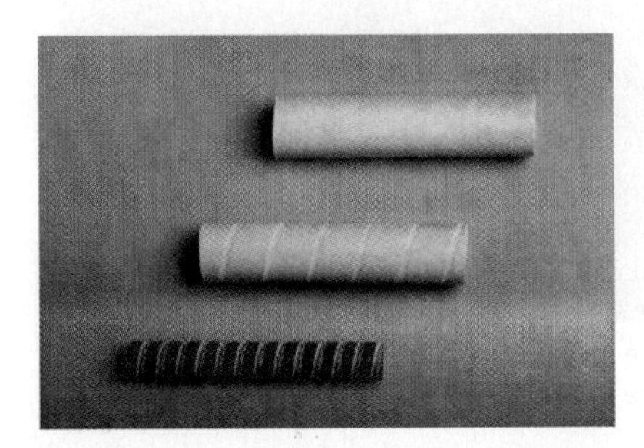

Fig. 9.12 *Tubular types of fibre glass reinforcement*

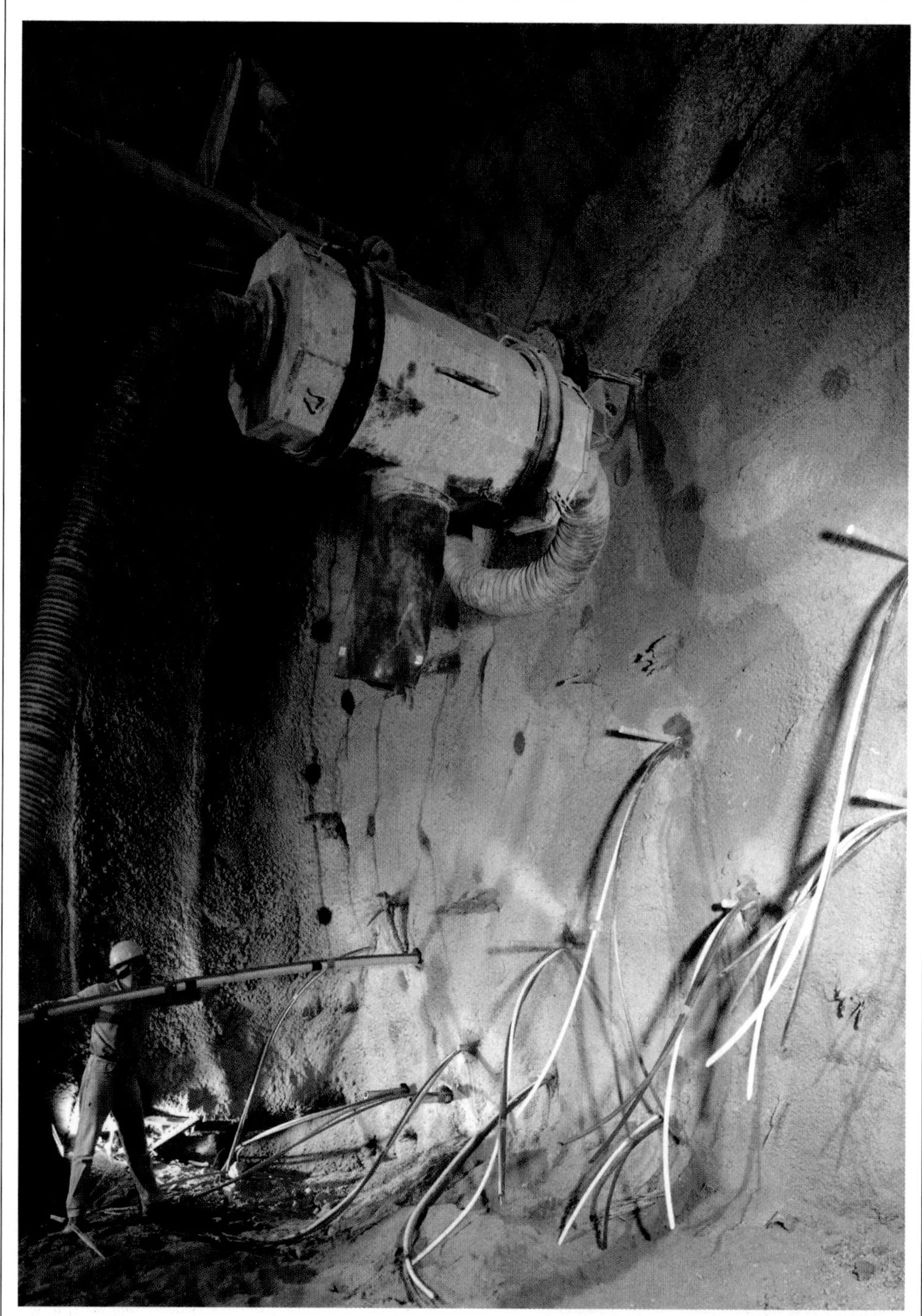

Cavity preconfinement after reinforcement of the core-face with fibre glass reinforcement: manual insertion of reinforcement in the core-face (Caserta-Foggia railway line, S. Vitale tunnel, 1993, ground: scaly clay, overburden: ~ 130 m)

The following is normally used:

- tubular reinforcement ∅ 60/40 mm, thickness 10 mm;
- structural elements assembled on site with a diameter of between 60 and 100 mm, consisting of three plates (40 mm × 7 mm or 40 mm × 5 mm) fitted on special plastic spacers.

The latter come in different forms (see Fig. 9.13). The plate type structural elements are assembled on site using special steel bands (Fig. 9.14). It is important at this stage to prevent the elements from coming into any contact with mud, clay, grease, oil or other materials which might compromise the friction grip of the reinforcement once they are cemented.

Since this type of intervention is performed mainly in sections of tunnel with *unstable face* or *stable face in the short term* behaviour, it is absolutely essential for the face to be

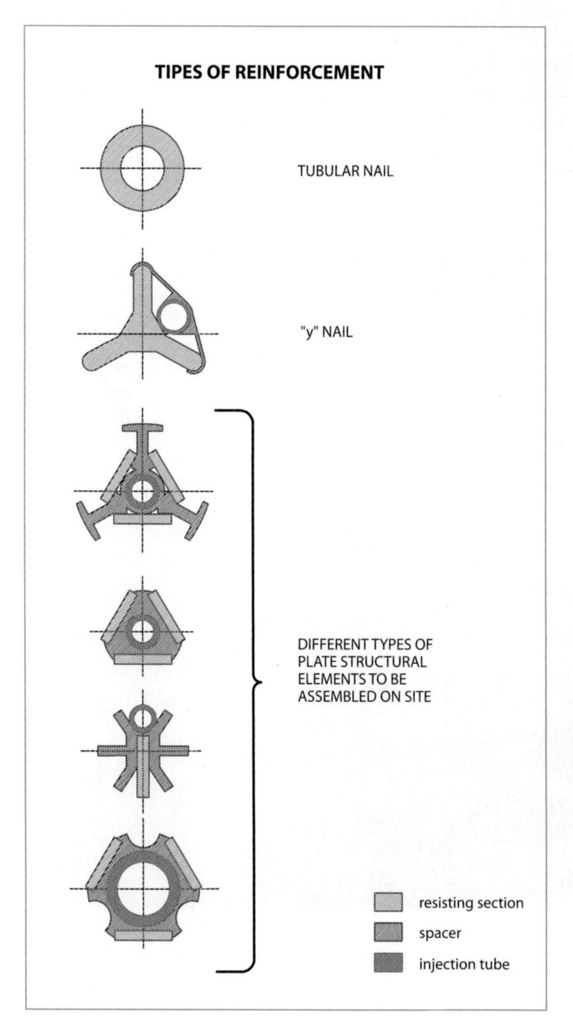

Fig. 9.12

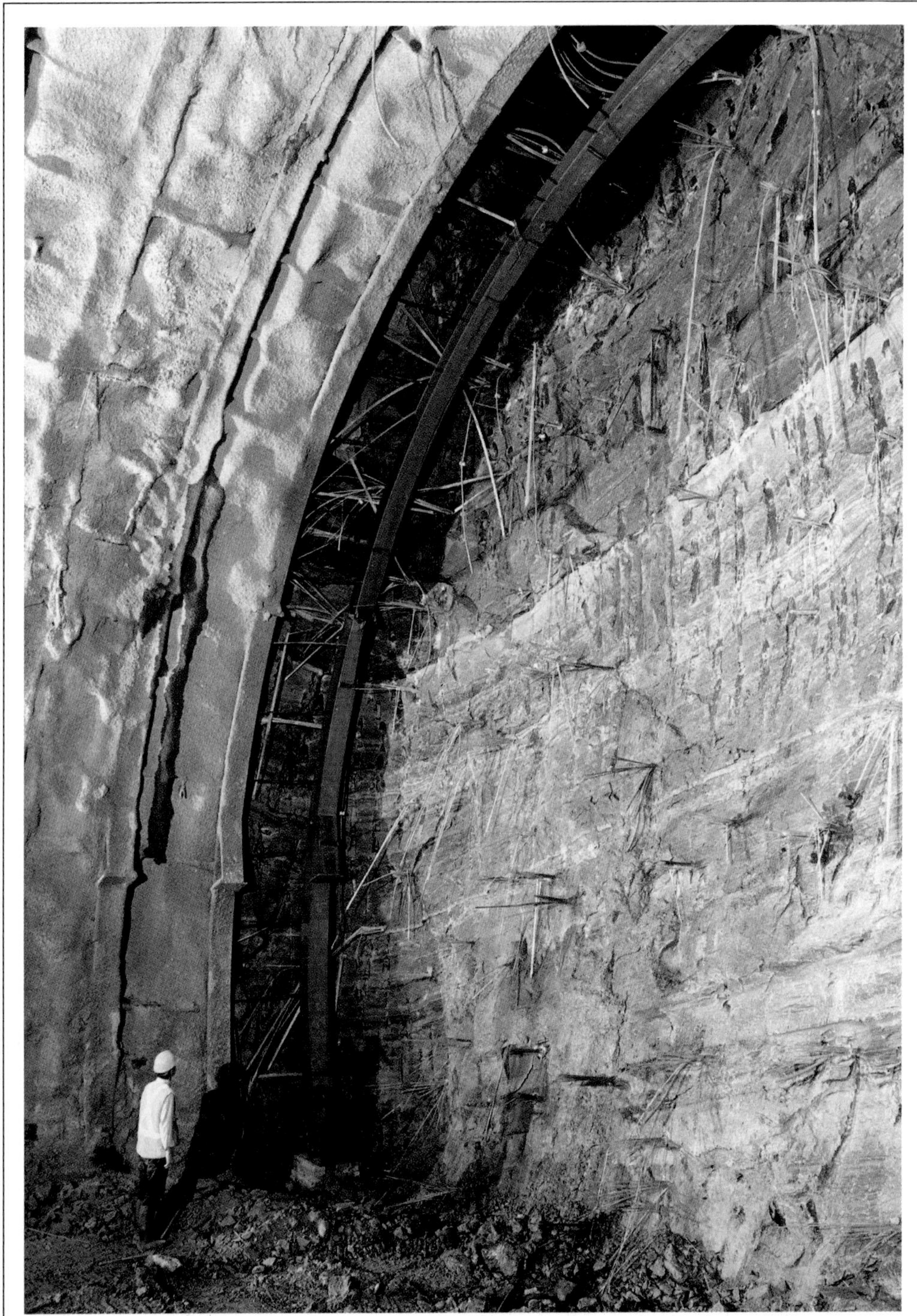

Cavity preconfinement after reinforcement of the core-face with fibre glass reinforcement: detail of the reinforced face (TGV Méditerranée, Lyon-Marseilles railway line, Tartaiguille tunnel, 1997, ground: clay, max. overburden: ~ 110 m)

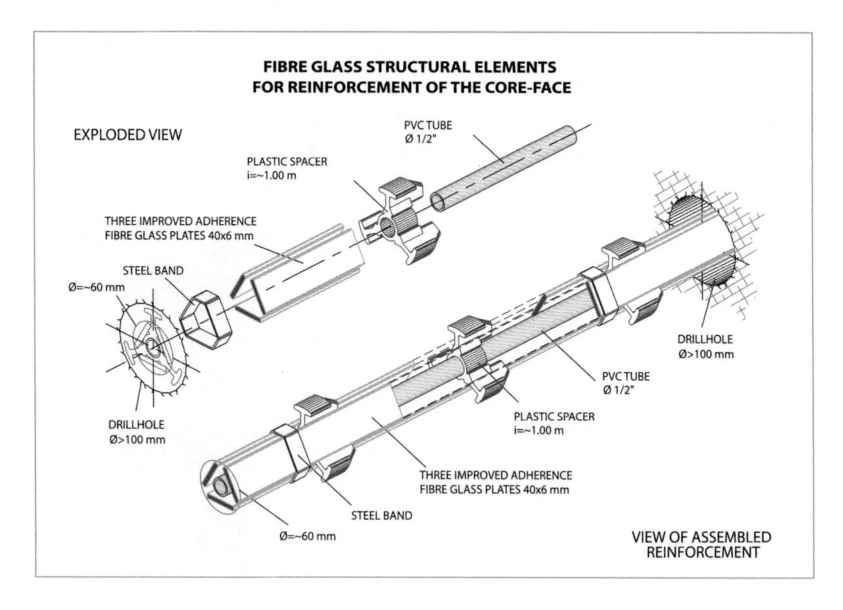

Fig. 9.14

given a concave shape and lined with shotcrete in order to encourage the channelling of stresses around the tunnel and to guarantee the necessary safety for site personnel.

Work to reinforce the core-face using fibre glass structural elements starts with the stage in which the holes for the reinforcement are drilled. It is a very delicate phase and in many cases it is actually decisive for the effectiveness of the technique.

The diameter of the hole must be as small as possible and at the same time allow the subsequent reinforcement and injection operations to be easily and properly performed. Whichever drilling system is adopted, it must be performed entirely dry, preferably with helical equipment, because water based drilling fluids would damage the ground around the borehole irreparably.

The reinforcement must be inserted immediately after the holes are drilled. In no case whatsoever may more than 4 or 5 holes be drilled before inserting the reinforcement, because the core-face would be irreversibly weakened.

Fig. 9.15 *Tubular fibre glass reinforcement*

FULL FACE ADVANCE AFTER REINFORCEMENT OF THE CORE-FACE

Operational stages

PHASE 1

Reinforcement of the core-face by means of fibre glass structural elements

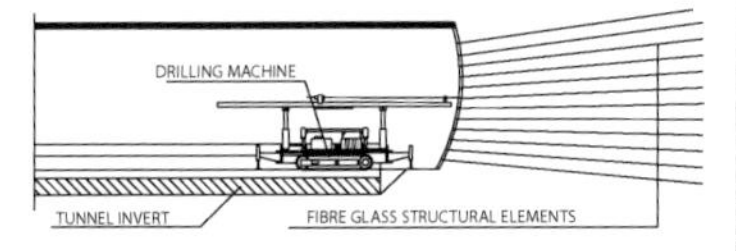

PHASE 2

Excavation step
(m. 0,70 ÷ 1,00 full face)

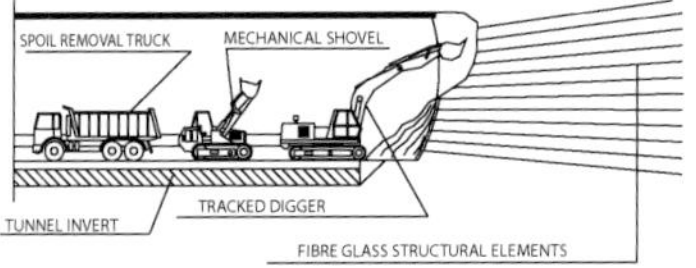

PHASE 3

Placing of shotcrete on the face and surface of the cavity to protect walls of the tunnel

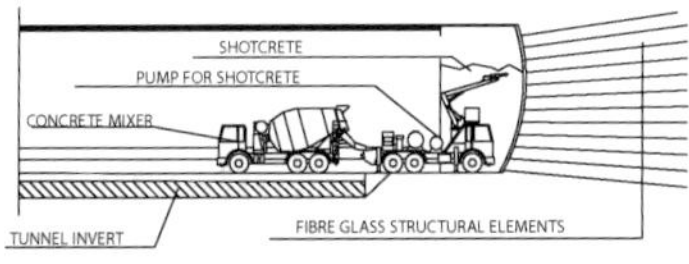

PHASE 4

Placing of the steel rib and connecting reinforcement in the section of tunnel excavated in phase 2

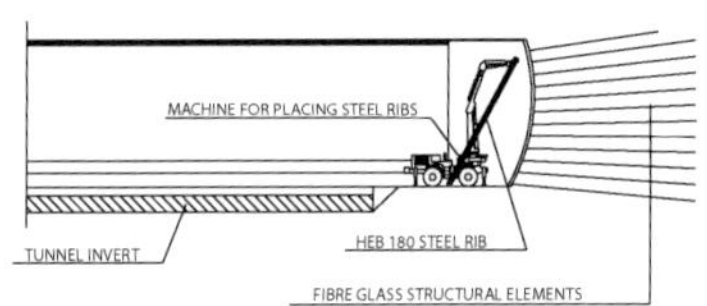

PHASE 5

Placing of the preliminary lining in fibre reinforced shotcrete

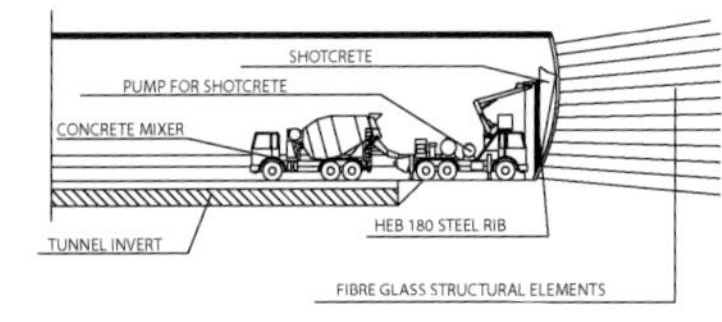

PHASE 6

Excavation and casting of side walls and tunnel invert close to the face

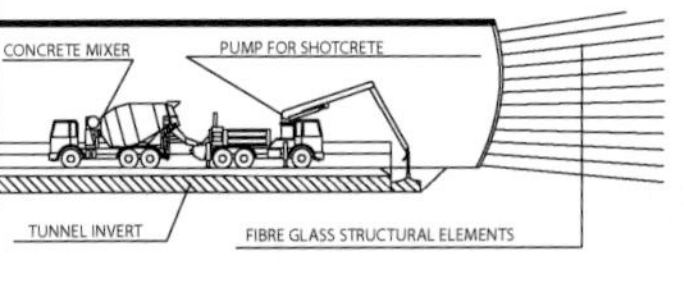

PHASE 7

Casting of the final lining at distance from the face of ≤ 4 ÷ 5 Ø

PUMP FOR SHOTCRETE
LINING
MOBILE FORMWORK
CONCRETE MIXER
TUNNEL INVERT
FIBRE GLASS STRUCTURAL ELEMENTS

Fig. 9.16

Fig. 9.17 *Detail of fibre glass reinforcement injected with cement grout*

The reinforcement must be injected with mortar immediately after it is inserted into the boreholes in order to make it active at once.

The cementation stage also requires particular care, especially when the reinforcement is inclined in an upwards direction. When simple cementation of a structural element in a hole is required, the mixtures most commonly adopted are of cement containing additives to accelerate setting times and to prevent shrinkage (e.g. *flowcable*).

The choice of cement is fundamental for fast setting times and special test beds should be prepared: the minimum strength after 48 hours should be greater than 5 MPa.

Pressure injections are performed using stable cement mixtures with fine or superaerated cements. The stability of the water/cement mixture is obtained by preparing modest quantities of bentonite to be mixed before the pressure injection.

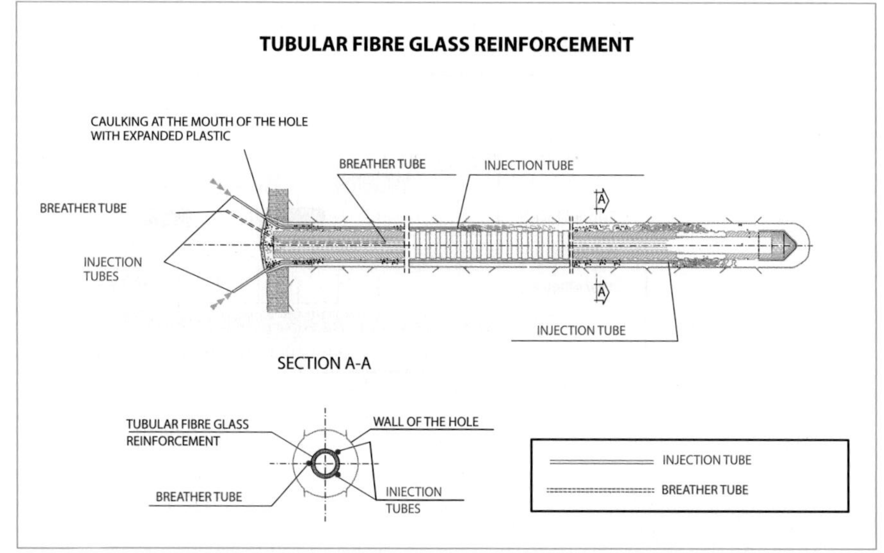

Fig. 9.18

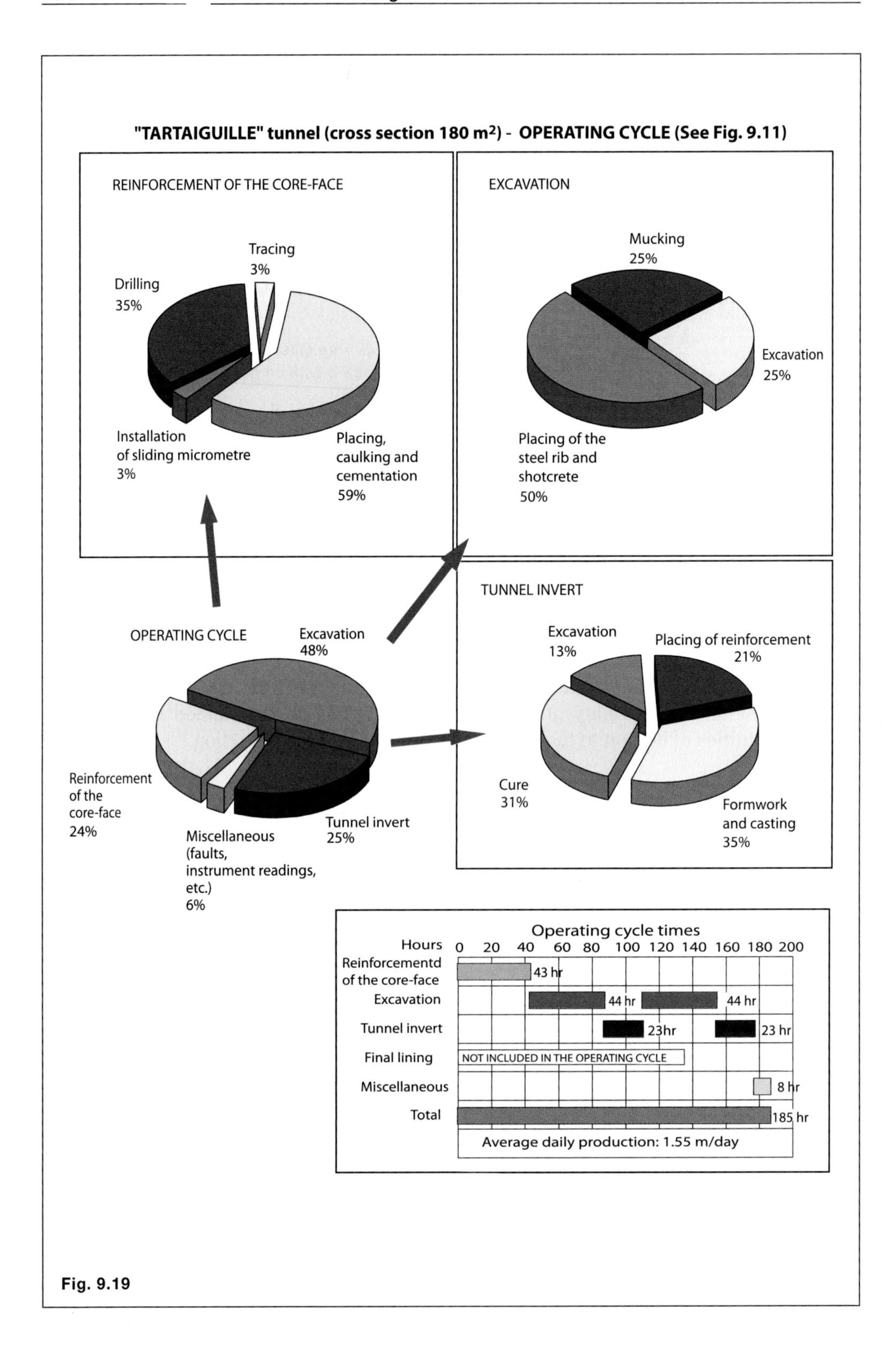

"TARTAIGUILLE" tunnel (cross section 180 m²) - OPERATING CYCLE (See Fig. 9.11)
REINFORCEMENT OF THE CORE-FACE
Tracing 3%
Drilling 35%
Installation of sliding micrometre 3%
Placing, caulking and cementation 59%
EXCAVATION
Mucking 25%
Excavation 25%
Placing of the steel rib and shotcrete 50%
OPERATING CYCLE
Excavation 48%
Reinforcement of the core-face 24%
Miscellaneous (faults, instrument readings, etc.) 6%
Tunnel invert 25%
TUNNEL INVERT
Excavation 13%
Placing of reinforcement 21%
Cure 31%
Formwork and casting 35%
Operating cycle times
Hours 0 20 40 60 80 100 120 140 160 180 200
Reinforcementd of the core-face
43 hr
Excavation
44 hr
44 hr
Tunnel invert
23hr
23 hr
Final lining
NOT INCLUDED IN THE OPERATING CYCLE
Miscellaneous
8 hr
Total
185 hr
Average daily production: 1.55 m/day

Fig. 9.19

Chemical mixtures are rarely used because they are complex to prepare and the chemicals may interfere with ground water.

Recently, expansive mixtures have been used (containing aluminate additives), which reinforce the ground with an effect equivalent to an injection under a pressure of 1–1.5 MPa, without the use of valved reinforcement or pressure injections.

The fibre glass elements must be cemented using special injection and breather tubes to prevent air bubbles from forming and badly cemented sections as a consequence. It is performed by pumping the cement mix into the hole via a delivery tube until it flows back through the breather tube, which is kept open. At this point the latter is closed (by simply bending it) and the pressure is raised to up to approximately 5 bar, which is maintained for a few seconds before pumping is stopped.

Injection must be performed:

- from the end of the borehole towards the wall of the face when the holes are inclined in a downwards direction;
- vice versa when they are inclined upwards.

If the design requires high pressure cementation, then a valved tube of the collar sleeve type must be inserted in the structural element (Fig. 9.20). When the cementation phase to form a “sheath” is complete, pressure injection is performed for each individual valve with a double shutter according to the design specifications.

In order to prevent injected mortar from leaking from boreholes, the heads of the fibre glass reinforcement must be sealed, with care always being taken to ensure that the breather tube is always positioned above the injection rod. Sealing (which is also referred to as calking) can be performed in three different ways:

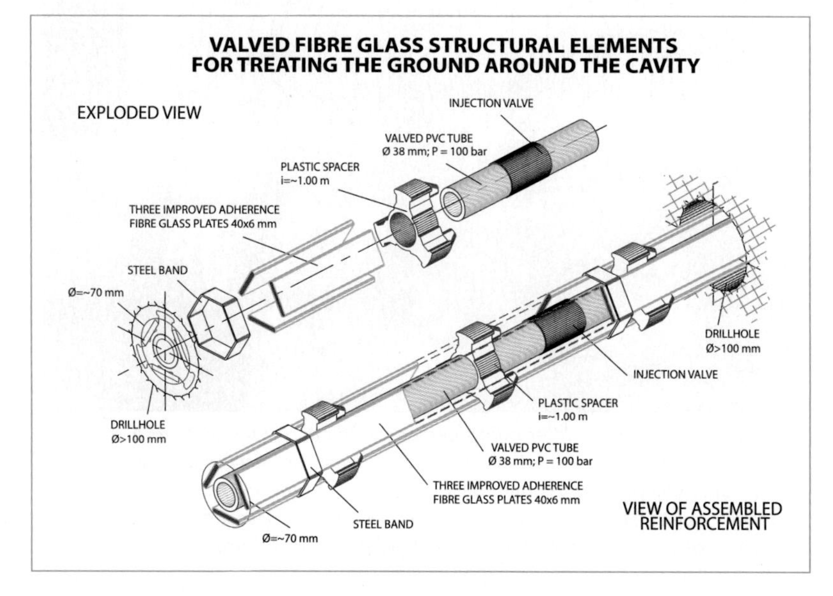

Fig. 9.20

Cavity preconfinement after reinforcement of the core-face with fibre glass reinforcement: view of the reinforced face with a span of greater than 20 m and a cross section of 194 m² (Rome ring road motorway, Appia Antica tunnel, 1999, ground: granular pyroclastites, overburden: ~4 m)

Cavity preconfinement after reinforcement of the core-face with fibre glass reinforcement: the face reinforced with fibre glass elements inserted into micro-columns of ground created by means of jet-grouting (high speed/high capacity railway line between Bologna and Florence, Firenzuola tunnel, 2000, ground: silt and silty sand with interbedded gravel, overburden: ~ 40 m)

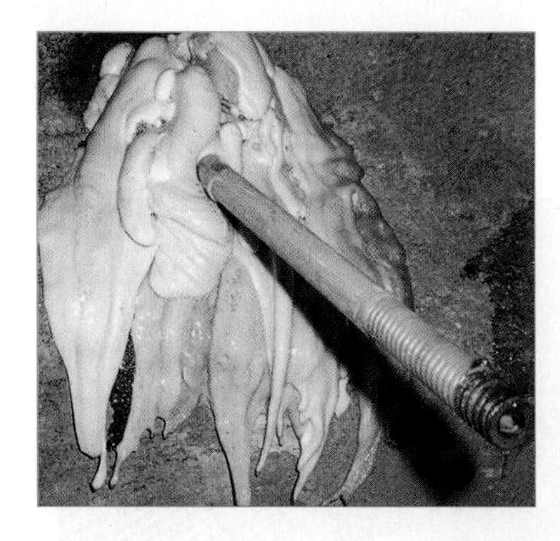

Fig. 9.21 *Caulking tubular fibre glass reinforcement with polyurethane foam*

- by using rapid setting cement;
- by using polyurethane foams (Fig. 9.21);
- by inserting a truncated cone plug of polystyrene in the borehole.

No plates or other devices are needed for sealing.

Simple controls must be performed systematically (for each set of structural elements) during construction to check that the treatment has worked properly. The most important are:

- the quality of the injection mixture is tested to ensure that it complies with design specifications, with regard above all to how rapidly it sets;
- checking the quantity of mixture injected for each element inserted;
- pullout tests to check element-mortar-ground adherence and the relative anchoring capacity.

9.4.4 Cavity preconfinement by means of truncated cone 'umbrellas' formed by sub horizontal columns of ground side by side improved by jet-grouting

Jet-grouting is a ground treatment system that consists of injecting controlled volumes of cement mix into the material to be improved at very high pressure (from 300 to 600 bar) through nozzles of appropriate diameter. Although there are various methods of performing it (monofluid, two fluid and three fluid) practically the only treatment used for horizontal work in tunnels is the monofluid system (injection of the cement mixture only). As opposed to conventional injections based mainly on the permeation and impregnation of fluids, which are therefore limited by the absorption capacity of the ground, jet-grouting is based mainly on hydrofracturing (*claquage*). This breaks up the ground by the mechanical action of a jet of fluid under very high pressure and therefore of high velocity, which mixes, compacts and consolidates it into a well defined form. The mechanical characteristics of the ground are increased by the treatment to the point where its permeability and strength properties are comparable to those of concrete.

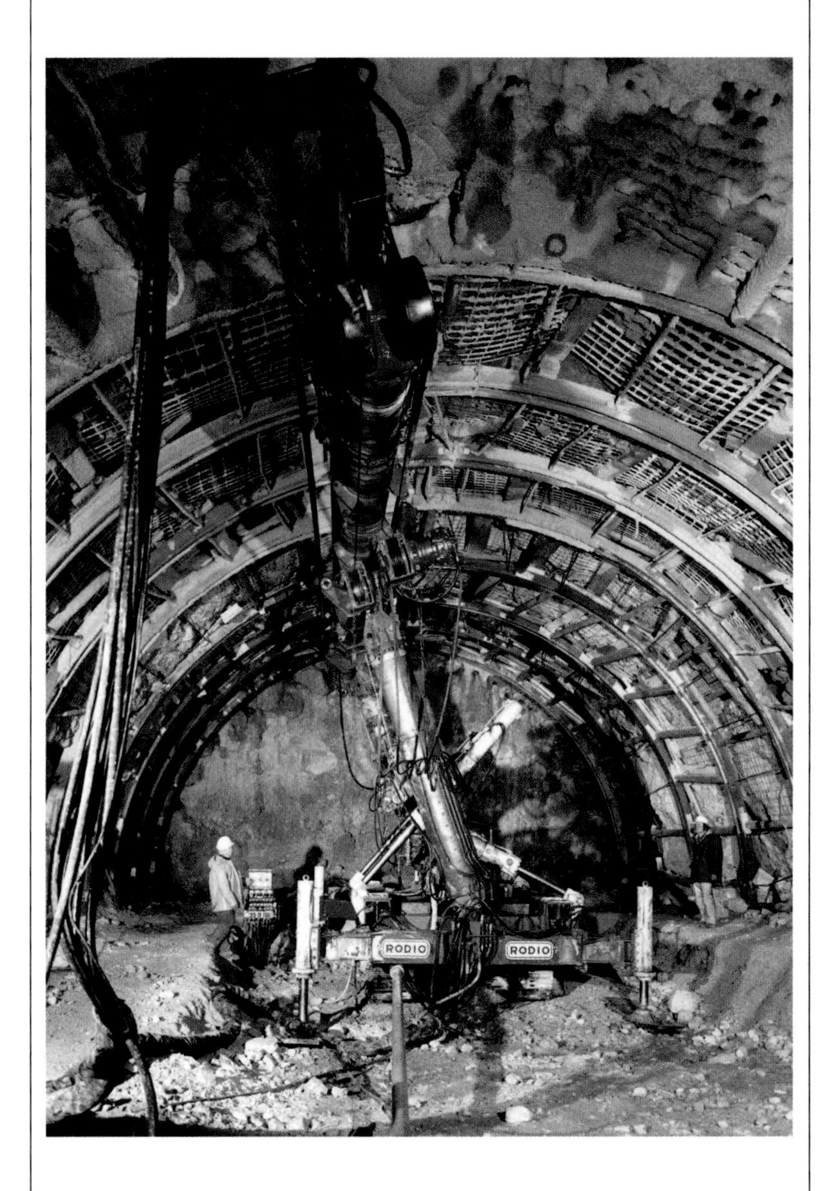

Cavity preconfinement by means of truncated cone 'umbrellas' consisting of sub horizontal columns of ground, side by side, improved by jet-grouting: a moment during the first ever use of the system in the world on the Campiolo tunnel (Udine-Tarvisio railway line, 1983, ground: rubble slopes, overburden: ~ 100 m)

The operating sequence for the procedure can be divided into two distinct stages:

- drilling to a depth not less than the diameter of the tunnel with a special injection device (*monitor*) fitted on the end of the drilling rod;
- injection through the monitor while simultaneously extracting and rotating it at the programmed velocity.

The volumes of improved soil obtained as a result of the operation have the shape of a column.

Experience indicates that this technology can be used in all granular soils and in cohesive soils with a shear strength that allows them to be broken up by the jet of the mixture under very high pressure. It is therefore also possible to treat heterogeneous soils and to guarantee uniform consolidation and impermeability. Provided the head is not greater than 5–6 m, the presence of a water table in a hydrostatic regime does not at all compromise the treatment, which remains possible even in a hydrodynamic regime up to velocities of around 1 cm/sec, if appropriate accelerators are used for grout setting.

The geomechanical strength of the ground treated depends primarily on the ratio of water to cement in the grout that is injected and on the granulometry curve of the *in situ* ground. Strength values generally vary between 12 and 18 MPa for sands and gravels, while the values for fine soils will fluctuate between 2 and 14 Mpa with high dosages of cement.

The dimensions of the improved ground that can be obtained depend on the characteristics of the *in situ* natural ground and on the operating parameters of the treatment (Fig. 9.22). It is normally possible to obtain columns with a diameter of between 0.40 and 0.80 m.

Sub horizontal jet-grouting is used in tunnelling to create a series of columns of improved ground side by side ahead of the face around the profile of the extrados of the tunnel to be excavated, with each sequence telescoping out of the previous one (Fig. 9.23). An arch of improved ground with considerable strength is created in this manner which:

- provides protection of the ground inside the advance core in a longitudinal direction lightening the load on it and giving it stability;

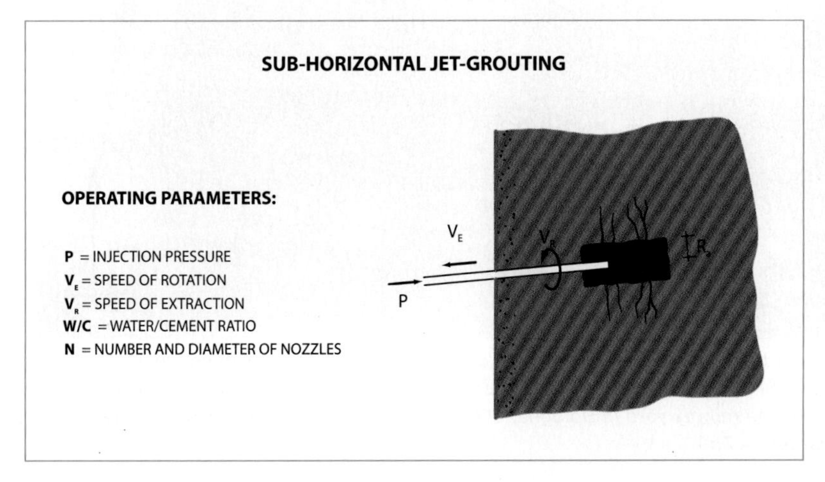

Fig. 9.22

Cavity preconfinement by means of truncated cone 'umbrellas' consisting of sub horizontal columns of ground, side by side, improved by jet-grouting: portal of the double bore underpass of the Ravone railway yard in Bologna (2000, ground: silty sand and graven in a sandy-silty matrix, overburden: ~ 7 m)

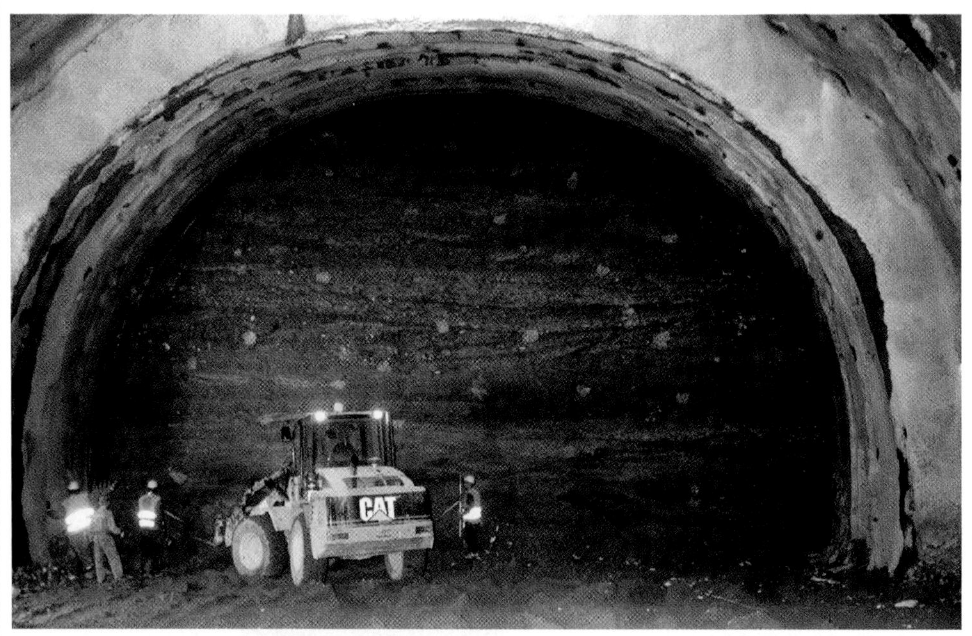

Cavity preconfinement by means of truncated cone 'umbrellas' consisting of sub horizontal columns of ground, side by side, improved by jet-grouting: view of the reinforced core-face (underpass of the Ravone railway yard in Bologna, 2000, ground: silty sand and gravel in a sandy-silty matrix, overburden: ~ 7 m)

- produces cavity confinement action in a transverse direction, sufficient to prevent the ground around it from decompressing and consequent deformation from occurring. It therefore allows subsequent tunnelling operations to proceed under the protection of an arch effect that is already operational and therefore in complete safety.

Naturally, less decompression and less deformation also means less pressure on the final lining, which therefore need not be so thick.

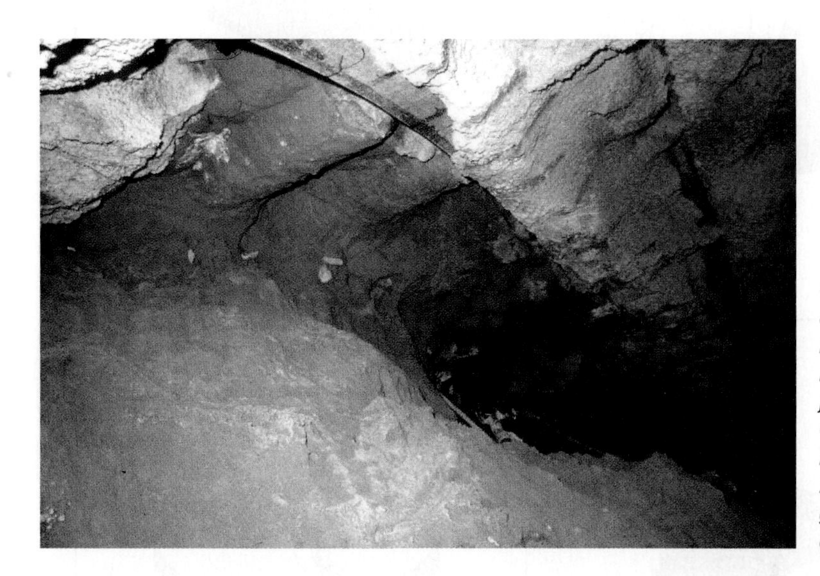

Fig. 9.23 *Detail of columns of ground improved by means of horizontal jet-grouting (Milan-Chiasso railway line, M. Olimpino 2 tunnel, 1984, ground: alluvial, overburden: ~ 50 m)*

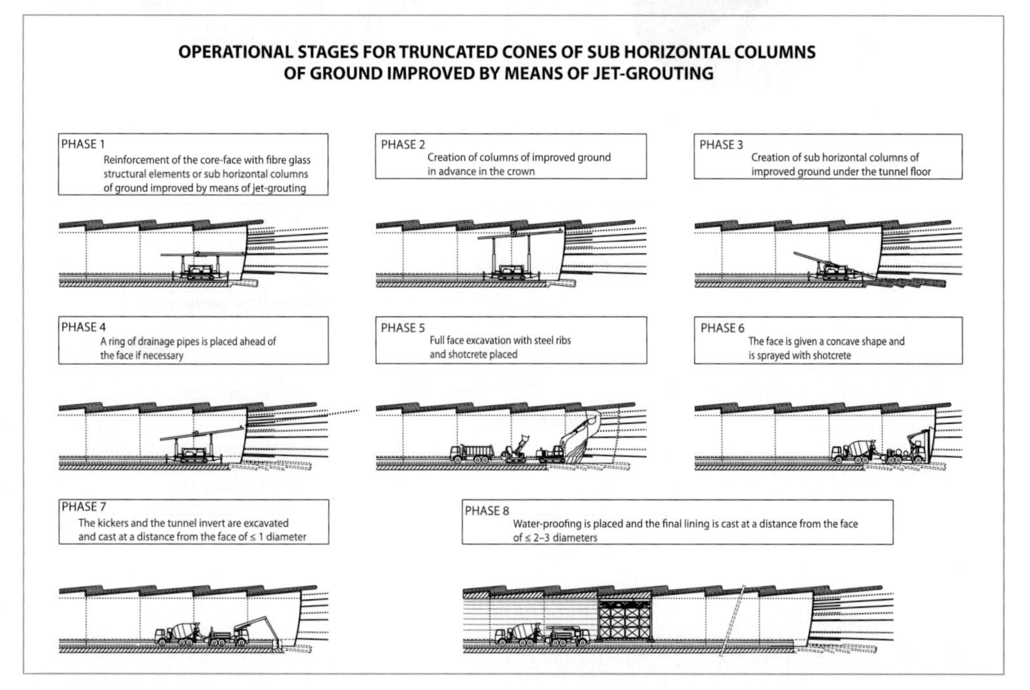

Fig. 9.24

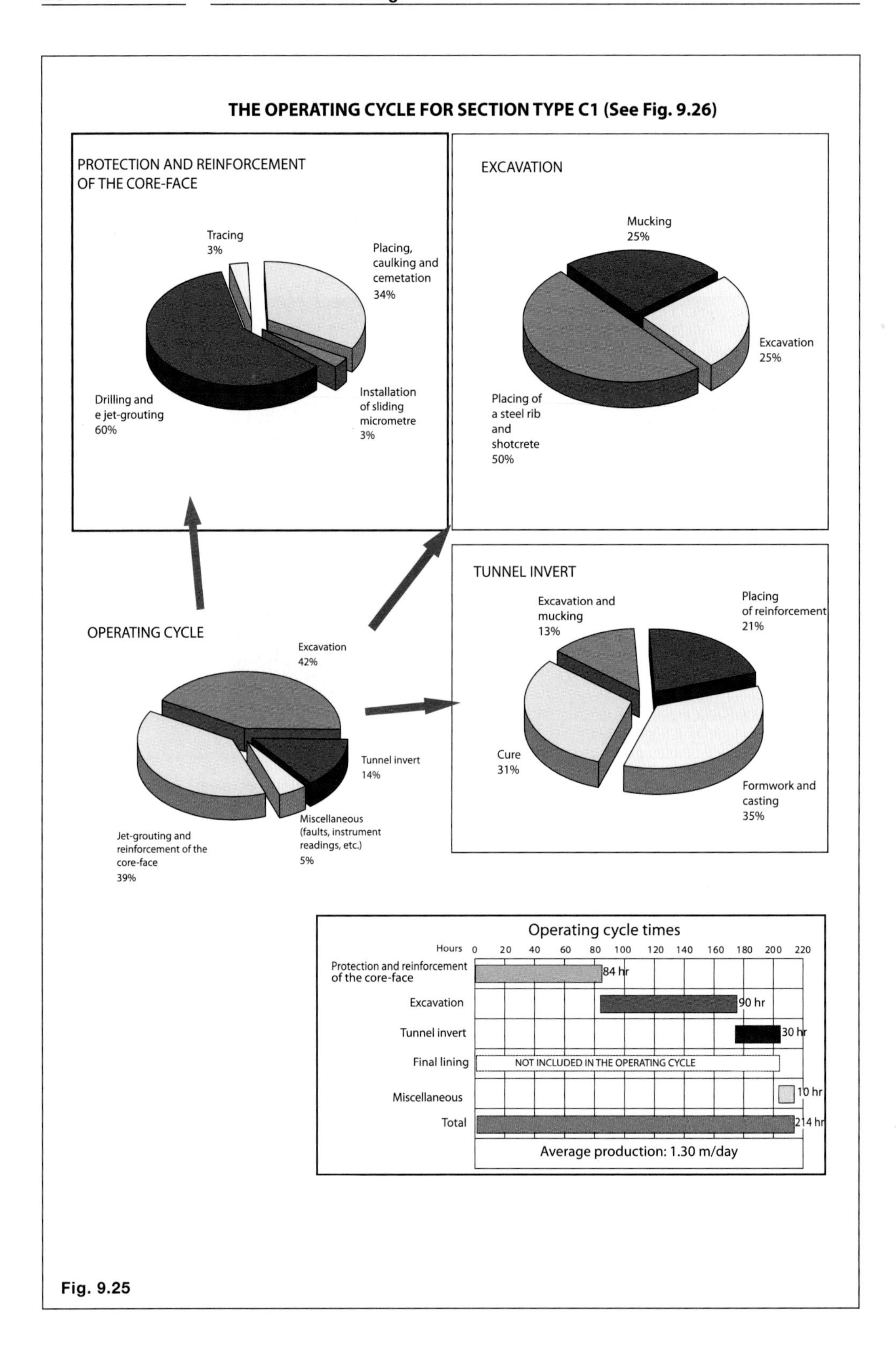
THE OPERATING CYCLE FOR SECTION TYPE C1 (See Fig. 9.26)
PROTECTION AND REINFORCEMENT OF THE CORE-FACE
Tracing 3%
Placing, caulking and cemetation 34%
Installation of sliding micrometre 3%
Drilling and e jet-grouting 60%
EXCAVATION
Mucking 25%
Excavation 25%
Placing of a steel rib and shotcrete 50%
TUNNEL INVERT
Excavation and mucking 13%
Placing of reinforcement 21%
Cure 31%
Formwork and casting 35%
OPERATING CYCLE
Excavation 42%
Tunnel invert 14%
Miscellaneous (faults, instrument readings, etc.) 5%
Jet-grouting and reinforcement of the core-face 39%
Operating cycle times
Hours 0 20 40 60 80 100 120 140 160 180 200 220
Protection and reinforcement of the core-face
84 hr
Excavation
90 hr
Tunnel invert
30 hr
Final lining
NOT INCLUDED IN THE OPERATING CYCLE
Miscellaneous
10 hr
Total
214 hr
Average production: 1.30 m/day

Fig. 9.25

In practice, ground improvement and excavation of the tunnel are performed in a cycle of operations as outlined in Fig. 9.24, so that excavation always takes place under the protection of an arch of improved ground. It may be advisable under very difficult conditions to create a few columns of improved ground with a reinforcement function inside the core-face.

It is important to pay attention to the structural continuity (overlap between truncated cones of columns) of the treatment in the crown and on the side walls of the future tunnel and to ensure that it extends below the level of excavation in order to be certain that the stresses channelled by the arch are properly transmitted to the natural ground.

It is best to wait for a mixture to set at least partially when performing jet-grouting treatment in order to prevent the injection of one column from damaging others already in place. This can be done by injecting columns in a non contiguous sequence.

The ground can usually be drilled using a rotary or rotary-percussion method and the same rod that is employed for the high pressure injection. The first method is preferred in soils with a medium to fine granulometry because lighter rods are required, while in coarse, non cohesive soils, or where stone blocks are present, rotary-percussion methods may be more efficient in terms of production rates, but will require bulkier equipment.

As opposed to improvement of the ground ahead of the face using conventional grout injections, the operating parameters of the jet-grouting treatment are not determined by the permeability of the soil, because the force with which the mixture is injected remixes the material completely and this also changes its permeability completely.

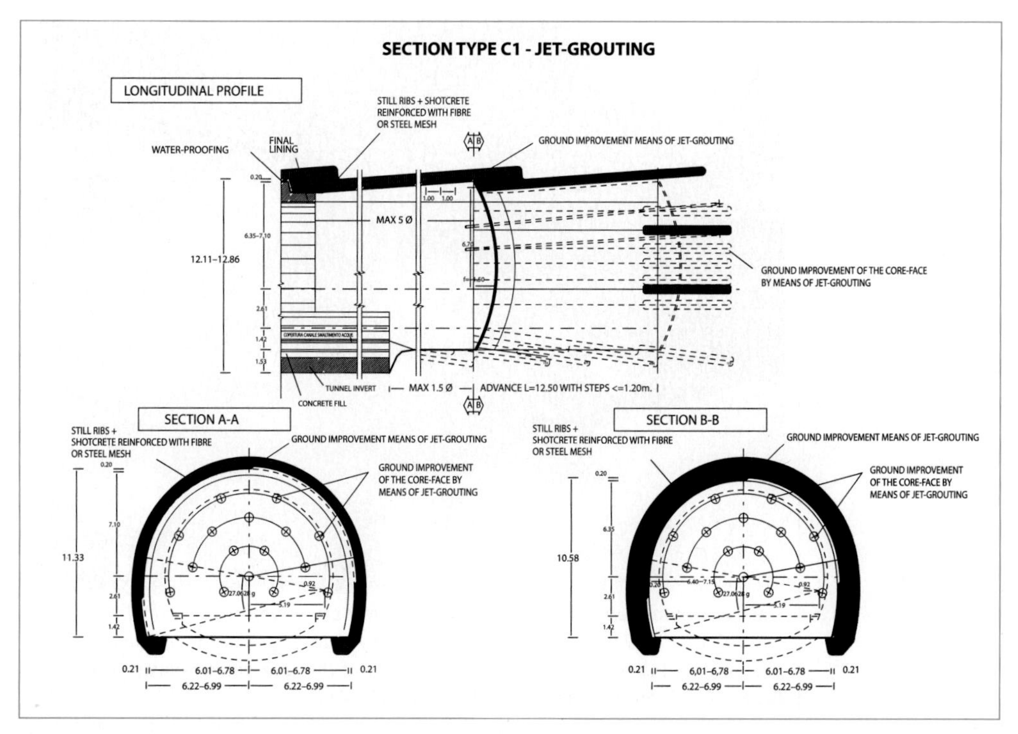

Fig. 9.26

Cavity preconfinement by means of truncated cone 'umbrellas' consisting of sub horizontal columns of ground, side by side, improved by jet-grouting: view of the construction site for the construction of the underpass of the Campinas railway yard in Brazil (1987, ground: sand, overburden: 2.5 – 4 m)

Cavity preconfinement by means of truncated cone 'umbrellas' consisting of sub horizontal columns of ground, side by side, improved by means of jet-grouting: detail of the treatment (underpass of the Campinas railway yard in Brazil, 1987, ground: sand, overburden: 2.5 – 4 m)

The main operating parameters for the treatment, which are normally decided by performing field tests, are as follows (Fig. 9.22):

- **injection pressure**: which is controlled by pressure gauges and determines the energy of the jet and therefore the radius of its influence. The limit to the pressure that can be achieved is dependent on the power of the pumps employed and the pressure that the line can withstand; working pressures of between 300 and 600 bar are generally used;
- **the number and diameter of the nozzles**: together with the injection pressure determine the flow rate of the mixture and, as a consequence, the velocity of the treatment. A larger nozzle cross section allows the power employed to be better exploited, while a greater number of nozzles gives lower performance for the same total flow rate, because there is a greater loss of concentrated load;
- **water to cement ratio**: this is the parameter which has most effect on the mechanical parameters of the improved column of ground. A low value with a mixture which therefore has a high specific weight is of great importance under the water table, especially under a hydrodynamic regime (danger of flooding); in this case it is important to use additives for rapid setting;
- **injection time:** this depends on the speed of withdrawal and of rotation of the rods. The former (normally one metre every 2 to 6 minutes), in 4 cm steps, is controlled by a timer fitted on the rod, which determines the time that the nozzles remain at a given depth. The injection time affects the size of the diameter of the improved ground, its mechanical characteristics and naturally also the time taken for the operation. The speed at which the rods rotate (normally 10 to 20 r.p.m) is governed by the withdrawal velocity and is designed to optimise the perforating capacity of the jet. If the speed of rotation falls below a given limit, the jet is reflected, which decreases its efficiency.

The drilling techniques, methods of injection and composition of the stabilisation mixture must be perfected in relation to the nature of the material treated and the local hydrogeological conditions, by performing *in situ* field tests, which must be the primary instrument used in deciding the mixture and the operating parameters, because theoretical tools do not yet provide valid support in this direction.

Controls must be carried out on columns to check that the results achieved comply with the design specifications. The following must be monitored:

- correct positioning and length of the columns and any deviations from the theoretical alignment, to be performed systematically as tunnel advance proceeds;
- the thickness of the arch of improved ground and its compressive strength after the preliminary shotcrete lining has been placed, by means of continuous rotary coring performed radially to the axis of the improved ground and drilled to a depth where it penetrates the natural ground;
- the effectiveness of the intervention in terms of the increase in the elastic modulus of the ground treated, by using seismic tests.

Cavity preconfinement by means of ground improved by conventional grout injections: treatment from a central pilot tunnel (Milan urban railway link line)

9.4.5 Cavity preconfinement by means of truncated cone 'umbrellas' of ground improved by means of conventional grouting

Cavity preconfinement by means of truncated cone 'umbrellas' of ground improved using conventional grouting methods consists of injecting the ground ahead of the face to create an arch of improved material with a predetermined length and strength and truncated cone geometry. A shell of material is created in this manner around the core-face with fair mechanical characteristics able to confine the surrounding ground radially and to prevent it from decompressing.

When the ground improvement is complete, the tunnel is excavated working under the protection of the arch of treated ground and for a length that is shorter than that section of ground. Excavation is therefore reduced to simply removing the ground enclosed inside the arch of treated ground.

Under very difficult stress-strain conditions, or when it is necessary in tunnels with a shallow overburden in an urban environment to prevent surface settlement completely, the treatment must be extended to include the ground in the core-face (Fig. 9.27).

In particular cases (shallow overburden, large diameter tunnels, etc.) conventional grout injections can be performed from a pilot tunnel or even from the outside (Fig. 9.27 and 9.28).

The technique using conventional grout injection is normally employed to drive tunnels with unstable face behaviour in non cohesive or pseudo cohesive soils with a permeability coefficient of not less than 10^{-6} m/s. The injections are performed from holes drilled into the ground in which valved tubes (*manchettes*) are fitted.

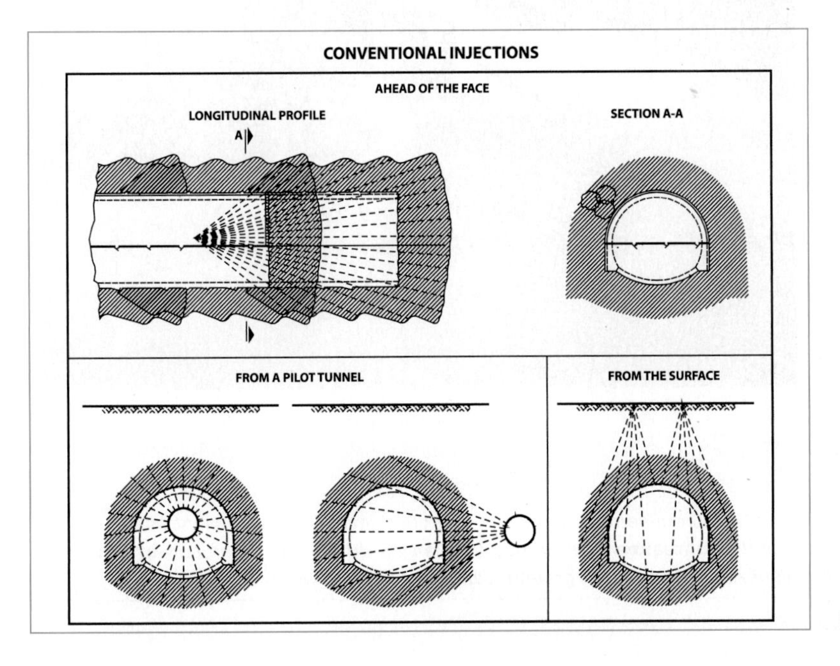

Fig. 9.27

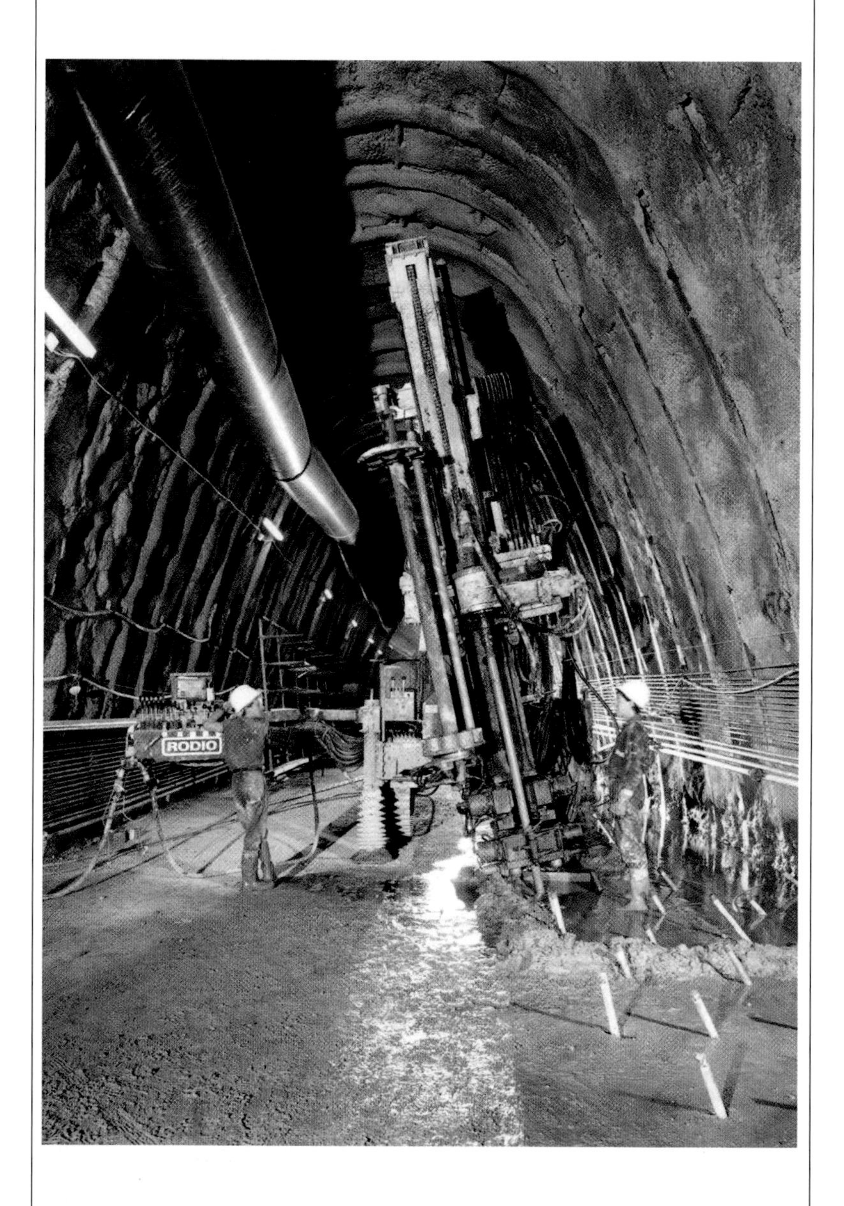

Cavity preconfinement by means of ground improved by conventional grout injections: treatment performed from a side drift (Milan urban railway link line, Venezia Station, 1988, ground: sand and gravel, overburden: ~ 4 m)

Fig. 9.28 *Conventional cement injections from a pilot tunnel*

Before the treatment starts the following must be accurately specified:

- the quantity and distribution of the injection tubes (as a function of the mechanical characteristics of the material to be treated and the degree of preconfinement it was decided must be exerted at the design stage);
- distance between valves (depending on the permeability of the soil);
- injection pressure (which must normally be less than the ultimate strength of the discontinuities in the rock mass (*claquage*).
- the fineness of the cement (which influences the penetrability of the mixture and how fast the grout sets);
- the composition of the mixture with particular regard to:
  - the viscosity (which must be low for rapid execution of the intervention);
  - the water to cement ratio (which has a predominant effect on the mechanical parameters of the final improved ground);
  - the use of appropriate additives (e.g. to slow setting speed or to fill large interstitial spaces).

To achieve this it is essential to have a large range of survey data, all obtained from a geological survey campaign, on the strength and deformability characteristics of the material to be treated and on its permeability for porosity and fracturation.

It is good practice to first verify that the intervention designed is adequate from a practical viewpoint and in terms of effectiveness by performing field tests from which valuable information can be obtained for defining the injection parameters (flow and pressure limit) and to establish the real improvement effect that can be achieved. Ground treatment in the tunnel should not commence until reasonable certainty has been acquired on the basis of the results of the tests performed that the design specifications can be met in terms of the thickness and strength of the band of improved ground.

Nevertheless all the necessary controls must be carried out during and after ground improvement actually performed for tunnel advance to verify that the results of the treatment are the same as those obtained in the field tests.

Cavity preconfinement by means of ground improved by radial injections: the effect of treatment performed from a central pilot tunnel (Line 3 of the Milan metro, 1985, ground: sand and gravels, overburden: 20 m)

Cavity preconfinement by means of truncated cone 'umbrellas' of ground improved by conventional grout injections (Ancona-Bari railway line, Vasto tunnel, 1996, ground: silty clay, overburden: ~ 35 m)

It is important in this respect for all the materials employed to conform to the design specifications as perfected during the field tests.

The equipment employed for the treatment must also enable all the work to be performed continuously and in a manner which guarantees the uniformity of the treatment. The mixing plant will therefore be fitted with equipment to weigh the cement and measure the water and the additive dosages and to count mixing cycles.

The injection unit will be equipped with high pressure pumps fitted with pressure gauges on the apertures of the bore-holes.

The composition of the grout mixture must be appropriate for the local penetrability and strength properties of the medium. This is achieved by broad use of microfines based cement mixes with grain dimensions of between 1 and 25 μm and a specific Blaine surface of up to 12,000 cm^2/g, which in addition to being atoxic are also very fluid and stable and at the same time they possess high permeation properties. The boreholes must be drilled using equipment most suited to the local characteristics of the medium. Since this intervention is normally performed in non cohesive soils, the walls of the boreholes may tend to collapse making it difficult to insert the injection tubes. It must therefore be possible to line the holes, if necessary, with a sleeve, which obviously must be withdrawn before cementation is performed.

Normally the tubes used for injections are in PVC fitted with valves located at constant predetermined intervals (normally 3 valves per metre). An injection needle slides inside the tube, fitted with a double shutter which is positioned at each valve for the time needed to perform the injections. The valved tubes must therefore be able to withstand the increase in the internal pressure that is generated up to the design values which may even be as high as 15 MPa.

Excavation of a tunnel along treated sections must be performed with caution, checking beforehand that the treatment has been performed successfully. Should the characteristics of the soils tunnelled change substantially from what was assumed when the system was set up, then the advance ground improvement parameters adopted must be promptly checked and recalibrated if necessary during construction.

The frequency with which quality controls are performed will depend on the local geomechanical characteristics of the medium (and must in any case be sufficient to furnish a clear idea of the quality of the ground improvement along the whole length of the sections treated) and they are designed to verify that the results achieved satisfy the minimum strength requirements and fall within the tolerances specified in the design. If possible they should be performed before the section of tunnel with the advance ground improvement is excavated.

The most important measurements are those for the thickness of the layer of improved ground and for its compressive strength at different stages of hardening. Both can be measured by continuous

Cavity preconfinement by means of truncated cone 'umbrellas' of ground improved by conventional grout injections (Sibari-Cosenza railway line, tunnel 4, 1998, ground: silty sand, overburden: ~ 40 m)

Cavity preconfinement by means of truncated cone 'umbrellas' of ground improved by conventional grout injections: effect of the treatment on the rock mass (Gran Sasso motorway tunnel, Valle Fredda fault, 1971, ground: milonite, overburden: 275 m)

rotary coring with a drilling rod and double core. Pieces are selected from each core for simple compressive strength tests. Pressure and expansion tests can be performed *in situ* inside the core boreholes.

If it is not possible to complete quality controls before tunnel advance takes place, they should in any case be performed subsequently after the preliminary shotcrete lining has been placed. This can be done with core boreholes drilled radially to the axis of the tunnel and sufficiently long to core the whole thickness of the improved ground and penetrate the natural ground behind it. They should be concentrated in points of greatest risk where it is felt that the presence of poorly improved ground is more likely.

9.4.6 Cavity preconfinement by means of truncated cone "umbrellas" of drainage pipes ahead of the face

Truncated cone "umbrellas" of slotted drainage pipes are launched ahead of the face to intercept the flow of water in the ground around the advance core of the tunnel, to prevent the circulation of water inside the core-face and to reduce the hydraulic pressure. This has the result of improving the natural strength and deformability characteristics of the advance core, enabling it to exert an appreciable cavity preconfinement action. To achieve this it is essential that the drainage pipes are positioned strictly outside the core-face, on the extrados of the perimeter of the cross section to be excavated with a truncated cone geometry, which is repeated in sequence when tunnel advance resumes so that the core is always completely protected. In practice the stages of tunnel advance and the placing of drainage pipes must always be performed in an alternating sequence so that a continuous succession of overlapping truncated cone 'umbrellas' is obtained with a length of not less than the diameter of the tunnel.

The length of the drainage pipes used will depend on the diameter of the tunnel, the permeability of the ground and the characteristics of the water table through which the tunnel passes. It will also be conditioned by the operating machinery available. The pipes are generally around three times the diameter of the tunnel.

They are placed while tunnel advance is halted after first drilling the borehole, with destruction of the core, in which the drainage pipe is inserted. The pipe is usually in PVC with an initial slotted section (bottom of the borehole, to capture water) (Fig. 9.30), covered with a sleeve of geotextile (non woven) to prevent clogging, and a second non perforated section, which is cemented in with the walls of the borehole by injecting grout (Fig. 9.31). The two sections are separated externally by a membrane which prevents the drainage water from hindering the injection operations. A 'packer' positioned outside the tube and filled with cement grout through a valve specially positioned on the wall of

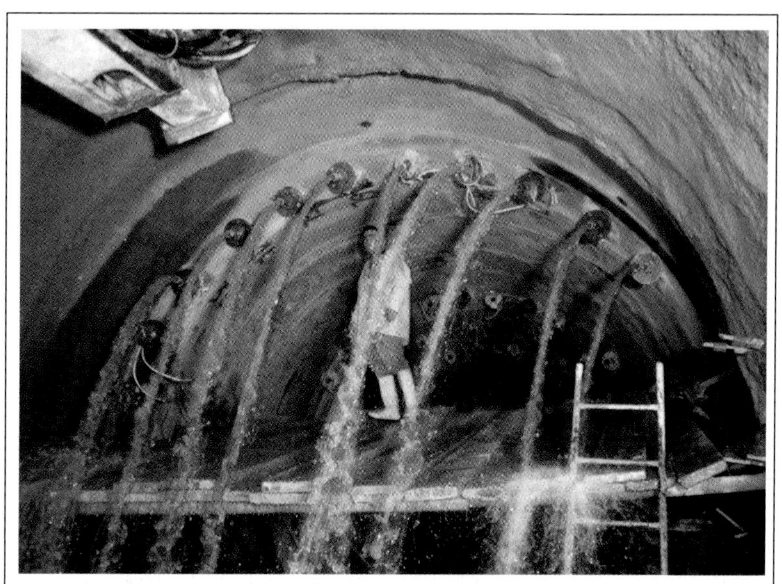

Cavity preconfinement by means of truncated cone 'umbrellas' of drainage pipes ahead of the face (Sibari-Cosenza railway line, tunnel 4, 1998, ground: silty sand, overburden: ~40 m)

Cavity preconfinement by means of truncated cone 'umbrellas' of drainage pipes ahead of the face: water drainage through NP jointed steel ribs (Gran Sasso motorway tunnel, construction site on Casale S. Nicola side, 1974, ground: limestones, overburden: ~800 m, water head: ~600 m)

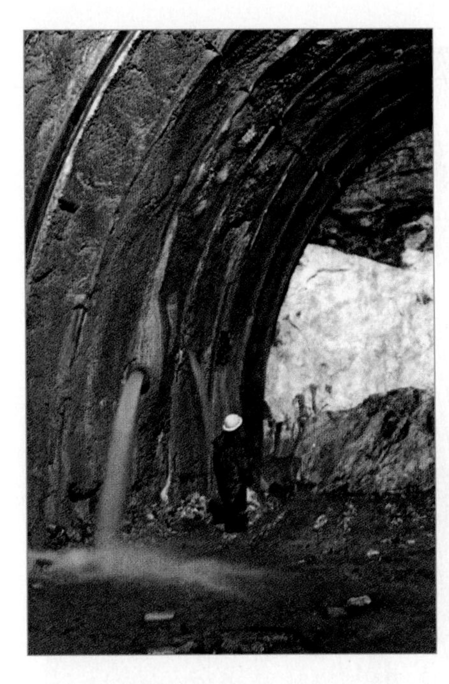

Fig. 9.29
Advance through an aquifer under 60 bar of hydraulic pressure (Gran Sasso motorway tunnel)

Fig. 9.30
Slotted pipe

the pipe immediately ahead of the membrane isolates the perforated section from the non perforated section when the grout injections are performed.

From an operational viewpoint, the installation of sub horizontal drainage pipes is performed in the stages described below.

1. Boreholes are drilled according to the geometry specified in the design after first protecting the wall of the face with a layer of shotcrete. The drilling methods used will depend on the nature of the material treated and on the local hydrogeological conditions. The equipment should, however, satisfy the following requirements:

   - possibility of drilling holes with a temporary lining at least 15 m in length, without the need to manoeuvre the rods, keeping faithfully to the truncated cone geometry as specified at the design stage;
   - a rotary head with through hole and not too bulky in terms of the distance between the extrados and the axis of the bore;
   - a sufficiently rigid design of the slide carriage, the guide devices for the rods and the positioning equipment in order to guarantee that drilling is performed within the specified geometrical tolerances.

   Drilling must be performed using tools appropriate for the design diameter and allow the subsequent operations for inserting the slotted pipes to be performed without any problems. If the walls of the boreholes are unable to support themselves long

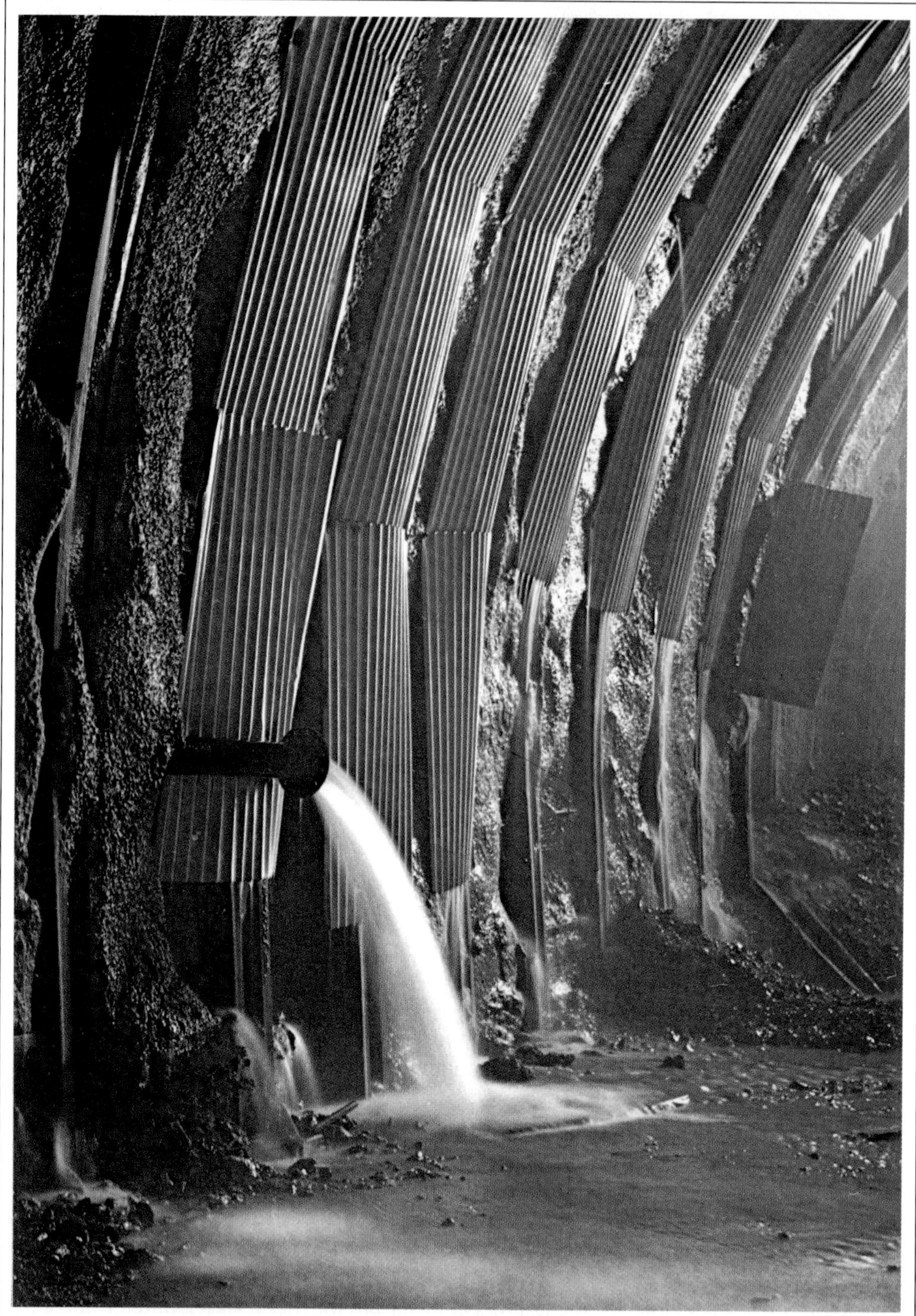

Cavity preconfinement by means of truncated cone 'umbrellas' of drainage pipes ahead of the face: effect of drainage (Gran Sasso motorway tunnel, construction site on Casale S. Nicola side, 1974, ground: limestones, overburden: ~ 800 m, water head: ~ 600 m)

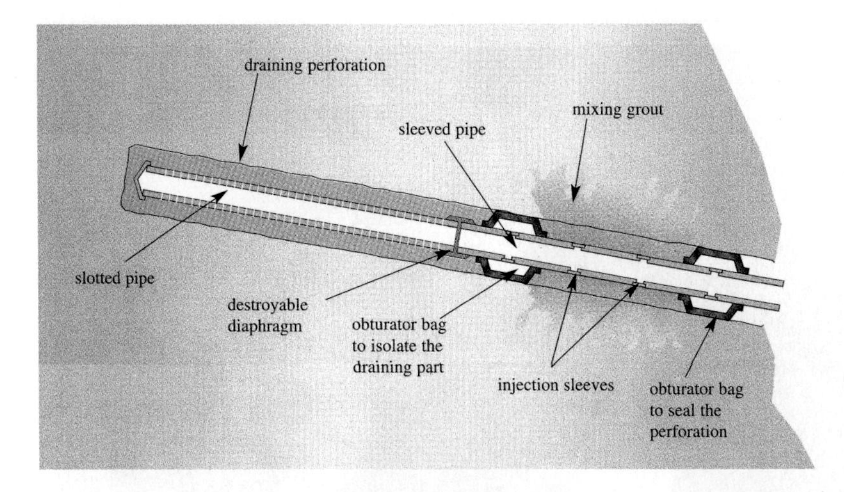

Fig. 9.31

enough for the drainage pipes to be inserted, then drilling sessions must interrupted at intervals for the insertion of a temporary lining, which is left in place until the drainage pipe is inserted. In this case, when the flow rate and the pressure of the water are high, the mouth of the borehole is tightly sealed by using a pipe fitting containing a stuffing box (which is connected using a water-proof joint to the drill tube) placed on the face at the axis of each borehole and an external discharge outlet which will allow the controlled outflow of the cuttings and prevent entrainment phenomena from being triggered. If necessary, this pipe fitting, which is known as a *"preventer"*, allows the borehole to be closed rapidly if uncontrolled flows of water and/or particles of soil occur. It is recommended that the *preventer* is set rigidly into the surrounding ground with appropriate cementation.

2. Insertion of the drainage pipe in the borehole enclose in a geotextile sleeve to protect the slots from clogging and activation of it.
   If the walls of the borehole are stable then the micro-slotted tube can be simply caulked and sealed on the outside after it has been inserted, while the water that is captured must be carried in pipes away from the tunnel face. Otherwise a temporary lining must be inserted, especially if working under high pressures, which involves the following operations:
   - insertion of the slotted tube inside the temporary lining;
   - retrieval of a length of the borehole lining corresponding to the length of the slotted section of the drainage pipe and the packer, so that the latter comes into direct contact with the walls of the borehole;
   - the packer bag is filled with cement mixture under controlled pressure, injected through a special valve. Once the sack has been filled, the injection pressure is maintained until the mortar mixture in the packer has set, at which point the lining tube is withdrawn completely and grout is immediately injected into the non perforated part of the drainage pipe starting with the most distant valve;

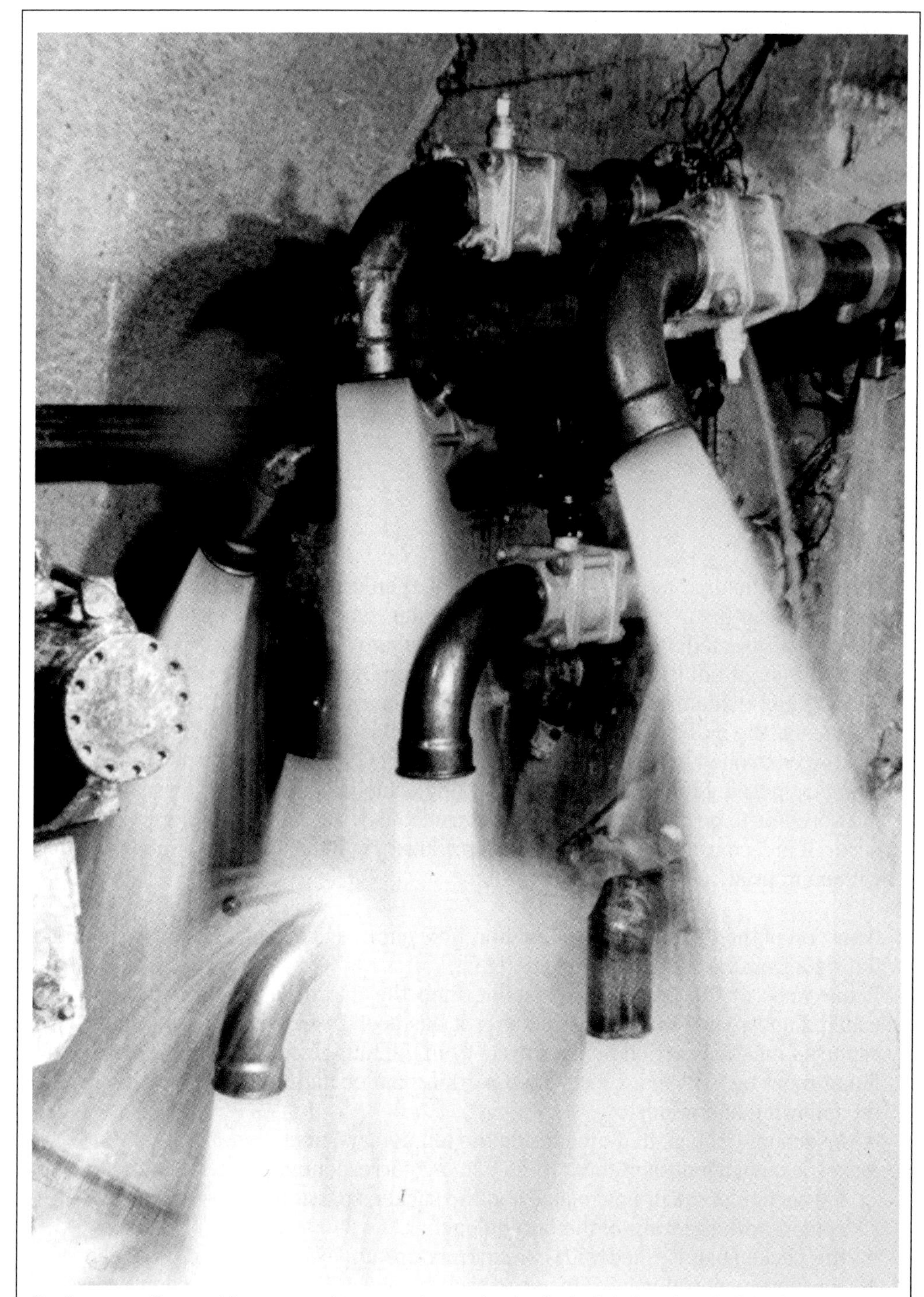

Cavity preconfinement by means of truncated cone 'umbrellas' of drains ahead of the face: effect of drainage passing through an aquifer under 20 bar of pressure (Gran Sasso motorway tunnel, Valle Fredda fault)

- the internal membrane located next to the packer is broken using a normal rigid rod inserted into the drainage pipe.

Depending on the type of ground and the likelihood of the slots in the drainage pipe becoming clogged, it may be necessary to unblock the pipe by pumping water into it at a constant rate and at a pressure greater than the external hydrostatic pressure for the time needed to clear the drainage section completely.

3. Inspection of the functioning of the drain.
This inspection must be performed immediately after the drainage pipe is placed and then periodically for the whole of its operational life. The function performed by drainage pipes with respect to intervention to stabilise tunnels is often fundamental to the safety of a tunnel in both the long and short term. The risk of a drainage pipe clogging must never be underestimated since it can occur as a result of countless different causes, even after a very long time. The geotextile filter plays a very important role and it must be chosen very carefully on the basis of the local characteristics of the ground. It is then placed with appropriate precautionary measures taken to avoid damaging it and to allow it to perform its function as well as possible.
Finally it is important to channel drainage water appropriately to prevent it from spreading in the tunnel where it might compromise subsequent stabilisation operations.

9.5 Cavity confinement intervention

9.5.1 Confinement of the cavity by means of radial rock bolts

Effective cavity confinement action can be achieved by inserting a series of rock bolts into holes in the surrounding rock mass drilled radially to the axis and perpendicularly to the walls of a tunnel. They are usually made of stainless steel, fibre glass or some other long lasting material with guaranteed tensile and shear strength.

The equipment used to drill the holes does not need to possess particular characteristics and can be selected according to the local conditions of the ground. Since this treatment is usually performed in ground of reasonably good quality, the walls of the boreholes rarely require lining.

There are two types of rock bolt, end anchored and fully bonded bolts. End anchored rock bolts consist of bars which only grip the ground at the ends, the end inside the rock mass (bottom of the hole) by means of an anchoring system which is usually of the expansion type (Fig. 9.32) and the end at the tunnel wall by means of a bearing plate to distribute the load. Consequently, this type of rock bolt can

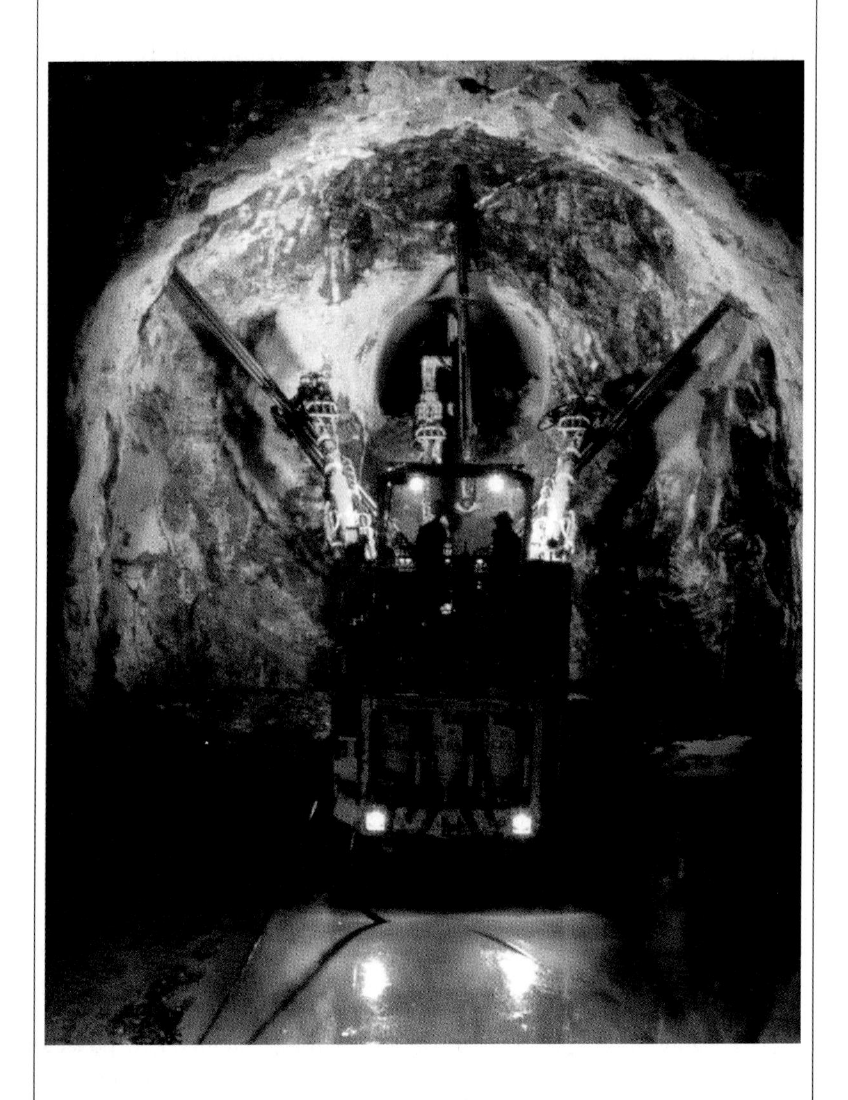

Confinement of the cavity by means of radial rock bolts (Udine-Tarvisio railway line, S. Leopoldo tunnel, 1986, ground: calcareous dolomite

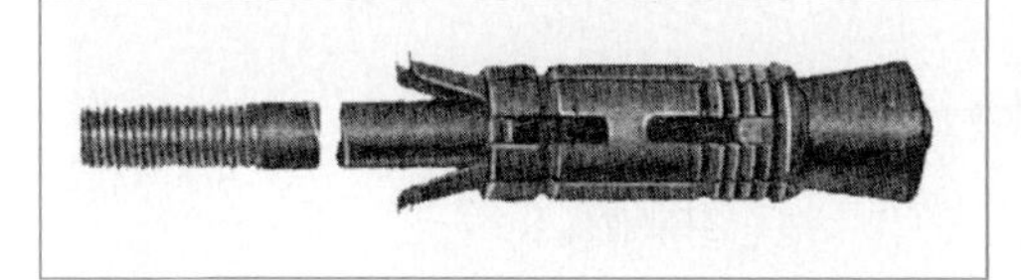
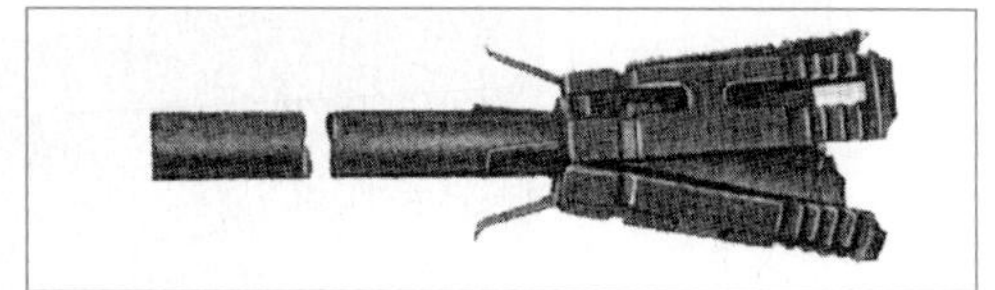

Fig. 9.32 *End anchor by expansion*

only be employed successfully in rock masses which may even be heavily fractured, but are of sufficiently good mechanical quality to guarantee a secure anchorage. They work exclusively by tensile stress and can be untensioned or tensioned. The former do not act until convergence of the cavity actually occurs, which they resist, while the latter are able to tighten the discontinuities present in the material and therefore increase its shear strength.

The cavity confinement action exerted by radial rock bolting using end anchored rock bolts may be direct or indirect, depending on the length of the rock bolts in relation to the thickness of the band of plasticised material around the cavity. If they are longer than that thickness then the radial confinement action is direct and equal to the pull exerted per unit of surface area. If however they are shorter than that thickness, then the cavity confinement action is indirect being equal to the radial confinement pressure that the band of ground confined between the two ends of the bolts is able to develop (see Fig. 9.33 and 9.34).

Grouted or fully bonded rock bolts, on the other hand, consist of bars or tubes in steel, or fibre glass, or some other material which are cemented to the walls of the hole along their entire length using either epoxy resin or cement mortar. As a consequence, their range of use can be extended effectively to include soft or poor quality rocks. They work mainly by means of shear and tensile strength and perform a true and genuine ground reinforcement function by significantly increasing the cohesion of the band of material treated. The cavity confinement action that they produce is therefore of the indirect type because it is developed by means of the ring of reinforced ground.

In order for grouted rock bolts to perform their ground improvement function effectively, care must be taken to make sure that they are grouted into the walls of the borehole as completely as possible so that the rock bolt acts immediately to resist any deformation of the ground that runs across its axis.

There is a particular type of fully bonded rock bolt which is not cemented to the walls of the hole by means of grout but which is anchored by friction. These are the Swellex type bolts which consist of steel tubes squeezed and bent in on themselves along their axis. They are simple and quick to place: after they are inserted they are injected to make them swell with water or cement mix under high pressure. As this expansion process takes place, the bolt changes shape to fit the irregularities of the hole perfectly (Fig. 9.35). This guarantees high resistance against withdrawal.

The operations for rock bolting consist of the following:

1. a hole is drilled using appropriate equipment of the diameter required for placing the bolt (which must be the minimum size necessary for fully bonded bolts). The cuttings are removed and the hole cleaned;
2. the rock bolt is inserted in the hole using the proper equipment;
3. it is anchored in the hole:

Confinement of the cavity by means of radial rock bolts: placing of a rock bolt (purifier in a cavern at Brunico, 1993, ground: quartziferous phyllites, max. overburden: ~ 125 m)

Confinement of the cavity by means of radial rock bolting: effect of rock bolting performed from a pilot tunnel (Udine-Tarvisio railway line, Malborghetto tunnel, 1986, ground: Ugovizza Breccia, overburden: ~ 70 m)

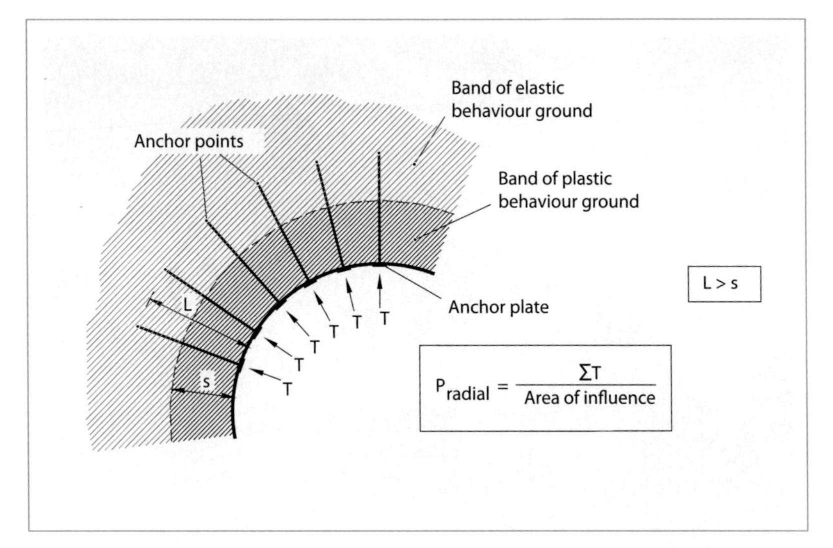

Fig. 9.33

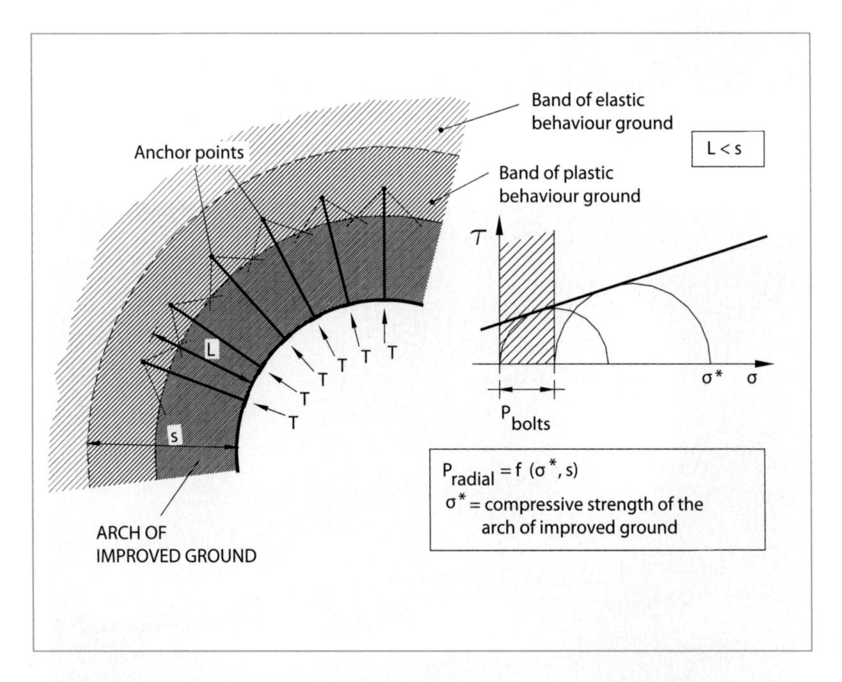

Fig. 9.34

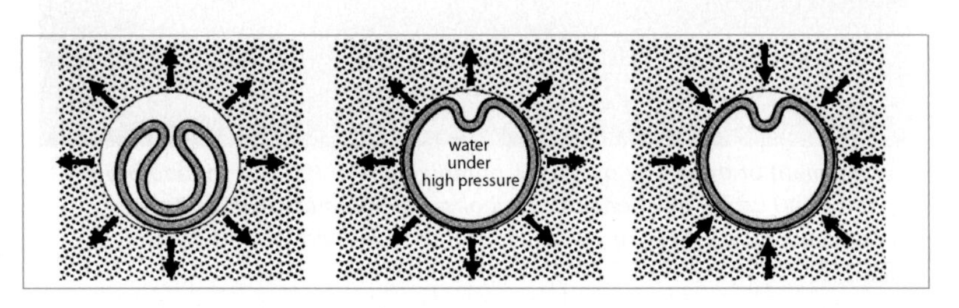

Fig. 9.35
Functioning of Swellex rock bolts

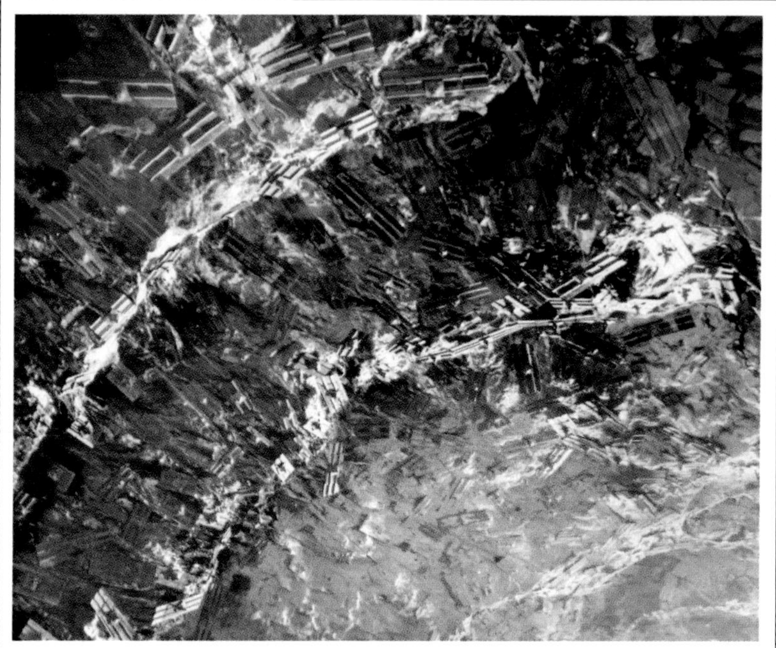

Confinement of the cavity by means of radial rock bolts: intervention using Ancral rock anchors (Frejus motorway tunnel, 1978, ground: crystalline schistose limestone, max. overburden: ~ 1700 m)

Confinement of the cavity by means of radial rock bolts: cavern-laboratory C, stabilised using fully bonded rock bolts and shotcrete (INFN laboratory of the Gran Sasso, 1984, ground: limestones, overburden: ~ 1400 m)

- by expansion of the anchor head at the end of the bolt in the bottom of the hole, for end anchored bolts;
- by the injection of cement mortar containing additives to fully bond the bolt to the walls of the hole along its entire length for grouted rock bolts. The bars of grouted rock bolts are fitted with small flexible PVC tubes for injection and to allow air to escape while the hole is filling. Care must be taken to avoid squeezing these tubes (e.g. by keeping them above the bar) when introducing the bar into the hole if injection is to be performed successfully. The injection mixture normally used is cement mortar containing accelerator and anti-shrink additives. When the drill holes deviate from the horizontal by more than 35° then the grouted rock bolts must also be fitted with an expansion head to anchor them at the bottom of the hole in order to secure them long enough for the mixture to set;
- by inflating them in the case of Swellex bolts;

4. the bolt is tensioned by tightening the nut on the threaded end of the bolt after first fitting the bearing plate in position (the nuts must be tightened according to the design specifications using a torque wrench);
5. steel mesh and lattice reinforcement is anchored to the head of the bolts, if required, to form the reinforcement for a layer of shotcrete that will be cast subsequently.

With some special types of self drilling rock bolts, insertion and drilling is performed in one operation.

Systematic quality controls should be carried out on rock bolts after they are placed based on the design specifications for the structural function they perform and on the materials selected. The most important of these for end anchored rock bolts is undoubtedly periodical pull testing to ensure that the specified working tension is guaranteed in the long term. At the same time, the integrity of the cementation must be checked periodically for grouted rock bolts. Testing devices exist for these purposes that are fitted directly on the bolts and allow the long term functioning of the bolts to be monitored easily and automatically.

9.5.2 Cavity confinement using a preliminary lining shell of shotcrete

The preliminary lining of the walls of a tunnel with a shell of shotcrete is widespread general practice as an effective means of confining the cavity. It consists of spraying a concrete mixture with a low water to cement ratio, which will develop high strength in a short time, at high velocity onto the walls of a tunnel. It adheres firmly to the surface to be lined by compacting under the force of the jet. If the composition is right and it is sprayed properly, the free surface of the shell of shotcrete has an attractive even appearance and for many tunnels driven through ground with rock type behaviour, it can even be considered for use as a final lining.

The method of use and the ease with which it can be placed make shotcrete an ideal lining instrument for smoothing irregularities in the surface of an excavation where overbreak has accidentally occurred. It can be reinforced where necessary, not only by steel ribs which are normally buried, but also by one or two layers of steel mesh or by

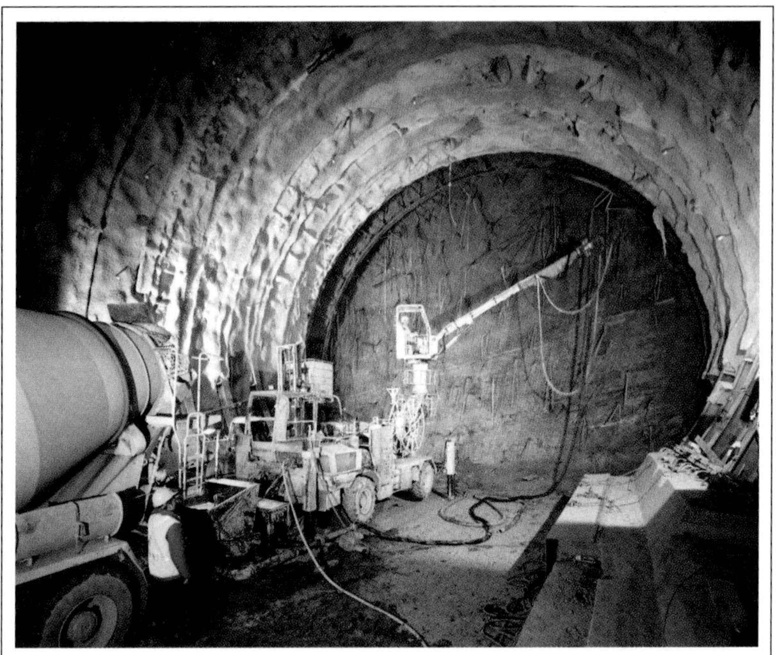

Confinement of the cavity by means of a preliminary lining shell of shotcrete: placing of the shell (TGV Méditerranée, Lyon-Marseilles railway line, Tartaiguille tunnel, 1997, ground: clay, max. overburden: ~ 110 m)

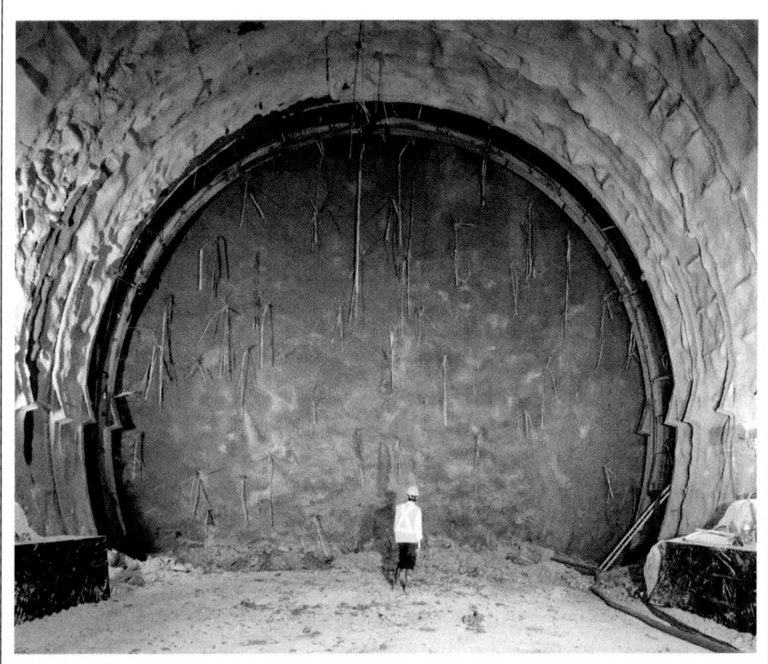

Confinement of the cavity by means of a preliminary lining shell of shotcrete: view of the sprayed face (TGV Méditerranée, Lyon-Marseilles railway line, Tartaiguille tunnel, 1997, ground: clay, max. overburden: ~ 110 m)

steel fibre (fibre reinforced shotcrete), which significantly increase its strength (tensile strength above all) and ductility. The degree of improvement obtained in the mechanical characteristics of shotcrete by fibre reinforcing it depends on volume and the length to diameter ratio of the fibres. The minimum quantity must comply with the design specifications (usually at least 20–30 kg per m^3 of mixture).

The thickness of the shell, which must in any case never be less than the design specification, normally ranges from a minimum of 5–10 cm to a maximum of 35–40 cm.

From an operational viewpoint shotcrete is usually placed wet, which is much healthier, more efficient and reliable than dry spraying. When prepared for wet spraying, all the components, including the water and steel fibres if used, are mixed beforehand and then conveyed by a pump to the spray gun. At this point the compressed air is added, which projects the shotcrete onto the surface to which it is applied. (Fig. 9.36).

Its composition is based on the same principles as normal concrete. The correct granulometry is essential for good results. Generally the curve for the percentage passing through against mesh size must be even with neither peaks nor jumps (Fig. 9.37). According to where it is sprayed, it is best for the granulometry of the mixture to lie towards the finer section of the granulometry curve for the arch of a tunnel, towards the centre of the curve for vertical application or toward the courser part of the curve for floor work. It is essential to add extremely fine material and cement to lubricate the flow and to prevent it from segregating. However, the percentage of these materials should not be too great or the tubing will clog, especially if it is long.

The aggregates (natural rounded aggregates are preferable to crushed quarry aggregates) must have a size of not more than 15 mm for compatibility with normal shotcreting equipment, while for the best strength properties, rapid setting and ease of spraying, the water cement ratio should not be greater than 0.5.

In order to achieve fluidity, cohesion and rapid setting without compromising the strength of the concrete, it is essential to use high strength cement in quantities higher than for normal concrete and also to use fluidifying and rapid setting additives appropriate for the type of cement employed (silicate or alkali free cement, 8–10% of the weight of the cement).

Not only should the shape of the fibres used in fibre reinforced shotcrete have an L/d ratio of >60 but they must have high tensile strength and minimum elongation characteristics. They are therefore made of low carbon content steel, are of small diameter (fractions of a millimetre) and a few centimetres in length; they are shaped at the ends to improve adherence (Fig. 9.38).

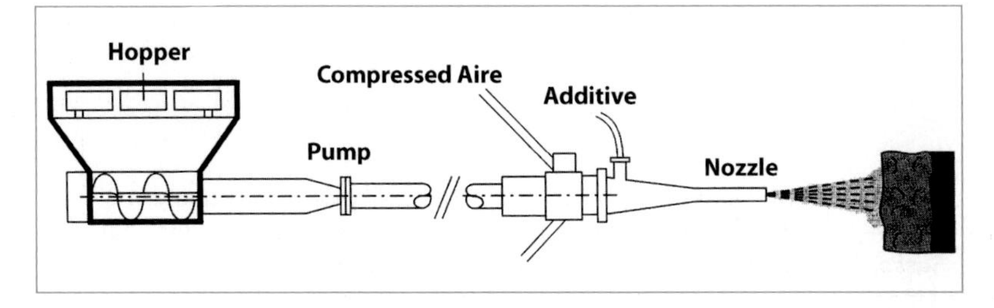

Fig. 9.36

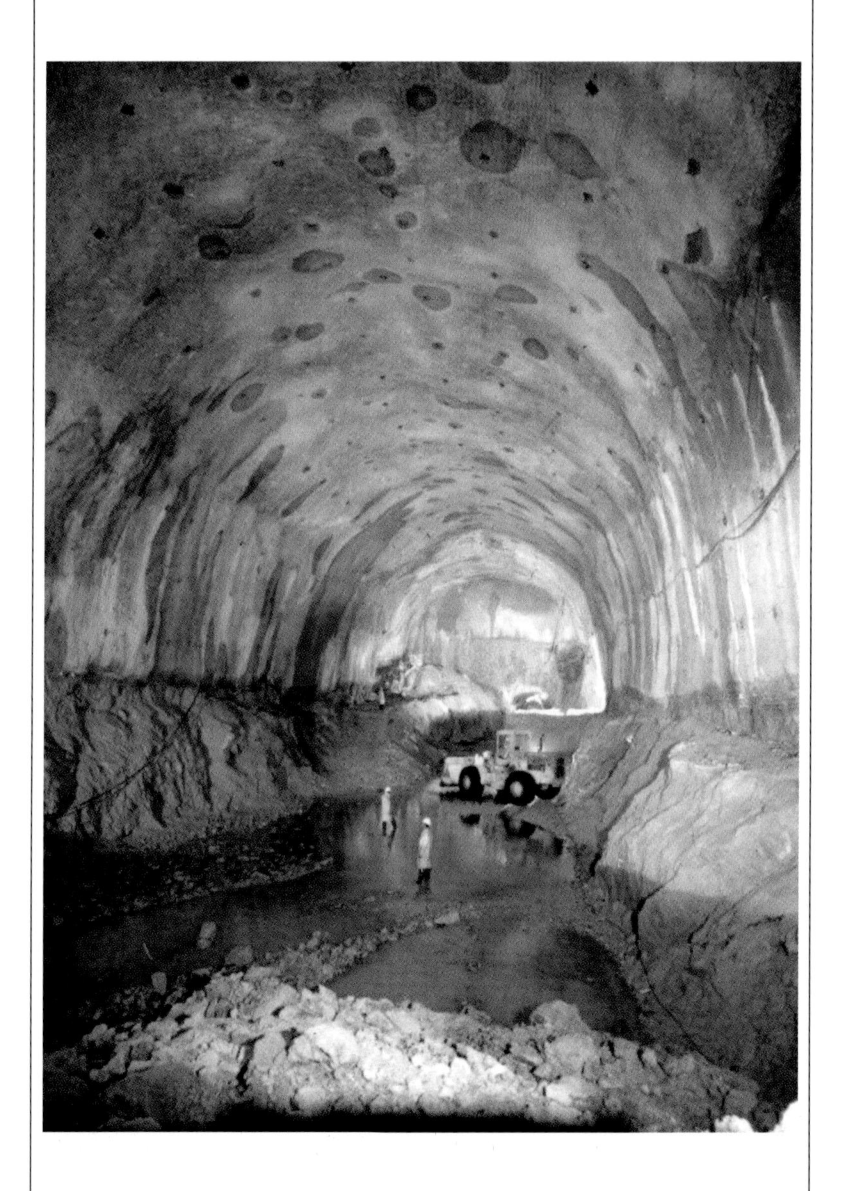

Confinement of the cavity by means of radial rock bolts and shotcrete: cavern-laboratory C, during bench excavation (INFN laboratory of the Gran Sasso, 1984, ground: limestones, overburden: ~ 1400 m)

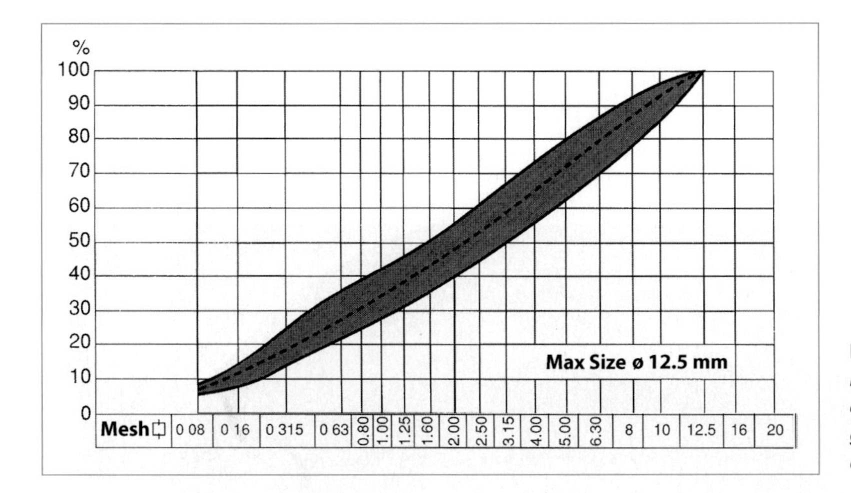

Fig. 9.37 *Example of shotcrete granulometry curve (AFTES)*

The composition and strength of the shotcrete must be tested carefully both when the mixture is prepared and after it is cast at different stages of hardening so that its strength can be plotted over time.

In addition to the usual simple or triaxial compression laboratory tests on sample cubes to check that the strength complies with the design specification, tensile, shear punching and bending tests must also be performed.

The quantity of fibres must also be checked when fibre reinforced shotcretes are used. This can be performed by comparing the percentage of fibres found in given volumes of mixture taken from the mixer with the design percentage. Energy absorption tests should also be performed to check the ductility.

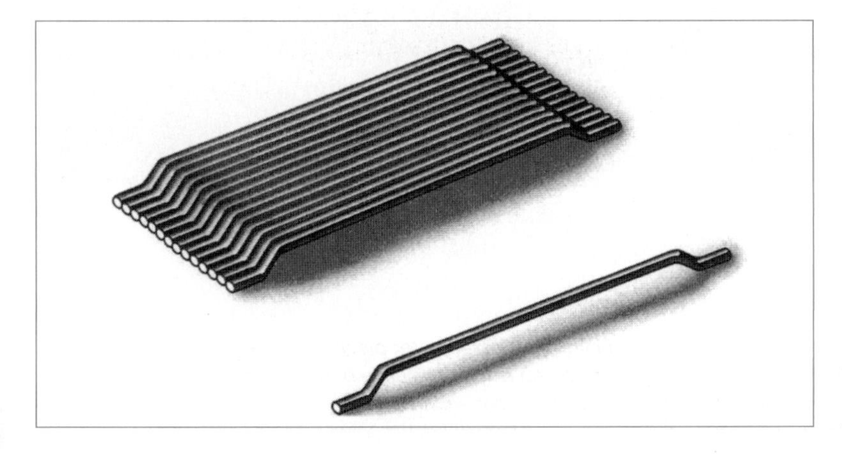

Fig. 9.38 *Steel fibre for fibre reinforced shotcrete*

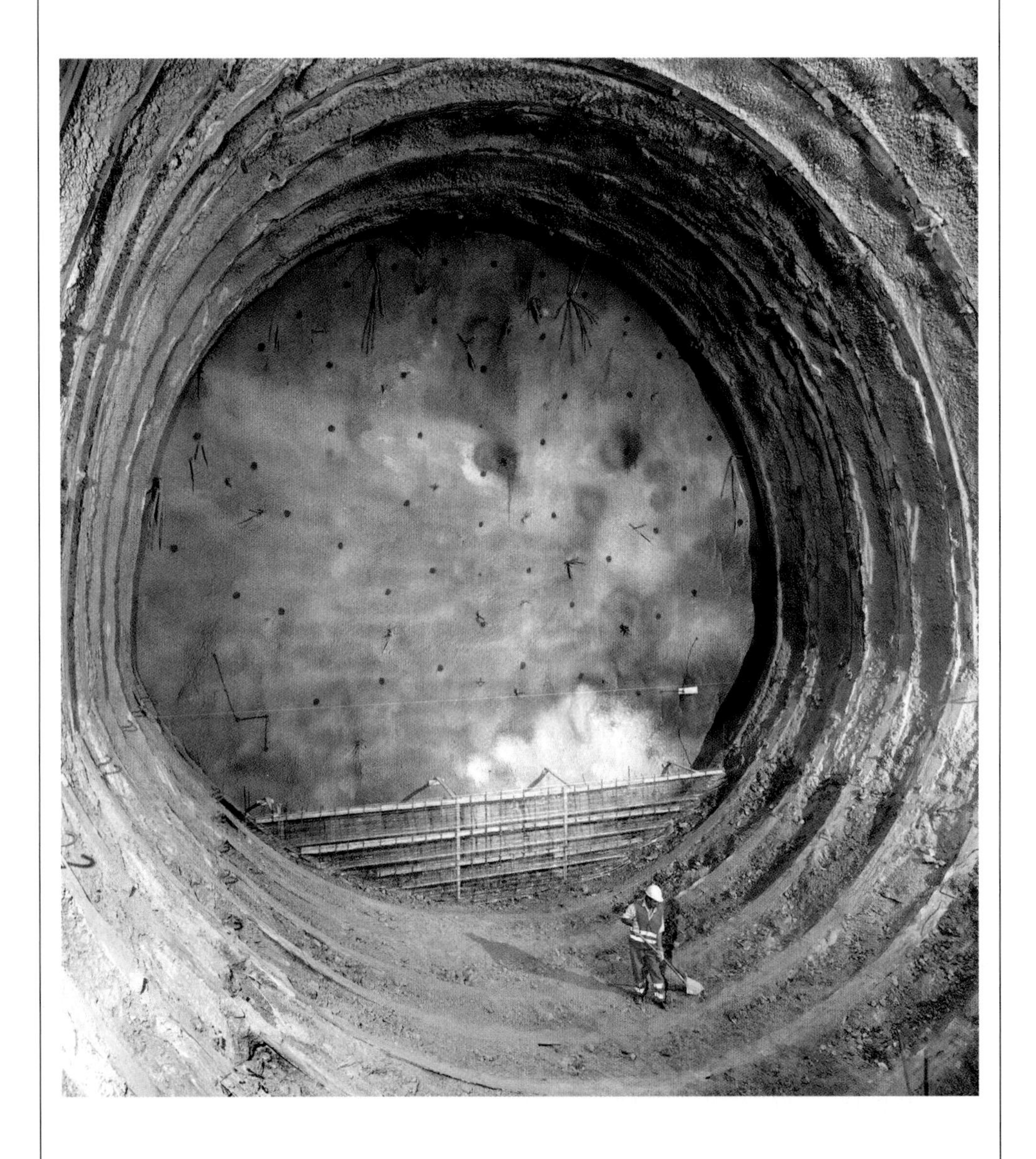

Cavity confinement by means of a tunnel invert: preparation for placing the tunnel invert immediately behind the face (high speed/high capacity railway line between Bologna and Florence, Raticosa tunnel, 2000, ground: scaly clay, overburden: ~ 600 m)

9.5.3 Confinement of the cavity by means of the tunnel invert

A tunnel invert is a structural element which closes the ring of the lining of a tunnel to give it greater rigidity and a very much greater capacity to exert cavity confinement action. Placing it is of particular importance in sections of tunnel with *stable core-face in the short term* or *unstable core-face* stress-strain behaviour, where it should be placed close to the face. In fact the worse the deformation response of the ground to excavation, the closer the tunnel invert should be cast to the face.

Although a sharp decrease in deformation phenomena in the tunnel is always produced after the tunnel invert is cast and better control of the deformation response is generally achieved, attention must nevertheless be paid to the possible increase in radial deformation that may manifest when the ground is excavated to house it. If this is not done properly and the necessary precautions are not taken, this excavation can lead to a dangerous weakening of the preliminary lining.

In order to prevent this from happening, especially when the tunnel invert is placed at a distance of not more than three tunnel diameters from the face, it is good practice to cast the kickers before this excavation takes place.

Great attention must be paid to the problem of the bonding between the kickers and the tunnel invert, where the maximum transmission of

Fig. 9.39 *Reinforcement for the tunnel invert (Tartaiguille tunnel)*

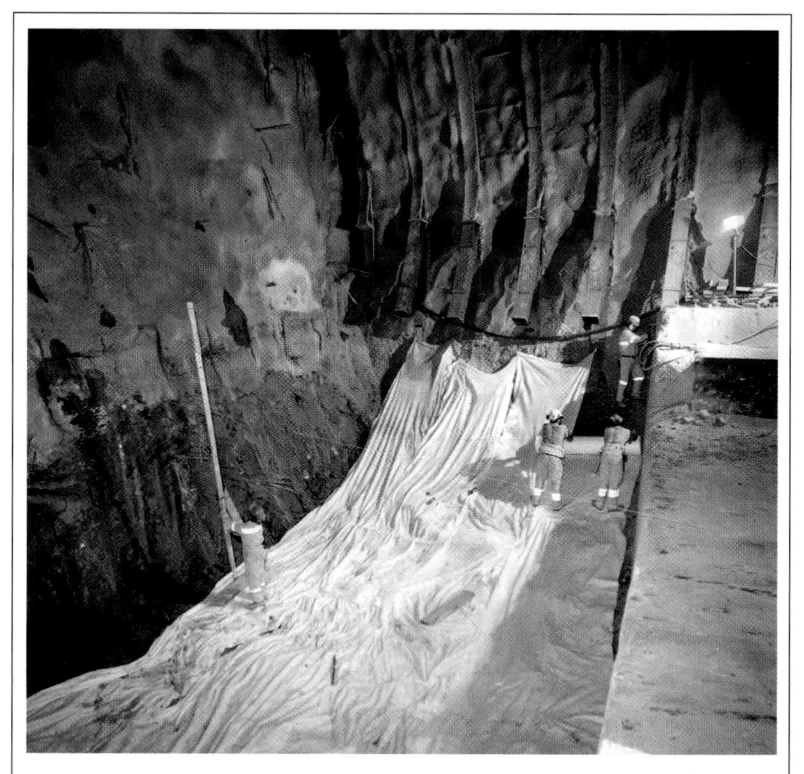

Cavity confinement by means of a tunnel invert: laying of the waterproofing for the tunnel invert (TGV Méditerranée, Lyon-Marseilles railway line, Tartaiguille tunnel, 1997, ground: clay, max. overburden: ~ 110 m)

Cavity confinement by means of a tunnel invert: casting the tunnel invert immediately behind the face (TGV Méditerranée, Lyon-Marseilles railway line, Tartaiguille tunnel, 1997, ground: clay, max. overburden: ~ 110 m)

stresses must be ensured. While it is important to select the most appropriate profile, it is also always useful to place reinforcement bars to join the two castings.

It is indispensable, in very difficult stress-strain conditions to place struts between the feet of steel ribs as soon as excavation of the invert is complete. This will in effect complete the lining structure without having to wait for the concrete to set.

The shape and thickness of the tunnel invert are decided in the therapy phase, according to the tunnel section type employed. If possible, it is preferable to avoid having to reinforce the tunnel invert, either by modifying its curve to minimise the tensile stresses acting on it, or by using a fibre reinforced concrete.

From a construction viewpoint it is best to cast the tunnel invert and its foundation on *in situ* ground and not on material that has been disturbed; any over excavation should be refilled and reshaped using lean concrete.

Quality controls to be performed during construction to verify the characteristics of the tunnel invert consist of the normal tests for plain or fibre reinforced concretes.

9.5.4 Confinement of the cavity by means of the final lining

The final lining is generally placed by *in situ* casting of a layer of either plain or reinforced concrete on the walls of the preliminary lining using appropriate formwork. Alternatively it can be also be constructed by placing prefabricated lining segments in RC or prestressed concrete (PRC).

The type and thickness of the lining, as well as the class of concrete to be used are decided beforehand at the design stage, depending on the tunnel section type employed.

It is very important for the thickness of the final lining actually placed to comply with the design specification and also for no cavities to be formed behind the extrados of the lining.

It is indispensable in the presence of water to place a layer of waterproofing anchored to the preliminary lining (which should not be damaged when formwork is placed or the concrete is cast) and to implement suitable drainage systems to ensure the long life of the lining.

Great care must be taken over the bonding between different pourings with thorough cleaning of contact surfaces. The free surfaces of the final lining must be perfectly even and shaped according to design drawings, with particular attention paid to sections on tunnel bends.

Standard quality controls on the properties of the concrete should be performed by testing samples taken when concrete is cast and when it has set by taking core samples.

Cavity confinement by means of a final lining: final lining of prefabricated segments in RC (high speed/high capacity railway line between Bologna and Florence, Ginori service tunnel, 2001, ground: limestones and marls, max. overburden: ~ 600 m)

The presence of gaps between the final and the preliminary linings can also be detected using non destructive methods such as radar surveys.

9.6 Waterproofing

Waterproofing is normally placed between the preliminary shotcrete lining and the final concrete lining, whenever the tunnel passes through an aquifer or when it is considered that water might enter, even if only rainwater.

Its normal function is to protect the final lining and to convey water not intercepted by drainage pipes to channels located at the foot of the side walls or under the tunnel invert. If, however, the tunnel must not be drained, because of questions of hydrogeological or local environmental equilibriums, then its purpose is to isolate the tunnel hydraulically from the ground in which it is located.

Waterproofing consists of a protective layer of non woven fabric and a waterproof layer of PVC sheeting at least 2 mm thick partially overlapping and heat welded (Fig. 9.40). It should be placed a short time before the final lining is placed when cavity convergence should be mostly finished. Any strong localised inflows of water should be intercepted and drained, because they might hinder proper placement

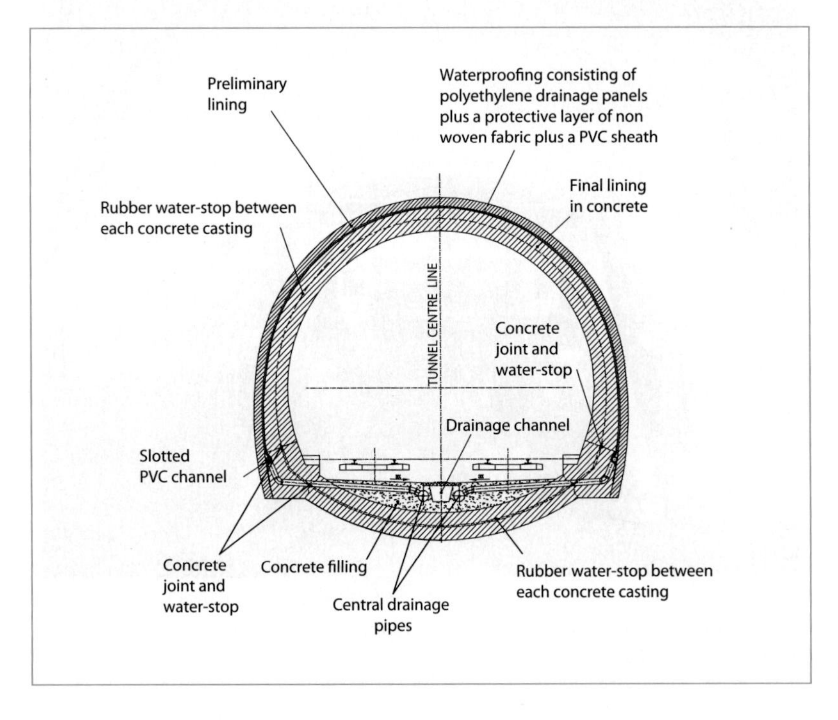

Fig. 9.40

Cavity confinement by means of a final lining: final lining in concrete (Ancona-Bari railway line, Vasto tunnel, 1996, ground: clayey silts, max. overburden: ~ 100 m)

Waterproofing: the PVC sheath placed just before the final lining is cast (Ancona-Bari railway line, Vasto tunnel, 1996, ground: clayey silts, max. overburden: ~ 100 m)

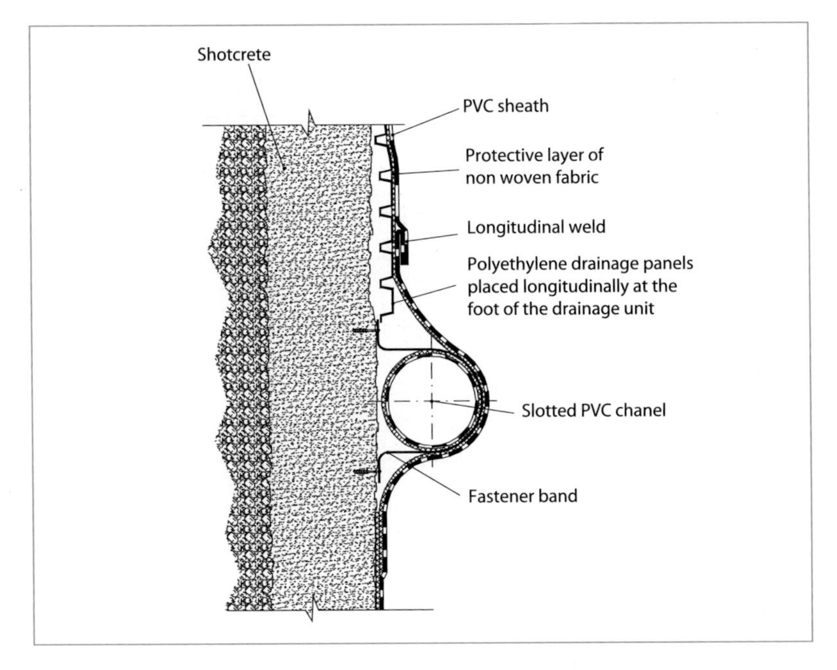

Fig. 9.41

of the protective sheeting and waterproofing if not controlled. This can be achieved by using plastic (polyethylene) panels protected by a layer of non woven fabric and fixed to the walls of the tunnel where the water is entering using cement mortar with rapid setting additives. The panels must then be connected to the main drainage channels located at the foot of the waterproof lining which then run to drains in the tunnel floor (Fig. 9.41).

All projecting metal parts such as the anchor heads of rock bolts, steel mesh, etc. must then be eliminated and the existing lining must be evenly finished with fine shotcrete to smooth over rough areas and parts such as steel ribs and chains in order to avoid damaging the geotextile and PVC sheeting.

The protective layer (consisting normally of non woven fabric usually in continuous filament polypropylene, preferably bound by needle punching) considered by some as of no use, is in fact of great importance because it performs and anti-perforation function and because a thin layer of high permeability acts as a medium through which water that comes into contact with the waterproofing will easily flow. This facilitates the interception and channelling of water to the drainage channels considerably. This layer must be placed by proceeding crosswise to the axis of the tunnel (Fig. 9.42); the sheets of non woven fabric with a gram weight of approximately $400\,g/m^2$, partially overlapping, are fixed to the preliminary lining by nails of 2–4 cm fitted with semi-rigid

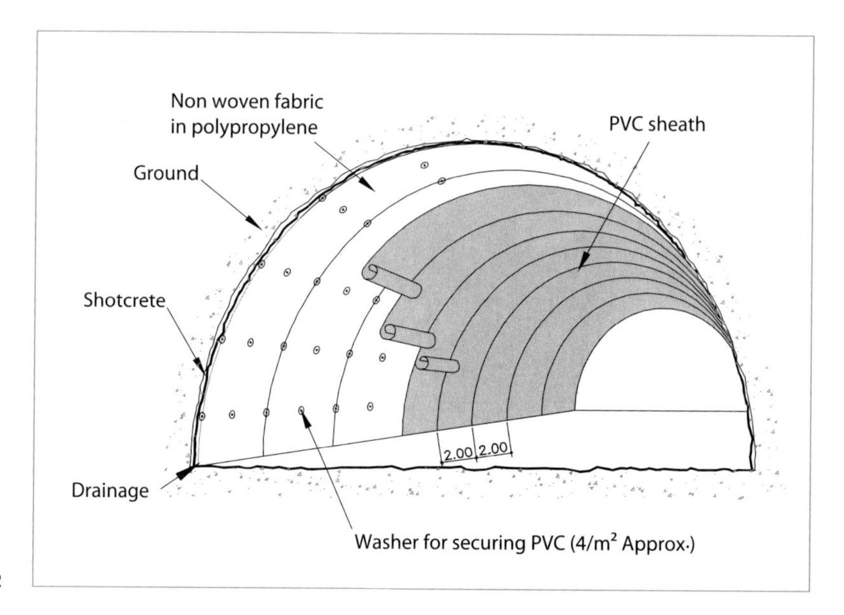

Fig. 9.42

PVC washers. The washers (8–9 cm diameter approx.) perform the dual function of supporting the non woven fabric and acting as an anchor to which the PVC waterproofing can be fixed. The latter is hot air welded to the washers, starting at the key of the crown and continuing down on both sides so that it is held in the correct position until the final lining is cast. The overlapping edge of each individual sheet is then welded to the adjacent sheets and then sealed by heat welding to form a joint consisting of a double beaded weld. The area between the two beads must then be pressure tested to verify that the joint is perfectly waterproof.

Before they can be used, both the protective and the waterproofing layers must pass quality controls and tests (thickness, compressive and tensile strength, resistance to thermal stress, rot resistance, etc.) to verify that they comply precisely with the design specifications.

Further controls, mainly visual inspection, should also be performed after placing the water proofing to check that all the specifications and precautions needed for the proper functioning of the waterproofing have been taken, even in the long term.

CHAPTER 10

The monitoring phase

10.1 Background

Once the design stage has finished, monitoring phase begins when excavation work (the construction stage) starts and it is in this phase that the reliability of the predictions of the deformation response of the rock mass to excavation made in the diagnosis and therapy phases are tested.

The particular nature of underground constructions, which normally consist of an artificial composite structure created inside an excavation, [which guarantees the stability and persistence over time of that excavation by interacting with the ground that stresses the structure in a manner that cannot be unequivocally determined, but which depends on the construction procedure followed and also on the rheological characteristics of the material involved (ground, stabilisation instruments)], requires an appropriate monitoring system to installed and activated as soon as construction begins and sometimes even before it begins, in order to be able to measure the deformation response of the ground systematically in absolute terms and in all its components.

This is required in order to achieve the following:

- to be able to verify, during the construction stage, the appropriateness of the design decisions made and to calibrate them during construction (intensity and distribution of stabilisation instruments);
- to make all the essential data available when commissioning an infrastructure in order to be able to reliably assess its structural safety and suitability for use. Commissioning a tunnel involves testing it to verify that the design predictions formulated in terms of the stress-strain response of the rock mass to excavation and of the stress state and deformation of the stabilisation structures and lining, were found to be accurate during the construction of the tunnel and after completion. This requires reconstructing the stress and deformation history of the tunnel in terms of its interaction with the surrounding ground by examining the deformation response of the ground to excavation and the stabilisation structures employed at different stages in the construction process. This reconstruction can only be achieved through detailed analysis of the monitoring data and investigations conducted *in situ* where necessary;
- to make it possible and easy to check the condition of the tunnel over time when it is in service and this must be possible in relation to the rheological behaviour of the rock mass and possible changes in the hydrogeological conditions surrounding the cavity in specific sections (fault zones, walled sections etc.).

This chapter will therefore illustrate the criteria that a design engineer must follow in order to prepare a detailed monitoring programme that is able to fully satisfy these requirements and which at the same time is custom designed for the specific tunnel to be built and which takes account of the specific geological context.

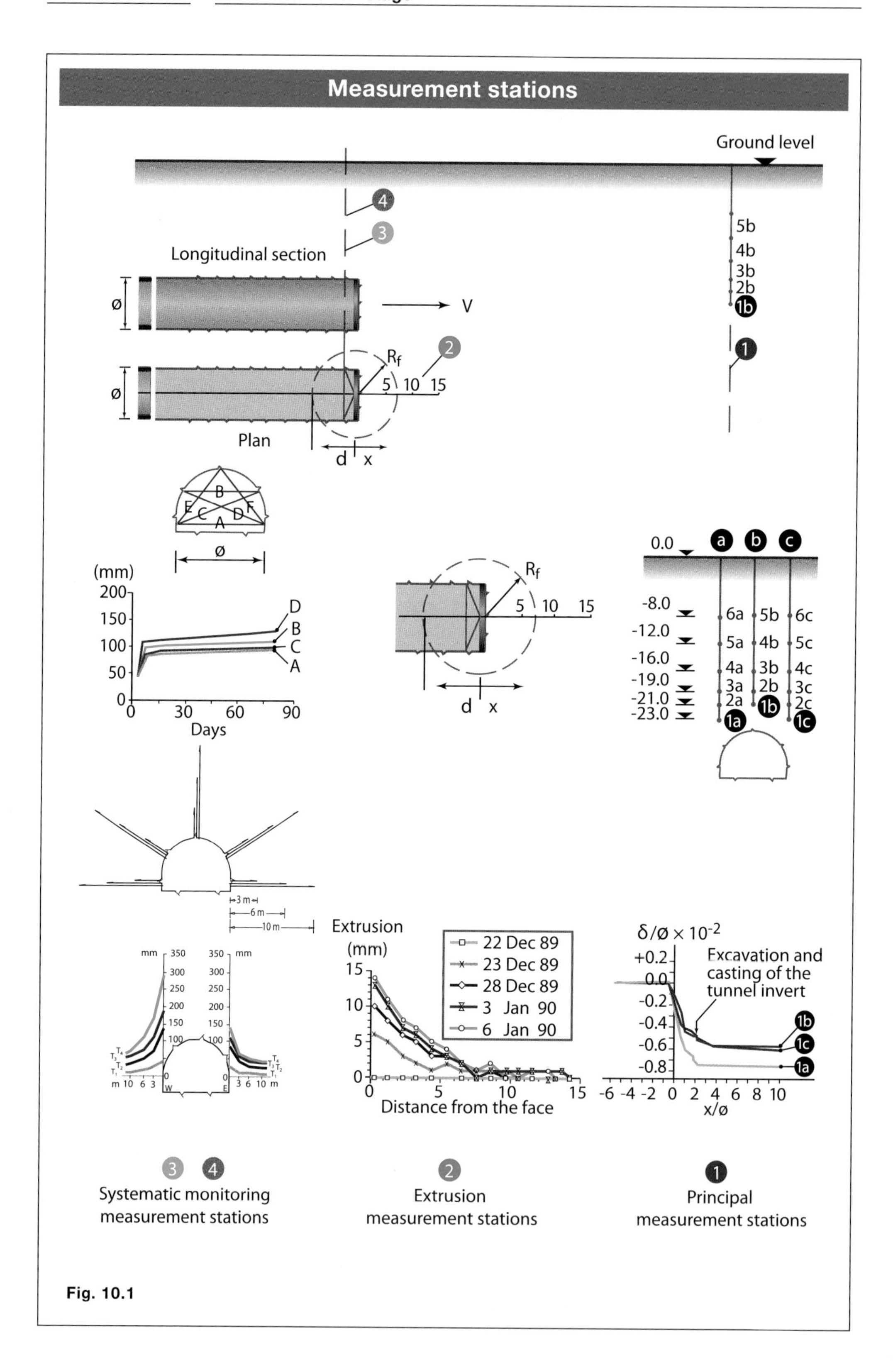

Measurement stations
Ground level
Longitudinal section
Plan
V
Rf
5 10 15
d x
Ø
5b
4b
3b
2b
1b
1
2
3
4
B
E C A D F
(mm)
200
150
100
50
0
0 30 60 90
Days
D
B
C
A
0.0
a b c
-8.0
-12.0
-16.0
-19.0
-21.0
-23.0
6a 5b 6c
5a 4b 5c
4a 3b 4c
3a 2b 3c
2a 1b 2c
1a 1c
3 m
6 m
10 m
Extrusion (mm)
22 Dec 89
23 Dec 89
28 Dec 89
3 Jan 90
6 Jan 90
Distance from the face
δ/Ø × 10-2
Excavation and casting of the tunnel invert
x/Ø
Systematic monitoring measurement stations
Extrusion measurement stations
Principal measurement stations

Fig. 10.1

The distinction which will subsequently be made between monitoring *during construction* and monitoring *when in service* is performed purely for the sake of description.

In service monitoring not only shares most of the necessary instrumentation with construction monitoring but is also compared to and interpreted in the light of the data provided by it. Long term monitoring of the stress-strain state of a tunnel would lose large part of its meaning if one did not have a fair knowledge of its stress-strain history since its birth, which is to say since tunnel advance first commenced.

10.2 Basic concepts

As we have already seen, the calculation methods available to a design engineer are capable in most cases of solving, with a fair degree of accuracy, both the problem of predicting the stress-strain response of a future tunnel and that of designing and testing the stabilisation intervention performed in the therapy phase.

However, the accuracy of the results they provide depends inevitably on the accuracy of the input data used in the calculation (strength and deformability of the medium, natural stress state, the rigidity of the confinement intervention etc.), for which a certain margin of error will always exist.

It is therefore indispensable to verify the hypotheses and predictions made in the diagnosis and therapy phases very carefully during construction. It is possible to obtain from the results of measurements and above all from comparing the predicted design deformation response with the actual response measured, the information needed to perfect the construction methods originally specified, to weight the calculation models used if necessary *(back-analysis)* and finally to estimate the real degree of safety for an underground construction by continued monitoring over time even when the tunnel is in service.

The output from monitoring activity must therefore guide the design engineer and the project manager in deciding whether to employ the tunnel section type specified, or to modify some of operational magnitudes (within the limits of the variabilities specified in the design and according to the criteria specified in it), or to proceed to the design of a new section type altogether to deal with particular conditions not identified in the survey phase and therefore not specified in the therapy phase.

In this context the monitoring phase consists of the following:

- during construction:
  - using appropriate instruments to measure the development, type and magnitude of the deformation phenomena produced in the medium in the form of extrusion of the core-face and convergence of the cavity, as a result of tunnel advance and also to measure the stresses and deformation within stabilisation structures themselves;
  - interpreting the results of measurements so that they can be compared with design predictions;
  - verifying the match between predictions and empirical observations;
- when in service:
  - monitoring the behaviour of the tunnel over time, in order to be able to programme indispensable maintenance prudently and guarantee safety over the whole of its service life.

Measurement instruments for the monitoring phase

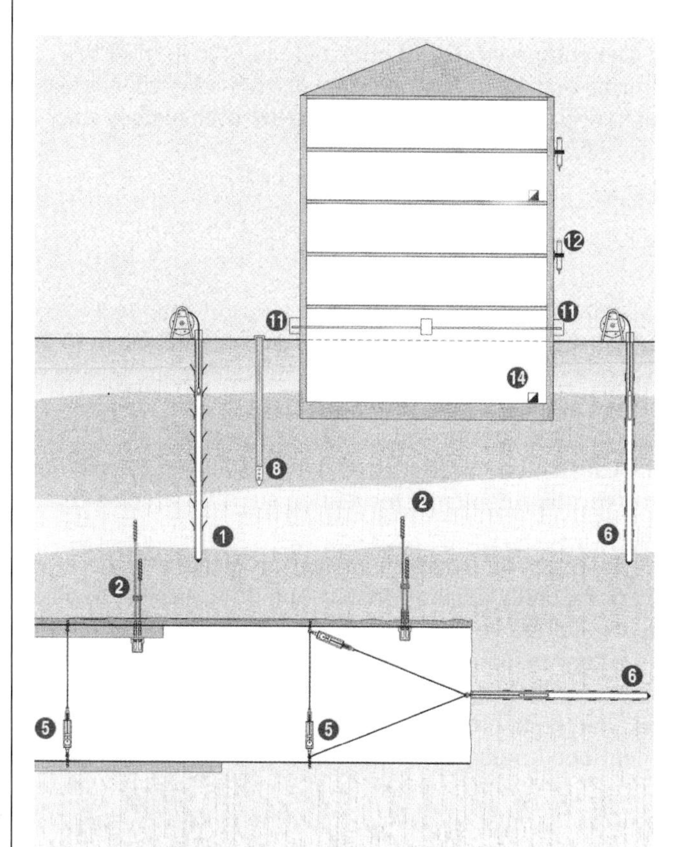

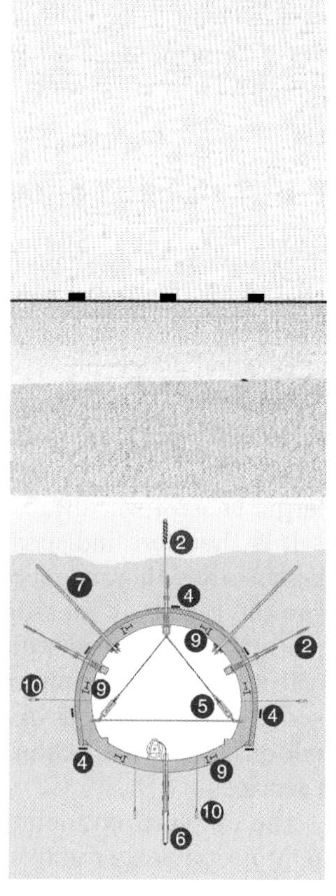

1 **Magnetic settlement meters**
Ground settlement

2 **Multipoint extensometers**
Deformation / plasticitation of the ground

3 **Smach accelerograph**
Vibrations during excavation

4 **Pressure cells**
Pressure on the ground-lining interface

5 **Tape distometers**
Convergence in the tunnel

6 **Incremental extensometers, inclinometers**
Ground extrusion and deformation

7 **Anchor load cells**
Rock bolts and anchors pull

8 **Piezometric-settlement column**
Piezometry and settlement

9 **Strain gauges**
Lining stress and strain

10 **Electrical piezometers**
To control neutral pressures

11 **DSM level measurement system**
Differential settlements

12 **Surface clinometers**
Vertical inclination of buildings

10.3 Measurement stations

In order to monitor the deformation behaviour of the rock mass to excavation, it is essential to set up measurement stations. The location of these along the tunnel alignment must be carefully studied to achieve the best possible compromise between conflicting requirements: the need to have a good knowledge of the stress-strain state in the rock mass and the need to guarantee safety for personnel against the need to keep costs low and not to interfere with excavation work and the use of the tunnel in service.

To achieve this, the ADECO-RS approach provides for four basic types of measurement station (Fig. 10.1):

- principal measurement stations;
- extrusion measurement stations;
- monitoring stations;
- systematic measurement stations.

Principle measurement stations are designed to give a complete map of the stress-strain state of the ground around cavity before, during and after the passage of the face through the section monitored. For technical and economical reasons, there use is limited to sections of shallow tunnel.

They are often used during construction in conjunction with **extrusion measurement stations,** which measure deformation of the ground in the advance core into cavity and allow the radius of influence of the face, **Rf,** to be determined.

The **monitoring stations** produce similar information to that provided by the principal measurement stations, but as they are installed inside the tunnel, they can only measure stress-strain phenomena that occurs after the passage of the face. They are particularly useful in the long term, because they provide indications of the size of bands of plasticised ground around the cavity.

The **systematic measurement stations** only provide information on tunnel convergence and are generally used on preliminary and final linings.

Monitoring and systematic measurement stations must be set up very close to the face and the zero reading taken immediately if they are to provide useful and accurate information.

10.3.1 Principle measurement stations

The principal measurement stations are set up before the passage of the face through the section measured, in order to also acquire information on deformation of the rock mass ahead of the face. It will be recalled that in the absence of ground reinforcement or improvement ahead of the face, radial deformation starts to manifest at a distance of a few tunnel radii from the face and at the face itself this can even amount to 30–40% of final convergence.

The principal measurement stations must also be able to provide clear and full information on the stress-strain response of the tunnel and the surrounding rock mass after stabilisation intervention.

To achieve this the following must be performed:

Measurement instruments for the monitoring phase

Downhole extensometers

Downhole extensometers are used to measure ground movement along the axis of a drillhole at a certain number of what are termed *measurement points* at predetermined depths along the drillhole. The profile of the deformation of the ground around a hole is obtained by measuring the change in the position of those points over time with respect to a reference point that is considered fixed.
These instruments are used widely inside tunnels above all to measure extrusion deformation deep into the core-face and convergence around the cavity deep into the ground.
Extensometers can be incremental and single or multipoint. The former measure relative displacements along the axis of the drillhole from a series of equidistant measurement points located on an inclinometer tube which lines the drillhole.
The latter, on the other hand, measure displacements in the direction of the axis of the drillhole from one or more measurements (up to a maximum of seven) with respect to the reference point at the head on the surface.
While the measurement device (probe) in incremental extensometers is only inserted in the inclinometer tube during measurement (more than one measurement station can therefore be operated using a single probe), multipoint extensometers are single piece units from this viewpoint.
One important feature of incremental extensometers is that measurements can be taken with them even when the inclinometer tube has been partially demolished. This characteristic makes them irreplaceable for measuring extrusion of the core-face during tunnel advance (dynamic measurement).

- monitoring of settlement and deformation and also of changes in the hydrogeology of the ground that occur before the passage of the face, as it simply approaches the section monitored;
- measurement of settlement and any changes in the hydraulic load that are produced following excavation, during and after the passage of the face through the section monitored;
- measurement of the extension of any plasticised zones before, during and after the passage of the face through the zones;
- assessment of the effect of ground reinforcement or improvement and of the preliminary and final lining on the stress-strain state of the walls of the cavity, firstly in the zone where they are placed, then subsequently when the face has moved away from the measurement station during the final phases when the ring of the lining is completed and closed and finally when the tunnel is in service;
- measurement of the stresses and strains in the stabilisation structures employed at every stage of the construction and life of the tunnel.

In order to achieve this it is essential to install appropriate instrumentation in the section it is intended to study that is able to function before and after the passage of the face.

At least three multipoint extensometers (the number and length of the individual bases will depend on the overburden and the local geotechnical and stratigraphic conditions) and one inclinometer must be inserted in vertical boreholes drilled from ground level before the passage of the face well outside its radius of influence. The inclinometer should be positioned a few metres to the side of the future tunnel and driven down below the level of the tunnel invert.

If the tunnel runs through an aquifer, the measurement station should also include the installation of a piezometer from ground level in order to measure variations in pore pressure.

By combining the information furnished by these instruments, a complete three dimensional picture can be obtained of the real deformation response of the rock mass to excavation.

Once the face has reached the section monitored, the instrumentation is completed to provide detailed measurement of the development of the stress-strain state in the ground and in the stabilisation structures.

To achieve this, between three and five distometer nails must be installed to measure cavity convergence and a few pairs of devices to measure radial and tangential stresses (e.g. pressure cells) should be installed in the preliminary and final linings in the crown, the side walls and the tunnel invert.

Where there are strong neutral pressures or when drainage or water proofing is employed around the cavity, a series of piezometers

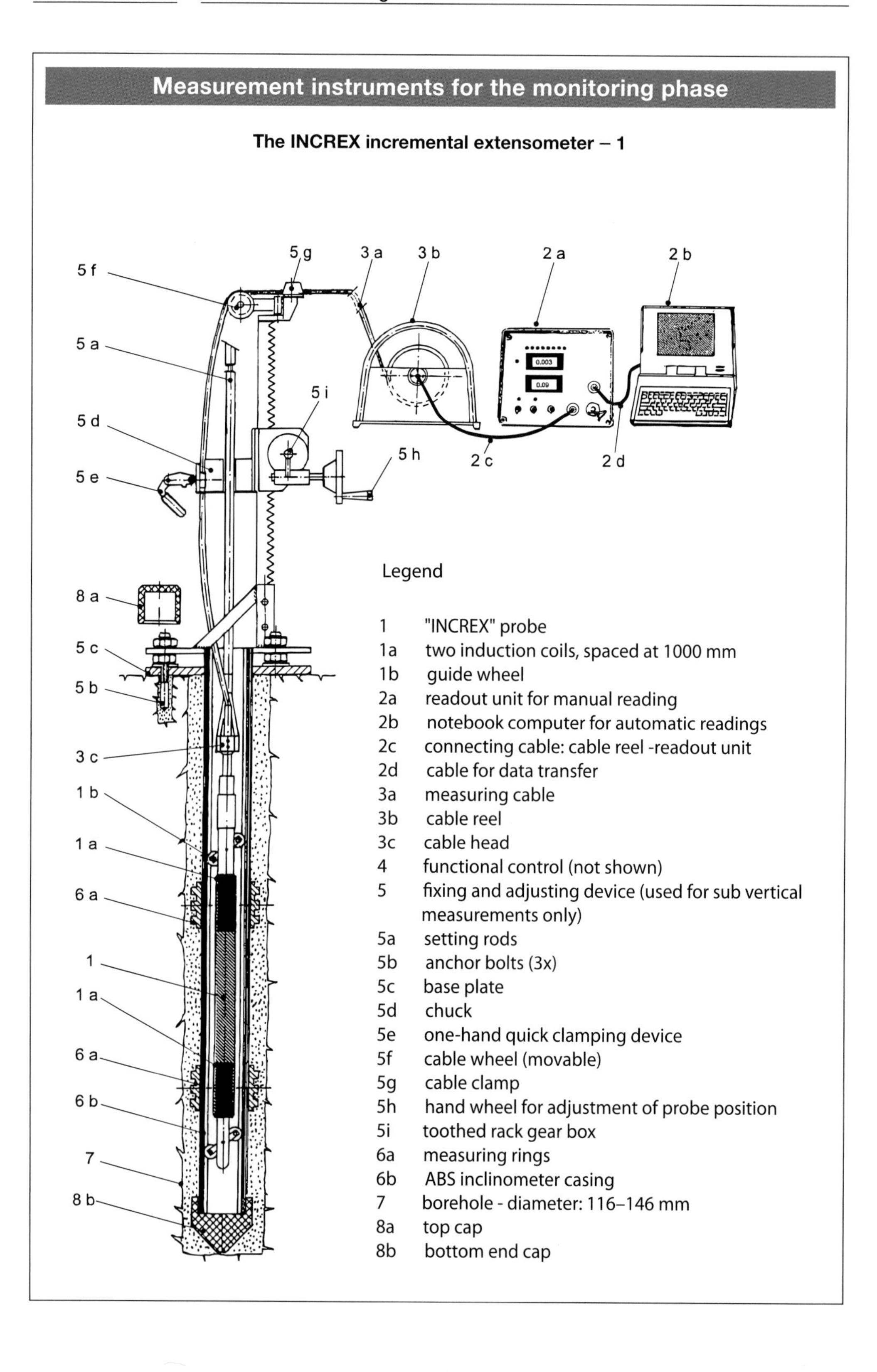
Measurement instruments for the monitoring phase
The INCREX incremental extensometer – 1
5 f
5 g
3 a
3 b
2 a
2 b
5 a
5 i
5 d
5 h
2 c
2 d
5 e
0.003
0.09
8 a
5 c
5 b
3 c
1 b
1 a
6 a
1
1 a
6 a
6 b
7
8 b
Legend
1 "INCREX" probe
1a two induction coils, spaced at 1000 mm
1b guide wheel
2a readout unit for manual reading
2b notebook computer for automatic readings
2c connecting cable: cable reel -readout unit
2d cable for data transfer
3a measuring cable
3b cable reel
3c cable head
4 functional control (not shown)
5 fixing and adjusting device (used for sub vertical measurements only)
5a setting rods
5b anchor bolts (3x)
5c base plate
5d chuck
5e one-hand quick clamping device
5f cable wheel (movable)
5g cable clamp
5h hand wheel for adjustment of probe position
5i toothed rack gear box
6a measuring rings
6b ABS inclinometer casing
7 borehole - diameter: 116–146 mm
8a top cap
8b bottom end cap

of varying lengths or a continuous piezometer must be installed, located on the wall not occupied by the extensometer, to monitor changes in the hydraulic gradient caused by tunnel construction.

10.3.2 Extrusion measurement stations

The purpose of extrusion measurement stations is to furnish information on the radius of influence of the face, $\mathbf{R_f}$, and on the magnitude of the deformation occurring in the ground that constitutes the advance core in order to assess its actual stability.

This therefore consists of the following:

- measuring the size of the relaxed zone with respect to the face;
- measuring the radial and longitudinal deformation of the advance core.

Figure 10.2 illustrates the typical configuration of an extrusion measurement station. It consists of:

- an incremental extensometer inserted into a hole drilled into the centre of the face with a length of at least two to three times the diameter of the tunnel with the head fitted for topographic sighting of its position;
- nine optical targets for topographic survey of the face set on three levels as indicated in Fig. 10.2, but only for periods when the face is at a halt for periods of not less than one week.

If the advance core is reinforced with fibre glass structural elements, it is useful to integrate extrusion measurements with measurements of the stress in those elements in order to acquire a more detailed understanding of both the relaxation of the ground

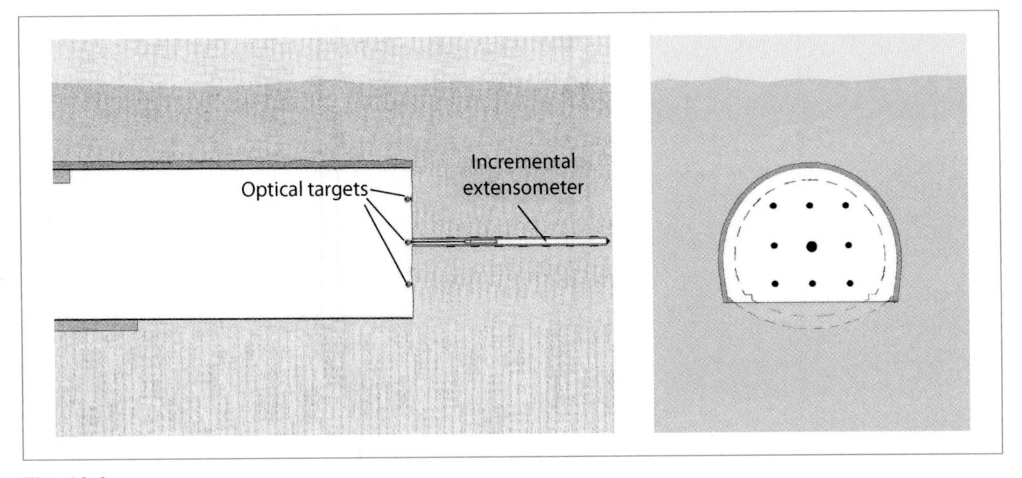

Fig. 10.2

Measurement instruments for the monitoring phase

The INCREX incremental extensometer – 2

The INCREX incremental extensometer is for measuring ground deformation along the axis of the inclinometer tube at regular 1.00 m intervals.
This is performed by first installing metal measuring rings at the measurement points on the tube (at regular intervals of 1.00 m) and then inserting it in a borehole of between 11 and 15 cm in diameter drilled along the axis to be measured. The rings are fixed into the surrounding ground by injecting cement grout into the space between the tube and the wall of the borehole.
The distance between the different pairs of rings can then be measured electronically at any time by inserting a special probe in the tube from the outer end of the borehole consisting of two electric inducers set at a distance of 1.00 m from each other.
When in the measuring position, the inducers interact with a pair of measuring rings to generate an electrical signal which is proportional to the distance that separates them. The distance between each pair of measuring rings can be measured by sliding the probe down the tube and in time a continuous curve of the axial deformation of the ground can be obtained.
The measurement technique described will guarantee accuracy of ±0,003 mm/m.

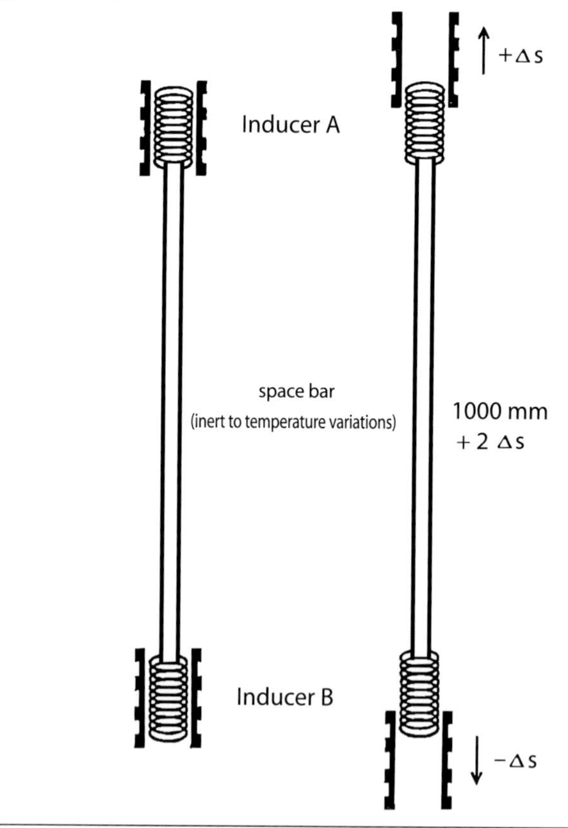

produced by tunnel advance and the stabilising effect produced by reinforcement of the advance core.

10.3.3 Monitoring stations

Monitoring stations are measurement stations that are placed inside the tunnel very close to the face for the purpose of monitoring the stress-strain state of the rock mass around the cavity and of assessing the importance of any plasticised zones there may be and how they develop over time.

The basic difference between monitoring and principal measurement stations therefore lies in that in the former there are no measurement instruments outside the tunnel and therefore there is no measurement of the deformation occurring in the ground before the passage of the face.

They are used where the overburdens or the morphological configuration will not allow drilling from the surface to position the necessary instrumentation in the zone monitored.

Monitoring stations consist of: a few multipoint extensometers placed radially in the crown and at the springline of the cavity immediately after the passage of the face in the section monitored; 2 or 5 distometer nails to measure cavity convergence; at least two load cells placed under the feet of the steel ribs and a few pressure cells to record radial and tangential stresses in the linings. For the latter purpose strain gauges may also be placed on the reinforcement.

It is essential in tunnels under the water table to install one or more piezometers.

10.3.4 Systematic measurement stations

The purpose of systematic measurement stations is to measure the magnitude of convergence and, if necessary, the stress state of the ground simply and economically, but with high frequency in time and space along the whole route of a tunnel.

They normally consist of distometer nails placed, as with monitoring stations, as close as possible to the face and optical targets installed on the final lining.

Where it is considered necessary, a few radial and tangential pressure cells may be added along with a few load cells where rock bolts or steel ribs are placed.

In cases of high convergence during construction, it is advisable to take topographic sightings of the feet of the steel ribs with the same frequency that convergence readings are taken.

Measurement instruments for the monitoring phase

The sliding micrometer

A *sliding micrometer* is an extremely versatile, precise and easy to handle incremental extensometer, very useful for systematic measurement, in any direction, of axial displacement of the ground over time at a series of measurement points set along the axis of a drillhole. It consists of a series of PVC pipes (each of which one metre in length) with an overall diameter of 60 m. The pipes are joined by sleeves which allow a series of them to be assembled and act at the same time as conical stops for the probe. The extensometric column assembled in this manner is then inserted and cemented with expansive cement in a drillhole with a diameter of at least 110 m.
The probe consists of a telescopic unit fitted with springs and grooved spherical heads on each end. These protect an invar steel tube connected to a displacement transducer which measures the distance between them.
Measurements are taken by connecting the probe to a modular guide rod and then inserting it in the sliding position in the inclinometer tube until it reaches the end of the hole. It is then rotated through 45° to the measurement position and carefully withdrawn until the spherical head **A** comes into contact with the first measurement point. At this point it is withdrawn further until the spherical head **B** reaches the second measurement point. The control monitor then displays the reference value *digitally* for that specific measurement point. The probe is then rotated to the sliding position to free it from the measurement point, when it is withdrawn until it reaches the next point where the reading operations are repeated. Once the procedure has been repeated for each measurement point, the readings are processed by special software and converted to numerical values which give the movement that has occurred at each measurement point with respect to the original reference zero value. The instrument is accurate to ±0,002 mm/m.

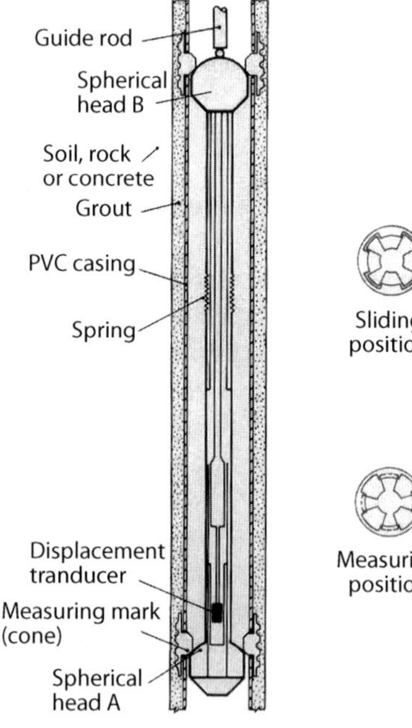

10.4 The design of the system for monitoring during construction

The design of the system for monitoring during construction will be based on the type and magnitude of the deformation phenomena that it is expected will have to be measured to implement the monitoring phase. Given the importance of the stability of the face, when an underground design engineer sets out to design a system he will draw up a plan that will above all allow the stress-strain response of the core-face to be monitored and assessed in terms of its correspondence to the predictions already formulated using the calculations performed during the diagnosis and therapy phases. As a consequence:

- in stable core-face conditions (rock type behaviour), in which the ground is stressed in the elastic range and the deformation response, at least in overall terms, is immediate and very small, it will not be necessary to specify very frequent systematic monitoring nor will the use of sophisticated instrumentation be required. Greater instrumentation will only be necessary for particular geostructural situations to monitor possible local instability (e.g. rock slide or wedges of rock being expelled) or, in very rigid material affected by discontinuity planes and fracturing sub parallel to the axis of the alignment, to prevent *rock burst*. As a consequence, it will normally be sufficient to place systematic measuring stations at a distance of one every 100 m on average, together with a few monitoring stations;
- in stable core-face in the short term situations (cohesive type behaviour), in which the ground is stressed in the elasto-plastic range and deformation phenomena is mostly deferred and not negligible in magnitude, it is important for the design engineer to monitor the development and magnitude of deformation and stress in the medium carefully, by placing extrusion, principal or monitoring (depending on the depth of the tunnel) and systematic measurement stations very frequently. This is the only way in which the necessary information can be obtained to assess the adequacy of the stabilisation intervention and to calibrate and perfect it.
  The frequency with which the various types of measurement station are installed will depend on the local geostructural and tectonic characteristics of the ground. Extrusion measurement stations should be installed whenever it is predicted that a substantial zone of ground with plastic behaviour will be formed ahead of the face, or when it is wished to calibrate and fine tune the excavation and stabilisation techniques employed. This is also recommended when tunnel advance is halted for more than seven days. Many experiences in the past have in fact demonstrated that if the core-face is not adequately protected in these circumstances, the relaxation of the ground in it spreads so much into the rock mass that it causes serious problems of instability when work resumes.
  It is best for extrusion measurement stations to be followed by systematic measurement stations installed at a distance of one quarter of the diameter of the tunnel in order to be able to reconstruct the three dimensional deformation of the ground on both sides of the core-face.
  Finally, principal or monitoring stations will normally be separated by systematic measurement stations installed every 20–40 m;

Measurement instruments for the monitoring phase

The sliding deformeter

The *sliding deformeter* is an economical version of the *sliding micrometer*. Extremely versatile, easy to handle, tough and reliable, it provides a valid alternative for all those cases in which, although a fair degree of accuracy is required, the high level of accuracy provided by the sliding micrometer and the INCREX extensometers is not necessary. *Sliding deformeters* are accurate to ±0,02 mm/m.

They consist of a series of PVC pipes (each of which three metres in length) with an overall diameter of 32 mm. In this case too the pipes can be joined by sleeves which allow a series of them to be assembled. The points for the probe to stop against are set at intervals of one metre. The extensometric column assembled *in situ* in this manner is then inserted and cemented with expansive cement into a drillhole with a diameter of 56 mm approx. As with the *sliding micrometer*, the probe is fitted with spherical heads with grooves cut into them to allow them to pass freely through the measurement points. The measurement procedure is also the similar.

The length M_0 is the distance between the head of the probe and the upper measurement points. The values are transmitted mechanically without hysteresis from a measurement rod to a displacement sensor located on the outside of the measurement tube. A subsequent measurement (M_1) gives the new distance between the head of the probe and the measurement point. The displacement ΔL of the section L_0 is given by the difference $M_1 - M_0$. The differential displacements along the line of measurement give the distribution of the deformation along the inclinometer tube. The total displacement along the measurement line can be calculated by summing all the deformations.

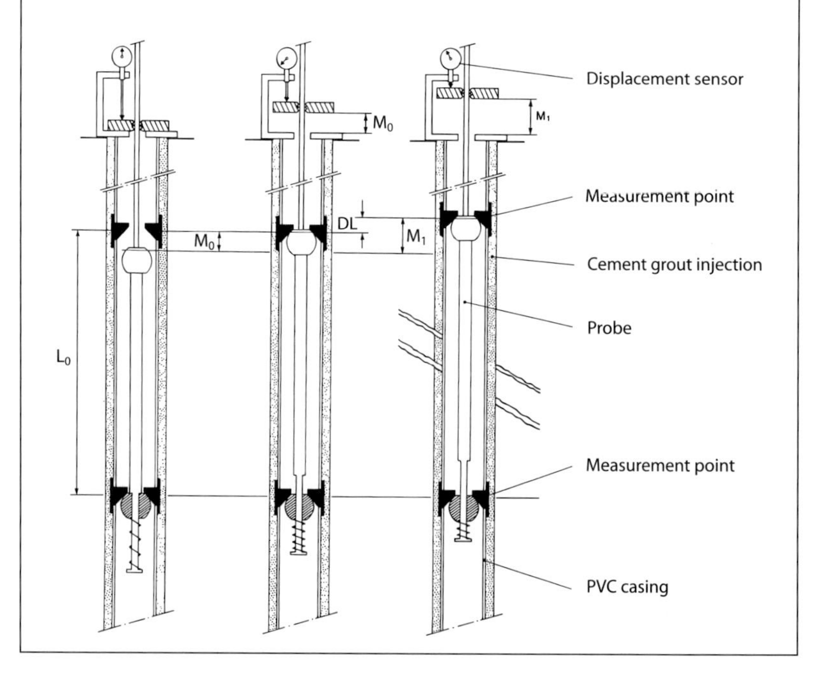

- in unstable core-face situations (loose type behaviour), where deformation of the ground would develop into the failure range if not properly controlled by intervention to preconfine the cavity in the operational phase, it is imperative for the design engineer to specify extremely frequent installation of extrusion, principal or monitoring and systematic measurement stations (the latter with a frequency of one every 10–20 m).
  It is very common in situations of this type to pass from conditions of stability to conditions of instability very suddenly following very small increases in deformation. Small changes in the characteristics of the medium or small deficiencies in the implementation of the works are sufficient to give rise to extremely serious and irreversible deformation, such as the failure of the core-face and the collapse of the cavity. Extrusion measurements are of fundamental importance in this context. In fact when the stress-strain situation is difficult to monitor, convergence alone is not sufficient to keep the deformation response of the rock mass under control and to prevent tunnel collapse. This is because convergence is the last stage of the deformation process and it is therefore an uncontrollable phenomenon, the result of the plasticisation of the ground around the cavity which, as is known, cannot be held back once it has been allowed to developed significantly. Extrusion, on the other hand, is the first stage of the deformation process. If it is properly monitored, it can even develop into substantial preconvergence and convergence phenomena, but not into tunnel collapse, because it allows the time needed to intervene effectively.

Monitoring must be carefully directed in certain contexts.

For example, in sections of tunnel affected by **large tectonic dislocations,** one may come across material of varying nature and competence organised in heavily disordered formations, which can develop in ways that are at times very difficult to predict. In these conditions material that is geomechanically very different, particularly with regard to deformability, is found close together. Modelling its behaviour can therefore be arduous: the only way to obtain precise quantitative information is by accurate measurement of the type and magnitude of deformation that occurs during and after excavation. It may also be useful to know the characteristics of the geostatic tensor, or more simply the ratio between the horizontal and the vertical pressures, remembering that it normally tends towards one as the overburden increases and the importance of structural discontinuities diminishes. Monitoring stations must therefore be installed to measure the deformation that is generated around the perimeter of the cavity and at depth inside the formation tunnelled and to measure the stress state inside the rock mass in order ascertain whether residual tectonic stresses are acting. This can be done by installing either principal or monitoring measurement stations depending on the overburden of the tunnel.

If there are fracture planes or faults along which it is feared that slip may occur, then it is wise to set joint measurement instruments across the discontinuities in different directions. The design engineer will decide the number and location of these and other instruments according to the specific circumstances given the very particular nature of these contexts.

It is of great importance to monitor the hydrogeology in intensely tectonised zones, because this is often characterised by sharp changes in permeability accompanied by equally sharp changes in neutral pressure. The strong hydraulic gradients that may be

Measurement instruments for the monitoring phase

The magnetic extensometer

The magnetic extensometer is widely used primarily for measuring vertical settlement of embankments for roads during construction and horizontal displacements of dams. It consists of an inspection tube protected by a corrugated outer sheath to which metal magnetic rings (measurement points) are fixed at regular intervals. Once the tube is inserted into a drillhole, the rings are fixed firmly into the ground. Measurement is performed by a probe which slides inside the inspection tube and measures the position of the magnetic rings and therefore also the movement that has occurred between one measurement and another. The measurements are accurate to around ±1 mm for each ring, **which is not, however, sufficient** for measuring total and differential extrusion of the core-face of a tunnel.

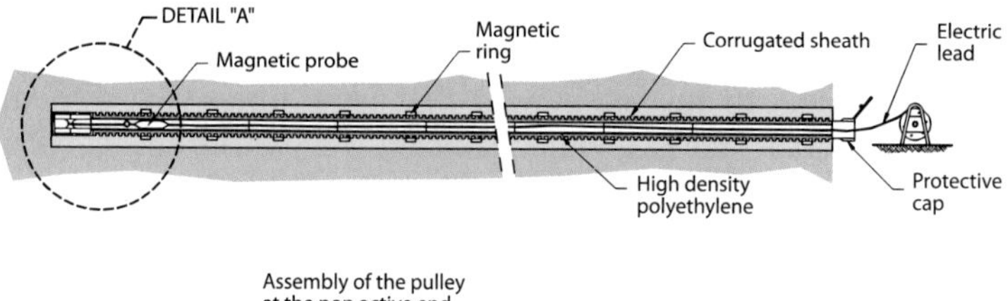

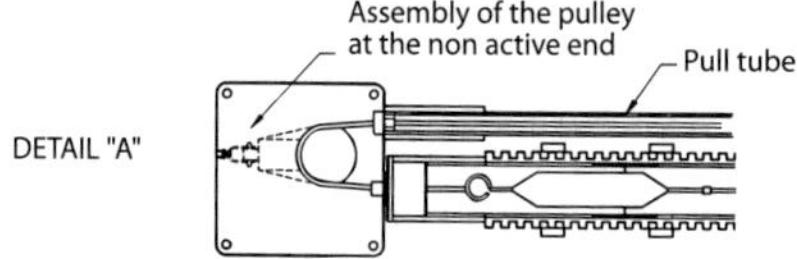

encountered can seriously compromise the long and short term stability of a tunnel if not brought quickly under control.

It is important **under the water table** for water intercepted during the advance and then conveyed into drains located in the final lining structure to be systematically analysed and the flow rate of drainage water measured. This must be performed in relation to changes in piezometric levels and to the possible effects of springs in the vicinity of the tunnel, in order to construct a hydrogeological model that is as close to the reality as possible and to assess the long term effects of the drainage exerted by the tunnel on the aquifer. The design engineer must also perform complete monitoring of ground deformation along **shallow tunnel** sections by installing principal measurement stations to measure deformation from the surface that develops at tunnel depth and settlement at ground level. This is particularly important if the tunnel passes under surface constructions of any consequence, such as roads or buildings or when excavation passes beneath a river bed. In the former case the surface constructions themselves must be monitored with appropriate instruments (surface clinometers, differential settlement gauges, etc.) designed to verify that original conditions do not change (absolute and differential monitoring of settlement, monitoring of any fissures that open and of disturbance caused by vibrations, etc.). In the latter case, hydrogeological monitoring must also be performed. This is indispensable for monitoring how bodies of surface water are drained by the tunnel and for ensuring that no statics or safety problems arise in the tunnel under construction.

In **urban areas** the design engineer will pay great attention to the functioning of all utility infrastructures which could be damaged by the passage of a tunnel. Leaks, which may even be large and dangerous, can be caused in aqueducts, sewers and gas lines as a consequence of settlement caused by the passage of a tunnel and high voltage power lines will tolerate only very slight deformation before they cease to function. It is indispensable in this respect to employ an integrated monitoring system so that measurements can be cross checked with iterative processing of the data in order to eliminate or at least substantially reduce the possibility of errors arising from potentially inaccurate measurements.

10.5 Monitoring the tunnel when in service

If the huge importance and the negative, or positive impact that an underground construction can have on its surrounding environment is considered then clearly its functioning and safety must be monitored continuously and this must continue after construction for at least the whole of its service life.

Until not many years ago monitoring of this type was not very common, partly because of the need to interrupt service to allow maintenance personnel the time and possibility to take the necessary measurements. Today, automatic devices are available with which remote measurement of most of the parameters needed to monitor the health of an underground construction can be performed. This means that a monitoring system can be activated during construction which is able to continue furnishing data for many years after a tunnel is commissioned without interfering with tunnel service.

With sufficient knowledge of the stress-strain behaviour of the ground-structure system since the start of construction together with up-to-date information on the condition

Measurement instruments for the monitoring phase

Single and multipoint extensometers

Single and multipoint extensometers are used to measure ground movement along the axis of a drillhole with respect to the mouth of the hole at a certain number of measurement points located at various depths. An extensometer is defined as single-point or multipoint depending on the number of measurement points installed in it (it can have a maximum of seven points).

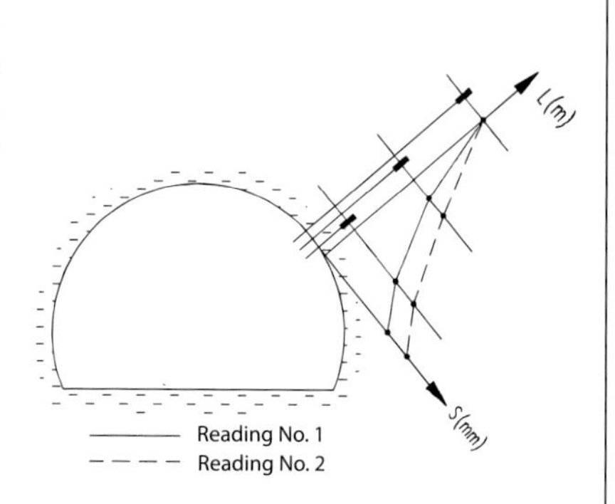

The installation of a single-point extensometer requires a drillhole with a diameter of 4.5 cm, approx., while a diameter of 10 cm approx. is needed for a multipoint extensometer. Typical use of single and multipoint extensometers is for measurement of deformation around the cavity.

The system consists of installing anchor points in the ground inside drillholes connected to the surface by steel, invar steel or fibre glass rods protected by a tough anti-friction sheath.

Invar steel is preferable to ordinary steel when large variations in temperature are expected, because it has a coefficient of expansion that is about one tenth that of ordinary steel. Fibre glass rods, now in very widespread use because they are practical (the extensometer arrives on site with the bases rolled up ready for installation), cannot be used to measure compressive movements nor when the length of the measurement points to be installed is greater than 70 –80 m.

The rods are free to slide inside the sheath and transfer the movement of the anchor point to the head of the instrument. Each movement can be measured using either a digital comparator or an electric displacement transducer.

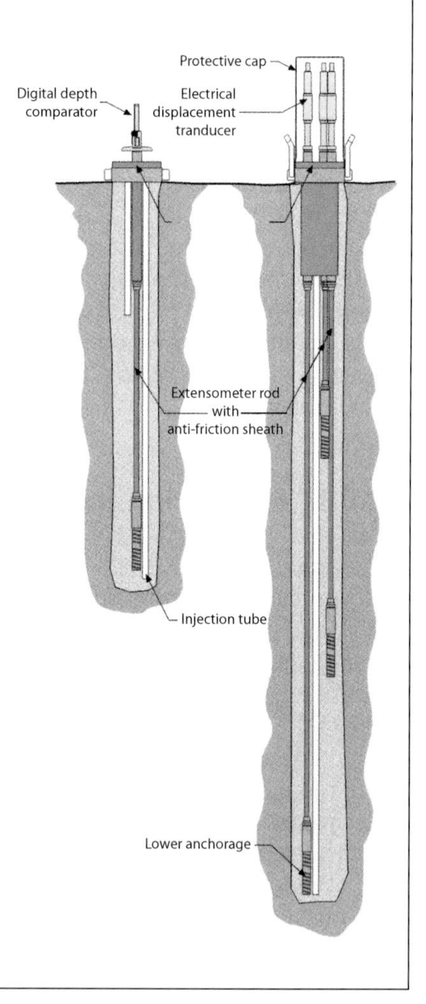

of a tunnel, the problem of maintenance and repair can be approached using scientific methods.

Most of the instrumentation installed in the measurement stations used for monitoring during construction can be used to advantage for this purpose. More particularly, interpretation of the measurements taken when the tunnel is in service is easier and more meaningful if they can be compared directly with similar measurements taken during construction.

Clearly those tunnels which require most careful monitoring when in service are those affected by situations which typically change over time, such as tunnels in landslide or heavily tectonised zones for example, or which in any case are subject to substantial viscous or creep phenomena.

The portal sections of tunnels with shallow overburdens should also be monitored for the whole working life of a tunnel because they can be affected by large movements in the surrounding ground (e.g. the result of seismic activity) at any moment even after construction is complete.

Which measurements are the most appropriate to take when a tunnel is in service? Before this question can be answered one must have a clear idea of the purpose of long term monitoring. It must definitely perform the following:

- verify that the functioning of the construction, understood as meaning all that intervention performed to ensure the stability and integrity of the underground cavity, is guaranteed with the safety coefficient specified in the design;
- monitor deformation of the cavity, which may occur over time;
- allow the impact of the construction on the surrounding environment to be monitored and assessed, with particular attention paid to pre-existing hydrogeological equilibriums (potential damage that should be monitored consists for example of a possible decrease in the fertility of agricultural soil on the surface as a result of the water table sinking, or difficulty in disposing of drainage water when it contains harmful substances either in solution or suspension).

Monitoring a tunnel in service should therefore consist of acquiring data on changes in both the stress-strain state of stabilisation instruments and the hydrogeological conditions around cavity.

The instrumentation already installed at the construction stage can be used to furnish directly, and not by deduction, the distribution of the pressures acting on the preliminary and final linings. The design engineer should pay special attention to asymmetrical loads that may arise as a result of particular pre-existing stress states in the ground, which are not normally predicted by normal numerical models.

Convergence measurements must occupy an important place in monitoring in service. They consist basically of measuring the internal profile of the lining and verifying that it does not deform over time

Measurement instruments for the monitoring phase

Distometer nails

Distometer nails are simple nails with a threaded or eyebolt heads used as reference points for the measurement of convergence in a tunnel.
This is performed by positioning them around the cavity and fixing them partially into the ground immediately after the passage of the face using a geometrical configuration which gives two, three or five measurement points.
The convergence for each measurement point configured in this way can be measured periodically using various systems. The oldest and most commonly used employs an invar steel tape graduated in centimetres (tape distometer) which, once it is connected at the ends to a pair of distometer nails, is brought to a constant tension by a special dynamometric device. The distance between each pair of nails is read using a mechanical or digital gauge incorporated in the

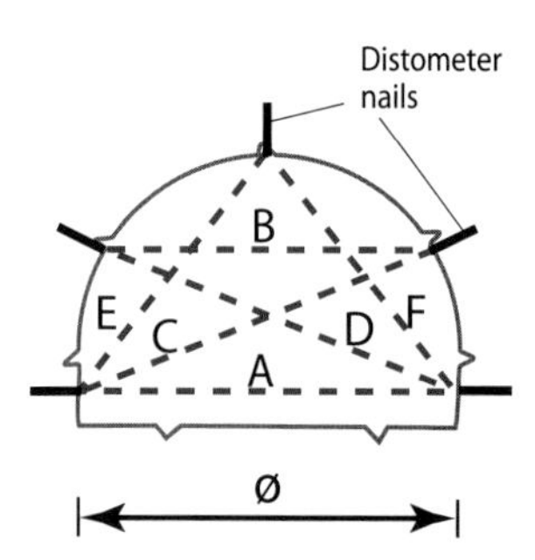

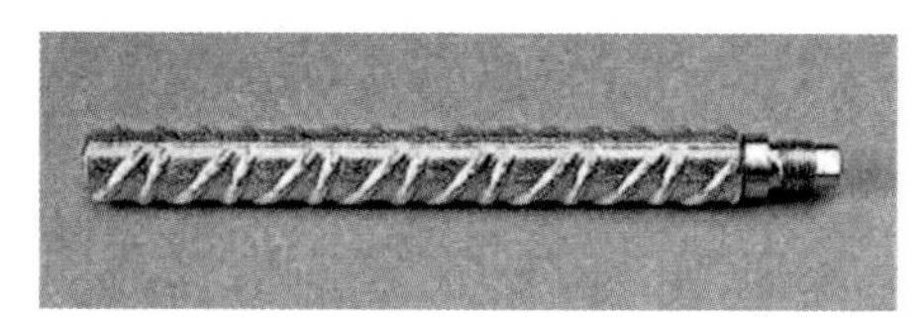

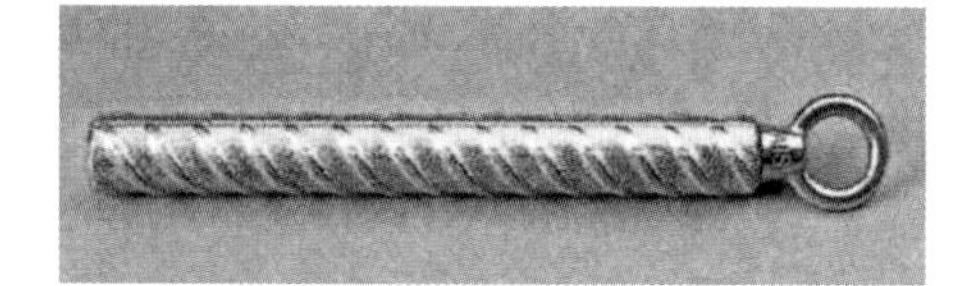

instrument.
In recent years the use of topographical survey systems has become widespread to avoid interference with construction site activity when measurements are taken. Appropriate optical targets are fitted on the distometer nails and distances are measured periodically using a theodolite.
Whichever method of measurement is used, it is important to be able to compare the measurements taken to a "zero" reading taken immediately after the distometer nails are placed.

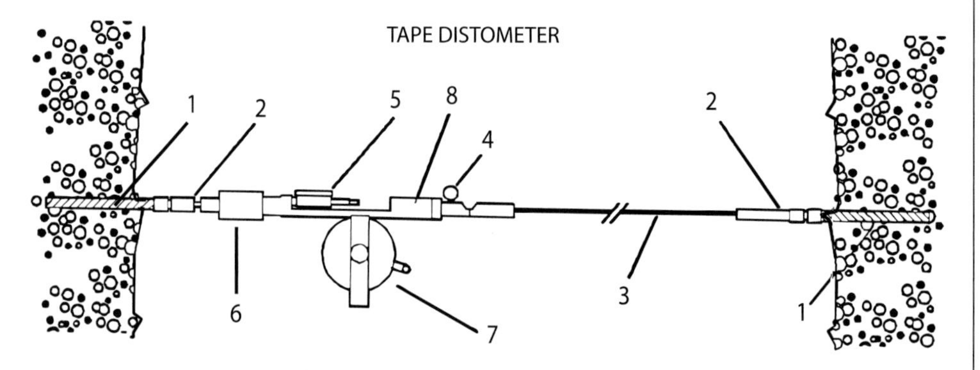

(1) Distometric nail
(2) Ball joint with connector
(3) Perforated steel invar tape
(4) Gudgeon pin
(5) Mechanical or digital gauge
(6) Tensioning device
(7) Tape winder
(8) Spring housing and tension indicator

(i.e. detecting possible ovalisation of the cross section of a tunnel). Normal distometers, convergence nails or topographic targets are sufficient for this type of measurement, but systems which interfere less with tunnel service such as radar or television can be used in particular situations.

Naturally if measurement campaigns should discover anomalous behaviour, the design engineer must make a detailed analysis of the situation, in order to identify the causes and therefore the best remedies for restoring the original conditions of safety.

If a tunnel runs through substantial aquifers, then it is very important for the design engineer to monitor the transition periods (generally long and occupying large physical areas) when the water table first dries and is then replenished. This can be done by installing a network of piezometers before construction actually commences, which should remain in operation for the whole working life of a tunnel. If tunnel advance takes place under hydrodynamic conditions, there may be very substantial subsidence along the tunnel alignment, which would include the whole area affected by ground consolidation phenomena.

Given the number of instruments used and the number of readings to be taken more or less simultaneously, not to mention the environment in which measurements must be taken, it is extremely advisable to employ automated methods as much as possible to acquire and process data. This is the only way in which regular and systematic measurements can be taken without causing even the slightest disturbance either to excavation activities during construction or to users when a tunnel is in service.

A typical data acquisition and transmission system consists of (Fig. 10.3):

- a series of **sub-systems**, each used to manage a measurement station by means of a computerised control unit located in a protected place inside a tunnel near the measurement station;
- a number of **central units** each controlled by a micro processor which controls and interrogates the various sub-systems via a telephone cable according to set schedules, taking readings of all the instruments in the different sections of the tunnel at once or, if necessary, a reading of a single instrument. These are positioned near tunnel portals and are connected by modem to an operating unit with which it can dialogue by receiving and transmitting commands, even in the form of alarm signals to report potential situations of danger;
- an **operating unit**, the real brain of the system, programmed to manage one or more central units via modem. All monitored sections of the tunnel can be controlled by it with the acquisition of data and all the necessary processing.

The centralisation and computerisation of measurements enables systems to be built that are constantly active and which can take safety

Measurement instruments for the monitoring phase

Inclinometers – 1

General information on the instrumentation

Inclinometers are widely used for monitoring vertical and horizontal displacements of the ground and for measuring variations in the position of structures.

They function by measuring variations from the vertical at significant points. Numerical integration is used to obtain the movements from the inclination.

Measurements can be taken using either removable or fixed equipment. The former are used when readings for the whole length of the inclinometer tubes is required and when ground movements are large in magnitude, but over a long period of time. If, however, movement must be monitored with very frequent observations or automatically, then fixed inclinometers are recommended, connected when necessary to a computerised data acquisition system. The basic instrumentation consists of the following parts:

1. **an inclinometer tube**, in ABS, aluminium or fibre glass with grooves (see figure);
2. **an inclinometer probe**, consisting of a cylindrical metal body with two carriages which by sliding in the grooves in the inclinometer tube ensure that the azimuthal direction is maintained. The distance between carriages determines the interval between the measurements (generally 50 cm). The measurement sensors are housed in the cylindrical body. They are of the digital or analogue, servo-accelerometer or magnetic-resistance type, which measure the inclination of the inclinometer tube in two reciprocally orthogonal planes;
3. **a measurement cable** used to move the probe and as an electrical connection. It has notches in it, generally every 50 cm, to facilitate positioning of the inclinometer probe during measurement operations;
4. **a control unit** with which inclinometer measurements can be read at different distances in the form most appropriate for processing. Digital probes can be connected directly to a portable computer.

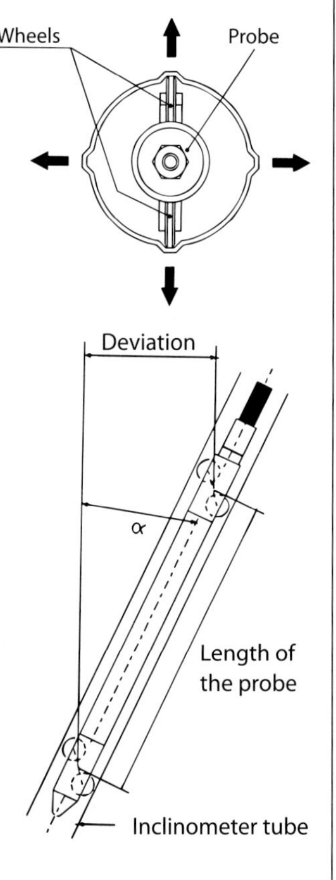

The following accessories are also advisable for proper and accurate measurement:

- **a test probe**, to check that the inclinometer tube is properly placed and functioning before the measurement probe is introduced;
- **a spiral probe**, to measure the angle of spiralling or twisting of the grooves in the inclinometer tubes that may occur when sections of tube (joined by sleeve couplings) are not perfectly aligned or as a result of operations to recover the temporary lining of the borehole used during installation;
- **a test bench** to calibrate the inclinometer probe.

Fixed inclinometers are similar to the removable types, but consist of a chain of inclinometer probes inserted permanently in an inclinometer tube and connected to automatic data acquisition systems.

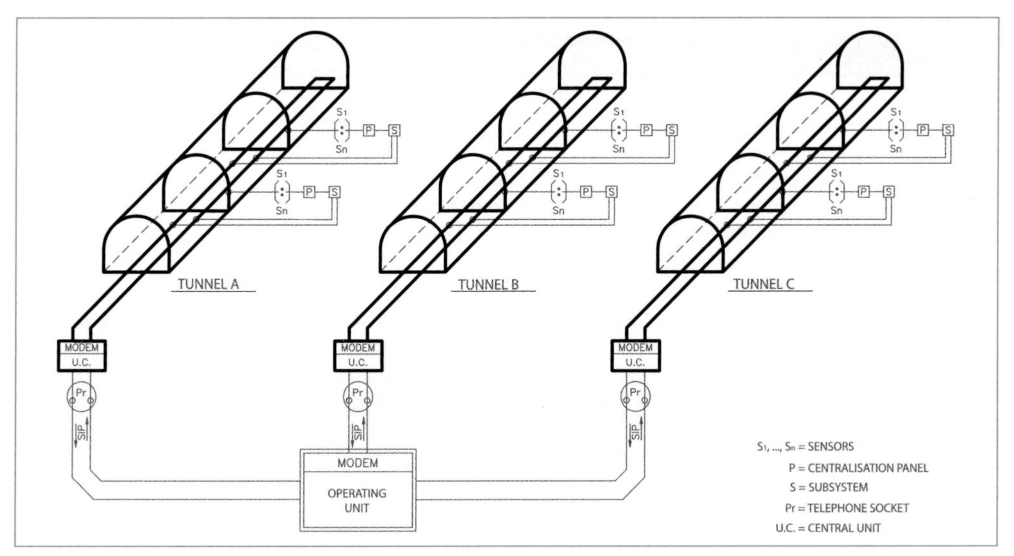

Fig. 10.3

and prevention measures by alerting tunnel management authorities when threshold stress and deformation values are exceeded.

10.6 The interpretation of measurements

10.6.1 Background

For monitoring to be genuinely useful in fine tuning the stages and rates of excavation and the type and intensity of the stabilisation instruments used, it is fundamental for the measurements to be properly interpreted. Interpretation is often not easy for a whole series of reasons, but mainly because of the following:

- **scattered distribution**: if the instruments used do not provide perfect repetition of the magnitude of the measurements, there may not be a sufficiently close cluster around a central value and this will room for doubt over which is the most accurate measurement to consider;
- **the reliability of the instruments**: the delicacy of many measuring devices, negligence in following instructions, wrong positioning or wrong use of an instrument in relation to the local situation, or malfunctioning of the equipment can lead to serious errors of interpretation and sometimes even to the loss of the instrument. It is always best to use less sophisticated equipment, more appropriate to an underground environment, and to install more than the minimum number considered necessary;
- **external influences**: the presence of still or running water, variations in temperature and humidity and so on with respect to the reference configuration can negatively

Measurement instruments for the monitoring phase

Inclinometers – 2

Execution and processing of inclinometer measurements
Measurements consist of reading the inclination of the inclinometer column with respect to the vertical at intervals of 50 cm along it on two orthogonal planes determined by the position of the guide grooves in the inclinometer tube.
The first series of measurements taken along the whole tube is termed the *zero reading* in the sense that it is the reference reading against which all movements of the inclinometer tube will be calculated from that moment onwards. It is important for readings to be always performed for all four guides because tubes often become oval in the course of time as deformation occurs and this prevents the probe from sliding along one pair of guides, while the other pair still remains free.
The guides should also be checked for spiralling during the zero reading. It is good practice to check the condition of a tube before inserting a probe in it, by using a *test probe*, which has the same external mechanics as an inclinometer probe, but without the costly sensors. The accessibility of the tube guides can be tested by running a test probe along the whole length of the tube to assess the risk of losing a probe because of obstructions inside the tube.
The readings can either be against the head of the inclinometer tube (the position of which is determined by optical measurements) or against the bottom (considered as anchored or at least firmly set in a bed of stable ground). The accuracy will depend on the quality of the instrumentation used and on how well measurement is performed. It can be affected by systematic and accidental errors.
The following may cause systematic errors:

- variation in the sensitivity of the measuring equipment (because of temperature changes or ageing sensors);
- variation in the position of the inclinometer sensors due to the mechanical design of the instrument;
- variation in the zero calibration of the sensors themselves;
- spiralling of the tube.

The following may cause accidental errors:

- changes in the position of the probe due to variations in the mechanical tolerances between wheels and guides due to deposits, encrustations, joints, etc.;
- inaccurate determination of the position of the probe.

In order to correct for systematic errors when interpreting readings, measurements must be performed twice in different positions by rotating the probe through 180°. Monitoring, archiving, interpretation and graphical presentation of measurements is normally performed automatically by using special software which is generally able to describe the deformation of the inclinometer tube installed in the ground on the basis of Cartesian or polar co-ordinates, in the following terms:

- **absolute** with respect to the vertical (i.e. the absolute position of the tube);
- **differential** with respect to the previous reading (the movements of the tube are relative to the zero reading or a previous reading.);
- **local**: "local" contributions are highlighted for each individual section of the differential curve.

affect the reliability of measurements. Equally negative consequences can be produced as a result of excavation operations, especially if tunnel advance is performed by blasting, or during stabilisation operations.

Naturally the more systematic monitoring is and the more accurately it is interpreted, the greater and more useful the information will be to the design engineer in verifying the accuracy of the predictions of the stress-strain response of the ground made in diagnosis phase and in critically analysing the decisions made in the therapy phase concerning stabilisation intervention. More specifically, he will be able to employ standard back-analysis procedures to calibrate the behaviour model of the ground calculated in the diagnosis and therapy phases and to perfect the design during construction as well as to verify whether it is appropriate in relation to the geotechnical conditions actually encountered.

It is important for the design engineer to consider all aspects that might influence the genesis and development of the deformation response of a tunnel during excavation when interpreting measurements. It should not be forgotten that structural discontinuities of greater or lesser magnitude can have a dominating effect in this respect. Rock falls and *rock bursts*, for example, are very frequently attributable to the particular position of the discontinuities in relation to the direction of tunnel advance.

A few criteria that may be useful for correct interpretation of core-face extrusion and cavity convergence measurements are given below.

10.6.2 Interpretation of extrusion measurements

Core-face extrusion measurements are normally plotted graphically to give the magnitude of deformation as a function of time or face advance (see Fig. 10.4). Analysis of these graphs generally provides all the information necessary for correct assessment of the stability of a tunnel in the face zone: increasing extrusive deformation is a warning sign which is very much more useful than the same trend for cavity convergence, because as opposed the latter, it usually allows sufficient time to quickly intervene and take the counter measures needed to save the situation before it is irremediably compromised.

If a point on an extrusion curve over time is considered on a graph, the ordinate indicates the **magnitude** of extrusion at the point in time considered, the angle of the tangent to the curve gives the **velocity** of the deformation, while the curve itself describes its **acceleration**.

For interpretation purposes, the magnitude and velocity of extrusion are dependent on the geomechanical characteristics of the material and are not particularly indicators of the onset of failure. Acceleration, which is independent of the geomechanical properties of the ground, is, however, a certain indicator of the onset of collapse if it remains positive. It is therefore essential for accelerations of extrusion to be constantly monitored in order to ensure that there is always time to intervene and quickly place adequate stabilisation structures or, if danger is imminent, to alert personnel and remove equipment, utilities and structures from harm's way.

Curves for the velocity and acceleration of the movement of the ground in the direction parallel to the longitudinal axis of the tunnel into it can be plotted for each individual measurement point on an incremental extensometer rod. More specifically the

Measurement instruments for the monitoring phase

Piezometers – 1

Piezometers are widely used for measuring the piezometric level of aquifers or pore pressures in the ground.
The main determining characteristic for the performance of different types of piezometer is the volume of water needed to obtain the measurement. The smaller the volume, the better the piezometer is for rapid measurement of pressure changes or for material with poor permeability. Given this basic consideration, there are two types of instrument: **open circuit** and **closed circuit.**

Borehole **open circuit** piezometers consist of a filter cell (or a slotted pipe) a few decimetres long connected to the surface by means of one or more blind tubes. In the latter case water can be circulated through them to clean the filter (flushing).
The functioning of this type of piezometer is very simple. The borehole is filled with sand and fine gravel around the tube in order to allow water to enter the tube through the filter. After a certain period of time (depending on the permeability of the ground), the water stabilises at a certain level which is the same as that of the surrounding water table.

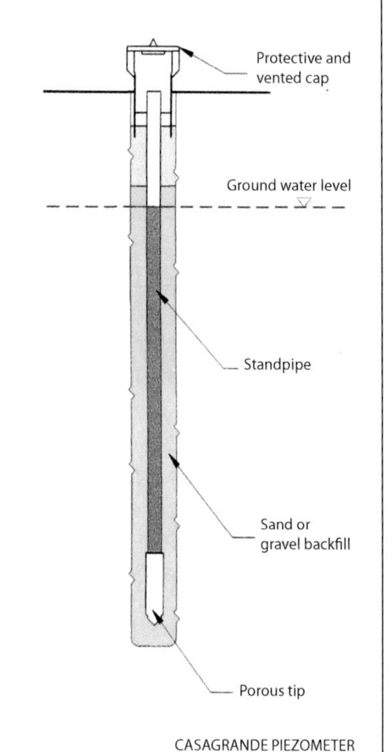

CASAGRANDE PIEZOMETER

If however the filter cell is isolated from the rest of the borehole, then the piezometer measures the water pressure in the stratum in which it is installed. This type of instrument is termed a **Casagrande** piezometer. Measurement in this case is performed using an electrical **dipmeter** consisting of a graduated cable fitted with a prod on the end which activates a light and/or sound signal when it comes into contact with water.
In soft soils "drive in" piezometers can be used consisting of a filter protected by a stainless steel point and a steel tube which is driven into the ground with a central plastic tube.

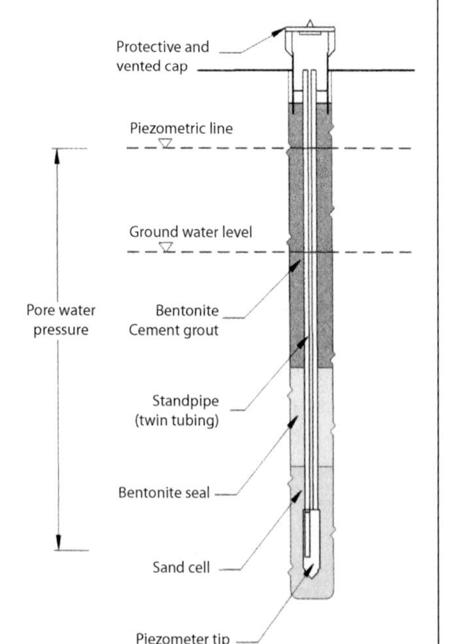

incremental curve, which relates to the entire length of the drill hole, and the curve for the individual sections into which the extensometer is divided should be plotted (Fig. 10.5). Curves of the latter type are particularly useful for comparing events and trends that manifest in different bands of ground within the core-face, or for making comparisons of the deformation behaviour of the ground along the same drill hole but at different times, or for correlating extrusion measurements with specific events, such as excavation activity or other external influences.

A design engineer can analyse core-face extrusion data to verify whether the behaviour category of a tunnel under construction actually corresponds to what was predicted in the diagnosis phase. If extrusion is practically nil, then the situation in question is definitely that of a stable core-face (behaviour category A); if there is significant extrusion, which, however, accelerates negatively then the core-face can be classified as stable in the short term (behaviour category B); if on the other hand there is clear extrusion characterised above all by lasting positive acceleration, then the situation is that of an unstable

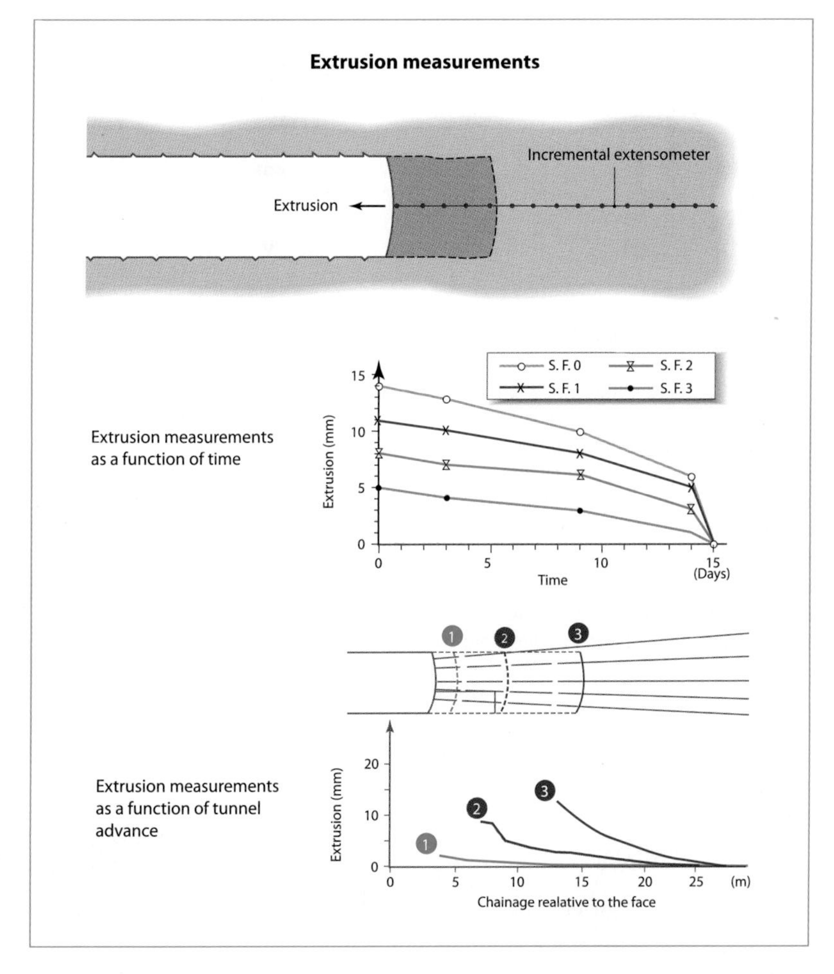

Fig. 10.4

Measurement instruments for the monitoring phase

Piezometers – 2

In **closed circuit** piezometers the water passes through a filter to enter a *hydraulic chamber* (pressure chambers), which is generally very small. Here it acts on a diaphragm causing it to deflect in proportion to the water pressure acting on it. Measurement can be performed by different electrical or pneumatic methods. The membrane (or diaphragm) in pneumatic piezometers acts as a separator between the water and a gas (nitrogen), and is controlled by a unit which balances the pressure between the water and the gas. The balanced pressure is displayed on the digital monitor of the control unit at the surface.
The main feature of this type of piezometer is that they have practically no zero drift over time and they are therefore highly reliable. The disadvantage is that measurements take a long time and connections of longer than 200 metres are not recommended, while automation is costly and not very reliable.

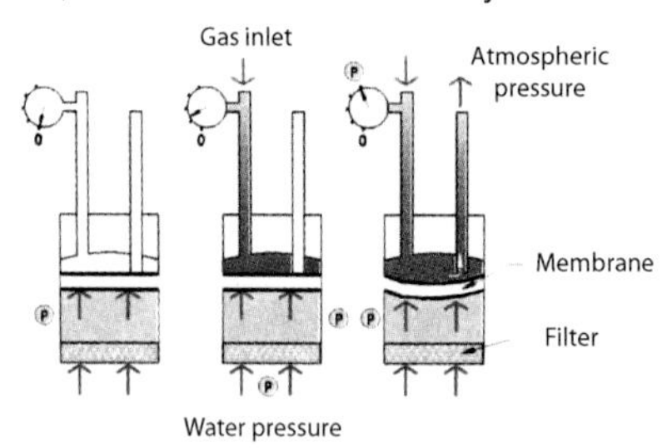

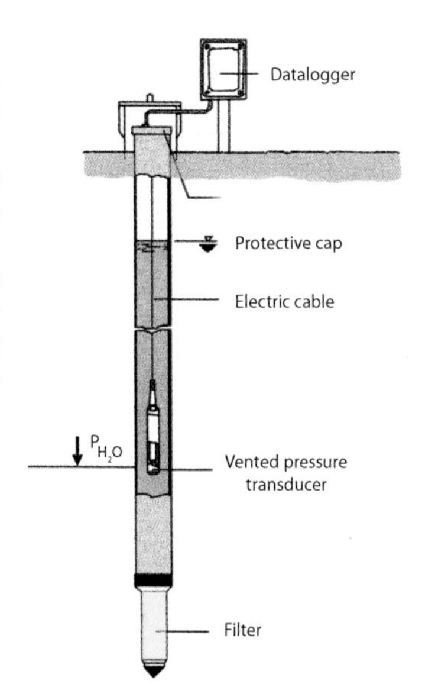

Electric piezometers may be of the vibrating wire or the ceramic (extensometer) type. Deformation causes a change in the tension of a steel wire stretched between the diaphragm and the body of the instrument in **vibrating wire type piezometers.** The deformation of the membrane, which gives the water pressure, can be obtained from the frequency at which the wire vibrates.
In **ceramic piezometers** the change in pressure is measured by a ceramic diaphragm in which extensometers are printed in a Wheatstone bridge configuration so that the electrical resistance varies as a function of the mechanical strain to which they are subjected. The change in the resistance is transformed into an easy to measure electrical signal. These piezometers give a particularly rapid response and are suitable for measuring very low pore pressures even in soils with low permeability.
Vibrating wire instruments are highly reliable for long term monitoring (10 years or longer), while ceramic piezometers can be used for dynamic measurement (hundreds of readings per second).

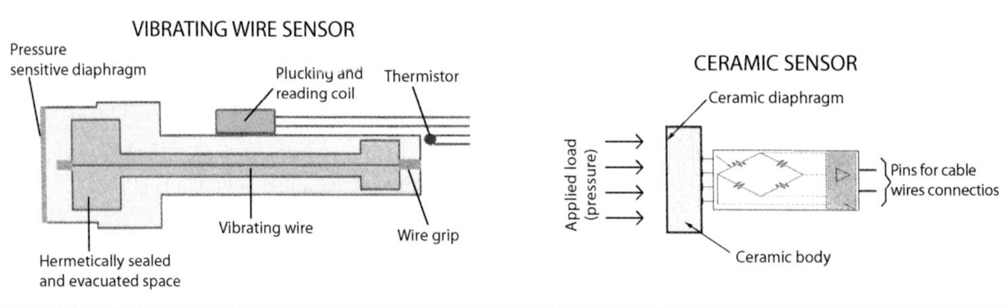

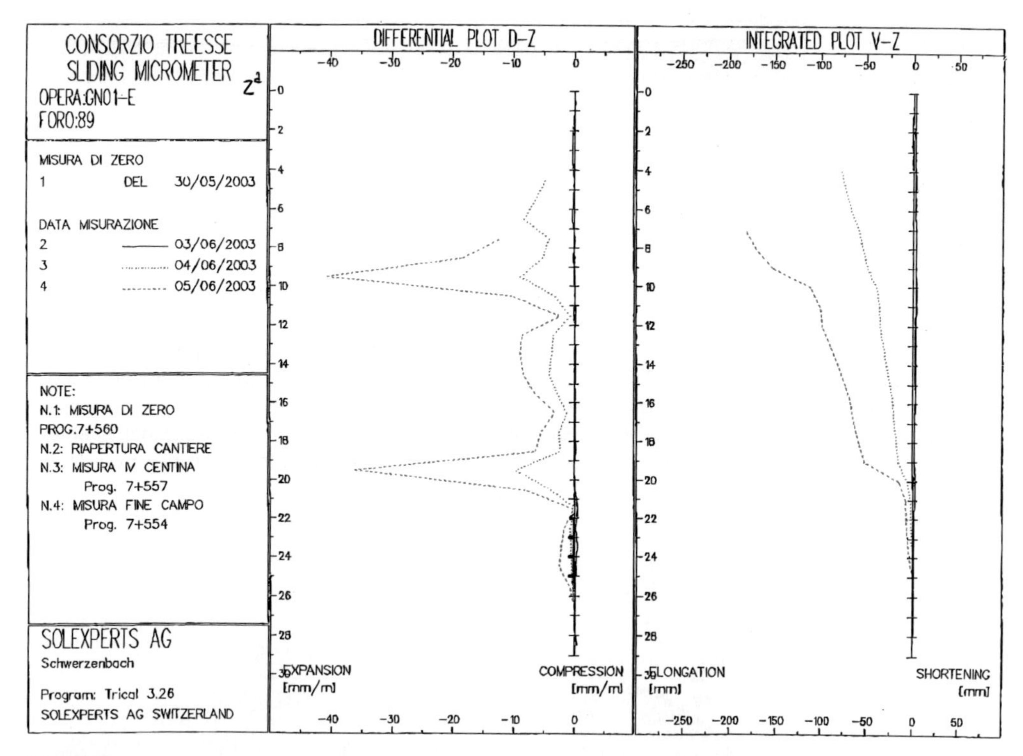

Fig. 10.5

core-face (behaviour category C) and excavation may only proceed if preconfinement of the cavity is performed to bring the overall stress-strain situation of the face-core back to at least conditions typical of the behaviour category B. Extrusion measurements in this case will be designed more than anything else to assess the effectiveness of the stabilisation intervention performed ahead of the face to limit extrusion.

Another variable that can be estimated from extrusion measurements is the radius of influence of the face, $\mathbf{R_f}$, intended as meaning the distance from the face of the source of stress turbulence induced by tunnel advance. This can also be estimated in sections of tunnel with shallow overburdens by taking extensometer measurements from ground level and starting to take readings before the face arrives.

10.6.3 The interpretation of convergence measurements

Convergence measurements taken during construction are basically of two types: surface and depth, which is to say they are either measurements of radial convergence of the profile of the cavity or they are measurements taken inside the rock mass at a distance from the axis of the tunnel. Measurements of the first type are made using distometers or topographic targets and single-point extensometers, while for the second type multipoint extensometers are used with the deepest anchor point located in a zone of the ground that is undisturbed by tunnel advance.

Measurement instruments for the monitoring phase

Pressure cells

Pressure cells are extremely useful for measuring the total pressure that builds up in the ground, in the lining or at the interface between the ground and the lining.
A large number of different types are available on the market which differ in size, shape, maximum working pressure, the characteristics of the pressure transducer used and the material of which it is made (galvanised steel or stainless steel), depending on the use to be made of it (measurement of radial/tangential stresses in a shotcrete lining or of the contact pressure between the ground and the lining, etc.).
They function on hydraulic principles and consist of two steel sheets welded around the whole of the edge and separated internally by a thin

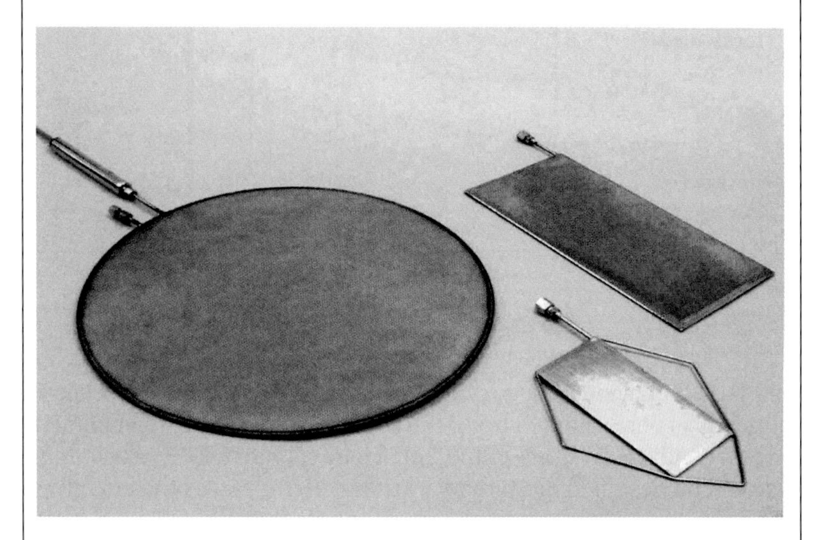

cavity which is filled under a vacuum with deaerated fluid to minimise its compressibility.
Any pressure whatsoever that is exerted on the cell is immediately transmitted by the fluid to a pressure transducer, which is usually either a ceramic sensor or a vibrating wire (both of which can be easily connected to an automatic data acquisition system) fitted either directly on the cell or remotely using a rilsan tube, which is also filled with oil.
When pressure measurements are taken inside shotcrete or concrete, a "pocket" may form while the mixture is setting which creates a gap between the cell and the surrounding material and prevents it from functioning properly. These problems can be solved by using special pumps which re-pressurise the hydraulic cell and restore the contact required by eliminating the gap.

The results of the measurements taken must be interpreted in the context of a whole series of qualitative observations which enable each particular situation to be assessed (e.g. prevalent formation of solid loads or of plasticised rings of ground) and the reliability of the measurement instruments themselves to be judged.

Convergence curves can be plotted as a function of time or of distance from the face depending on the aspect it is wished to focus on most (Fig. 10.6). If a point on a convergence curve over time is considered on a graph, the ordinate indicates **magnitude** and the angle of the tangent to the curve gives the **velocity** of the deformation, while the curve itself describes its **acceleration** (Fig. 10.7).

Measurements taken using multipoint extensometers can be used to plot the curve for radial deformation as a function of the distance from the walls of the excavation (Fig. 10.8). If the extensometer is sufficiently long, the displacement of the anchor at its end inside the rock mass can be considered to be nil. The head of the instrument positioned on the wall of the tunnel will therefore give the total radial displacement, while the intermediate measurement points will give the radial displacement at those points. What is termed the **deformation gradient** can be obtained from this curve, which, as

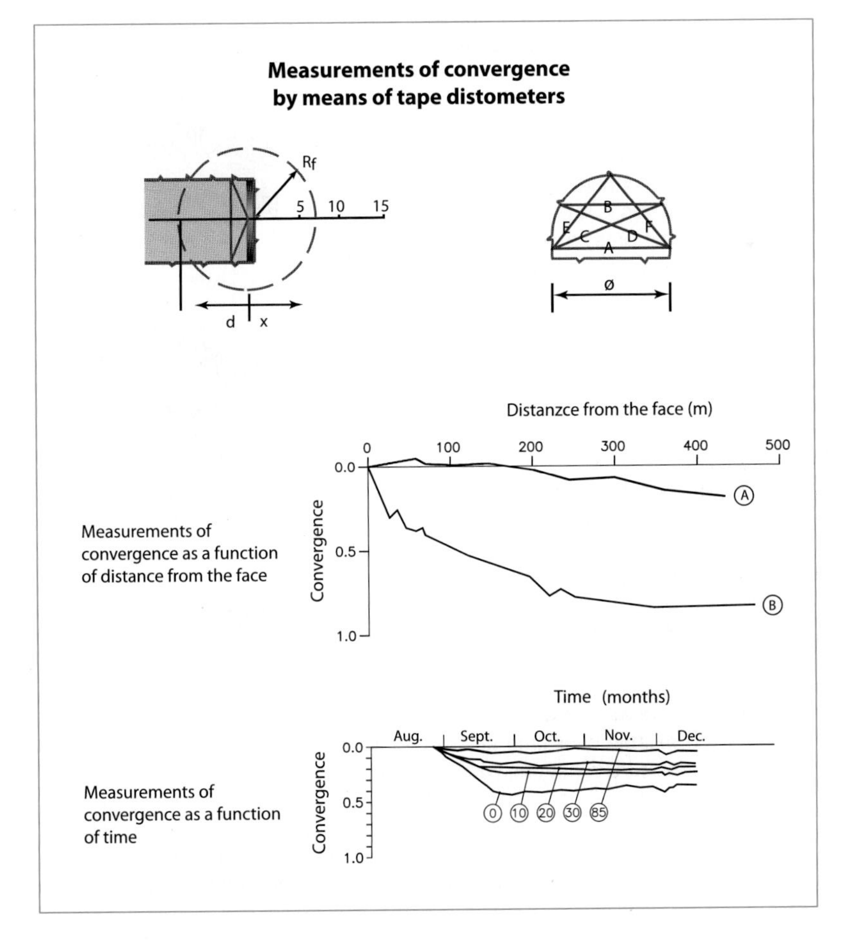

Fig. 10.6

Measurement instruments for the monitoring phase

Load cells

Load cells are devices that are used to measure loads transmitted by structural elements such as steel ribs, tie rods, etc.
The measurements most commonly performed are of:

- the load transmitted by steel ribs to the ground of the foundation;
- the load that tie rods in rock bolts are subjected to when they are tensioned or that they exert on the rock mass when in operation.

Hydraulic load cells are used for the first type of measurement consisting of a stainless steel cylindrical body, very similar in appearance and the principles by which they function to the pressure cells already described, but different above all in the maximum working load capacity and in the calibration of the transducer, which must translate the load in terms of stress rather than pressure. They are fitted using special load distribution plates under the feet of steel ribs.

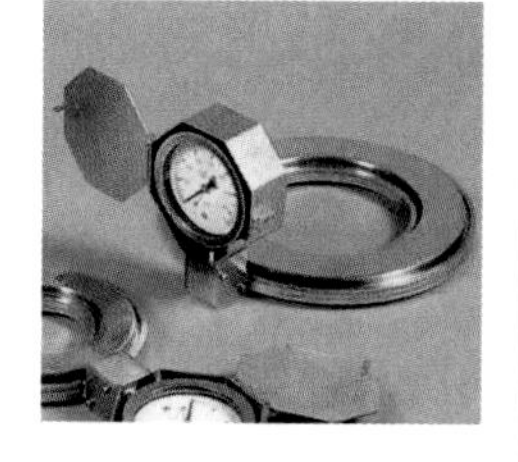

Either hydraulic or electric toroidal load cells are used for the second type of measurement. The hydraulic cells consist of a stainless steel body, toroidal in shape, with a pressure chamber inside filled with deaerated hydraulic fluid under a vacuum. The volume of the cell changes when it is subjected to a load and this is reflected immediately by a change in the pressure of the fluid, which is measured by a manometer graduated in kN.
Electric toroidal load cells also consist of a stainless steel body, *toroidal* in shape, which, however, measures loads by means of a resistance type strain gauge, which makes it very sensitive to eccentric loads. When it is subjected to a load, the cell deforms and this is immediately detected by the extensometers which generate an electrical signal proportional to the load applied.

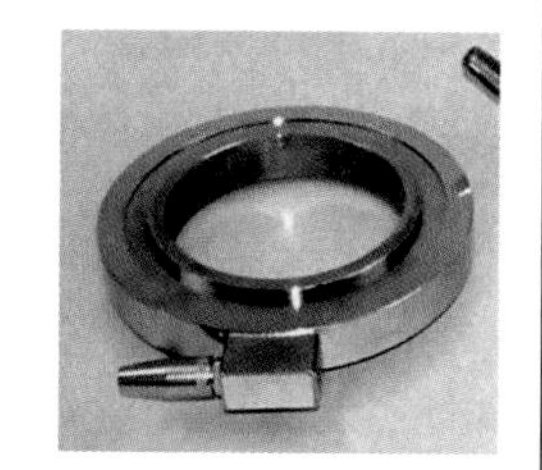

If they are to produce reliable measurements, toroidal cells should always be installed between a plate to distribute the load and a bearing plate, both made of steel in order to ensure even distribution of the load on the pressure chamber. Hydraulic cells give reasonable accuracy at a low cost, but they are not too reliable over time and are sensitive to temperature variations with respect to electric cells. They are also difficult to automate.

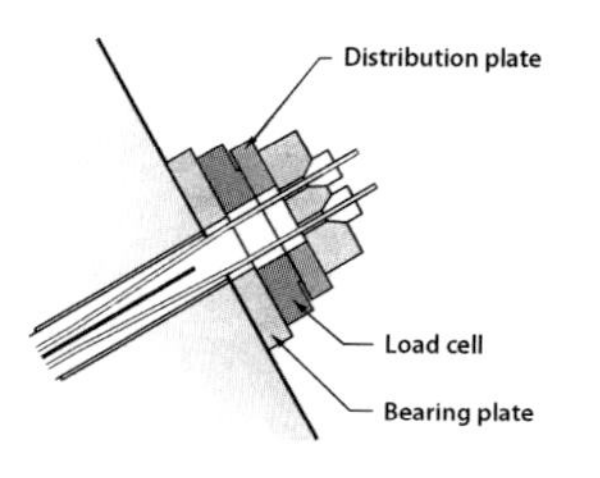

While electric cells on the other hand are more costly, they guarantee better compensation of eccentric loads and they are very robust, as well as allowing measurements to be automated. However, they are not recoverable.

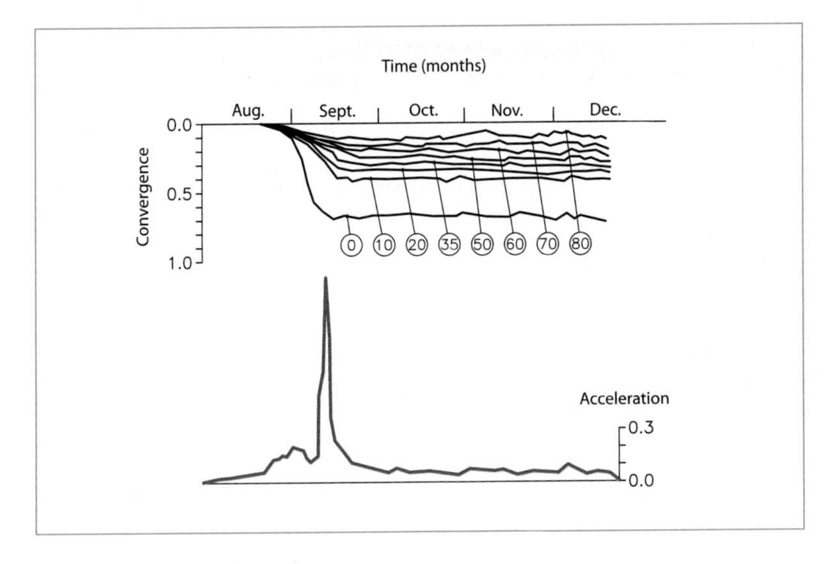

Fig. 10.7

we will see later in this section, together with the magnitude, velocity and acceleration, constitutes the set of parameters of greatest interest; studying them will lead to correct interpretation of convergence phenomena.

A comparison of the magnitude of actual convergence with the design prediction of it allows the validity of the initial hypotheses and the calculation methods used to be verified along with the effectiveness of the stabilisation structures implemented.

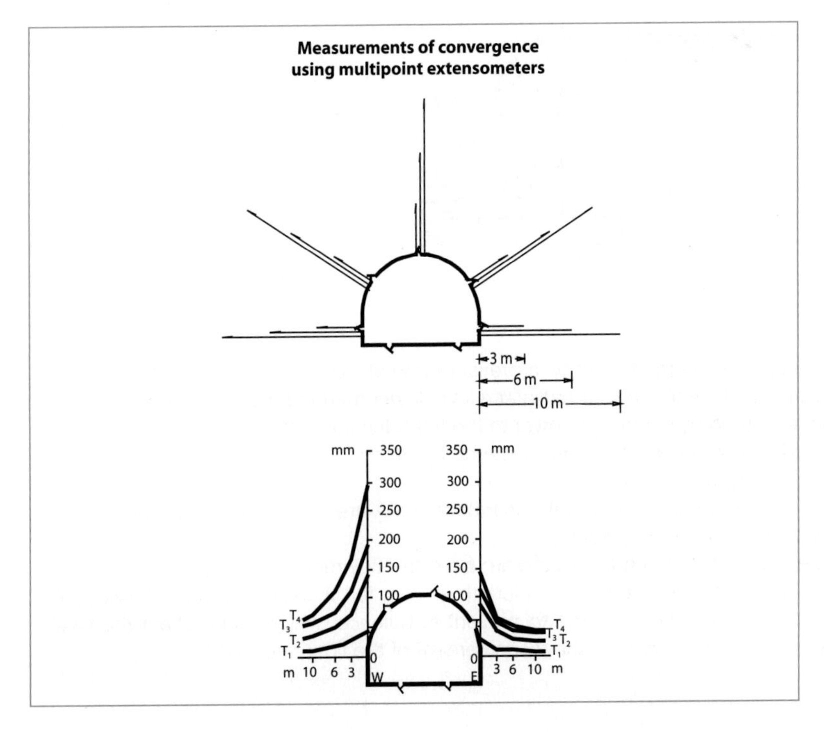

Fig. 10.8

Measurement instruments for the monitoring phase

Flowmeters

When the face advance occurs under hydrodynamic conditions, it is important to have precise knowledge of the flow rate of the water that is penetrating into the tunnel, both to assess the effectiveness of any controlled drainage that may have been implemented and to be able to plan intervention to reduce interference with and to safeguard the hydrogeological equilibrium of the aquifer affected by tunnel advance.
To achieve this water must be conveyed into a channel conveniently positioned in order to avoid interference with construction site activities and a flowmeter or other device for measuring flow rate inserted in it at an appropriate point.
It is a system that consists of the following (see figure):

- a weir basin;
- a weir (triangular, rectangular or other);
- a water level metre.

The measurement assumes that a cross section structure of known dimensions (weir) has been created to cause the level of the water in the 'weir basin' immediately upstream from it to change as a direct function of the change in the flow rate. This generally consists of a rectangular sheet of metal with an opening on one edge.

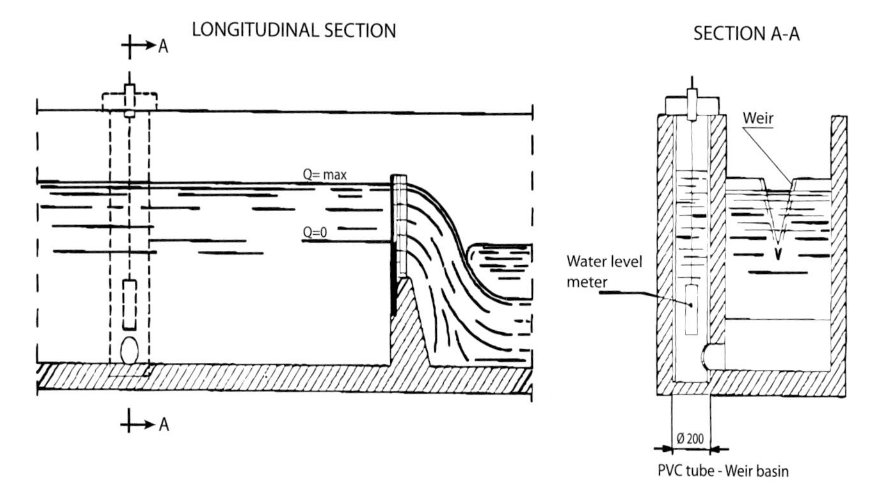

Changes in the water level are measured by a pressure transducer housed inside a plastic box placed in a pressure inlet beside the weir basin at a level lower than the top of the weir.
The water level is measured by supplying power to the transducer and measuring the return current from the transducer subjected to water pressure. Systems for both manual and automatic readings exist for this purpose.
Finally the measurements obtained from the transducer in mA are processed using appropriate mathematical formulas to give the flow rate.
For accurate measurement, the channel upstream from the flowmeter must have a fair gradient (1–3%) and a constant cross section for a length at least ten times longer than its width. A drop in level is useful downstream in order to prevent turbulence, which might affect the level upstream (hydraulic jump) and therefore the measurement of the flow rate also.

The behaviour type of the ground tunnelled can be assessed from combined assessment of the magnitude and velocity of convergence. Both depend on the geomechanical characteristics of the rock mass in relation to the natural stress state. Generally, they are not therefore indicators of possible instability. In any case, with the exception of particular situations such as in urban tunnels where convergence which exceeds certain thresholds would generate unacceptable surface settlement, there is no reason to attribute a maximum acceptable value to the magnitude of convergence alone, because taken by itself, it gives no indication of changes in the stability of the cavity.

On the other hand, as it is mainly independent of the characteristics of the rock mass, positive acceleration will lead to failure sooner or later and that is why convergence must present a certain degree of deceleration: a merely slight slowdown is only acceptable for one or two days following the opening of a cavity. Acceleration is normally at a maximum during and immediately after the passage of the face and at times during excavation of the tunnel invert; however when this is cast and has set, convergence must stabilise rapidly. As has already been seen for extrusion of the core-face, for convergence too, acceleration is the parameter which can be used to clearly evaluate the long and short term stability of a cavity.

The deformation gradient in the ground around a tunnel identifies the extension of the ground that has relaxed and the radius of influence of the excavation. In cases of ground with elasto-plastic behaviour a comparison of that extension with the value predicted for it by calculation gives useful indications for verifying the accuracy of the assumptions made with regard to the geomechanical characteristics of the rock mass and to its residual strength in particular.

In cases of ground with elastic behaviour, knowledge of the deformation gradient allows an idea to be gained of the dimensions of load solids that might weigh on the preliminary lining of a tunnel, and this will assist in deciding the best use of rock bolts to stabilise them.

Finally the anisotropy of deformation can be assessed from the percentage ratio of convergence measured along two different lines perpendicular to each other. Generally speaking, anisotropy is very much more pronounced where convergence is lowest and vice versa.

10.7 Back-analysis procedures

Once the measurements of the stress-strain response of a cavity have been read, collected together and interpreted, it is good practice to verify whether what was predicted at the design stage corresponds to the actual situation and to then proceed to fine tune the design.

The importance of employing an integrated system of design and construction which takes the deformation of the medium in which underground construction takes place as the only reference parameter is consequently quite clear here. It is worth remembering that deformation behaviour is intended as meaning the response of the medium through which a tunnel advances to the combined action of excavation and stabilisation.

The ADECO-RS approach, which possesses this characteristic, allows one to verify whether the design predictions made using numerical models in the non linear, elasto-plastic range actually correspond to the real behaviour of the ground-underground

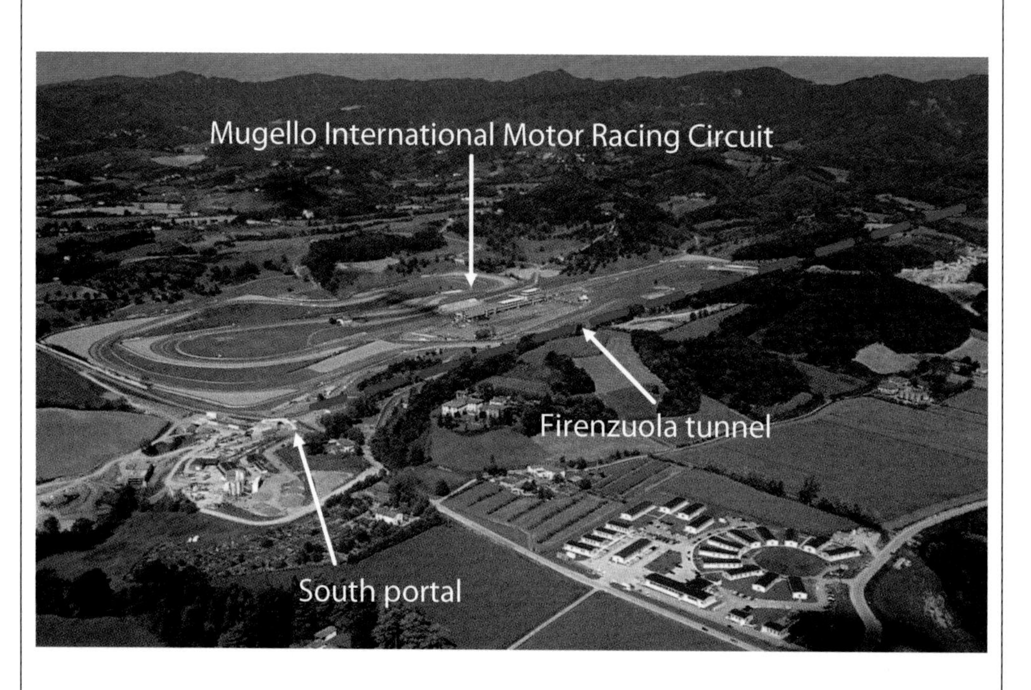
Mugello International Motor Racing Circuit
Firenzuola tunnel
South portal

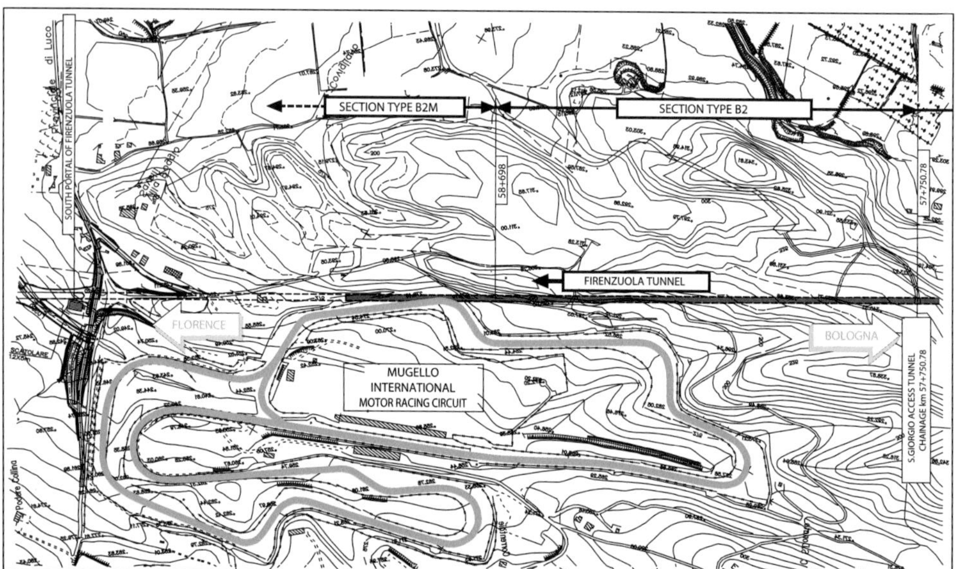
SECTION TYPE B2M
SECTION TYPE B2
SOUTH PORTAL OF FIRENZUOLA TUNNEL
58+698
57+750.78
FIRENZUOLA TUNNEL
FLORENCE
BOLOGNA
MUGELLO
INTERNATIONAL
MOTOR RACING CIRCUIT
S.GIORGIO ACCESS TUNNEL
CHAINAGE km 57+750.78

Fig. 10.9

construction combination both during construction and when the tunnel is in service.

If this condition is not satisfied, it is indispensable for the design engineer to remodel the situation with the assistance of the same calculation methods, but assuming more accurate geomechanical parameters and if necessary more appropriate underlying principles to then test and fine tune the design again.

In the spirit of the new legislation on design in Italy, together with more recent regulations, the design engineer is therefore also actively involved in the construction stage and not just to purely monitor the progress of the works systematically, but also and above all each time behaviour is observed that is different from that predicted, whether at the face or around the perimeter of the tunnel.

One example described below has been drawn from one of the authors' most recent and important tunnelling experiences. It shows perhaps most effectively how monitoring procedures should be performed.

10.7.1 Fine tuning of the design during construction of the tunnel beneath the Mugello international motor racing track with a shallow overburden

The problems connected with passing, with a very shallow overburden (~20m), beneath the track of the Mugello motor racing circuit (Fig. 10.9), which, as is well known, is used for motor cycle and automobile races, as part of the works for the Firenzuola Tunnel on the new high speed Bologna to Florence railway line were very particular.

10.7.1.1 The survey phase

From a geological viewpoint the ground affected by the works belong to the Mugello clays formation (aBM), consisting of silty-clayey, fluvial-lacustrian, Pleistocene deposits with a sub horizontal stratification (Fig. 10.10).

From a geotechnical viewpoint this formation can be classified as inorganic clays with medium-low plasticity (CL), good consistency and slightly overconsolidated (OCR between 2 and 5). The strength and deformability parameters of the material were investigated in the survey phase under drained and undrained conditions using simple and triaxial compression tests (CD, CU and UU) which gave values for undrained cohesion that increased with depth and fell in any case within the range of 1 to 5 kg/cm^2 (0.1–0.5 MPa). Under drained conditions the strength of the intensely stressed material decayed in two distinct phases:

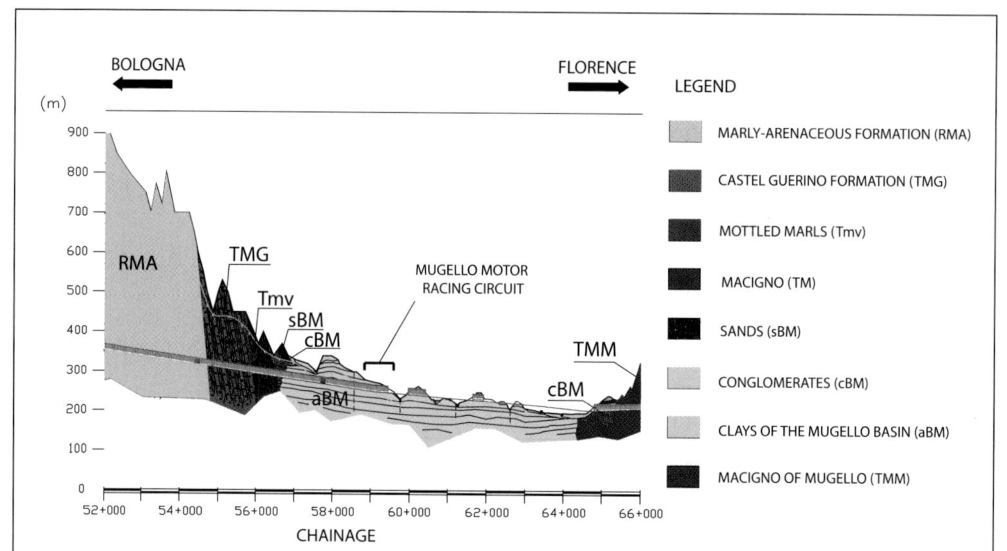

Fig. 10.10

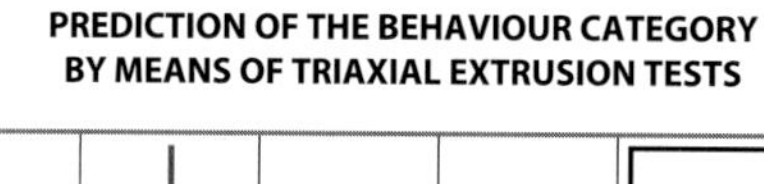

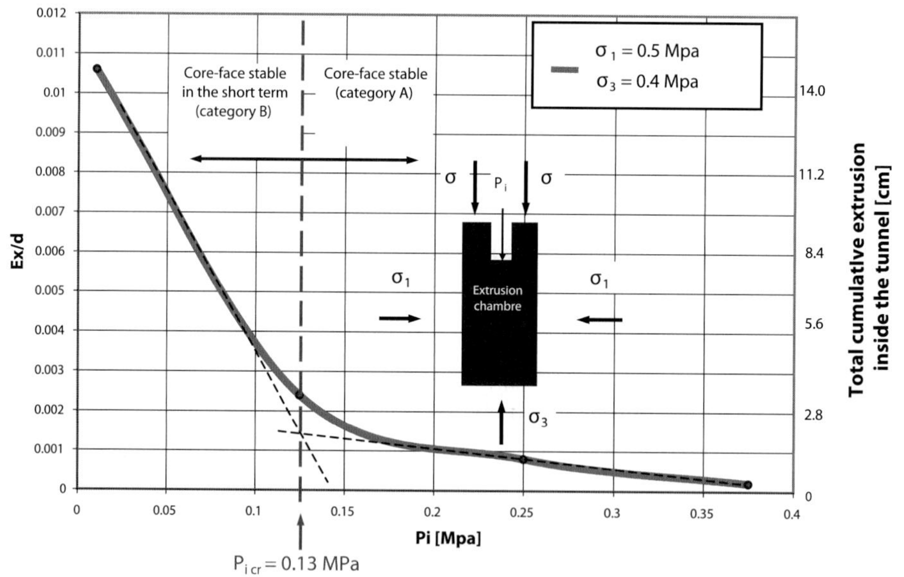

Fig. 10.11

1. on reaching peak strength values, after small relative slips, the particle bonds which give the material its effective cohesion c', were destroyed, while there was no change in the effective angle of friction with respect to the peak value;
2. after greater relative slip, the effective angle of friction decayed to the residual value.

To summarise, on first analysis the investigations attributed the following geotechnical values to the ground:

unit weight:	1.9–2 t/m^3 (19.0–20 kN/m^3)
peak effective cohesion:	0.2–0.3 kg/cm^2 (0.02–0.03 MPa)
residual effective cohesion:	0
peak effective angle of friction:	24°–28°
residual effective angle of friction:	15°–18°

The elastic modulus, calculated on the basis of the pressuremeter tests, was found to vary linearly with depth according to the law:

$$E(z) = 115 + 30.2\ z\ [kg/cm^2]$$

The consolidation coefficient (Cv), obtained from consolidation tests, also varied with depth, lying between 5×10^{-7} and 3×10^{-7} m/s.

Finally triaxial cell extrusion tests were performed, the results of which are given in Fig. 10.11.

The hydrogeological study identified a basically impermeable formation in which levels and lenses of sand up to 3 metres thick could nevertheless be found, the possible sites of artesian aquifers. A series of piezometers installed along the alignment of the future tunnel had found a water table with a piezometric line which tended to rise, roughly following the contact surface between the upper alluvial formation and the clays below. At tunnel level, the piezometric head was approximately 40 m at the points of the greatest overburdens.

10.7.1.2 The diagnosis phase

Analysis of the deformation response of the rock mass to excavation in the absence of stabilisation intervention in the diagnosis phase had revealed *stable core-face in the short term* behaviour (behaviour category B) (Fig. 10.11). As is known, this condition occurs when the stress state in the ground at the face and around the cavity during tunnel advance is sufficient to exceed the strength of the medium in the elastic range. Deformation phenomena therefore develops in the elasto-plastic range and, as a consequence, an arch effect is not created around the cavity very close to it, but at distance from it which will depend on the size of the band of ground which is subjected to plasticisation phenomena.

In this situation the stability of the core-face would be affected strongly by the speed of tunnel advance and, with the shallow overburden, the deformation in the advance core, in the form of extrusion, would have given rise to unacceptable surface settlement if not adequately contained.

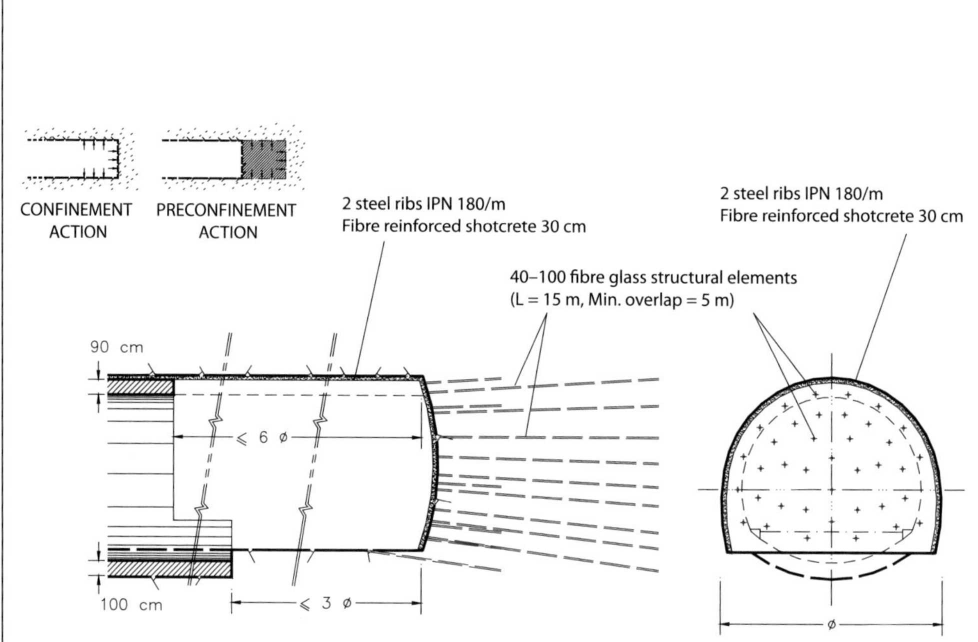

Fig. 10.12

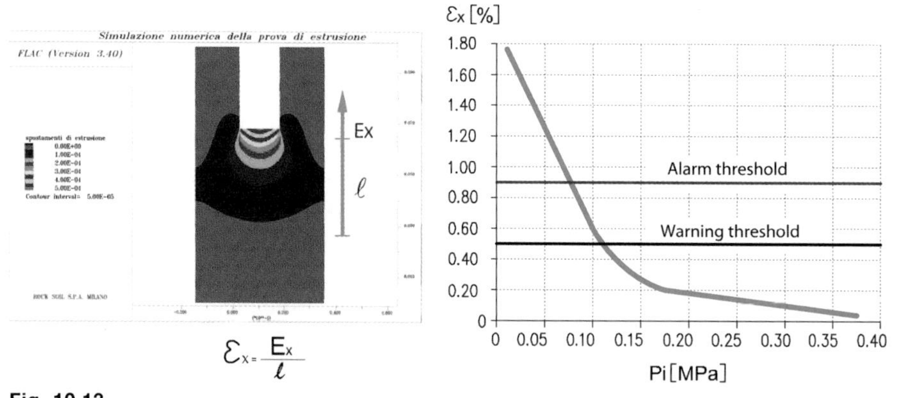

Fig. 10.13

10.7.1.3 The therapy phase

Consequently the design problem to be solved in the therapy phase was to devise design and construction action which would adequately contain settlement induced on the surface by excavation and the possible resultant damage to the road surface of the track, given the shallow overburdens (varying between 20 m and 60 m), the particular geo-morphological characteristics of the area and the geological and geotechnical nature of the alluvial deposits of the Mugello basin. Account also had to be taken of the race track's calendar of races and events of various types and of the absolute need for the race track to meet the deadline for official approval before the coming world motorcycle championships (April–May 2000).

Having taken account of all the requirements, the following tunnel advance method was decided at the design stage (see Fig. 10.12), to then be calibrated during the operational phase on the basis of monitoring data:

1. full face advance after first stiffening the advance core by placing 45–100 fibre glass structural elements 15 m long with 5 m overlap, grouted using a shrink controlled or an expansive mixture;
2. preliminary lining consisting of 2IPN 180 steel ribs placed a intervals of 1.00–1.40 m and a 30 cm layer of fibre reinforced shotcrete;
3. a tunnel invert, 100 cm thick, cast at the same time as the side walls at a maximum distance from the face of 3 Ø;
4. final lining in concrete, 90 cm thick in the crown, cast at not more than 6 Ø from the face.

The deformation specified and considered acceptable at the design stage was 5 cm for total cumulative extrusion and 8 cm for average radial convergence. The following threshold values were set for total differential extrusion: 0.5% for the warning threshold and 1% for the alarm threshold (see Fig. 10.13).

10.7.1.4 The monitoring programme

Given the delicacy of the situation, as the tunnel approached the race track a programme of monitoring in real time was started in June 1999, as specified in the design, but also with the purpose of guaranteeing the level of safety required for the race track to continue its programme of events.

The deformation response of the rock mass to excavation was carefully monitored ahead of the face both, from inside the cavity and from the surface (Fig. 10.16). It was analysed by taking systematic measurements of face extrusion (using both sliding micrometers and topographic survey measurements) and of cavity convergence from

Fig. 10.14

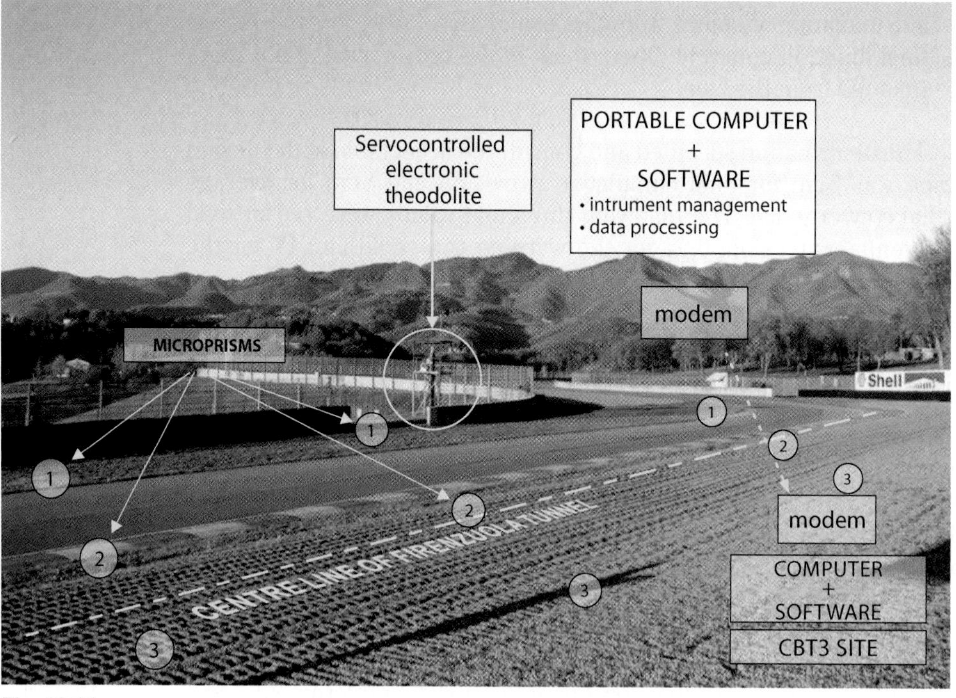

Fig. 10.15

inside the cavity and by setting up a real time topographic system on the surface for measuring surface settlement induced by the passage of the face.

The surface monitoring system was designed so that sightings could be taken automatically at hourly intervals from a specially constructed *light tower,* located near the zone of interference between the tunnel and the race track, of a series of optical targets placed in a grid configuration of approximately 10 × 5 m across the area of ground lying over the alignment of the future tunnel before and after the zone of interference with the track on the Borgo San Lorenzo bend (Fig. 10.14 and 10.15). The data acquired was transmitted in real time to a control unit, which immediately processed it and transmitted it to the design engineers on the construction site.

10.7.1.5 Final calibration of the design based on monitoring feedback

Monitoring showed that settlement phenomena had started approximately 60 m before the arrival of the face and then increased to values significantly higher than those predicted for the tunnel section type employed, recording settlement of approximately 14 cm along the tunnel alignment at chainage km 58 + 690, which then rose to 22 cm after the passage of the face (Fig. 10.17). A plot of surface settlement on a curve as a function of distance from the face shows that 60% of total surface settlement occurred before the arrival of the face. A further 30% occurred in the 20 m following the passage of the face (before the tunnel invert was cast) and the remaining 10% in the next 40 m.

At the same time, total cumulative extrusion of around 10 cm was measured inside the tunnel with average radial convergence of around 10–12 cm. Total differential extrusion was well above the alarm threshold (Fig. 10.18) that had been set.

These differences between predicted and measured values arose mainly because the actual geotechnical conditions were different from those that had been forecast by the geological survey on which the predictions were based.

The problem was then posed of how to calibrate the intensity and distribution of the specified stabilisation intervention between the face and the cavity in order to optimise the design from the viewpoint of settlement (damage to the track), cost and time (passing through the race track zone by February 2000).

The following steps were taken to solve the problem (see Fig. 10.19):

1. reconstruction of the situation by means of numerical modelling and back analysis procedures using the data acquired from

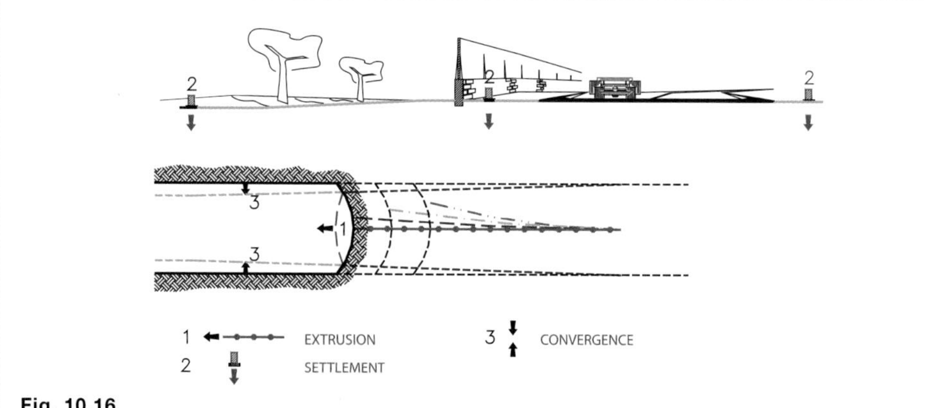

Fig. 10.16

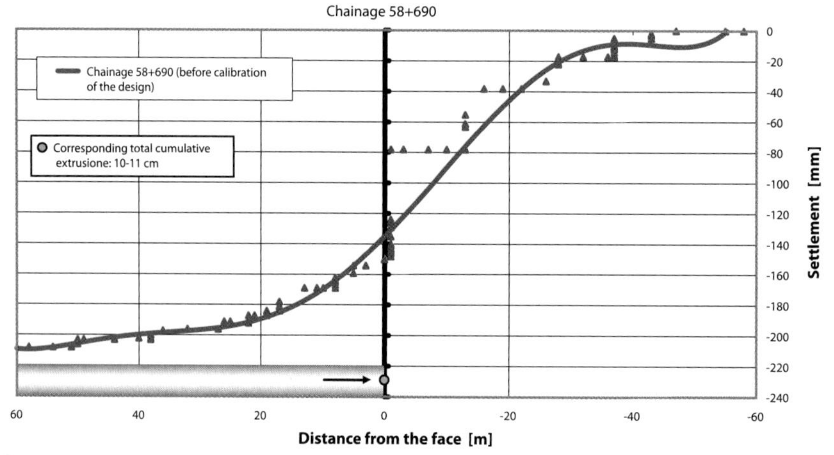

Fig. 10.17

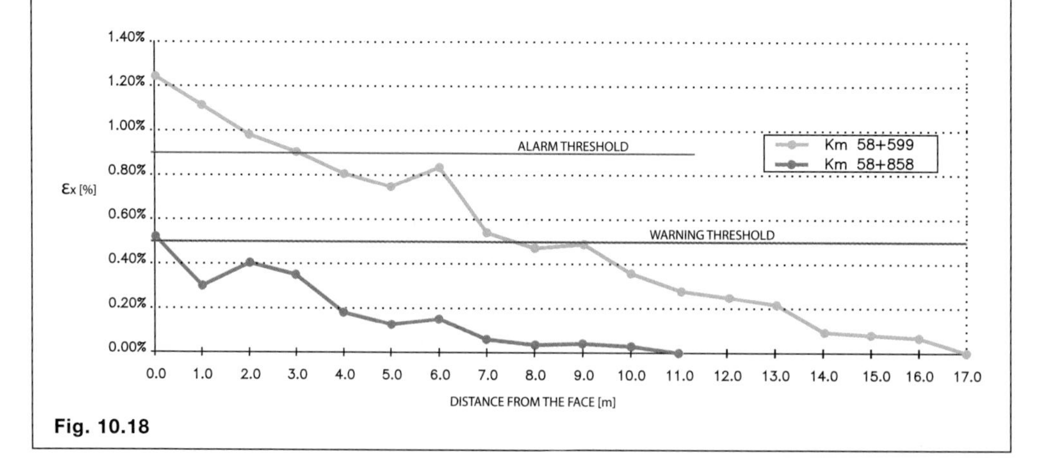

Fig. 10.18

monitoring (extrusion, convergence, surface settlement) the real situation in order to identify more accurate strength and deformability parameters and the most appropriate numerical procedures to model them;
2. use of the numerical models from the previous step to decide the cavity preconfinement and confinement intervention to be implemented to limit extrusion and settlement within the desired thresholds. A variety of finite difference numerical models were employed to reconstruct the real situation numerically, consisting of 12 axial symmetric and 32 plane models constructed using the version 3.40 of the FLAC software and a strain softening failure criteria which simulates real stress-strain response of the ground rather faithfully. The new geomechanical parameters that were calculated are summarised in the table below.

Parameter	After the survey phase	After back-analysis
Weight by volume	19.9–20 kN/m^3	19.9–20 kN/m^3
Peak effective cohesion	0.02–0.03 MPa	0.015 MPa
Residual effective cohesion	0	0
Peak angle of friction	24°–28°	22°
Residual angle of friction	15°–18°	15°
Modulus of elasticity	E(z) = 11.5 + 3.02 * z [MPa]	E(z) = 11.5 + 0.5 * z [MPa]
Overconsolidation ratio	2–5	2–5

Once the models had been weighted to numerically reproduce the monitoring data (Fig. 10.20), they were then used to perform the following:

1. to reconstruct the deconfinement curve for a given section as a function of distance from the face. Knowledge of this is essential for conducting reliable analyses using plane models;
2. to study the effect of possible cavity preconfinement and confinement intervention (Fig. 10.21).

Plane models on the other hand were used to study the effects obtained, in terms of the deformation response of the ground, from controlling how the core-face extruded when the ring of confinement inside the tunnel was closed in different ways (Fig. 10.22):

- model A: presence of strut in tunnel invert, which is cast at intervals of 4 m, and casting of the roof arch 40 m from the face;
- model B: presence of strut in tunnel invert, which is cast at intervals of 12 m, and casting of the roof arch 40 m from the face;

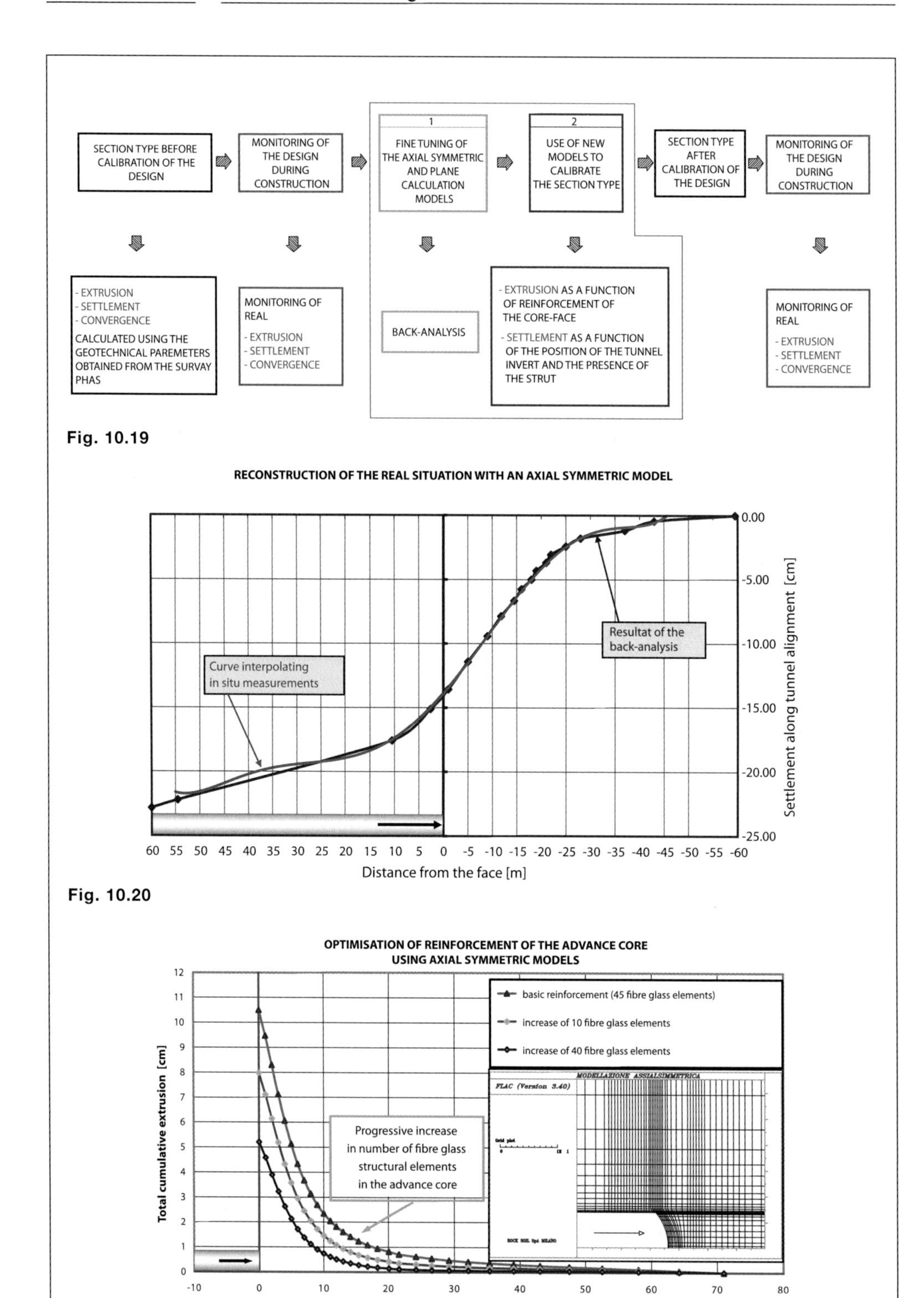

Fig. 10.19

Fig. 10.20

Fig. 10.21

- model C: absence of strut in tunnel invert, which is cast at intervals of 12 m, and casting of the roof arch 40 m from the face.

The following action was decided to calibrate the design on the basis of the results of the numerical modelling (see Fig. 10.23):

- use of 93 fibre glass structural elements 24 m long with minimum overlap of 12 m to reinforce the advance core;
- strut in the tunnel invert;
- tunnel invert cast at intervals of 12 m;
- use of expansive mixtures to grout the structural elements in the advance core.

10.7.1.6 The operational phase

Tunnel advance under the race track took place according to the decisions made in the therapy phase (Fig. 10.24) following the calibration of the design on the basis of monitoring data in faithful compliance with the design specifications, both with regard to deformation and construction times.

The passage under the critical zone of the Borgo San Lorenzo bend was successfully completed on schedule with an average advance rate of approximately 2 metres/day of finished tunnel.

10.7.1.7 The monitoring phase

Figure 10.25 summarises the monitoring data recorded during the passage under the track following action to calibrate the design. It is interesting to observe how the progressive decrease in total cumulative extrusion was followed by a corresponding decrease in both tunnel convergence and vertical settlement of the lining itself. Total differential extrusion also fell rapidly below the warning threshold. Figure 10.26 shows that, as a consequence, settlement along the tunnel alignment also fell significantly, with maximum settlement of around 13 cm, below the required threshold, during the passage beneath the track. This value, which in itself could be considered high, could in fact have been reduced further without difficulty by taking appropriate action to stiffen the advance core (dotted green line in the graph in Fig. 10.26). This, however, would not have allowed the tunnel to complete the passage under the track before the start of the racing season. The solution adopted not only guaranteed the stability and safety of the excavation, but was also excellent from the viewpoint of construction times and costs.

The case described, taken from experience acquired during the construction of tunnels on the new high speed railway line between

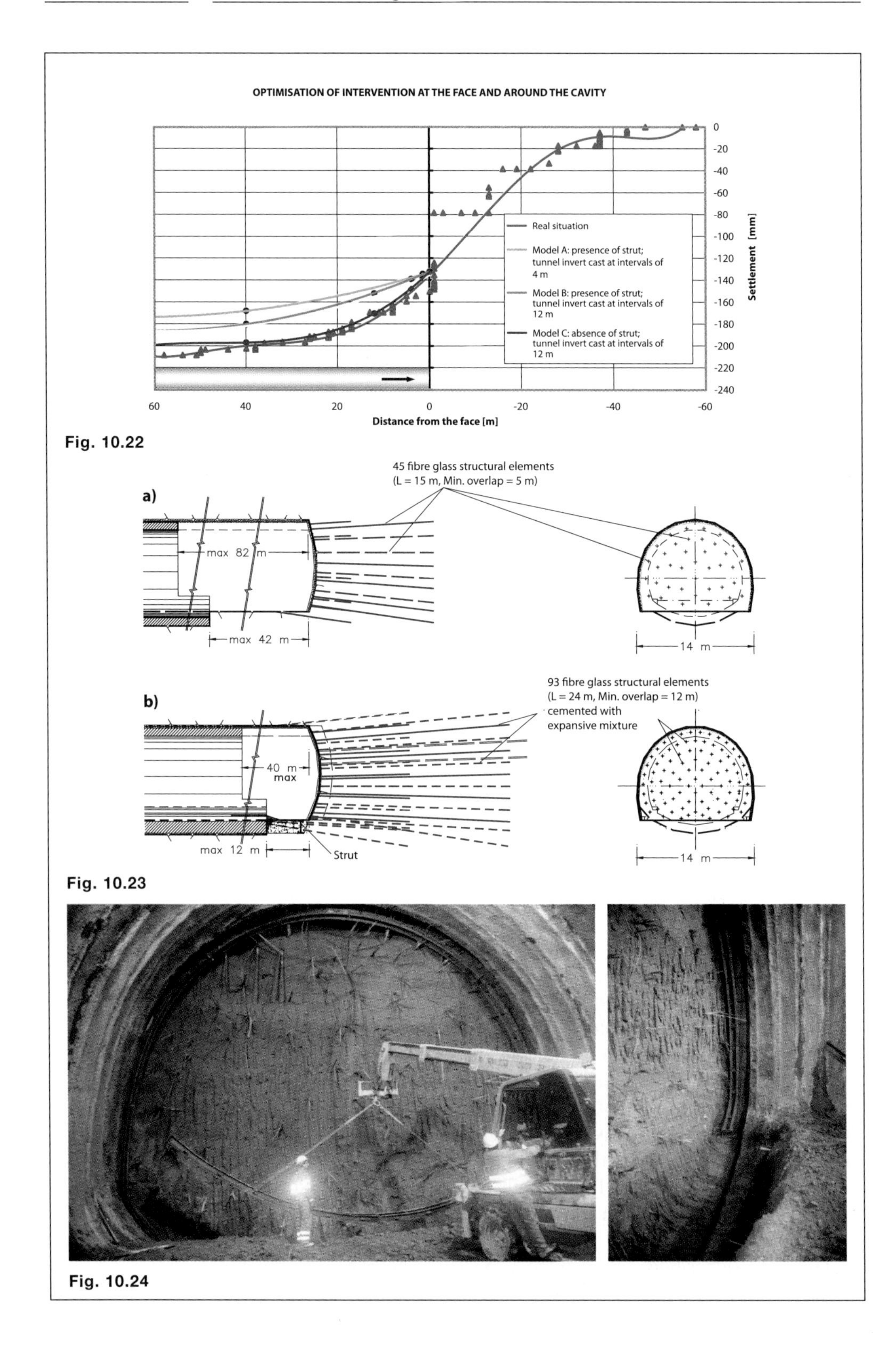

Fig. 10.22

Fig. 10.23

Fig. 10.24

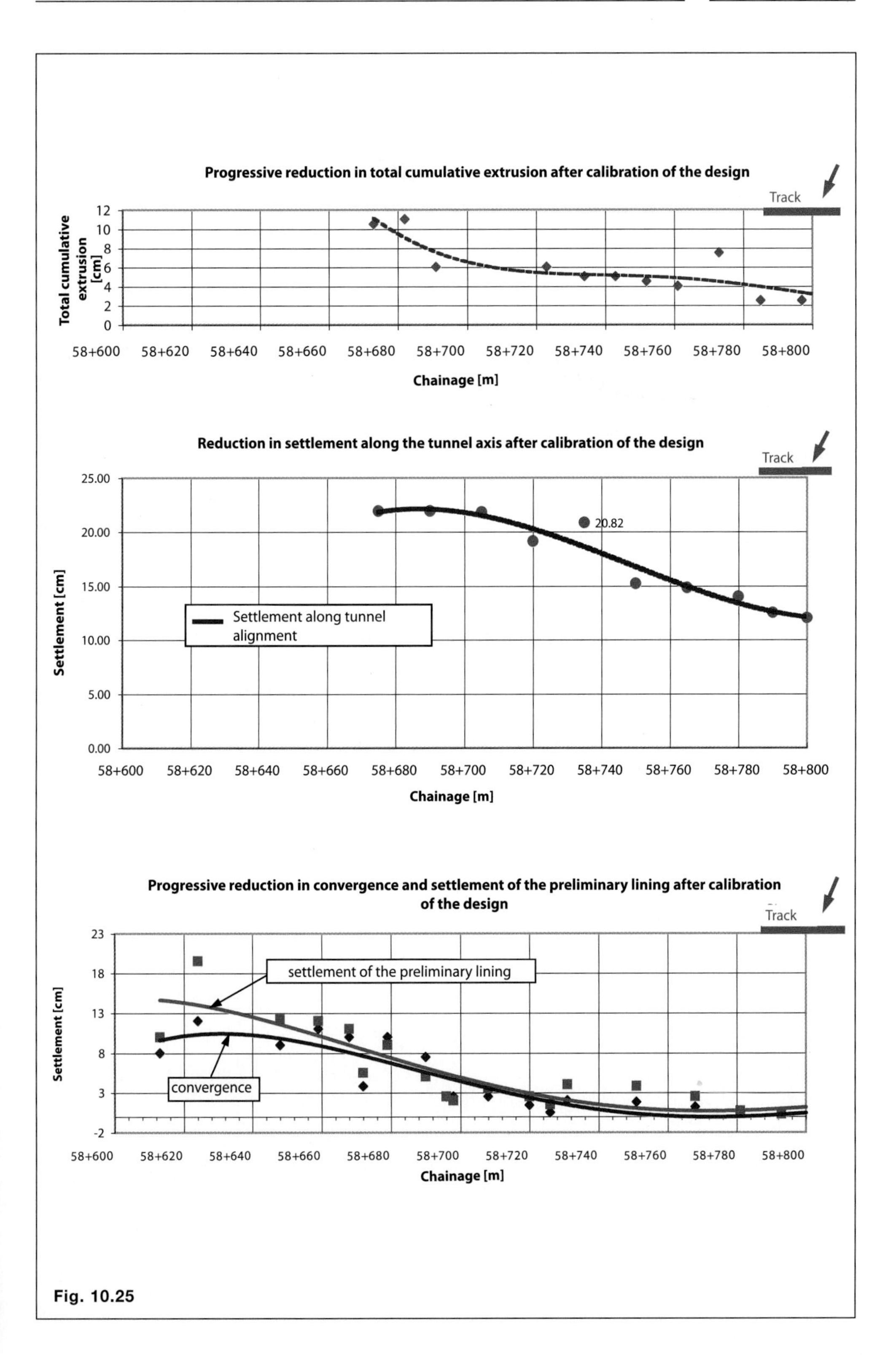
Progressive reduction in total cumulative extrusion after calibration of the design
Track
Total cumulative extrusion [cm]
12
10
8
6
4
2
0
58+600 58+620 58+640 58+660 58+680 58+700 58+720 58+740 58+760 58+780 58+800
Chainage [m]
Reduction in settlement along the tunnel axis after calibration of the design
Track
Settlement [cm]
25.00
20.00
15.00
10.00
5.00
0.00
20.82
Settlement along tunnel alignment
58+600 58+620 58+640 58+660 58+680 58+700 58+720 58+740 58+760 58+780 58+800
Chainage [m]
Progressive reduction in convergence and settlement of the preliminary lining after calibration of the design
Track
Settlement [cm]
23
18
13
8
3
-2
settlement of the preliminary lining
convergence
58+600 58+620 58+640 58+660 58+680 58+700 58+720 58+740 58+760 58+780 58+800
Chainage [m]

Fig. 10.25

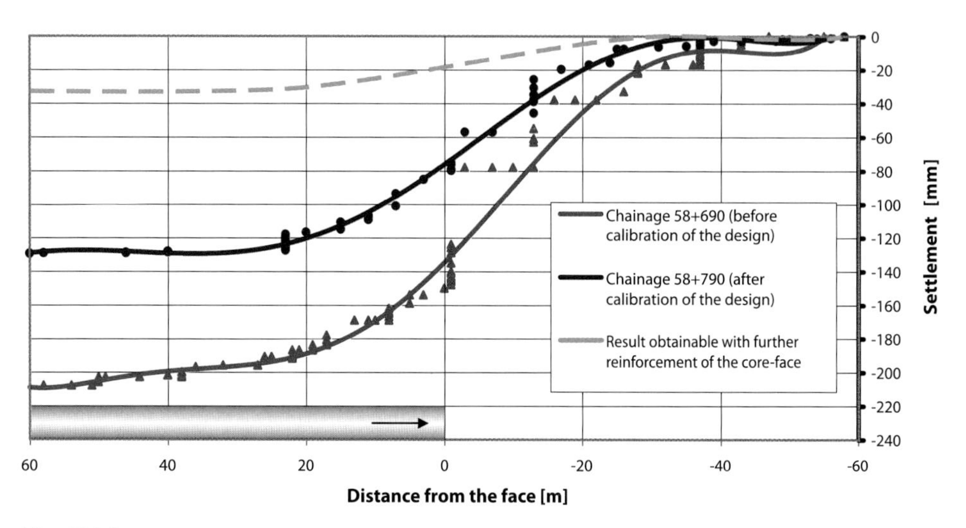

Fig. 10.26

Bologna and Florence, shows how the ADECO-RS approach also gives a new and more appropriate interpretation of monitoring to test and calibrate the design during construction, which will no longer normally constitute an occasion for compulsory radical revision of the original design, but will provide the chance to optimise it and to balance the use of stabilisation instruments between the face and the cavity within the context of the tunnel section type specified at the design stage.

Final considerations

If the deformation that is normally observed inside a tunnel while it is advancing is interpreted in terms of a process of cause and effect, it would seem perfectly reasonable to identify the cause as the action that is exerted on the medium and the effect as the deformation response of the medium that results from the action.

Given this assumption, while the cause (at least until just a few years ago) was never considered worthy of attention nor of detailed study and therefore remained only apparently determined, the effect was immediately identified as convergence of the cavity and became the subject of study (Fig. 1). Theories, design approaches and construction systems have been developed on the basis of these studies which assume that all the problems of tunnel construction can be solved in terms of simple radial confinement action.

The former include the very well-known *theory of characteristic lines* and the *convergence-confinement method* [11], [12], which, although they were the first to recognise the beneficial effect of the presence of the advance core for the stability of a cavity, they nevertheless furnished no effective suggestions on how to exploit that effect, nor did they indicate how do deal with an unstable face.

Among the latter, approaches like the NATM, based on geomechanical classifications (often used for purposes other than those for which they were conceived), undoubtedly constituted considerable progress with respect to the past when they were introduced. The principal merits of the NATM were that it:

- considered the ground as a construction material for the first time;
- introduced the use of new simple cavity confinement technologies with an active action such as shotcrete and rock bolts;
- underlined the need to measure and interpret the deformation response of the ground systematically.

Today, however, to consider the problem of the statics of tunnels exclusively as a two dimensional problem with all the attention focused on cavity convergence only, is to demonstrate (and this applies to all the approaches derived from it) serious limitations:

- it is an incomplete and partial classification system because it is not applicable in all types of ground and under all statics conditions;
- it completely neglects the importance of the advance core and the need to use it as a stabilisation instrument under difficult stress-strain conditions;

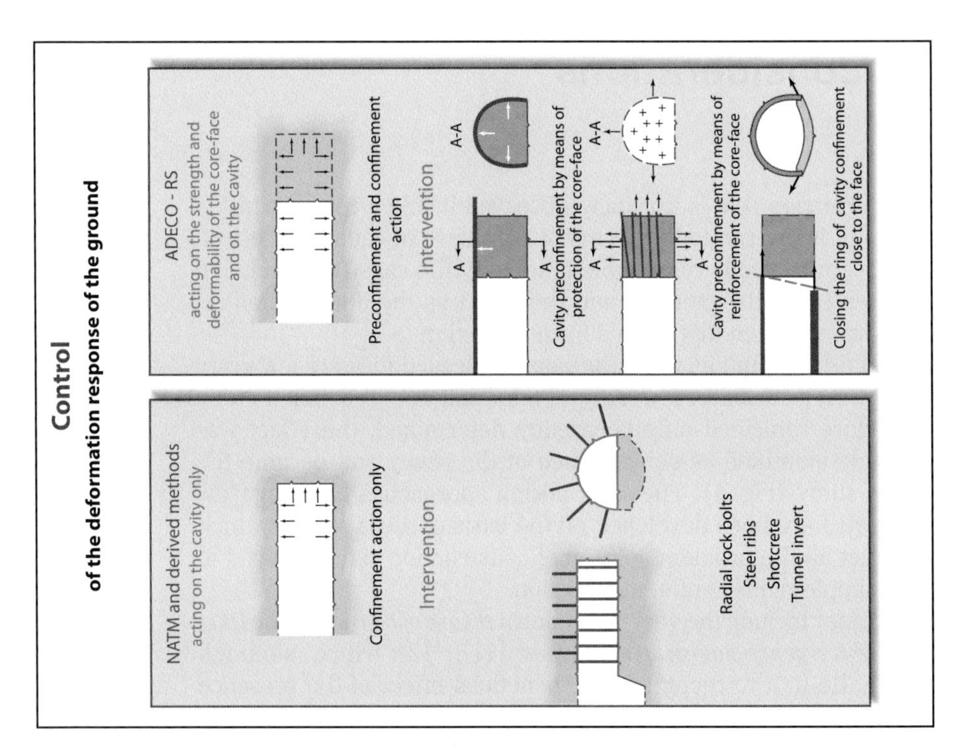
Control
of the deformation response of the ground
NATM and derived methods
acting on the cavity only
Confinement action only
Intervention
Radial rock bolts
Steel ribs
Shotcrete
Tunnel invert
ADECO - RS
acting on the strength and deformability of the core-face and on the cavity
Preconfinement and confinement action
Intervention
A-A
A
A
Cavity preconfinement by means of protection of the core-face
Cavity preconfinement by means of reinforcement of the core-face
Closing the ring of cavity confinement close to the face

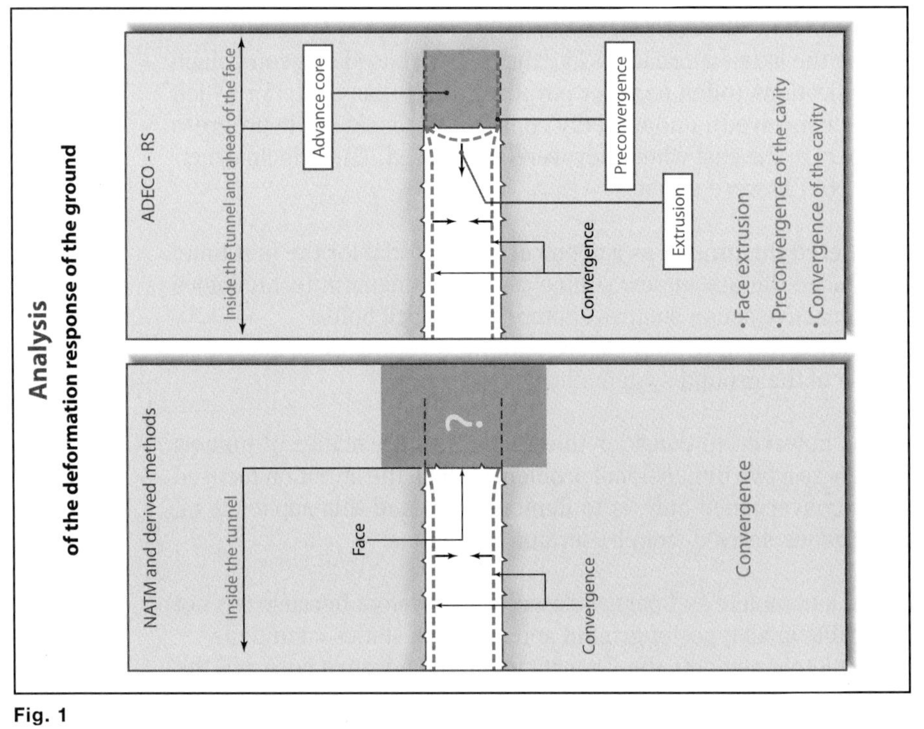
Analysis
of the deformation response of the ground
NATM and derived methods
Inside the tunnel
Face
?
Convergence
Convergence
ADECO - RS
Inside the tunnel and ahead of the face
Advance core
Preconvergence
Extrusion
Convergence
• Face extrusion
• Preconvergence of the cavity
• Convergence of the cavity

Fig. 1

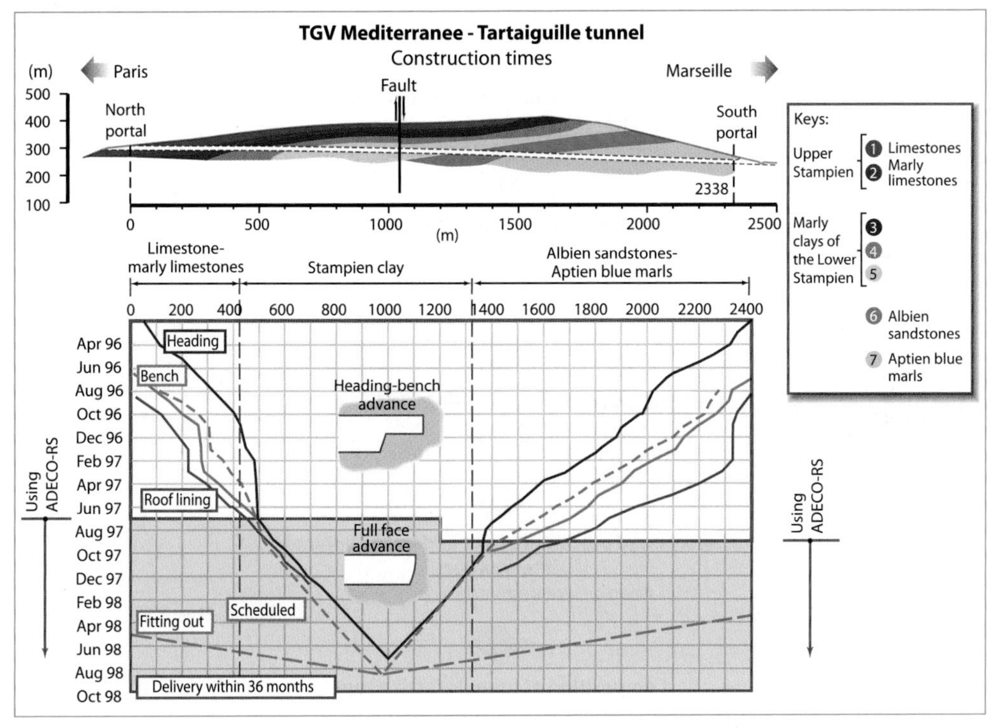

Fig. 2

- it ignores new technologies and continues to propose simple cavity confinement action as the only intervention to stabilise a tunnel;
- it does not make a clear distinction between the design stage and the construction stage;
- it solves the problem of how to monitor the adequacy and correct dimensions of the design adopted in an indisputably non scientific manner by quite happily comparing geomechanical classes with the magnitude of the deformation response of the ground.

The erroneous conviction that the effect of the action exerted on the medium during tunnel excavation was to be seen solely in terms of convergence of the cavity has led entire generations of engineers in Italy and abroad down the wrong path for decades. What was taught by the design and construction approaches in use (NATM and derived methods which still today act according to this erroneous conviction) induced them to concentrate on dealing with the effects (containing convergence of the cavity by simple confinement action) instead of on the causes of tunnel instability [7], [36].

This approach to the problem was successful for driving tunnels when stress-strain conditions were not particularly difficult but it

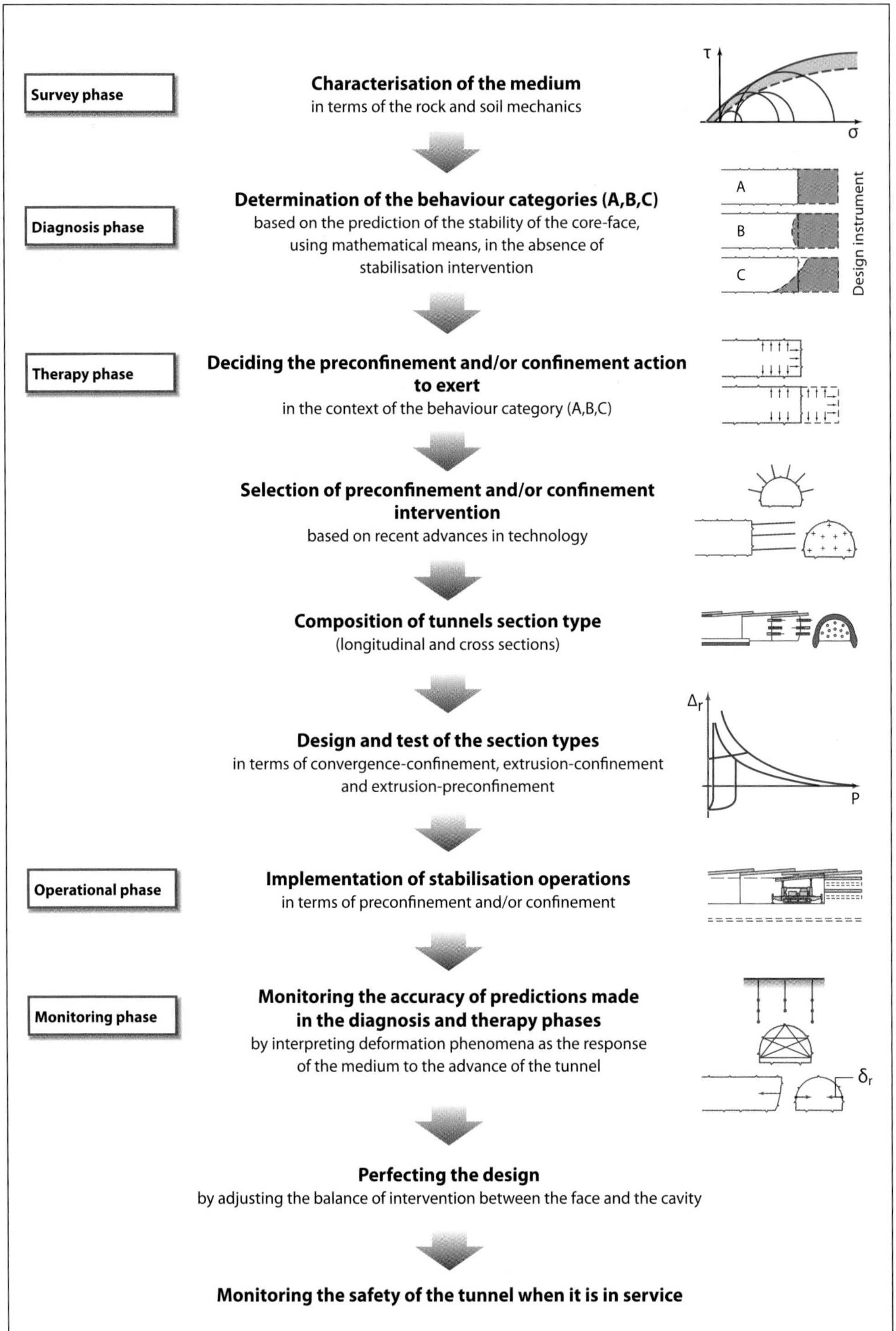

Survey phase

Characterisation of the medium
in terms of the rock and soil mechanics

Diagnosis phase

Determination of the behaviour categories (A,B,C)
based on the prediction of the stability of the core-face, using mathematical means, in the absence of stabilisation intervention

Therapy phase

Deciding the preconfinement and/or confinement action to exert
in the context of the behaviour category (A,B,C)

Selection of preconfinement and/or confinement intervention
based on recent advances in technology

Composition of tunnels section type
(longitudinal and cross sections)

Design and test of the section types
in terms of convergence-confinement, extrusion-confinement and extrusion-preconfinement

Design instrument

Operational phase

Implementation of stabilisation operations
in terms of preconfinement and/or confinement

Monitoring phase

Monitoring the accuracy of predictions made in the diagnosis and therapy phases
by interpreting deformation phenomena as the response of the medium to the advance of the tunnel

Perfecting the design
by adjusting the balance of intervention between the face and the cavity

Monitoring the safety of the tunnel when it is in service

Fig. 3

revealed its limitations when it came to more complex situations because:

1. of an incapacity to make reliable forecasts of tunnel behaviour during advance and therefore because of the absence of a diagnosis phase in the design procedures;
2. measures to confine deformation that could not be predicted in advance were improvised;
3. of a lack of effective stabilisation systems, capable of addressing the cause of instability (deformation of the core) and not just the effect (convergence);
4. of the inability to make a preliminary assessment of a project in terms of forecasting risks, time schedules and advance rates.

Given this situation, the rapid and constantly growing demand for tunnels of all types, including those with very difficult stress-strain conditions, urgently required the formulation of theories and procedures capable of controlling the deformation response of the medium under all possible *stress-strain conditions and not just under ordinary conditions.*

In order to find a way out of this stalled situation, it was necessary to bring the problem back to reality and treat it as a *three dimensional problem*, which in effect it is, and to consider the entire dynamics of tunnel advance and not just the last part of it.

It was with this philosophy that the theoretical and experimental research studies commenced from which the foundations of a new approach were to spring based on the *Analysis of COntrolled DEformation in Rocks and Soils*. It has been successfully employed over the last 10–15 years in a very wide variety of types of ground and stress-strain conditions, including the most difficult, and it has been used to solve problems in many particularly difficult situations (see the table on page 377), where the application of old concepts (NATM and derived methods), which do not reveal their limitations or intrinsic defects in simpler situations, had furnished disappointing and at times even catastrophic results.

It is certainly of significance in this respect to recall, in conclusion, what happened in France during the construction of the Tartaiguille tunnel for the new *TGV Méditerranèe*, Marseilles to Lyon high speed rail line (see also Appendix C).

Tunnel advance with a 180 m². cross section, began in February 1996 and proceeded with varying success according to NATM principles until September of the same year. At this point, when the tunnel started to pass through the heavily swelling argile du Stampien *formation (75% montmorillonite) growing difficulties were encountered until it became practically impossible to continue tunnel advance. In order to solve the problem, at the start of 1997 the SNCF (*Societè National du Chemin de Fer*) set up a study group (*Comité de Pilotage*) consisting of engineers from the French Railways, engineers from the G.I.E. Tartaiguille*

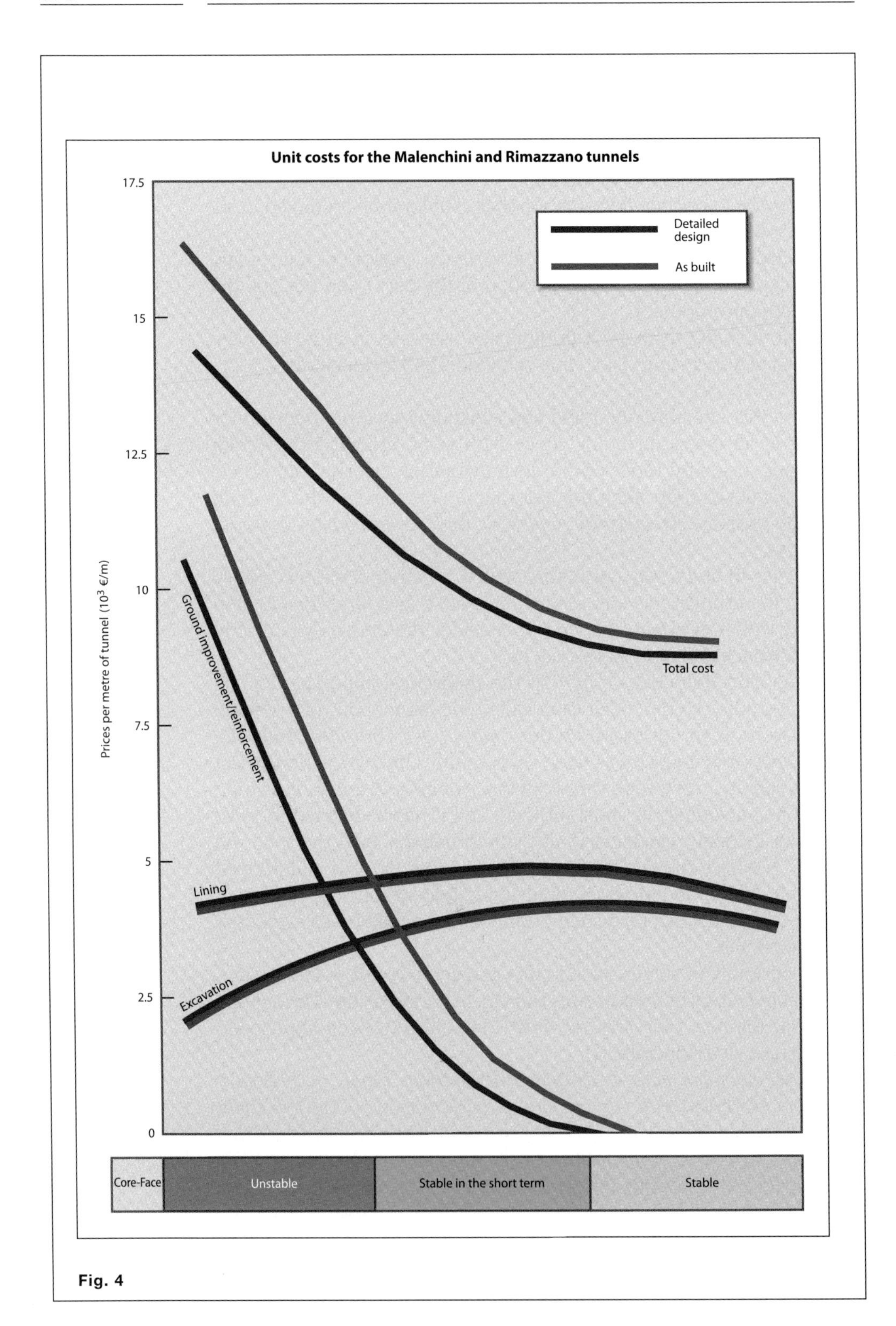

Unit costs for the Malenchini and Rimazzano tunnels
Detailed design
As built
Prices per metre of tunnel (10³ €/m)
17.5
15
12.5
10
7.5
5
2.5
0
Total cost
Ground improvement/reinforcement
Lining
Excavation
Core-Face
Unstable
Stable in the short term
Stable

Fig. 4

The application of the design and construction principles of the ADECO-RS approach has produced notable successes including the recovery and rescue of tunnels where tunnel advance had had to be abandoned using other principles of tunnel advance. These include the following:

TUNNEL	YEAR	ø [m]	GROUND	OVERBURDEN max [m]	PRODUCTION [m/day] mid – max
Tasso (Florence-Arezzo railway line)	1988	12.20	Sandy silts	50	2.0–3.2
Targia (Bicocca-Syracuse railway line)	1989	12.00	Hyaloclastites	50	2.0–3.3
San Vitale (Caserta-Foggia railway line)	1991	12.50	Scaly clays	100	1.6–2.4
Vasto (Ancona-Bari railway line)	1993	12.20	Silty sand and clayey silt	135	1.6–2.6
Tartaiguille (TGV Méditerranèe, Marseille-Lyon railway line)	1996	15	Swelling clay	110	1.4–1.9
Appia Antica (Rome motorway ring road)	1999	20.65	Sandy-gravelly pyroclastites	18	2.3–3.3

consortium, the consulting engineers of the railways, Coyne et Bellier and CETU, and the consulting geotechnical engineers from the Terrasol and Simecsol consortium. This group then consulted major European tunnelling experts inviting them to propose a design solution to pass through the clayey formation safely and on schedule.

After examining several proposals, none of which offered the guarantees of safety and reliability requested by the client, above all with regard to completion times, the SNCF were attracted by the proposal put forward by the author containing hypothesised construction times and costs guaranteed by the consulting engineer on the basis of similar cases successfully solved. In March 1997, Rocksoil S.p.A. was awarded the contract for executive design of the 860 m of tunnel still to be completed.

Tunnel advance resumed in July 1997 after radical revision of the design according to ADECO-RS principles (full face tunnel advance, see Fig. 5) and was finally able to continue without interruptions and with growing success as the site operators gradually gained confidence in the use of the new technologies. Exceptionally constant average advance rates were recorded (Fig. 2), which as guaranteed by the consulting engineers were higher than 1.4 m per day allowing the tunnel to be completed in July 1998 after only one year since work began with the new system and one and a half months ahead of schedule [69], [70], [71].

In the light of the considerable experience acquired over the last ten years [19], [21], [22], [71], [74], [75], it can be confidently stated that the ADECO-RS approach to the design and construction of tunnels can be used to produce virtually linear advance rates independently of the type of ground tunnelled and the contingent stress-strain conditions. It follows that while it was once only possible to talk of mechanisation under conditions that could be dealt with by simple confinement of the cavity or of the face (shields, TBMs), today mechanisation can be spoken of even under more complex and difficult conditions

Fig. 5 *Tartaiguille tunnel (France, TGV Méditerranée, Lyon-Marseilles railway line, 1998, ground: Swelling clay [75% montmorillonite], overburden: 100 m, diameter: 15 m).View of the face (180 m² cross section) reinforced with fibre glass structural elements*

Fig. 6 *Pianoro Tunnel (new high speed Milan-Rome-Naples railway line between Bologna and Florence, 1999, ground: cemented silty sand, overburden: 150 m, diameter: 13.30 m). View of the face reinforced with fibre glass structural elements*

which require preconfinement action. Tunnel excavation can finally be industrialised (constant advance rates, accurate forecasting of times and costs) independently of the type of ground and the size of the overburden involved.

To summarise, by making full use of the most recent knowledge, calculation processing power and advance technologies (Fig. 3) the ADECO-RS approach offers design engineers a simple guide with which tunnels can be classified in one of three fundamental behaviour categories. To do this, reference is made to the stability of the core-face which is predicted by means of in-depth stress-strain analysis performed theoretically using mathematical tools. For each section of tunnel with uniform deformation behaviour identified in this manner, the design engineer decides the type of action (preconfinement or simple confinement) to produce in order to control deformation and, as a consequence, he selects the type of stabilisation technique and the longitudinal and transverse tunnel section type most appropriate to each given situation, by using the instruments most suited to develop the necessary action. Tunnel section types are available for all types of ground and stress-strain conditions. Costs and construction times (per linear metre of tunnel) can be automatically calculated for each of these.

By using this approach:

- importance is given to stabilisation techniques as indispensable instruments for controlling deformation, and therefore as "structural elements" for the purposes of the final stability of a tunnel (tunnels are seen in terms of, and paid for in proportion to how much they deform). It is worth considering in this respect that careful examination of the budgets and costs of underground projects show that stabilisation and ground improvement works are now the only items which vary significantly as compared to excavation and lining items which increasingly tend to remain constant for all types of ground (see Fig. 4);
- with a complete and reliable design, a main contractor is induced to industrialise tunnel advance operations in all types of ground, even the most difficult;
- given the ability to plan construction times and costs, disputes which until very recently arose between the project managers and contractors are avoided;
- by employing one single reference parameter, common to all types of ground (the stress-strain behaviour of the core-face), that can be easily and objectively measured during tunnel advance, the problem of the clearest and most evident defect of previous classification systems (comparing geomechanical classes with the deformation response of the ground), which until today fuelled disputes between project managers and contractors, is solved.

As a result of these important features, the ADECO-RS approach has aroused considerable interest and rapidly established itself as an advantageous alternative to those used until now. The decision to use it in this respect for drawing up the design specifications on the basis of which contracts were awarded and then for the actual design of the new Bologna to Florence High Speed Rail line was particularly significant. It is certainly at present the largest tunnel construction project ever implemented in the world: approximately 84.5 km of tunnel with a cross section of 140 m^2. on a route with a total length of 90 km through notoriously difficult ground due to its variability and often very poor geomechanical properties (see also Appendix B). Despite the difficult conditions, the contract for the project was awarded on a "turnkey" basis, in which the contractor evidently felt that the design was sufficiently complete and reliable to accept all the risks, including geological risks. Full face tunnel advance (Fig. 6) took place simultaneously on a maximum of 26 faces to achieve average advance rates of around 1,600 metres/month of finished tunnel [72], [73], [74] and was completed in October 2006 on time and as budgeted in the executive design specifications (see Appendix B).

While the art of designing and constructing tunnels has perhaps lost some of its fascination with the demand to plan, it has certainly gained in efficiency and functionality with the introduction of the ADECO-RS approach, while neither restricting nor conditioning the imagination of the design engineer.

Appendices

Introduction to the appendices

It is said that a good example is worth more than any long theoretical discourse. I have therefore brought together some of the most important examples of tunnelling performed in recent years in these appendices. They are in seven parts.
The first three parts (A, B and C) concern three particularly significant projects:

A. the design and construction of tunnels for the new Rome-Naples high speed/capacity railway line. The application of the ADECO-RS approach is illustrated at both the design and the construction stages in terms of the reliability of the design;
B. the design and construction of approximately 73 km of running tunnels for the new Bologna-Florence section of the new high speed/capacity Milan-Rome-Naples railway line, where the considerable length of the alignment that runs underground, the heterogeneity of the ground tunnelled and the extremely difficult stress-strain conditions to be overcome constituted a severe test of the effective capacity of the new ADECO-RS approach to meet expectations in terms of the industrialisation of tunnelling;
C. the construction of the "Tartaiguille" tunnel in France. It proved too difficult to drive this tunnel using conventional means, while the application of the new advance method in the presence of a rigid core enabled it to be completed on time and within the budgeted cost.

The next four parts of the appendices (D, E, F and G) are dedicated to the most successful of the tunnelling technologies designed by the author over the last twenty years:

D. the cellular arch with which it is possible to construct large underground cavities (up to a span of 60 m) using bored tunnel methods without causing surface settlement even in ground of poor geomechanical quality (reference project: Venezia Station on the Milan urban railway link line);
E. artificial ground overburdens (AGO), to drive underground tunnels even where the necessary overburden is lacking, with considerable savings on times and costs which would be required to construct the tunnel itself artificially (reference project: the new Rome-Naples high speed/capacity railway line, tunnels Piccilli 1, Piccilli 2, Castagne, Santuario, Caianello and Briccelle);
F. shells of improved ground by means of vertical jet-grouting and the same technique applied horizontally for the construction of tunnel portals in difficult grounds (reference project: portal on the Messina side of the S. Elia tunnel of the Messina-Palermo highway);
G. the widening of road, highway and railway tunnels without interrupting traffic, which solves the problem of modernising old tunnels on major transport routes (reference project: widening of the Nazzano tunnel on the Milan-Rome motorway from two to four lanes in both directions).

I believe that reading these seven examples should be very helpful both for acquiring an in-depth understanding of how the ADECO-RS approach is applied in different stress-strain contexts and for fully appreciating the progress that has been made in tunnelling in the last twenty years.

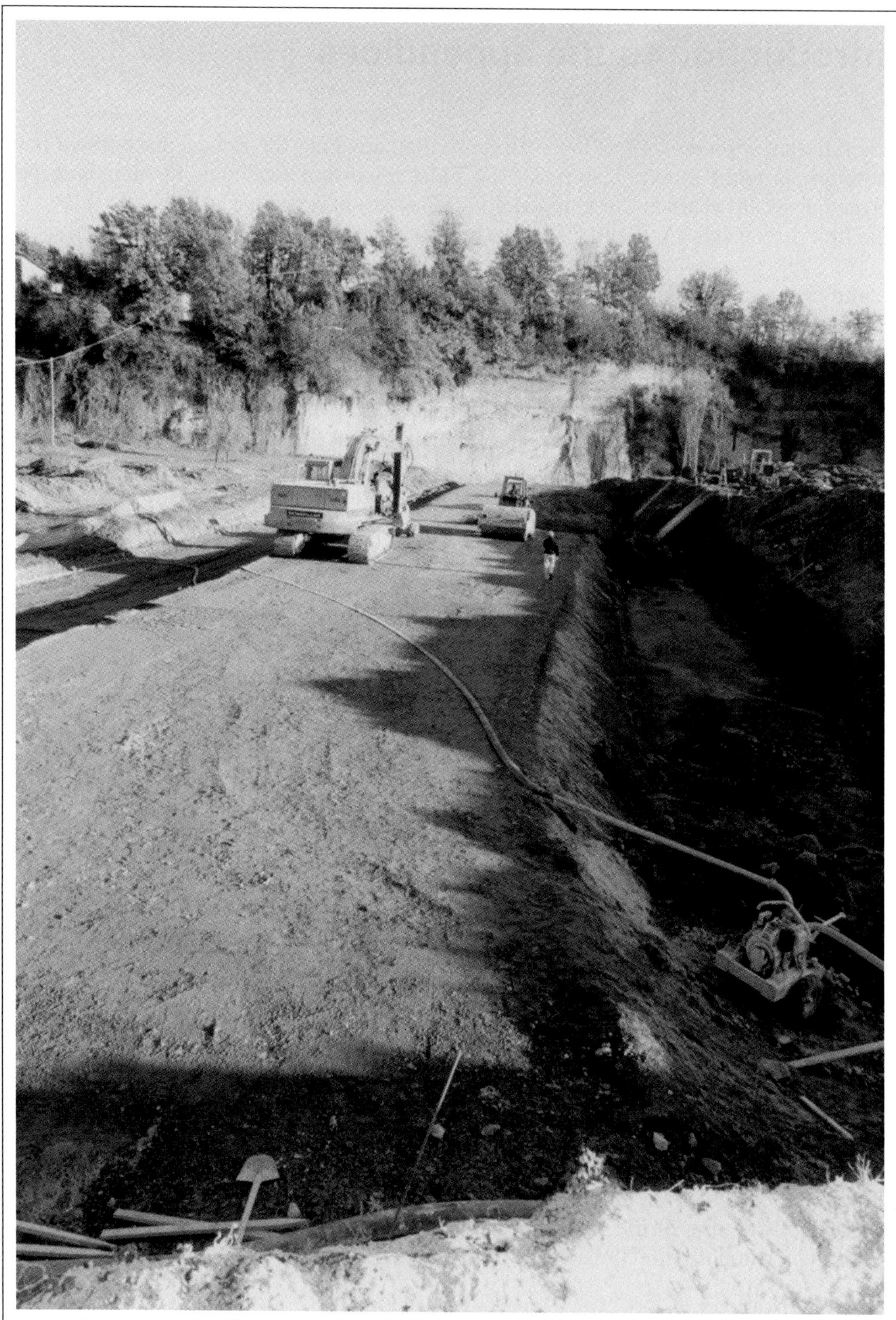

Rome-Naples high speed/capacity railway line, Piccilli 2 tunnel. Construction of the "artificial ground overburden (AGO)" in the section with no overburden, ground: pyroclastites

The design and construction of tunnels for the new Rome-Naples high speed/capacity railway line

The new Rome-Naples railway is part of the High Speed Train Milan-Naples line which in turn is a southern terminal of the European High Speed network.

The line has been divided into lots with differing costs. Construction contracts were awarded to five contractors belonging to the IRICAV UNO consortium (the general contractor for this part of the line), namely Pegaso, Icla, Italstrade, Vianini and Condotte.

The total length of the line to be constructed was 204 km and 28.3 km of it (13% of the total length) runs through bored tunnel.

The problems encountered in the design and construction of the parts of the tunnel designed by the author are discussed below. They amounted to 77% of the underground works for a total of 21.8 km divided into 22 tunnels.

Colle Pece Tunnel. North portal

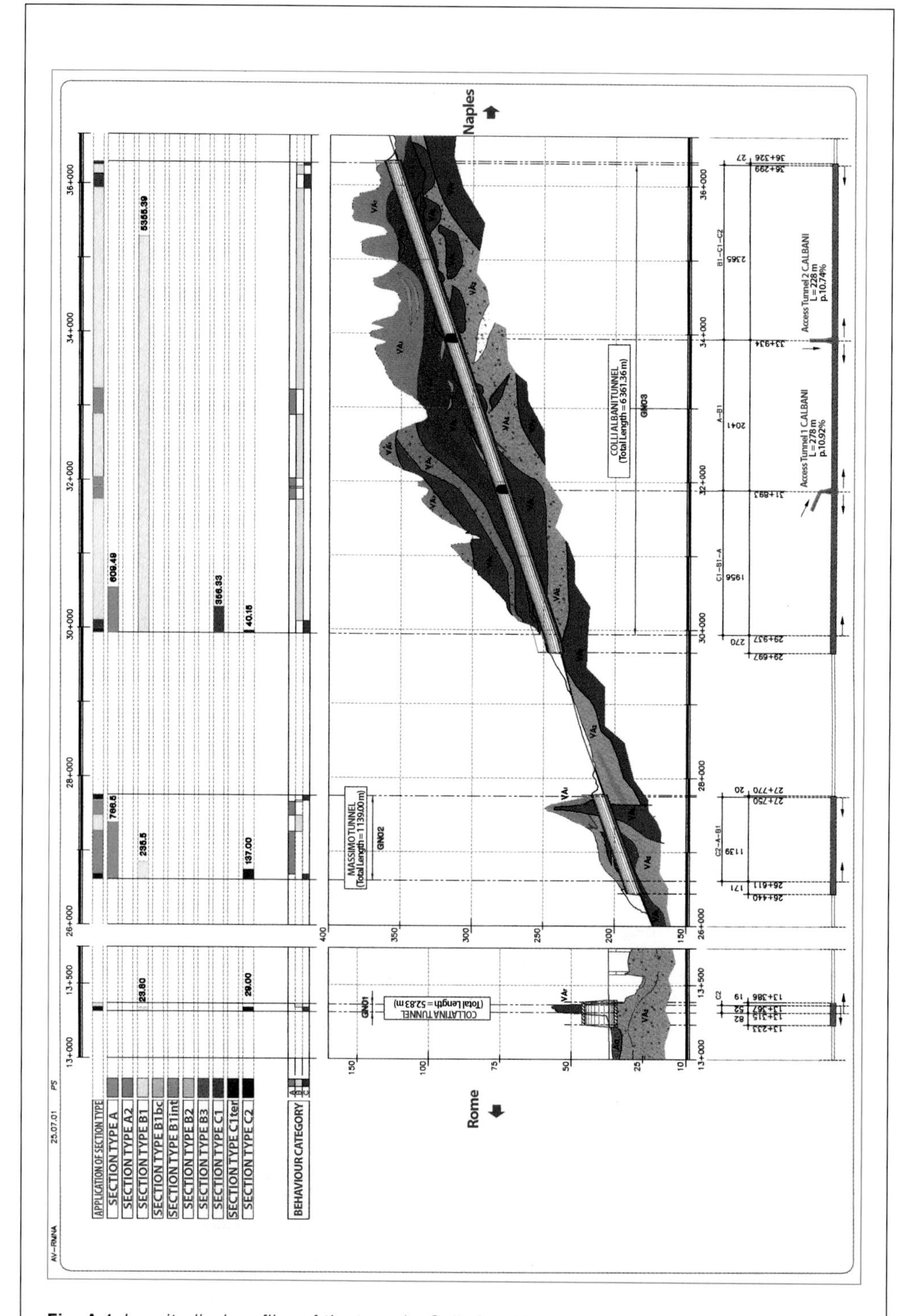

Fig. A.1 *Longitudinal profiles of the tunnels: Collatina, Massimo, Colli Albani*

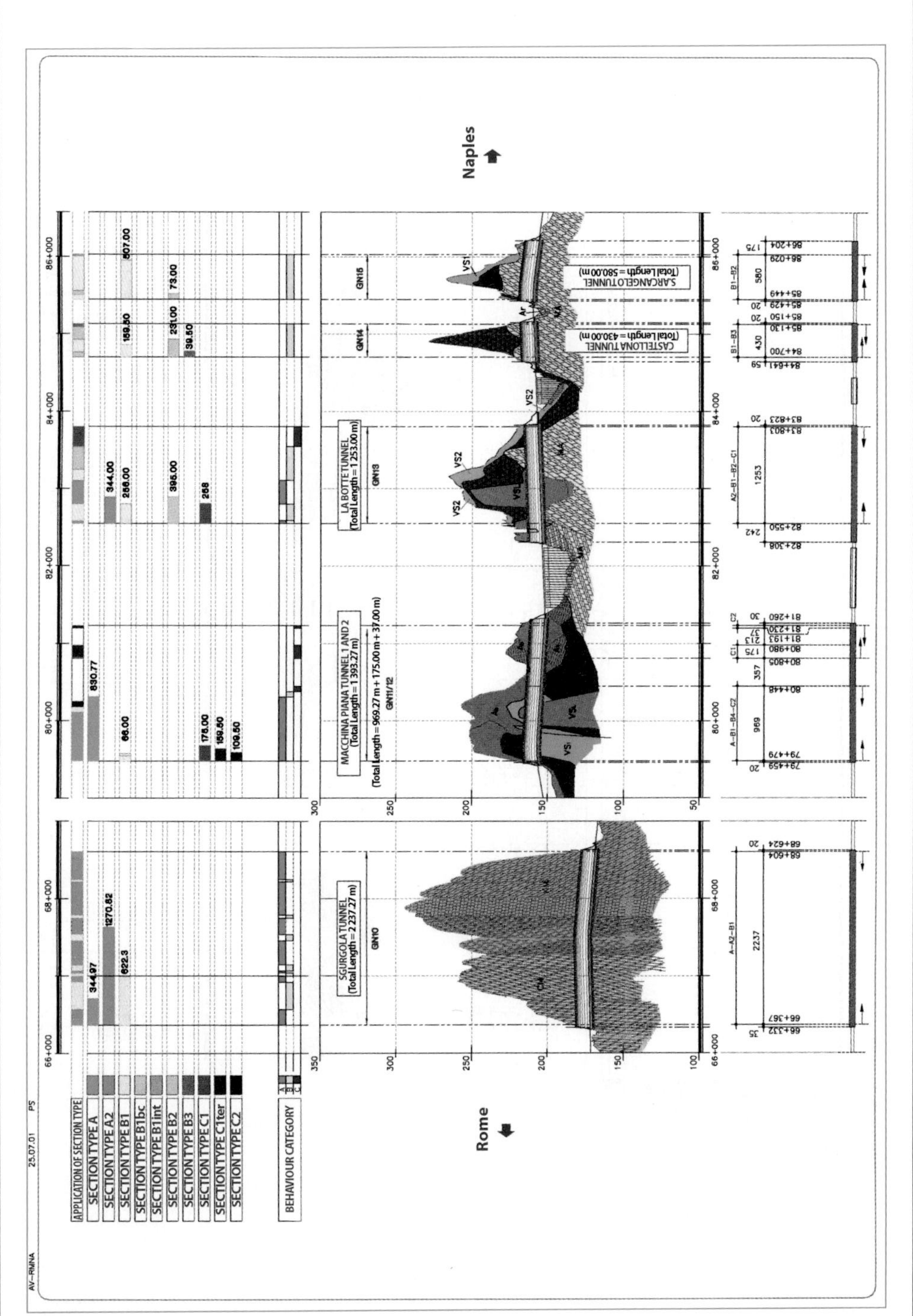

Fig. A.2 *Longitudinal profiles of the tunnels: Sgurgola, Macchia Piana 1 and 2, La Botte, Castellana, S. Arcangelo*

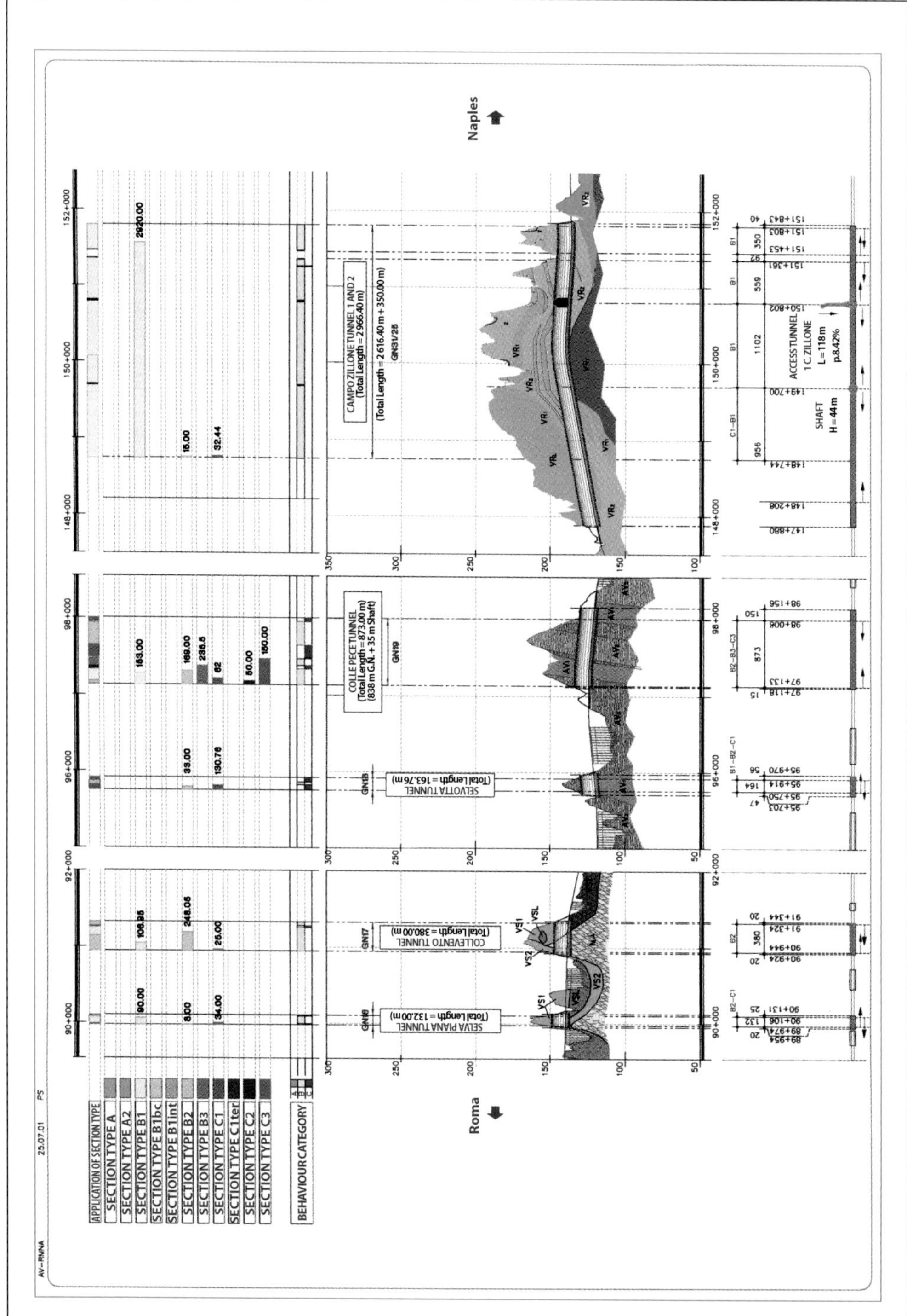

Fig. A.3 *Longitudinal profiles of the tunnels: Selva Piana, Collevento, Selvotta, Colle Pece, Campo Zillone 1 and 2*

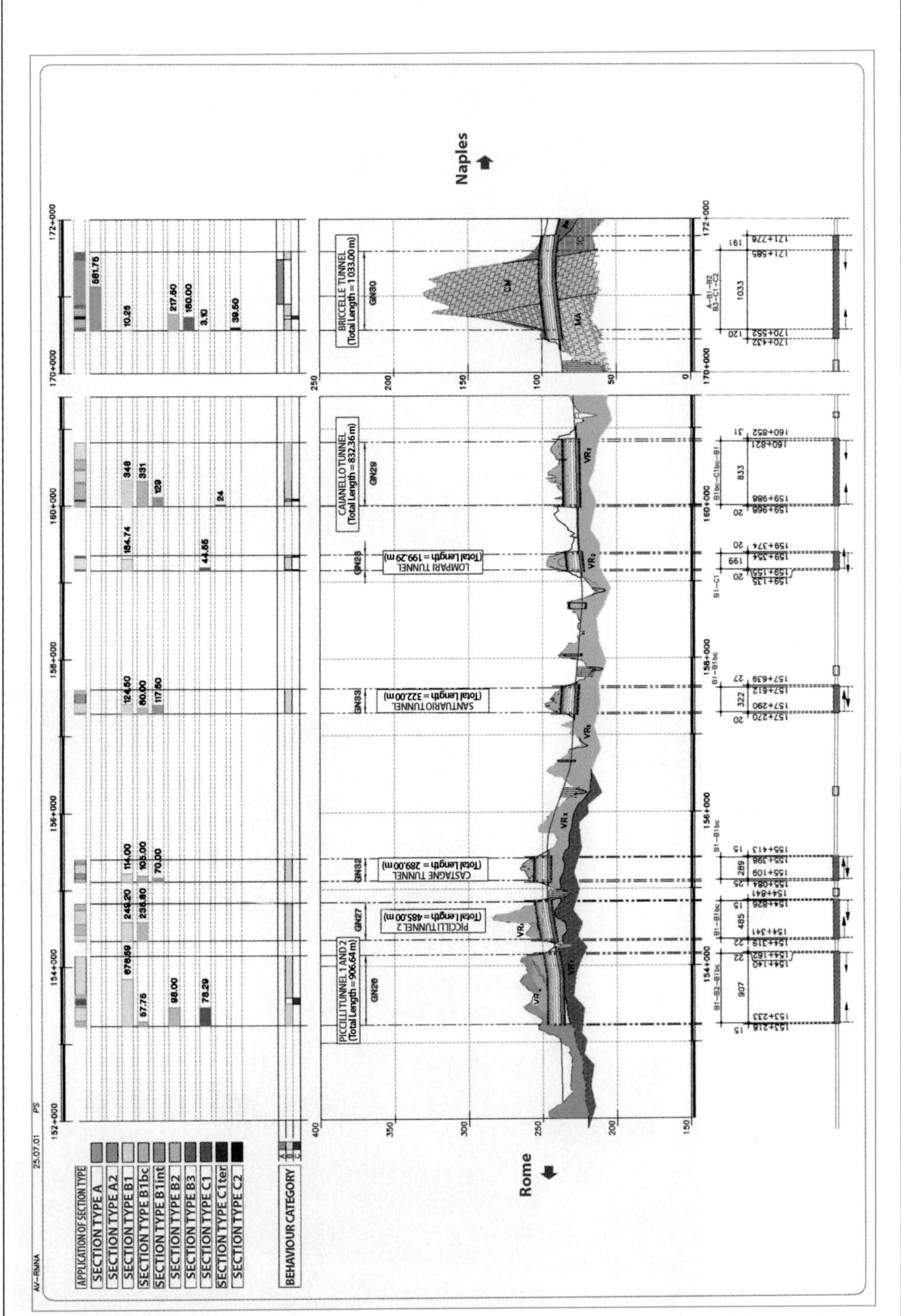

Fig. A.4 *Longitudinal profiles of the tunnels: Piccilli 1 and 2, Castagne, Santuario, Lompari, Caianello, Briccelle*

Colle Pece Tunnel. Face reinforced with fibre glass structural elements. Ground: scaly clays, overburden: ~ 25 m

Colle Pece Tunnel. Positioning a steel rib. Ground: scaly clays, overburden: 25 m

Figures A.1 to A.4 show the longitudinal geological profiles of the 22 tunnels: the *Colli Albani* tunnel (Pegaso lot, 6361 m) is the longest tunnel on the entire line, while the *Collatina* tunnel (Italstrade lot, 53 m) is the shortest.

The underground alignment runs through ground which can basically be classified as having two different types of origin:

- pyroclastic ground and lava flows, generated by eruptions of the volcanic complexes of *Latium*, *Valle del Sacco* and *Campania*;
- sedimentary rocks of the flyschoid and carbonatic type (marly and limy argillites) belonging to the Apennine system.

The overburdens vary greatly but never exceed 110 m, while they are often very shallow at the portals.

Geological and geotechnical background (survey phase)

As mentioned above, many of the tunnels pass through ground of volcanic origin. The activity of the volcanic bodies concerned began in the Pliocene period and developed mainly in the Pleistocene period from upper *Latium* down to the *Vesuvian* region dying out about 100,000 years ago. The volcanological development of the active centres passed through three different phases, which occurred in almost the same order in all the centres of activity: the phase of the volcano-stratum, the phase of great ignimbritic expansion and the phase of the construction of ash and lava cones.

Two important centres of eruption are identifiable along the alignment, one in the *Latium* area (*Colli Albani*) and the other (*Roccamonfina*) in the *Campania* area near the end of the route. There is a smaller volcanic body near the *Sacco* river (*Valle del Sacco*). Between the two major volcanic "domains", in the central part of the alignment, often interdigitating with the volcanic products of the *Valle del Sacco* volcanic body, there are outcrops of sedimentary rocks of the Apennine backbone of the Cretaceous and Miocene periods, in carbonatic, flyschoid and clayey-marly facies.

From a hydrogeological viewpoint, the route lies practically entirely above the regional water-table and consequently the tunnels are not subject to large heads of pressure; there are, however, some localised exceptions, for example in the *La Botte* tunnel of the Italstrade lot, and the *Castellona* tunnel of the Vianini lot, where the marly-arenaceous complex provides an impermeable bedrock to the overlying pyroclastites, favouring the formation of suspended water-tables with modest heads of pressure or at the northern portal of the *Colli Albani* tunnel where a water-table that supplies a fountain in the Vetrice area is intersected.

Massimo Tunnel. Passage under the crater of a volcano, overburden: ~ 15 m

Colli Albani Tunnel. A typical face running through vulcanites, overburden: ~ 55 m

	VOLCANIC COMPLEXES (Vulcaniti dei Colli Albani, Vulcaniti della Valle del Sacco, Vulcaniti di Rocca Monfina)			CARBONATIC COMPLEX (Calcari dei Monti Lepini)	FLYSCHOID COMPLEX (Argille Varicolori, Complesso Marnoso-Arenaceo)	
	Pyroclastites	**Tuffs**	**Lavas**	**Stratified limestones**	**Scaly clays**	**Clayey marls with arenaceous layers**
γ [t/m³]	1.4-1.6	1.6-2	2.6-2.7	2.5-2.7	2.0-2.1	2.2-2.4
C [Mpa]	0-0.1	0.1-0.5	0.5-5	0.5-1	0.01-0.05	0.2-0.4
φ [°]	25-35	20-25	30-35	35-40	18-23	28-33
σ_{gd} [Mpa]	1-5	-	-	1-4	-	-
E [Mpa]	1-5	300-600	2,000-5,000	10,000-12,000	50-100	200-400
υ	0.35	0.3	0.25	0.25-0.3	0.35	0.3

From a geotechnical viewpoint the lithotypes given in table below were identified in the geological complexes. The parameters for them with the variation in the geotechnical characteristics are given in the same table.

Stress-strain behaviour predictions (diagnosis phase)

It became immediately clear in the diagnosis phase that the tunnels to be driven, either because of the geotechnical characteristics of the ground or because of the varying overburdens, would be subject to extremely different stress-strain conditions.

The geological, geotechnical and hydrogeological information acquired and the results of calculations performed using analytical and/or numerical methods were therefore employed to divide the underground alignment into sections of uniform stress-strain behaviour as a function of core-face stability in the absence of intervention to stabilise the tunnel:

- stable core-face (behaviour category A);
- stable core-face in the short term (behaviour category B);
- core-face unstable (behaviour category C).

Category A comprised all those sections of tunnel in which calculations predicted that:

Colli Albani Tunnel. Detail of the ribs in the cross vault of the junction with the Finestra 1, access tunnel, ground vulcanites, overburden: ~ 50 m

- the stress state of the ground at the face and around the cavity would not have exceeded the natural strength of the medium;
- an “arch effect” would have been created close to the profile of the tunnel;
- deformation phenomena would have developed in the elastic range, having an immediate effect and a magnitude of a few millimetres;
- as a consequence, the face as a whole would have remained stable.

This category was found in lava, lithoid tuff and slightly fractured limestone sections, materials which generally present good strength properties in relation to the stresses mobilised by driving tunnels with the design overburdens.

Category B, however, included all those sections of tunnel where mathematical calculation predicted that:

- the stress state at the face and around the cavity during tunnel advance would have exceeded the natural strength characteristics of the medium, in the elastic range;
- an “arch effect” would not have been formed close to the profile of the tunnel, but at a distance equal to the size of the band of plasticised ground around the cavity;
- deformation phenomena would have developed into the elasto-plastic range with the effect deferred in time and measurable in centimetres;
- as a consequence, the core-face would have remained stable in the short term at normal tunnel advance rates, with contained extrusion of the core-face but not sufficient to affect the short term stability of the tunnel since the ground would still be able to generate sufficient residual strength.

This category was found in the flyschoid complexes (*Argille Varicolori* or marly-arenaceous ground) or in the stratified pyroclastites, as long as the overburdens were sufficient to make the natural formation of an arch effect possible.

Finally category C included all those sections of tunnel where numerical calculation predicted that:

- the state of stress in the ground would have exceeded the strength characteristics of the material considerably even in the zone around the face;
- an “arch effect” would have been formed naturally neither at the face nor around the tunnel since the ground would not have possessed sufficient residual strength;
- deformation phenomena would have developed into the failure range, with the effect deferred in time and be measurable in decimeters, giving rise to serious instability such as the failure of the face and the collapse of the cavity.

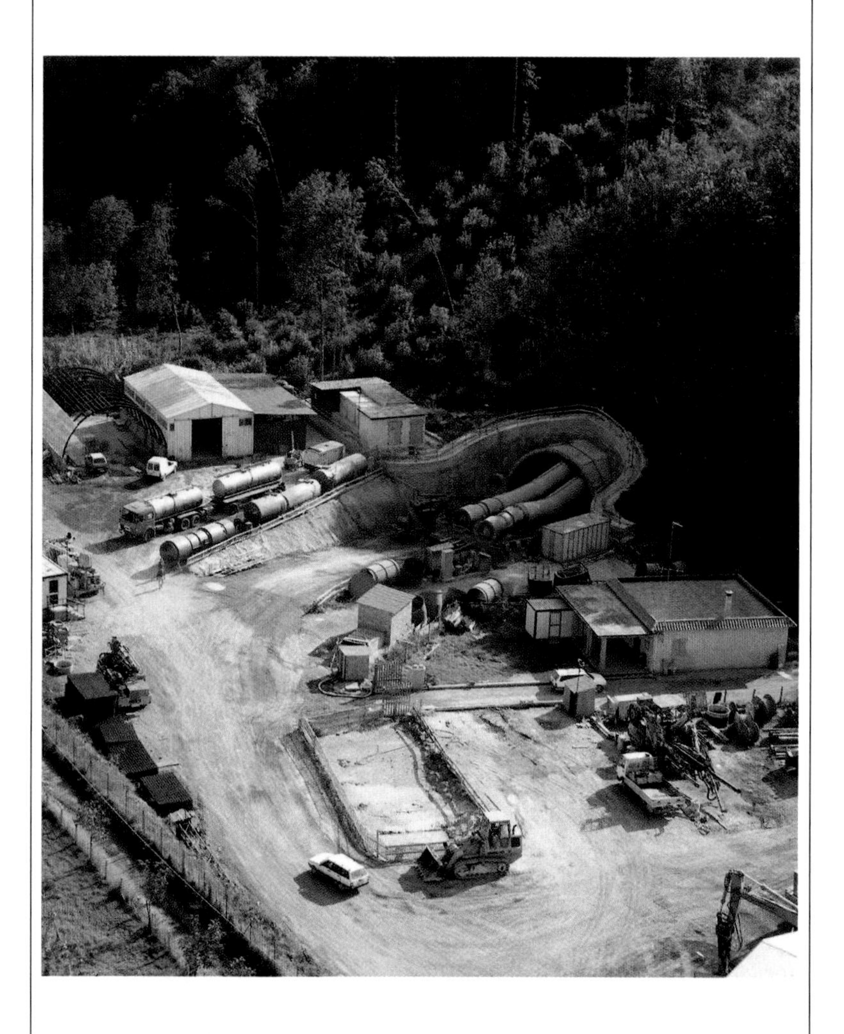

Colli Albani Tunnel. The portal of the Finestra 1 access tunnel

- as a consequence, in the absence of intervention to stabilise it, the core-face would have been completely unstable.

This category was found most frequently at portals and in general in sections with shallow overburdens, as well as in sections of clayey ground with a scaly structure belonging to the flysches of the *Argille Varicolori* with geotechnical characteristics close to the lower limits of the range identified (residual values). There is no way in which an arch effect can be formed in these cases except artificially.

Construction methodology (therapy phase)

After formulating reliable predictions of the stress-strain behaviour of the ground as a result of excavation, the most appropriate stabilisation techniques were chosen to control, contain or actually anticipate and eliminate deformation for each section of tunnel with uniform stress-strain behaviour.

The guiding principles on which the design of the tunnel section types was based are essentially as follows:

1. full face advance always, especially under difficult stress-strain conditions;
2. containment of the alteration and decompression of the ground around the tunnel by immediate application of effective preconfinement and/or confinement of the cavity (sub-horizontal jet-grouting, fibre glass structural elements in the core and/or in advance around the future tunnel and, if necessary, fitted with valves for cement injections, shotcrete, etc.) of sufficient magnitude, according to the case, to absorb a significant proportion of the deformation without collapsing or to anticipate and eliminate the onset of any movement in the ground whatsoever;
3. placing of a final concrete lining, reinforced if necessary, and completed, where necessary to halt deformation phenomena, with the casting of the tunnel invert in steps at a short distance from the face.

The following tunnel section types were actually designed (Fig. A.5):

- **for sections of tunnel belonging to behaviour category A** (*stable core-face*) a type A section was designed consisting of a simple preliminary lining in sprayed concrete reinforced with simple steel ribs and a final lining in concrete 60 cm thick, closed with a tunnel invert also 60 cm thick;
- **for sections of tunnel belonging to behaviour category B** (*stable core-face in the short term*) three main tunnel section types were designed:

Colli Albani Tunnel. Work to reinforce the core-face, ground: vulcanites, overburden: ~ 30 m

Santuario Tunnel. Treatment of the ground with lime to create Artificial Ground Overburden (A.G.O.), type of ground: pyroclastites

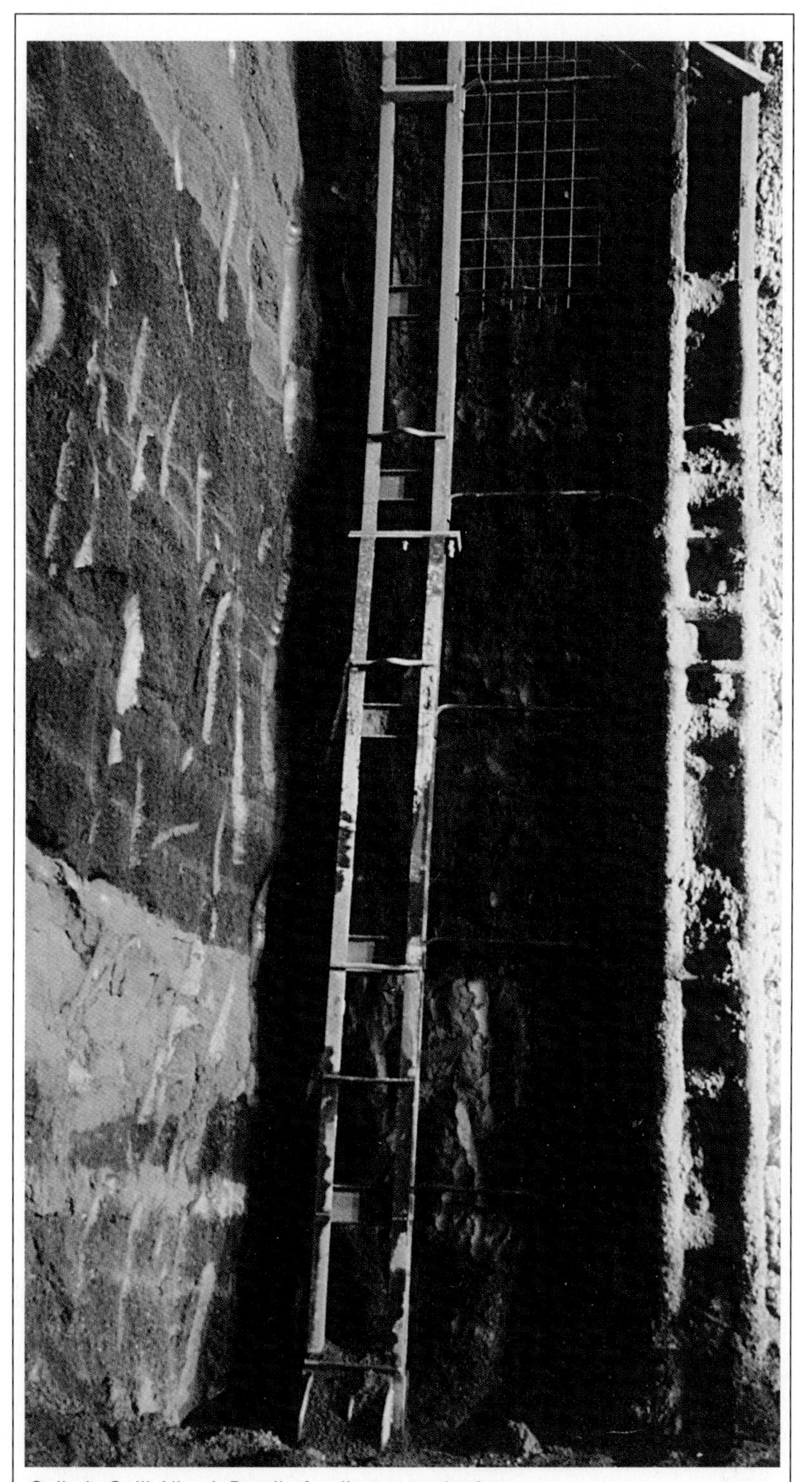

Galleria Colli Albani. Detail of a rib next to the face

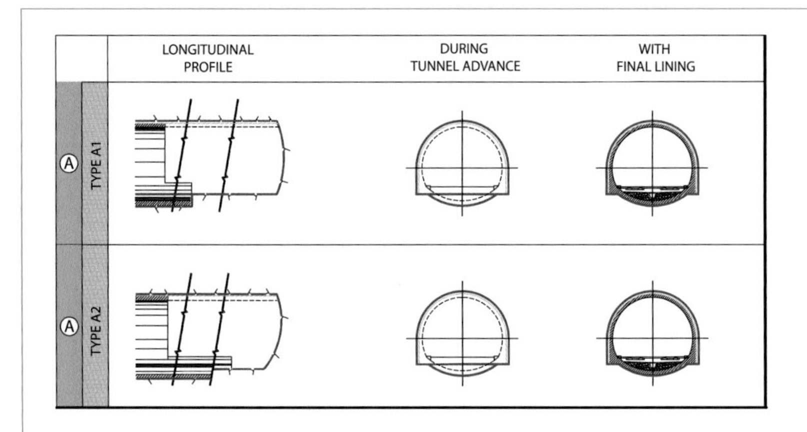
LONGITUDINAL PROFILE
DURING TUNNEL ADVANCE
WITH FINAL LINING
Ⓐ TYPE A1
Ⓐ TYPE A2

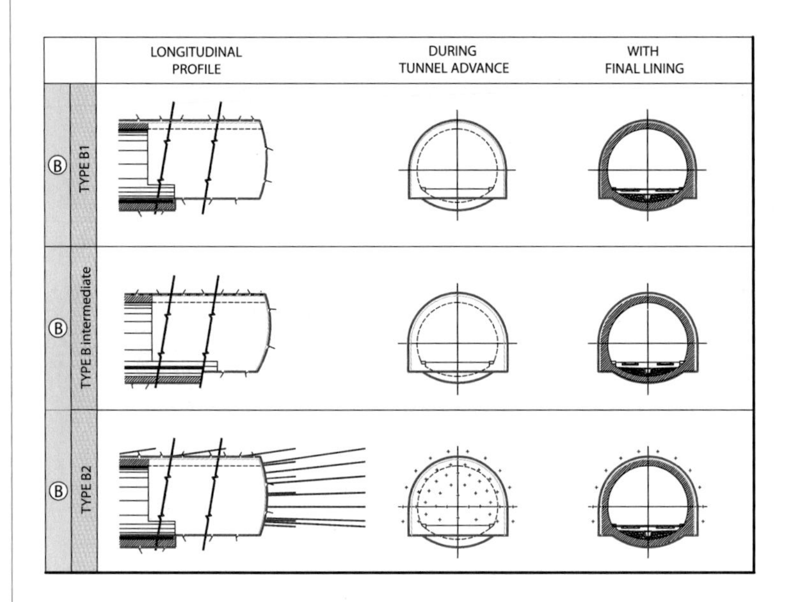
LONGITUDINAL PROFILE
DURING TUNNEL ADVANCE
WITH FINAL LINING
Ⓑ TYPE B1
Ⓑ TYPE B intermediate
Ⓑ TYPE B2

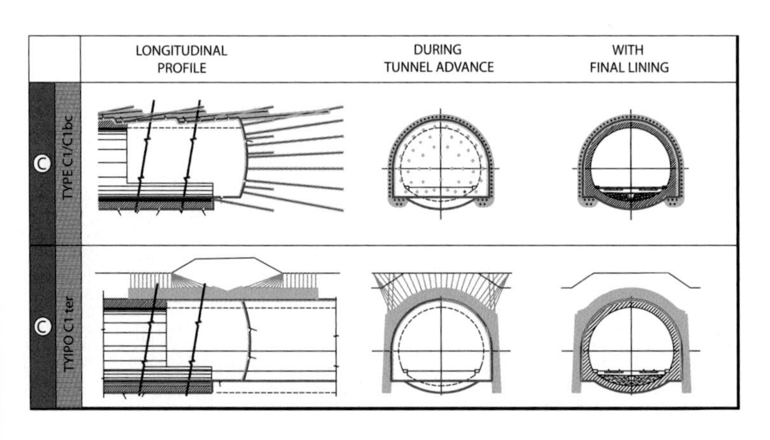
LONGITUDINAL PROFILE
DURING TUNNEL ADVANCE
WITH FINAL LINING
Ⓒ TYPE C1/C1bc
Ⓒ TYPO C1 ter

LEGEND
GROUND IMPROVEMENT/REINFORCEMENT

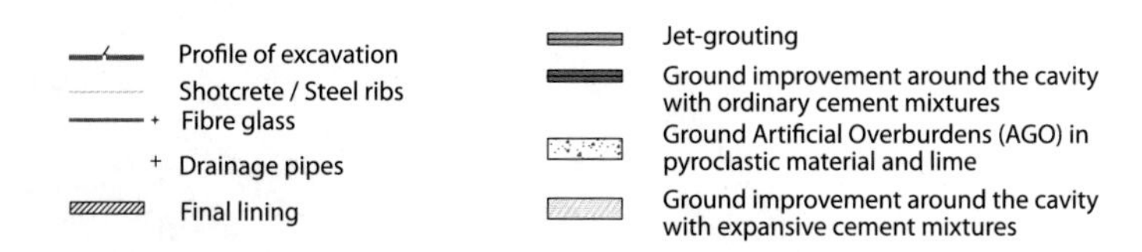

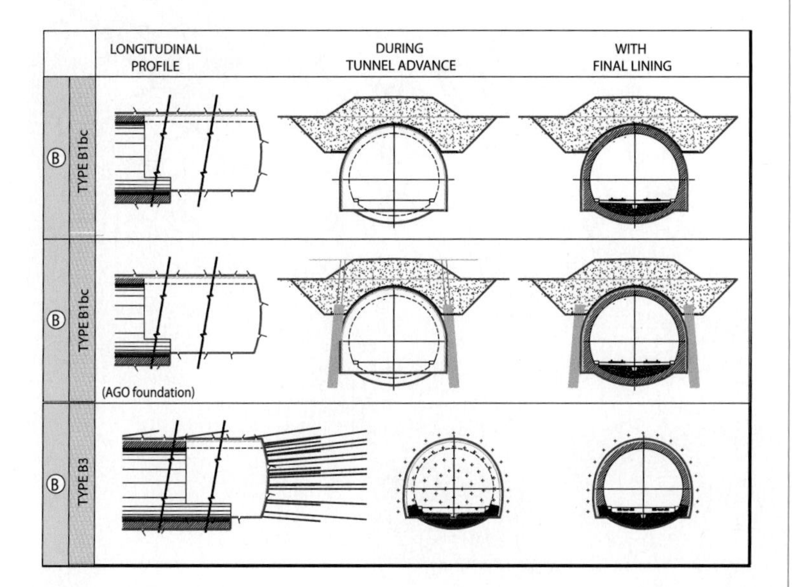

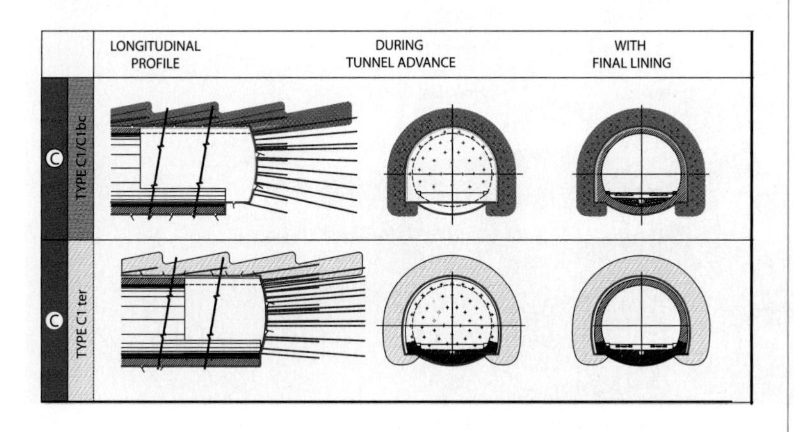

Fig. A.5. *Tunnel section types*

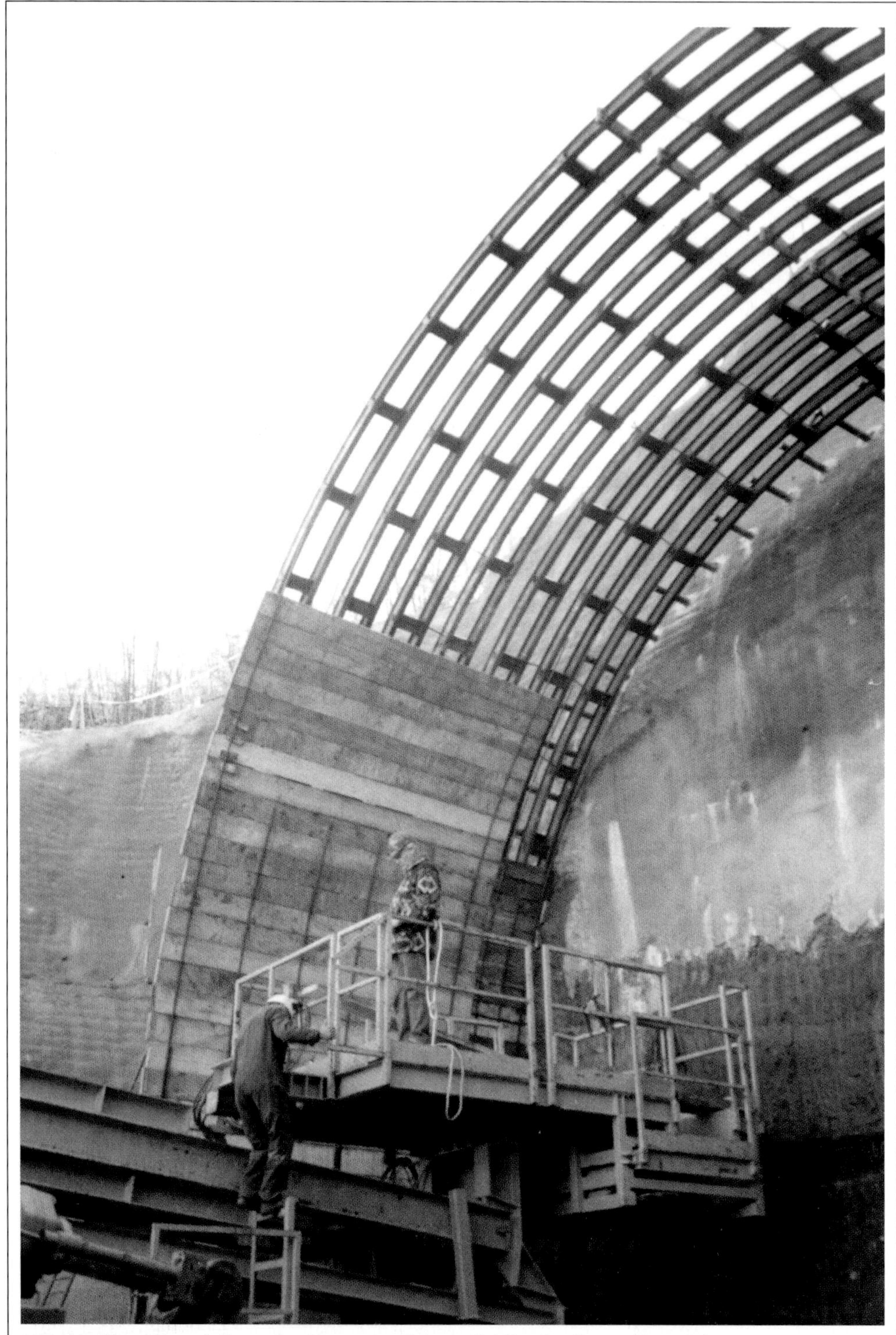

Macchia Piana Tunnel. Preparing the excavation scaffolding for the north portal

  - section type B1, consisting of a preliminary lining in shotcrete reinforced with double steel ribs + a final lining of 80 cm closed with a tunnel invert of 90 cm cast within 3 tunnel diameters from the face;
  - section type B2, for which reinforcement of the advance core is specified, performed using fibre glass structural elements + a preliminary lining in shotcrete reinforced with double steel ribs + a final lining of 90 cm closed with a tunnel invert cast within a distance of 1.5 tunnel diameters from the face;
  - section type B3, for which a geometry with curved tunnel walls is specified in order to withstand horizontal thrusts more effectively along with reinforcement of the advance core (more intensely than for B2 with more reinforcement and a longer overlap) again using fibre glass structural elements + a preliminary lining in shotcrete reinforced with double steel ribs + a final lining of 90 cm closed with a tunnel invert 100 cm thick cast in one piece with side kicker and floor slab in steps of 4–6 m from the face;
- **for sections of tunnel belonging to behaviour category C** (*unstable core-face*) two main types of tunnel section were designed:
  - section type C1, consisting of advance reinforcement of the ground around the tunnel using the technique of sub-horizontal jet-grouting + microcolumns of improved ground created using the same technique, but in the advance core and reinforced with fibre glass structural elements (15.5 m in length with overlap of 3 m) + a preliminary lining in shotcrete reinforced with double steel ribs + a final lining varying in thickness in the crown from 40 to 130 cm closed with a tunnel invert 100 cm thick, cast in steps 6–12.5 m in length, 1.5 tunnel diameters from the face;
  - section type C2, consisting of advance ground improvement around the future cavity using high pressure grout injections by means of valved fibre glass tubes + reinforcement of the advance core with grouted fibre glass structural elements (15 m in length, with an overlap of 5 m) + a preliminary lining in shotcrete reinforced with double steel ribs + a final lining of 90 cm closed with a tunnel invert 100 cm thick cast in steps 6–12.5 m in length, 1–1.5 tunnel diameters from the face.

Finally when the detailed design specifications were completed the percentages of different tunnel sections types with respect to the total length of the tunnels was as follows:

- section type A: 28.9%;
- section type B: 60%;
- section type C: 11.1%.

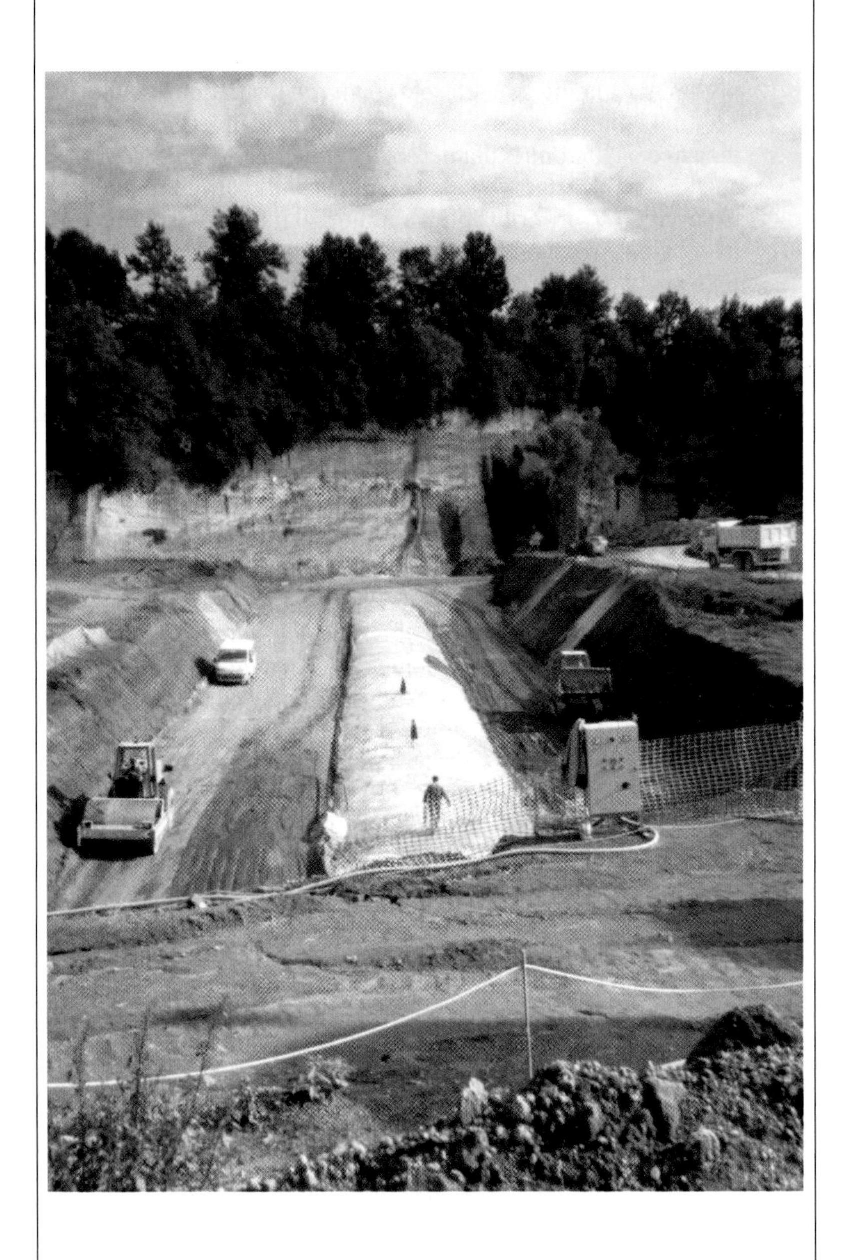

Piccilli 2 Tunnel. Construction of the "Artificial Ground Overburden (A.G.O.)" in the section with no overburden, ground: pyroclastites

In addition to the running tunnels, the detailed design also involved a few indispensible accessory works. The following are worthy of mention:

- three access tunnels for the passage of vehicles were driven to accelerate advance rates for the longest tunnels, working contemporaneously on several faces: two access tunnels on the *Colli Albani* tunnel (over 6 km in length) and a third access tunnel to the *Campo Zillone* tunnel (around 3 km in length);
- works for portals, which, depending on the morphology and the nature of the ground in question, were designed using the most appropriate technologies (shells of ground improved using jet-grouting methods, "berlin" walls, etc.).

Statics calculations

The statics and deformation behaviour of tunnels, both in the construction phases and the final service phase were analysed and verified by a series of calculations on three and two dimensional finite element models in the elasto-plastic range.

The models were developed to simulate the behaviour of the ground-tunnel system in the different construction phases, as realistically as possible. Particular attention was paid: to the effect of preconfinement treatments of the cavity and reinforcement of the advance core specified in the design; to deformation values to be expected at the constructions stage; and to stresses in the preliminary and final linings. The mathematical models were processed on computers using version 6.0 of the ADINA software package.

Contracts

Contracts for all the tunnels, just as for all the other surface works, required for construction of the line were of the "all-in, lump sum" type (2.844.644.600 euro, of which 324.231.600 for tunnels only) and awarded on the basis of detailed design specifications. With this type of contract the General Contractor IRICAV UNO accepted all risks including the geological risk.

Operational phase

Construction design began at the same time as excavation work (May 1995, *Colli Albani* tunnel) immediately after the contract was awarded.

Given the additional survey information available and direct in the field confirmation, the validity of the detailed design was substantially

Colli Albani Tunnel. Horizontal jet-grouting to improve the ground for the South portal

Macchia piana Tunnel. The scaffolding template for the North portal

confirmed at the operational phase and only a few minor refinements were made during construction design:

- in order to specifically tackle the scaly clays, characteristic of the *Colle Pece* tunnel, a section type C3 was introduced that is different from C2, due above all to the use of expanding cement mixes for grouting of the fibre glass structural elements;
- a section type "$B1_{bc}$" for shallow overburdens and a section type "$C1_{ter}$" were designed for tunnel sections with particular design characteristics;
- section types A2 and an "$B1_{intermedia}$", very similar to A1 and B1 described above, were developed to optimise works in a few particular circumstances;
- admissible variations (e.g. intensity of reinforcement) were specified for each tunnel section type according to the actual deformation behaviour measured during construction as compared to that predicted by design calculations. This was done in order to apply quality assurance norms.

Finally, when the construction design was completed, the percentages of different tunnel section types were as follows:

- section type A: 22.5 %;
- section type B: 69.4 %;
- section type C: 8.2 %.

At the end of December 1999, after 1,100 working days (1,700 total days), approximately 21.6 km of tunnel had been completed, almost all fully lined, equal to about 99% of the underground sections.

Average advance rates were about 20 m/day, not counting excavation performed to open access tunnels, shafts and other accessory works.

Figure A.6 shows production graphs for the *Colli Albani* and *Sgurgola* tunnels from which it can be seen that production rates were not only high (around 100 m/month per face) but above all very constant, a sign that the construction design matched actual conditions excellently.

All the civil engineering works were completed in March 2001. Table A.1 gives a clear comparison of the differences in the distribution of tunnel section types between the detailed design, the construction design and "as built". These differences did not result in any significant change in the overall cost of the works, since the greater percentage of B section types was compensated for by a decrease in A and, above all, in C (by far the most costly) section types.

Monitoring phase

Monitoring during construction

The adequacy of the design hypotheses was verified by geostructural mapping of the face and monitoring stress-strain behaviour of the face and the cavity observed during construction for each stage and sequence specified in the design.

HIGH SPEED TRAIN - Milan to Naples Line - Rome to Naples Section - SECTION TYPES DISTRIBUTION																										
		DETAILED DESIGN									CONSTRUCTION DESIGN								AS BUILT (March 2001)							
WBS	TUNNEL	TUNNEL LENGTH [m]	SECTION TYPES (Lenght in metres)							TUNNEL LENGTH [m]	SECTION TYPES (Lenght in metres)							Builtsection		SECTION TYPES (Lenght in metres)						
			A	B1	B1bc	B2	B3	C1	C2		A/A2	B1	B1 bc	B2	B3	C1	C2	(%)	[m]	A/A2	B1	B1 bc	B2	B3	C1	C2
GN01	COLLATINA	55							55	53		24					29	100.0	53	0	24					29
GN02	MASSIMO	1149	1092						57	1139	664	355					120	100.0	1139	766	236					137
GN03	COLLI ALBANI	6357	1462	3115				1780		6361	541	5425				355	40	100.0	6361	610	5355				356	40
GN10	SGURGOLA	2240	560	1680						2237	19887	250						100.0	2237	1615	622					
GN11	MACCHIA PIANA 1	1150	920	230						969	834	63					72	100.0	969	831	66					72
GN12	MACCHIA PIANA 2	540	540							212						191	21	100.0	212						175	37
GN13	LA BOTTE	1185	687			379			119	1253	300	348		342		263		100.0	1253	344	256		395		258	
GN14	CASTALLONA	470	249	221						430		178		173	79			100.0	430		159,5		231	39.5		
GN15	S.ARCANGELO	580				116	406		58	580		380		200				100.0	580		507		73			
GN16	SELVA PIANA	170	119	51						132		49		49		34		100.0	132		90		8		34	
GN17	COLLEVENTO	387				348			39	380		100		229,5		50,5		100.0	380		107		248		25	
GN18	SELVOTTA	173				173				164				31		133		100.0	164				33		131	
GN19	COLLE PECE	873				611	175		87	873		253		389	149	42	40	100.0	838		153		169	254	62	200
GN31	CAMPO ZILLONE 1	3163		1423		1740				2617		1565		801		251		100.0	2617		2586		15		15	
GN25	CAMPO ZILLONE 2	350				350				350		334				16		100.0	350		334				16	
GN26	PICILLI 1	845	127	591				127		907		553	58	296				100.0	907		677	58	98		74	
GN27	PICILLI 2	229		195				34		485		266	219					100.0	485		249	236				
GN28	LOMPARI	235		189			47			199		155				44		100.0	199		155				44	
GN29	CAIANELLO	210		168			21	21		832		474	334			24		100.0	832		477	331			24	
GN30	BRICCELLE	1033	424	609						1033	581	10		218	180	4	40	100.0	1033	582	10		218	180	4	39
GN32	CASTAGNE	0								289		189	100					100.0	289		184	105				
GN33	SANTUARIO	0								322		242	80					100.0	322		242	80				
TOTAL LENGHT [m]		21394	6180	8472		3717	649	1962	415	21817	4907	11212	791	2729	408	1408	362	100.0	21782	4748	12490	810	1488	474	1219	544

% SECTION TYPES USE	A	B1	B1bc	B2	B3	C1	C2	A/A2	B1	B1 bc	B2	B3	C1	C2	A/A2	B1	B1bc	B2	B3	C1	C2
	28.9	39.6	0.0	17.4	3.0	9.2	1.9	22.5	51.4	3.6	12.5	1.9	6.5	1.7	21.8	57.3	3.7	6.8	2.2	5.6	2.5

TOTAL SECTION TO BE TUNNELED (%/m): 0.0 | 0

Table A.I *Percentage use of different tunnel section types: a comparison between detailed design, construction design and "as built"*

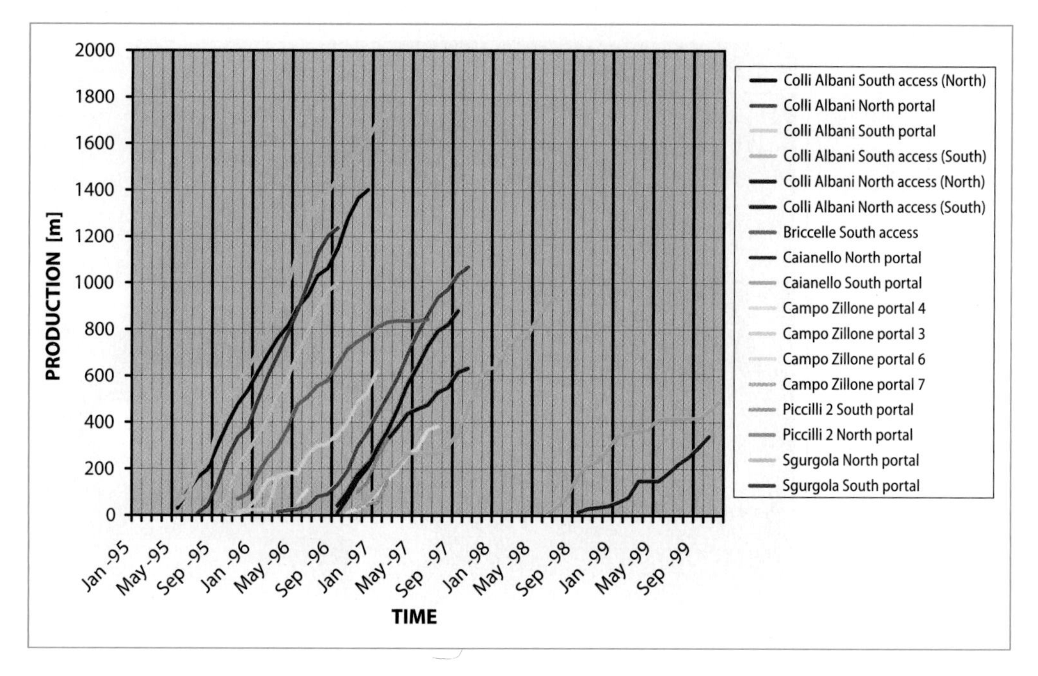

Fig. A.6 *The very distinct linear nature of the curves is a result of the high level of industrialisation achieved*

To achieve this a complete monitoring system was devised that included:

1. geological and structural mapping of the face;
2. face extrusion measurements;
3. convergence measurements;
4. extensometer measurements from the surface;
5. surface levelling measurements (on sections with shallow overburdens);
6. inclinometer measurements (at portals);
7. laboratory tests on samples taken from the face.

Final monitoring

Twenty two completely automated permanent monitoring stations for digital recording of data were located along the rail tunnels.

Their purpose was to:

1. verify design hypotheses;
2. monitor the behaviour of the tunnels over time;
3. provide information for maintenance purposes.

The following was performed as part of monitoring during construction:

1. convergence measurements to determine radial deformation;

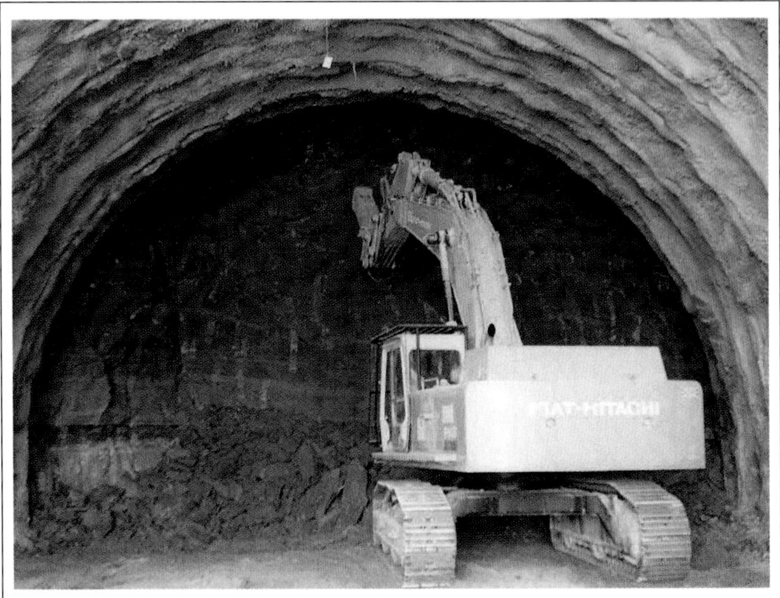

Macchia Piana Tunnel. Demolition of the core-face using a demolition hammer (ground: tuff, overburden: ~ 3 m)

Macchia Piana Tunnel. Work on the north portal (ground: rubble)

Colli Albani Tunnel. The junction with the south access tunnel, ground: vulcanites, overburden: ~ 40 m

2. extensometer measurements to assess the development of deformation inside the ground;
3. measurements of total pressure of the ground on the lining;
4. measurements of pore pressure in the ground;
5. measurements of stress in structural members;
6. measurements of the temperature of the concrete in the final lining;
7. measurements of vibrations induced in structures by the passage of trains.

Conclusions

On the basis of the final data reported above it can be stated that the design predictions made for the tunnels on the High Speed Rome-Naples line using the ADECO-RS approach were found to match the reality very closely despite the geological difficulty of the sites tunnelled.

When account is taken of the ground involved and of some objectively difficult conditions that had to be tackled, the fast average advance rates achieved constitute a good indicator of both the high standard and the reliability of the design using ADECO-RS principles and the high degree of industrialisation of the tunnel advance operations achieved on construction sites, as a consequence.

The underground works were completed on schedule in 2001. Construction costs too differ by only a few percentage points from forecasts and there were never any serious disputes between the contractors and the client.

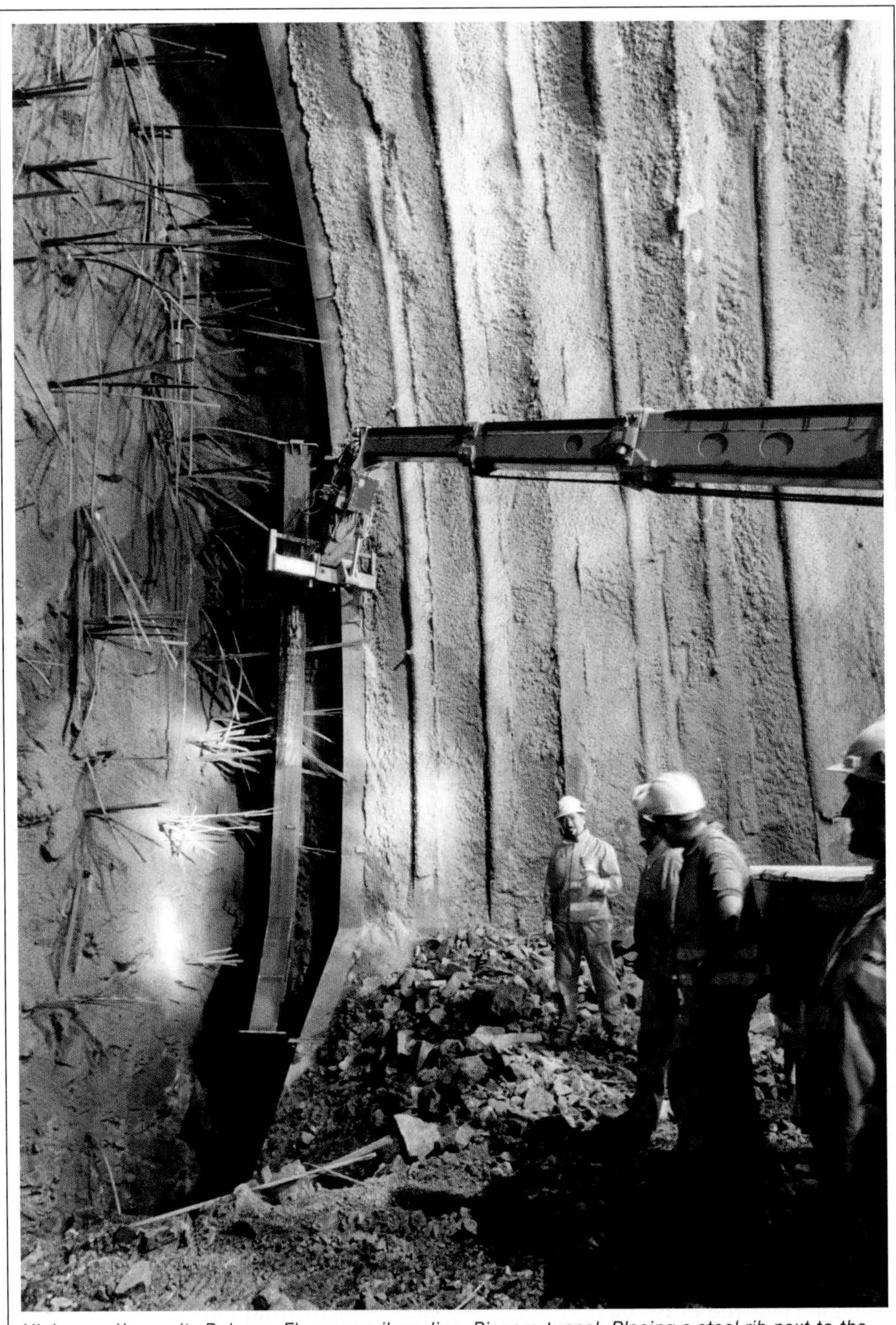

High speed/capacity Bologna-Florence railway line, Pianoro tunnel. Placing a steel rib next to the face

The design and construction of tunnels for the new Bologna-Florence high speed/capacity railway line

■ A look at the past

It is interesting to consider how the history of the railway connection between Bologna and Florence across the Tuscan Emilian Apennines has gone hand in hand with the emergence of railways as the main system of land transport for goods and passengers in a context of socio-political development which saw exponential growth in trade between southern and central Italy in the period between the unification of the country and its recent entrance into the European Union.

Work started on the first connection between Bologna and Florence in 1856 and it went into service in the November of the 1864. The alignment which follows the design

"Porrettana" Line: Construction of the South portal of the Signorino tunnel (1862 ca.)

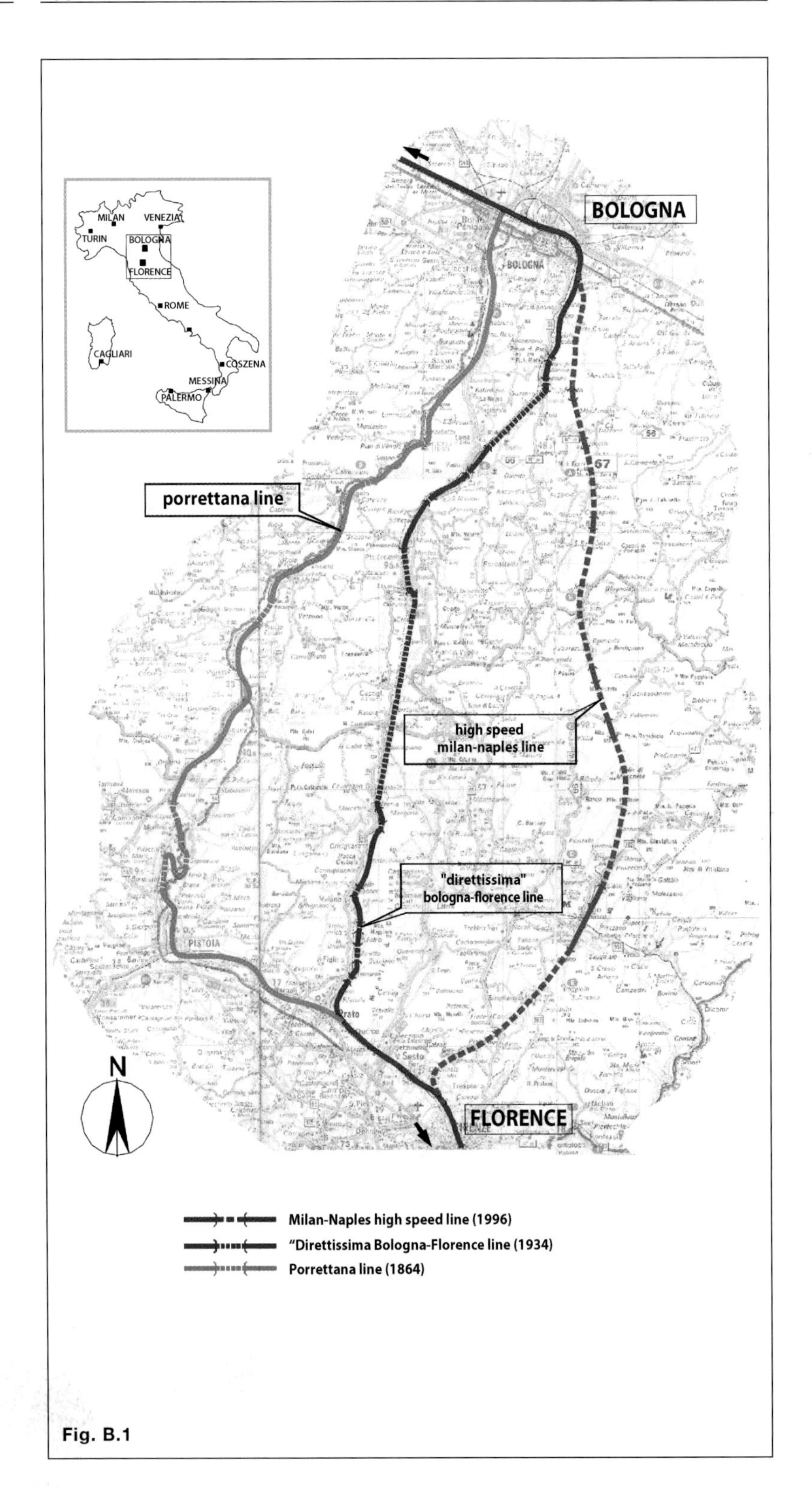
MILAN
VENEZIA
TURIN
BOLOGNA
FLORENCE
ROME
CAGLIARI
COSZENA
MESSINA
PALERMO
BOLOGNA
porrettana line
high speed
milan-naples line
"direttissima"
bologna-florence line
FLORENCE
N
Milan-Naples high speed line (1996)
"Direttissima Bologna-Florence line (1934)
Porrettana line (1864)

Fig. B.1

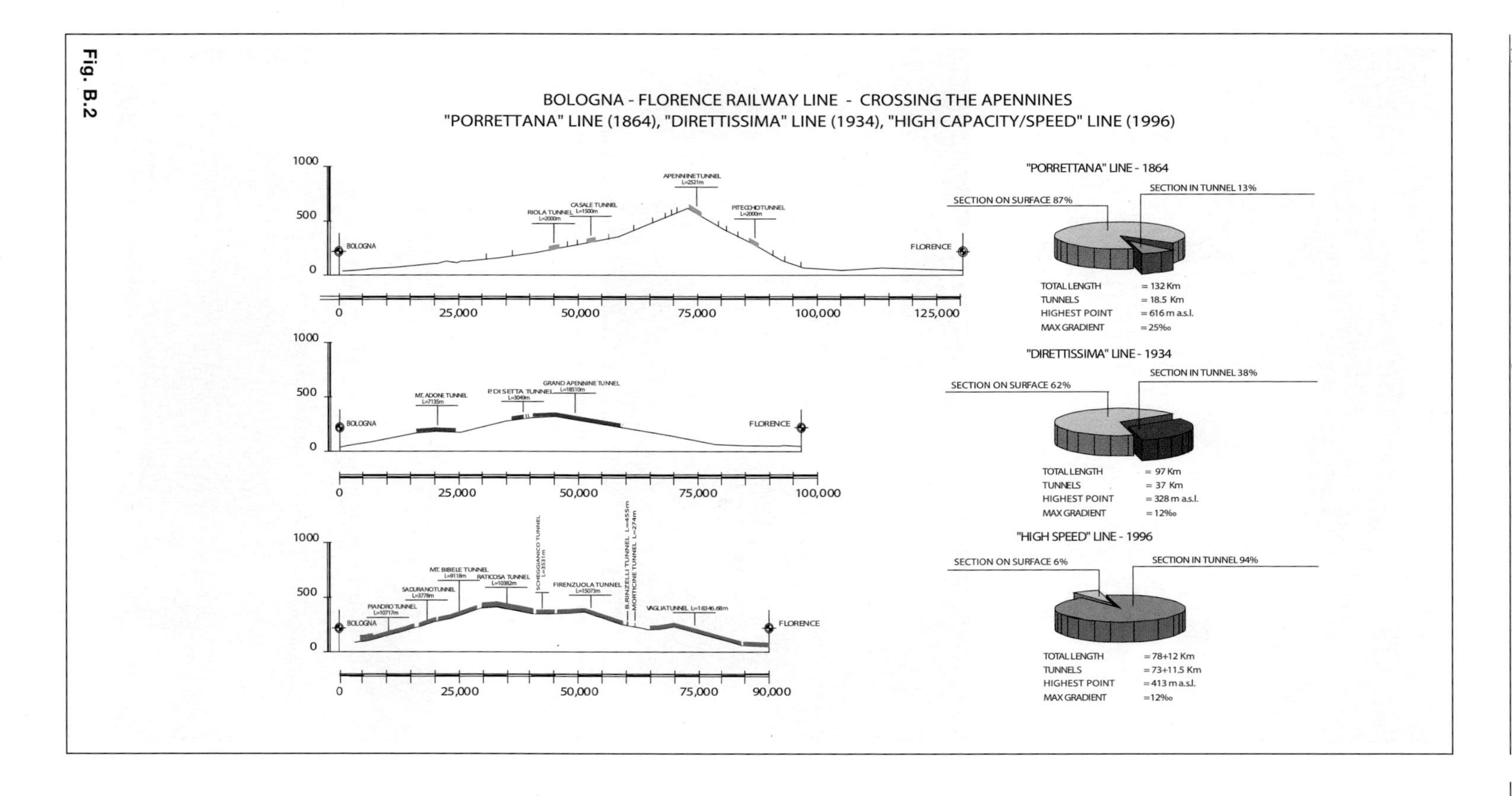
BOLOGNA - FLORENCE RAILWAY LINE - CROSSING THE APENNINES
"PORRETTANA" LINE (1864), "DIRETTISSIMA" LINE (1934), "HIGH CAPACITY/SPEED" LINE (1996)
1000
500
0
BOLOGNA
FLORENCE
RIOLA TUNNEL L=2000m
CASALE TUNNEL L=1500m
APENNINE TUNNEL L=2521m
PITECCHIO TUNNEL L=2000m
0
25,000
50,000
75,000
100,000
125,000
MT. ADONE TUNNEL L=7135m
P. DI SETTA TUNNEL L=3049m
GRAND APENNINE TUNNEL L=18510m
SCHEGGIANICO TUNNEL L=3531m
B.RINZELLI TUNNEL L=455m
MORTICINE TUNNEL L=274m
FIRENZUOLA TUNNEL L=15073m
VAGLIA TUNNEL L=18346.68m
90,000
"PORRETTANA" LINE - 1864
SECTION ON SURFACE 87%
SECTION IN TUNNEL 13%
TOTAL LENGTH = 132 Km
TUNNELS = 18.5 Km
HIGHEST POINT = 616 m a.s.l.
MAX GRADIENT = 25‰
"DIRETTISSIMA" LINE - 1934
SECTION ON SURFACE 62%
SECTION IN TUNNEL 38%
TOTAL LENGTH = 97 Km
TUNNELS = 37 Km
HIGHEST POINT = 328 m a.s.l.
MAX GRADIENT = 12‰
"HIGH SPEED" LINE - 1996
SECTION ON SURFACE 6%
SECTION IN TUNNEL 94%
TOTAL LENGTH = 78+12 Km
TUNNELS = 73+11.5 Km
HIGHEST POINT = 413 m a.s.l.
MAX GRADIENT = 12‰

Fig. B.2

Bologna-Florence Direttissima Line, excavation of the Grand Apennine Tunnel: reinforcement in the crown with steel ribs

Bologna-Florence Direttissima Line, excavation of the Station chamber from the previous Grand Apennine Tunnel: steel ribs in the crown and side drifts

Bologna-Florence Direttissima Line, the Grand Apennine Tunnel: the portal on the Florence side

by *ing.* Protche of Bologna is that of the Porrettana line (Fig. B.1), still in existence today, which winds along the Reno valley to pass over the Apennines at Pracchia (altitude 616 m a.s.l.) and reach Florence passing through Pistoia.

It was realised before work on the construction of this railway had even finished that because of its high gradients it would soon be insufficient to handle all the vigorously growing traffic between the Po river valley and the capital. A new more direct connection, and above all one with greater capacity, had therefore to be established between Bologna and Florence as soon as possible. The problem, examined by the top civil engineers of the period, gave rise to a series of proposals with different vertical and horizontal alignments. What they had in common was that in their desire to limit the length of the main tunnel, they were all induced to maintain the maximum altitude at around 500 m a.s.l. accepting gradients only a little lower than those of the existing line and hardly improving capacity.

In 1882 the authorities concerned appointed *ing.* Protche to examine the design drawn up by *ing.* Zannoni in 1871 and to propose any improvements he might see necessary. Protche's proposals were for a line that followed the Setta and Bisenzio valleys to connect with the existing line at Sasso Marconi and Prato; the Apennines would be crossed by a large 18,032 m tunnel with the highest point at 328 m a.s.l., which would give a maximum gradient of 12‰.

In 1902 the Upper Council of Public Works set up a special commission to examine the studies that had been presented at that time. The commission chose the Protche design because the gradients and the greater stability of the terrain along the alignment came closest to meeting requirements for the line. The ministerial commission chaired by the senator *ing.* G. Colombo, did not limit itself to selecting the Protche design but also studied detailed alternatives and chose the best.

In 1908 the government enacted Law No. 444/1908 on the basis of the Colombo commission report and the final design studies were to commence, with 150 million lire authorised for the construction of the Bologna-Florence *Direttissima*.

In reality construction work didn't start until 1913 and then it proceeded very slowly and intermittently because of the First World War and subsequent events which greatly upset the life of the country in that period. Continuous work didn't start until 1921 and the "Direttissima" was finally inaugurated in 1934 with total expenditure amounting to £ 1,122 million, 460 of which was spent on the Grand Apennine Tunnel alone.

Fifty years after the Bologna-Florence Direttissima was commissioned new demands for greater capacity arose. It was above all the demand to bring the national railway network up to the same standards as the European model to make it an integral part of the continental high speed network that led to the development of the high speed/capacity train project.

Bologna-Florence Direttissima Line, excavation of the Station chamber from the previous Grand Apennine Tunnel: access stairways on the right

This project redesigns the Italian railway network by quadrupling lines with new high speed/capacity routes that run East-West across the Po valley and North-South down the peninsula. It is because of this latter development that the Bologna-Florence section constitutes the greatest design and construction commitment.

Without going into the details of the TAV (*Treno ad Alta Velocità* – High Speed Train) project (described later) in this brief history, let it suffice to say that the State Railways granted the concession for the design, construction and management of the new lines for a period of 50 years to the company, TAV S.p.A., formed in 1991.

TAV then selected a number of large groups of companies as general contractors responsible for the design and construction of these lines. The group selected for the Bologna-Florence section was the FIAT group which sub contracted the design and construction to FIAT Engineering S.p.A. and the CAVET consortium. Rocksoil S.p.A of Milan was selected for the design of the underground works.

Study of the environmental impact and of the executive design of the works began at the beginning of 1992 when TAV delivered the general design for the alignment to FIAT. An alignment to the east of the existing route was decided on the basis of those studies with a total length of 78 km, of which 73 km underground and 5 km on the surface; the maximum altitude was 413 m a.s.l..

To conclude this brief history, Fig. B.2 illustrates the development of the routes from the 1864 "Porrettana" line to the future high speed/capacity line. It perhaps summarises the technological progress that has been made and the quality of the infrastructure of the latest line better than other descriptions. While the progress made by the Porrettana compared to the *Direttissima* line consisted of improving the vertical and horizontal geometry of the alignment (decreasing the maximum gradient, increasing the radii of curvature and shortening the length of the alignment, with the final result that it allowed greater speed), what changed with the new high speed/capacity line was not only the much broader geometrical parameters, but above all the design mentality of giving due importance to environmental factors and therefore resorting to the underground option as much as possible, thanks to the development of new construction technologies.

The new Bologna-Florence high speed/capacity railway line

The new Bologna-Florence high speed/capacity railway line across the Apennines consists of an alignment 78.5 km in length of which a good 70.6 km (90% approx.) passes through twin track underground tunnels driven through the insidious and complex terrain of the hills and mountains in those places.

The project involves the construction of:

- 9 nine running tunnels with a cross section of 140 sq.m and lengths of between 528 m and 16,775 m;
- 14 access tunnels for a total of 9,255 m;
- 1 service tunnels for a total of 10,647 m;
- 2 connecting tunnels passage way tunnels for a total of 2,160 m

The works are at an advanced state of completion with excavation of all the running tunnels completed in October 2005.

Design

The geological-geotechnical context (survey phase)

The exceptional complexity of the ground involved was well known. It had already been tackled with great difficulty for the construction of the "*Direttissima*" railway line inaugurated in 1934 still in service today. A sum of € 84 million, 2% of the total cost of the project, was therefore invested in the geological survey campaigns required for final design. This provided a geological-geomechanical characterisation of the ground to be tunnelled that was very detailed and above all accurate. As is shown in Fig. B.3, it consisted primarily of flyschioid formations, clays, argillites and loose soils, at times with extensive water tables, which covered more than 70% of the underground alignment, with overburdens varying between 0 and 600 m. Some of the formations also presented the problem of gas, always insidious and delicate to deal with. During the survey phase, the route was divided into sections with similar geological and geomechanical characteristics on the basis of the information acquired. Strength and deformation parameters most representative of the sections were attributed to these for the subsequent diagnosis and therapy phases.

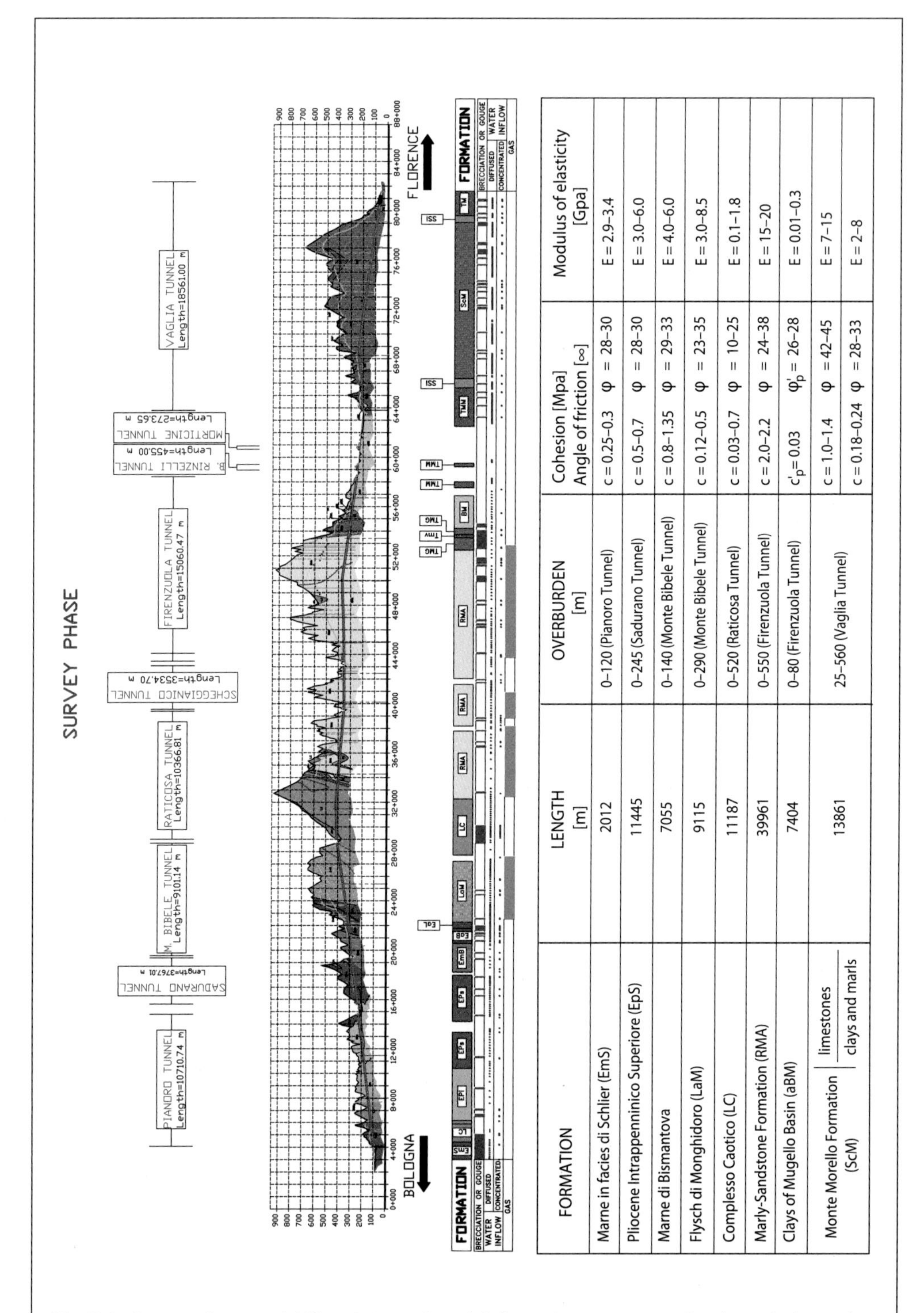

FORMATION		LENGTH [m]	OVERBURDEN [m]	Cohesion [Mpa] Angle of friction [∞]	Modulus of elasticity [Gpa]
Marne in facies di Schlier (EmS)		2012	0–120 (Pianoro Tunnel)	c = 0.25–0.3 φ = 28–30	E = 2.9–3.4
Pliocene Intrappenninico Superiore (EpS)		11445	0–245 (Sadurano Tunnel)	c = 0.5–0.7 φ = 28–30	E = 3.0–6.0
Marne di Bismantova		7055	0–140 (Monte Bibele Tunnel)	c = 0.8–1.35 φ = 29–33	E = 4.0–6.0
Flysch di Monghidoro (LaM)		9115	0–290 (Monte Bibele Tunnel)	c = 0.12–0.5 φ = 23–35	E = 3.0–8.5
Complesso Caotico (LC)		11187	0–520 (Raticosa Tunnel)	c = 0.03–0.7 φ = 10–25	E = 0.1–1.8
Marly-Sandstone Formation (RMA)		39961	0–550 (Firenzuola Tunnel)	c = 2.0–2.2 φ = 24–38	E = 15–20
Clays of Mugello Basin (aBM)		7404	0–80 (Firenzuola Tunnel)	$c'_p = 0.03$ $\varphi'_p = 26–28$	E = 0.01–0.3
Monte Morello Formation (ScM)	limestones	13861	25–560 (Vaglia Tunnel)	c = 1.0–1.4 φ = 42–45	E = 7–15
	clays and marls			c = 0.18–0.24 φ = 28–33	E = 2–8

Fig. B.3. *Survey phase: variability of strength and deformation parameters for the main formations*

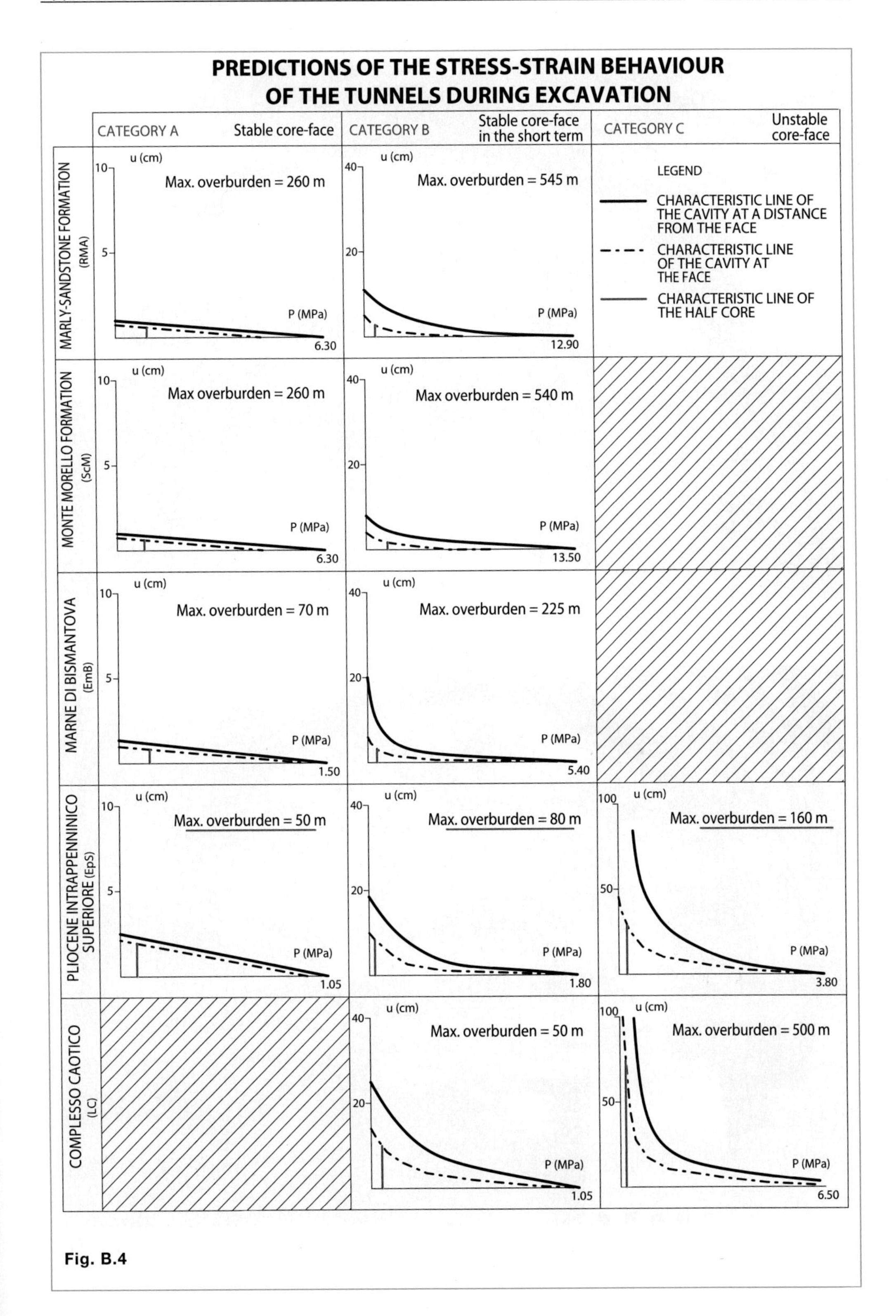
PREDICTIONS OF THE STRESS-STRAIN BEHAVIOUR
OF THE TUNNELS DURING EXCAVATION
CATEGORY A
Stable core-face
CATEGORY B
Stable core-face in the short term
CATEGORY C
Unstable core-face
MARLY-SANDSTONE FORMATION (RMA)
MONTE MORELLO FORMATION (ScM)
MARNE DI BISMANTOVA (EmB)
PLIOCENE INTRAPPENNINICO SUPERIORE (EpS)
COMPLESSO CAOTICO (LC)
u (cm)
P (MPa)
Max. overburden = 260 m
Max. overburden = 545 m
Max overburden = 260 m
Max overburden = 540 m
Max. overburden = 70 m
Max. overburden = 225 m
Max. overburden = 50 m
Max. overburden = 80 m
Max. overburden = 160 m
Max. overburden = 50 m
Max. overburden = 500 m
6.30
12.90
13.50
1.50
5.40
1.05
1.80
3.80
6.50
LEGEND
CHARACTERISTIC LINE OF THE CAVITY AT A DISTANCE FROM THE FACE
CHARACTERISTIC LINE OF THE CAVITY AT THE FACE
CHARACTERISTIC LINE OF THE HALF CORE

Fig. B.4

Monte Bibele Tunnel. View of the face while tunnelling through the Bismantova Marls, overburden: ~ 40 m)

Monte Bibele Tunnel. View of the face while tunnelling through the Monghidoro Flysch, overburden: ~ 200 m)

Predicting the behaviour of the rock and soil masses in response to excavation (diagnosis phase)

Whether because of the different geotechnical and geomechanical characteristics of the ground or the different overburdens, the tunnels to be constructed would very definitely be driven under extremely different stress-strain conditions. The underground route was therefore divided, in the diagnosis phase, into sections with uniform stress-strain behaviour. This was performed using the stability of the core-face in the absence of stabilisation measures as the only and universal parameter, which can be predicted according to the criteria illustrated in chapter 7, and also on the basis of the geological, geotechnical, geomechanical and hydrogeological information acquired in the survey phase. The sections were as follows:

- the core-face would in all probability be stable (behaviour category A; deformation phenomena in the elastic range, prevailing manifestations of instability: rock fall at the face and around the cavity);
- the core-face would in all probability be stable in the short term (behaviour category B; deformation in the elasto-plastic range; prevailing manifestations of instability: spalling at the face and around the cavity);
- the core-face would in all probability be unstable (behaviour category C; deformation phenomena in the failure range: consequent manifestations of instability: failure of the face and collapse of the cavity).

It was found from this analysis that 17% of the route would pass through ground which would have reacted to excavation with behaviour in category A, while it was predicted that 57% would be affected by deformation phenomena in the elasto-plastic range of behaviour category B and finally approximately 26% would have been characterised, in the absence of appropriate intervention, by serious instability of the core-face typical of behaviour category C.

Definition of excavation methods and stabilisation measures (therapy phase)

Once reliable predictions of the stress-strain response of the ground to excavation had been formulated, the action required (preconfinement and/or ordinary confinement) to guarantee the formation of an arch effect as close as possible to the profile of the excavation, in each situation hypothesised, was identified for each section of tunnel with uniform stress-strain behaviour. The advance methods (method of actual excavation, length of tunnel advances) and the most appropriate techniques for producing the required action and to guarantee, as a consequence, the long and short term stability of the excavation were then designed.

The variable character of the ground, present to a greater or lesser extent in all the tunnels, meant that totally mechanised technologies were not advisable with the exception of the service tunnel for the Vaglia tunnel. The main principles on which the design of the tunnel section types was based were therefore as follows:

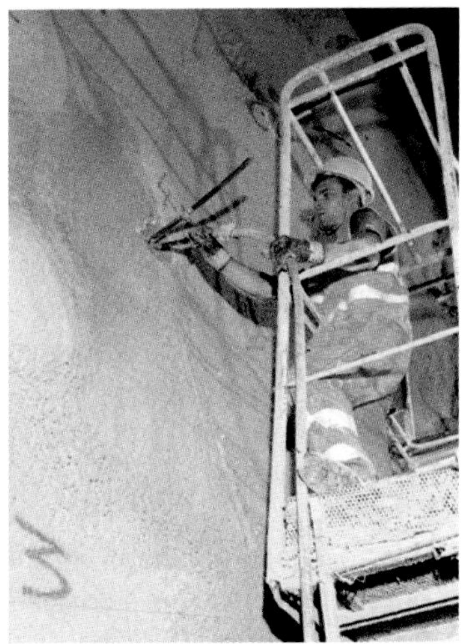

Raticosa Tunnel. Two moments in the reinforcement of the advance core using fibre glass structural elements. Drilling and insertion (left); cementation (right)

Raticosa Tunnel. Reinforcement of the core-face by means of fibre glass structural elements in the scaly clays of the Complesso Caotico (overburden: ~ 550 m)

1. full face tunnel advance always, especially under difficult stress-strain conditions: due to its peculiar static advantages and because large and powerful machines can be profitably used in the wide spaces available, it is in fact possible to advance in safety with excellent and above all constant advance rates even through the most complex ground by using full face advance after core-face reinforcement, when necessary;
2. confinement, where necessary, of the alteration and decompression of the ground caused by excavation, with the immediate application of effective cavity pre-confinement and/or confinement (sub-horizontal jet-grouting, fibre glass structural elements in the core and/or in advance around the cavity, fitted, if necessary, with valves for pressure cement injections, shotcrete, etc.) of dimensions sufficient, according to the case, either to absorb a significant proportion of the deformation without collapsing or to anticipate and neutralise all movement of the ground from the outset;
3. placing a final lining in concrete, reinforced if necessary, complete with the casting of a tunnel invert at short intervals immediately behind the face when the need to halt deformation phenomena promptly was recognised.

The follow design decisions were then made.

- **Pliocene Intrappenninico Superiore** (Pianoro tunnel): the just reasonable geomechanical characteristics of the ground in this formation and the stress states found resulted in the prediction of behaviour varying from stable core-face in the short term to unstable core-face (categories B and C). No intervention to improve the ground seemed necessary in the first case (which was prevalent), as long as a fast advance rate was maintained (higher than 2 m/day). The section types B0 and B2, considered appropriate for the expected operating conditions, were specified. Given the average strength of the matrix, an excavator hammer was specified as the means for excavating the ground. It was therefore considered that an advance step of 1.5 m, with the face profiled to give it a concave shape, would allow a steel rib to be placed on each eight hour shift and ensure the necessary speed of tunnel advance required.
- **Marne di Bismantova** (M. Bibele North tunnel): the consistency of the ground is that of rock with little fracturing, consisting of approximately two kilometres of marls followed by sandstones interbedded with marly strata. The diagnosis study found mainly stable core-face behaviour (category A), generally without the need for improvement of the rock mass. The section types Ab, Ac, B0, were therefore specified. Given the characteristics of the rock (σ_c >30 MPa), conventional excavation was specified by blasting with one round of shots per day to give tunnel advances of 4.5–5 m.

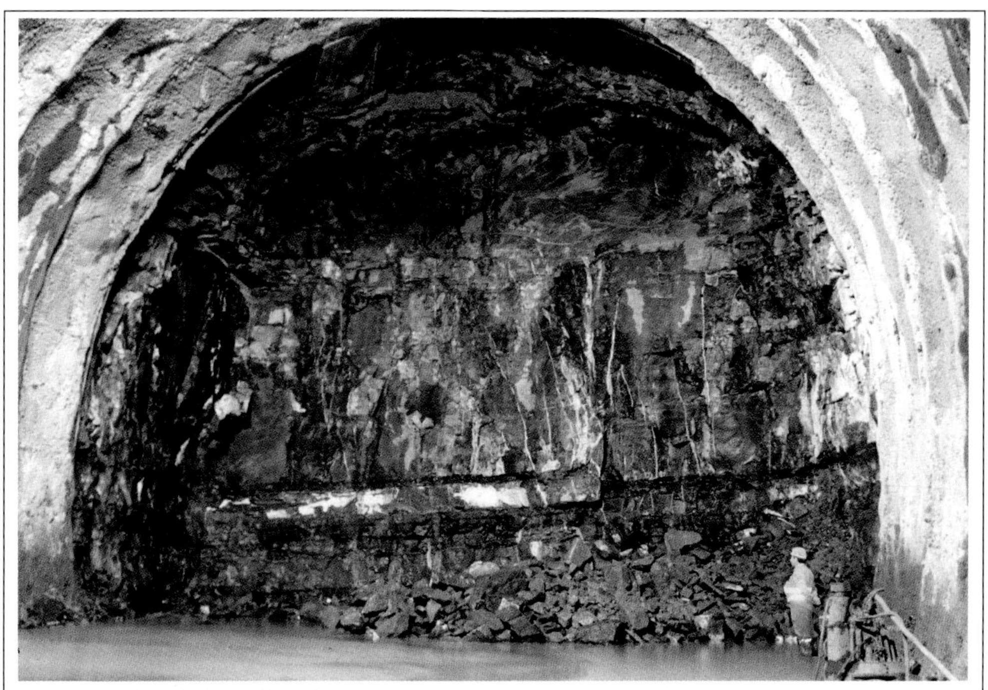

Firenzuola Tunnel. View of the face while tunnelling through the Marnoso-Arenacea formation (overburden: ~ 500 m)

Vaglia Tunnel. View of the face reinforced using fibre glass structural elements while tunnelling through the Clays of the Mugello basin (overburden: ~ 15 m)

- **Flysch di Monghidoro** (M. Bibele South tunnel): consists of densely alternating, heavily tectonised and intensely fractured argillites, clayey marls, limestones and calcarenites. Since the prevalent behaviour type found in the diagnosis phase was type B (stable core-face in the short term), with possible transformations to C (unstable core-face), the section types B0 and B2 were designed. Consequently, excavation would take place with and without intervention to improve and reinforce the ground. Given the marked fracturing of the ground, excavation using an excavator hammer was opted for, despite the rock consistency of the matrix, with maximum tunnel advances of 2,5 m in the better quality rock and up to 1 m in the ground consisting mainly of argillites.
- **Complesso Caotico** (Raticosa Tunnel): this consists of intensely fractured and tectonised scaly clays of poor geomechanical quality, to be excavated under high stress states (behaviour category C: unstable core-face). Full face advance after first reinforcing the advance core and the ground around the future tunnel with fibre glass structural elements was specified given the difficult conditions (section types C4R, C4V). The walls of the tunnel were then to be immediately lined with a ring of steel ribs closed in the tunnel invert with a strut to brace it and with shotcrete. The sidewalls and the tunnel invert were then to be cast within one tunnel diameter from the face. A ripper, with which it was easier to profile the face with a concave shape, was specified to excavate this soft and rather inhomogeneous ground. Finally, a tunnel advance of approximately 1.2 m was considered adequate.
- **Marly-Sandstone Formation** (Scheggianico and Firenzuola tunnels): this is an aquifer consisting of alternating marls and sandstones in sub-horizontal banks ranging from a few decimetres to some metres. Given the good strength and deformability characteristics of the material, it responds to excavation in the plastic range even under the largest overburdens (category A: stable core-face). Consequently section types Ab and Ac were specified, while the good mechanical quality of the rock made blasting the best method of excavation even if this would cause difficulties in profiling the tunnel cross section.
- **Clays of the Mugello Basin** (Firenzuola Tunnel): these consist of clayey silts with fine sandy and saturated interbedding. It was predicted that tunnel advance would take place under behaviour category C (unstable core-face). Appropriate intervention was therefore taken to improve and reinforce the ground (section types C1, C4R, B2). A ripper excavator was specified for excavation with immediate placement of the lining (sidewalls and tunnel invert cast within one tunnel diameter of the face).
- **Monte Morello Formation** (Vaglia Tunnel): this is a flysch consisting of limestones, marly limestones and poorly fractured

Vaglia Tunnel. View of the face while tunnelling through the Monte Morello formation (overburden: ~ 600 m)

S. Giorgio access tunnel. Construction of the junction with the Firenzuola running tunnel (ground: Clays of the Mugello Basin, overburden: ~ 30 m)

marls in compacted banks of a few decimetres. Given the good geomechanical quality, the main behaviour found was category A (stable core-face). Section types Ac or B0 were therefore specified. Blasting was the only choice for excavation considering the characteristics of the material.

Table B.I summarises the design specifications described.

The range of geological and geomechanical and stress-strain (extrusion and convergence) conditions within which they were to be applied was clearly defined for each type of longitudinal and cross section types as well as the position in relation to the face, the intensity and the phases and intervals for placing the various types of intervention (advance ground improvement, preliminary lining, tunnel invert etc). Very reliable work cycles based on a considerable number of previous experiences was drawn up from which precise predictions of daily advance rates could be made. Figure B.5 shows the main section types adopted (there were 14 in all), grouped according to the type of behaviour category, A, B and C. The 'variabilities' to be applied were designed for each section type for statistically probable conditions where, however, the precise location could not be predicted on the basis of the available data (see section 8.9).

It is essential to identify the variabilities for each tunnel section type that are admissible in relation to the actual response of the ground to excavation, which will in any case always be within the range of deformation predicted by the ADECO-RS approach. This is because it allows a high level of definition to be achieved in the design and also at the same time the flexibility needed to be able to adopt ISO 9002 quality assurance systems during construction to advantage. By employing this method, non

FORMATION	TUNNEL	σ_C [MPa]	BEHAVIOUR CATEGORIES	H_2O	SECTIONS TYPE	FACTORS AFFECTING EXCAVATION DESIGN			
						Ground improvement or reinforcement intervention	Excavation system	Excav. step [m]	Production predicted
Pliocene Intrappenninico (EPs -EPi)	Pianoro South	12	B, (C)		B0, B2 (C1, C2)	sporadic	hammer	1.5	2
Marne di Bismantova (EmB -EaB -EaL)	Monte Bibele North	30	A, (B)		Ab, Ac (B0)	sporadic	blasting	4.5–5	5
Flysh di Monghidoro [LaM]	Monte Bibele South	8	B, (C)		B0, B2 (C4)	sporadic	hammer	1–2.5	2
Complesso Caotico [LC]	Raticosa North	2	C		C4R, C4V	systematic	ripping	1.2	1
Marly-Sandstone [RMA]	Firenzuola from Rovigo	40	A, (B)	yes	Ab, Ac (B0)	systematic drainage	blasting	2.5–5	5
Clays of Bacino del Mugello [aBM]	Firenzuola from S. Giorgio	2	C, (B)	yes	C1, C4R (B2)	systematic	ripping	1–1.2	1,3
Monte Morello [ScM]	Vaglia	30	A, (B, C)		Ac (B0)	sporadic	blasting	2.5–5	5

Table B.1

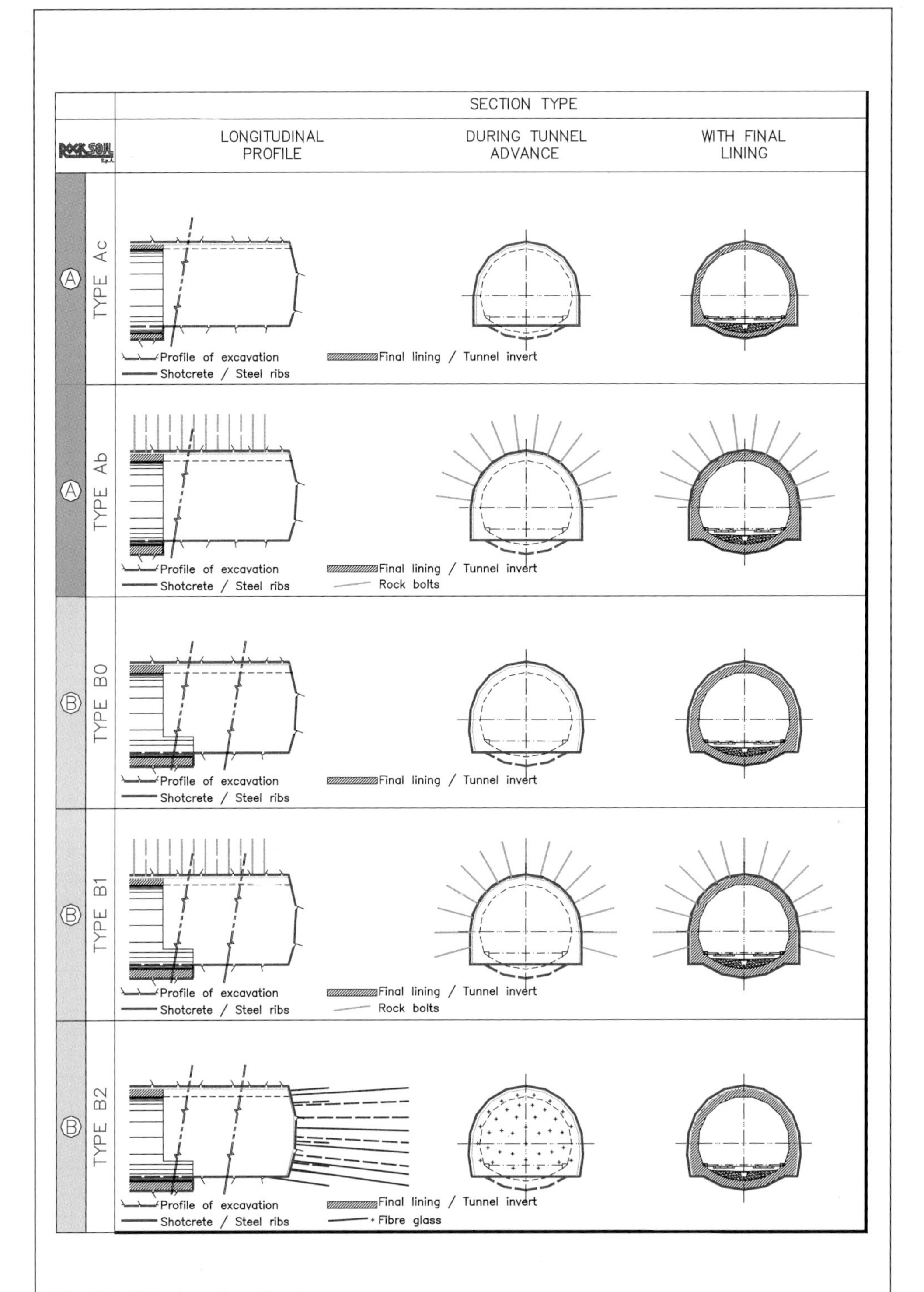

Fig. B.5 *Main tunnel section types*

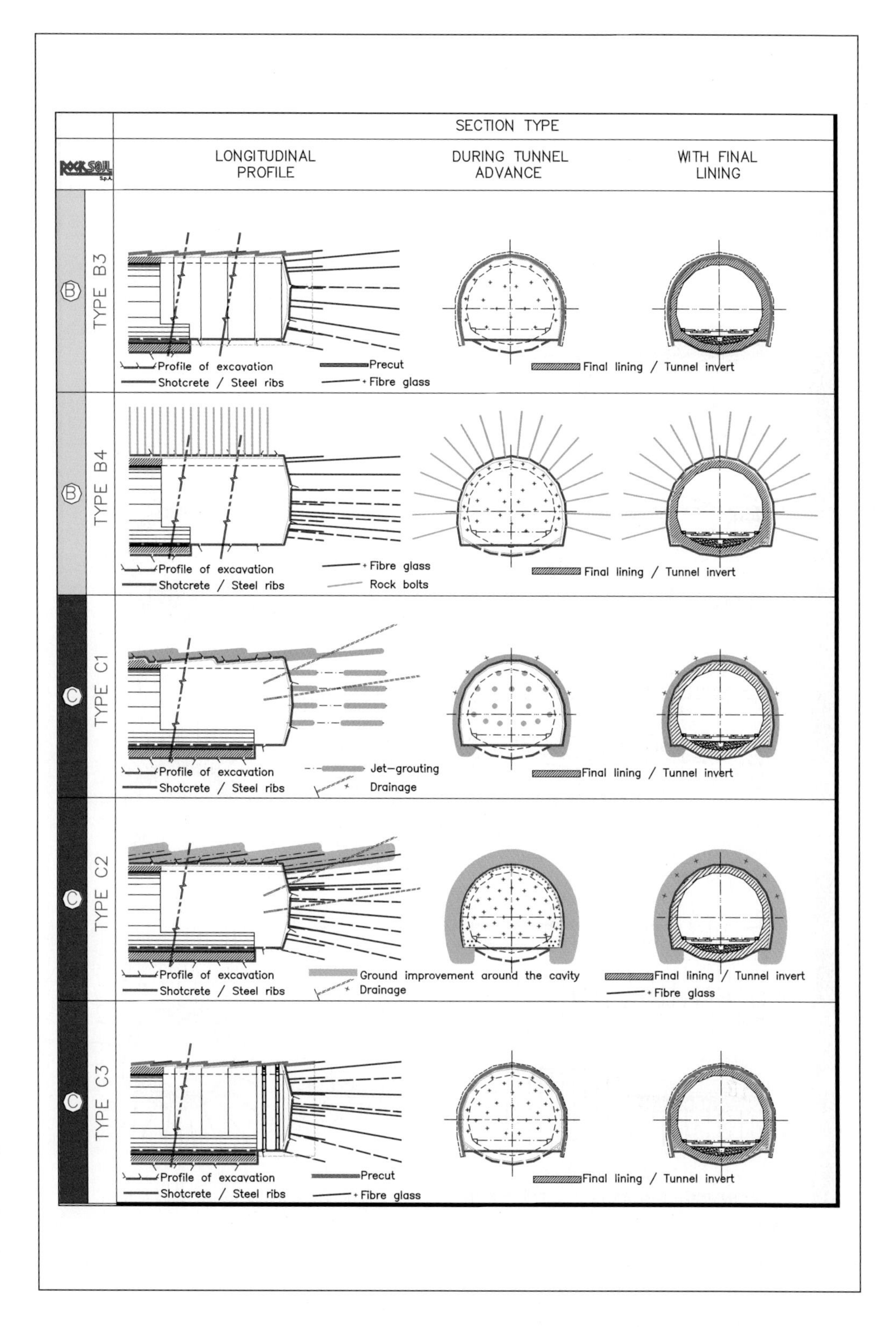
SECTION TYPE
ROCKSOIL S.p.A.
LONGITUDINAL PROFILE
DURING TUNNEL ADVANCE
WITH FINAL LINING
B
TYPE B3
Profile of excavation
Shotcrete / Steel ribs
Precut
Fibre glass
Final lining / Tunnel invert
B
TYPE B4
Profile of excavation
Shotcrete / Steel ribs
Fibre glass
Rock bolts
Final lining / Tunnel invert
C
TYPE C1
Profile of excavation
Shotcrete / Steel ribs
Jet–grouting
Drainage
Final lining / Tunnel invert
C
TYPE C2
Profile of excavation
Shotcrete / Steel ribs
Ground improvement around the cavity
Drainage
Final lining / Tunnel invert
Fibre glass
C
TYPE C3
Profile of excavation
Shotcrete / Steel ribs
Precut
Fibre glass
Final lining / Tunnel invert

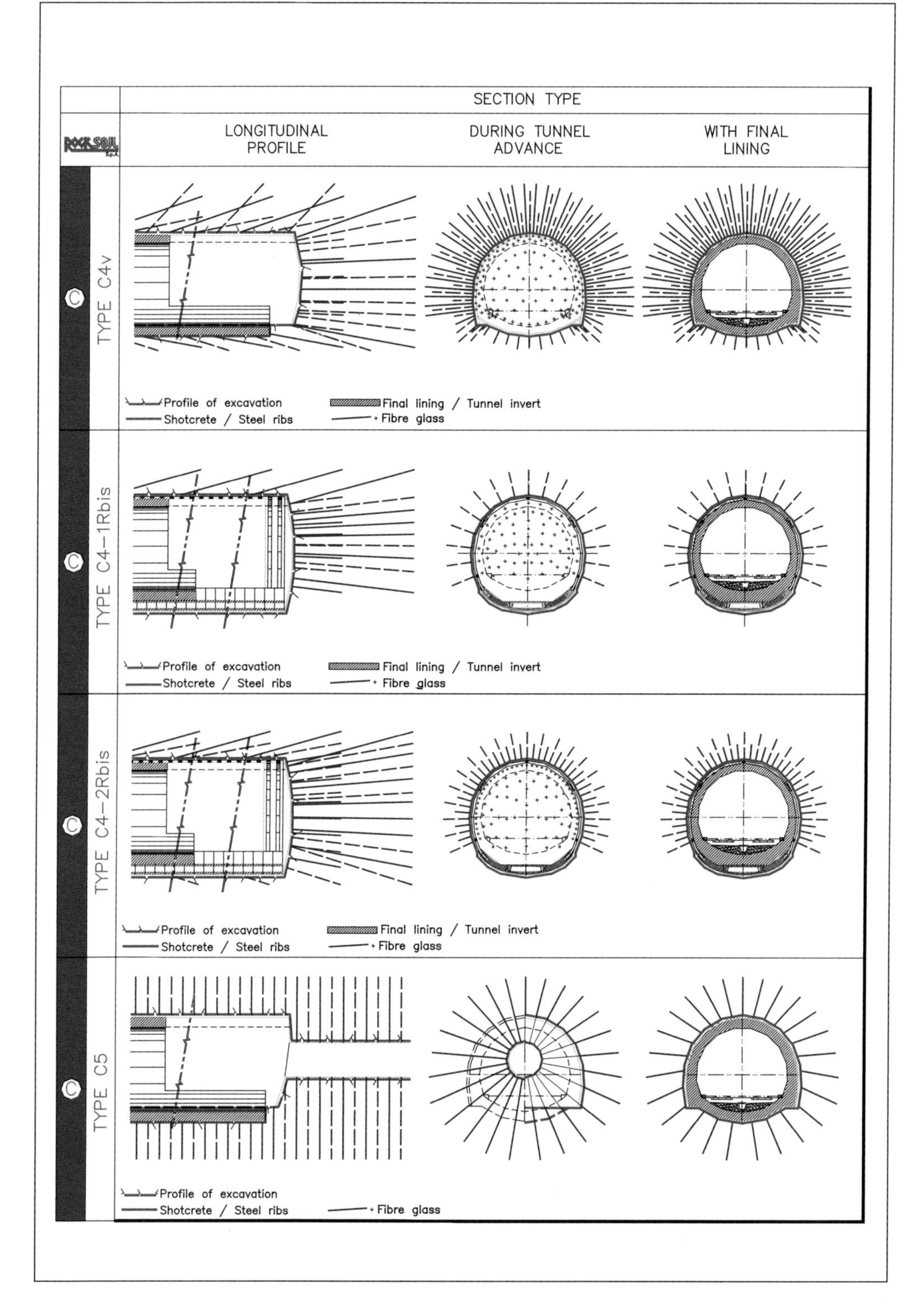
SECTION TYPE
LONGITUDINAL PROFILE
DURING TUNNEL ADVANCE
WITH FINAL LINING
ROCKSOIL S.p.A.
C
TYPE C4v
Profile of excavation
Shotcrete / Steel ribs
Final lining / Tunnel invert
Fibre glass
C
TYPE C4–1Rbis
Profile of excavation
Shotcrete / Steel ribs
Final lining / Tunnel invert
Fibre glass
C
TYPE C4–2Rbis
Profile of excavation
Shotcrete / Steel ribs
Final lining / Tunnel invert
Fibre glass
C
TYPE C5
Profile of excavation
Shotcrete / Steel ribs
Fibre glass

Scheggianico Tunnel. South tunnel under the state road (ground: Marly-Sandy formation)

Castelvecchio access tunnel. The construction of 9 tunnels on the new high speed/capacity Bologna-Florence railway line required the construction of 14 access tunnels for a total length of 9,255 m

Firenzuola Tunnel. View of the works for the construction of the S. Pellegrino chamber at the North portal (ground: Marly-Sandstone formation)

Firenzuola Tunnel. Construction of the artificial tunnel of the S. Pellegrino chamber at the North portal (ground: Marly-Sandstone formation)

conformities (i.e. differences between actual construction and design), which oblige partial redesign each time a change in conditions is encountered even if it involves only a minor change to the design, are avoided.

Each section type was analysed in relation to the loads mobilised by excavation as determined in the diagnosis phase, with regard to both the various construction phases and the final service phase by employing a series of calculations on plane and three dimensional finite element models in the elastic-plastic range.

Finally, precise specifications were formulated for the implementation of a proper monitoring campaign, which, according to the different types of ground tunnelled, would both guarantee the safety of tunnel advance, test the appropriateness of the design and allow it to be optimized in relation to the actual conditions encountered.

The philosophy which guided the design and subsequently the implementation of the monitoring system proposed can be summarised by the following basic concepts:

- *attention to detail*: each instrument must be selected and positioned to provide answers to a specific question;
- *monitoring "cause and effect"*: for the design engineer a change in an important parameter is governed by a law of cause and effect. It is only by measuring both phenomena that a relationship can be established between them and the necessary corrective action taken to remove the causes which resulted in undesired effect;
- *the principle of redundancy in measurements*: most measurements are of the highly localised type and as such can be influenced by the particular characteristics of the portion of the structure or rock in the immediate vicinity of the instrument.
  Single measurements may not therefore be representative of the phenomena on a larger scale. This problem can be solved by setting up measurement stations with a large number of points of measurements;
- *the principal of the plurality and the modularity of measurement systems and control parameters:* one single type of instrument may furnish information which is difficult to interpret or partial in determined contexts.

The use of instruments which measure different physical magnitudes which are correlated with each other may validate and/or complete the information on what is measured. Similarly, if the quantity and type of sensors can be adjusted in sections that are monitored according to how critical the situation encountered or predicted is, this guarantees appropriate dimensioning of the monitoring system and consequently also of the cost benefit ratio of the system itself. This last concept, together with those that precede it guarantees, amongst other things, a significant reduction in the subjectivity involved in the interpretation of monitoring phenomena and, secondly, allows

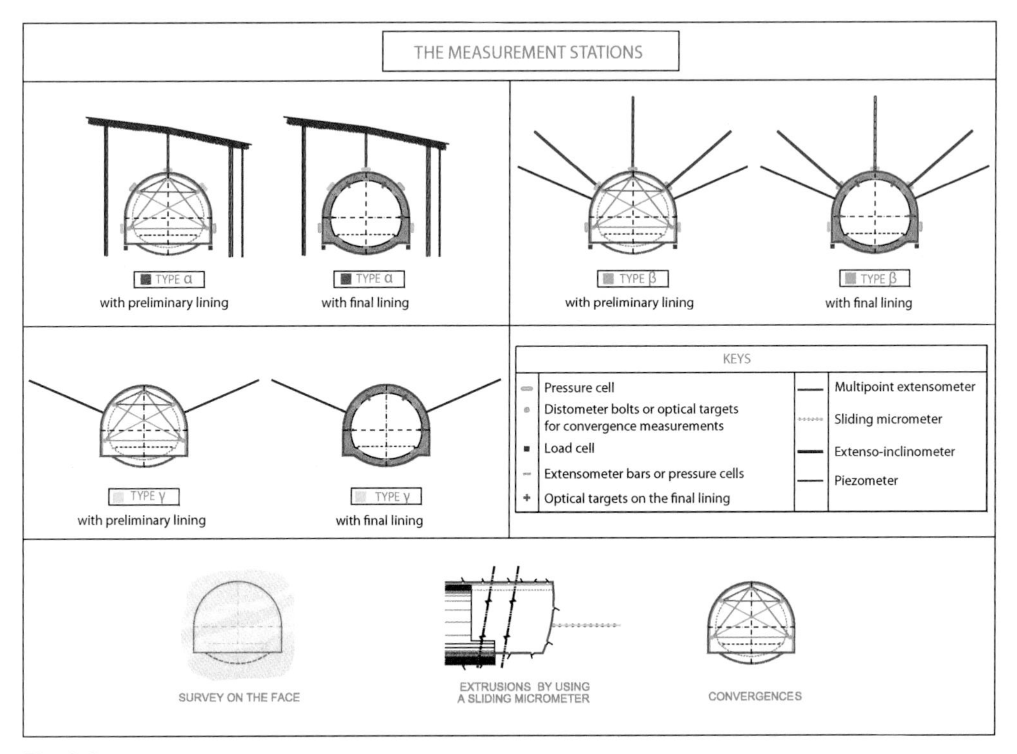

Fig. B.6

sufficient information to be acquired for statistical analysis to be performed. Geotechnical monitoring stations have been designed (Fig. B.6), which are based on these principles and on the parameters identified as the references for optimising the design during construction. They have been employed individually or in combination according to different criteria and frequency as a function of the three basic stress-strain behaviour categories (A, B and C).

Information on the implementation and use of monitoring is given later in this appendix in the section on monitoring during construction. However, before concluding the design stage and starting to discuss what happened in the construction phase, it should perhaps be stated that the detailed design illustrated here obviously also included the design of the necessary accessory works, such as portals, chambers, access and service tunnels which are not described here in order to avoid excessive detail.

Tunnel construction

Type of contract

The contract for the entire section of the railway between Bologna and Florence was awarded on a rigorous lump sum basis (€ 4.209 billion) by FIAT S.p.A., the general contractor, which accepted responsibility for all unforeseen events, including geological risks on the basis of the

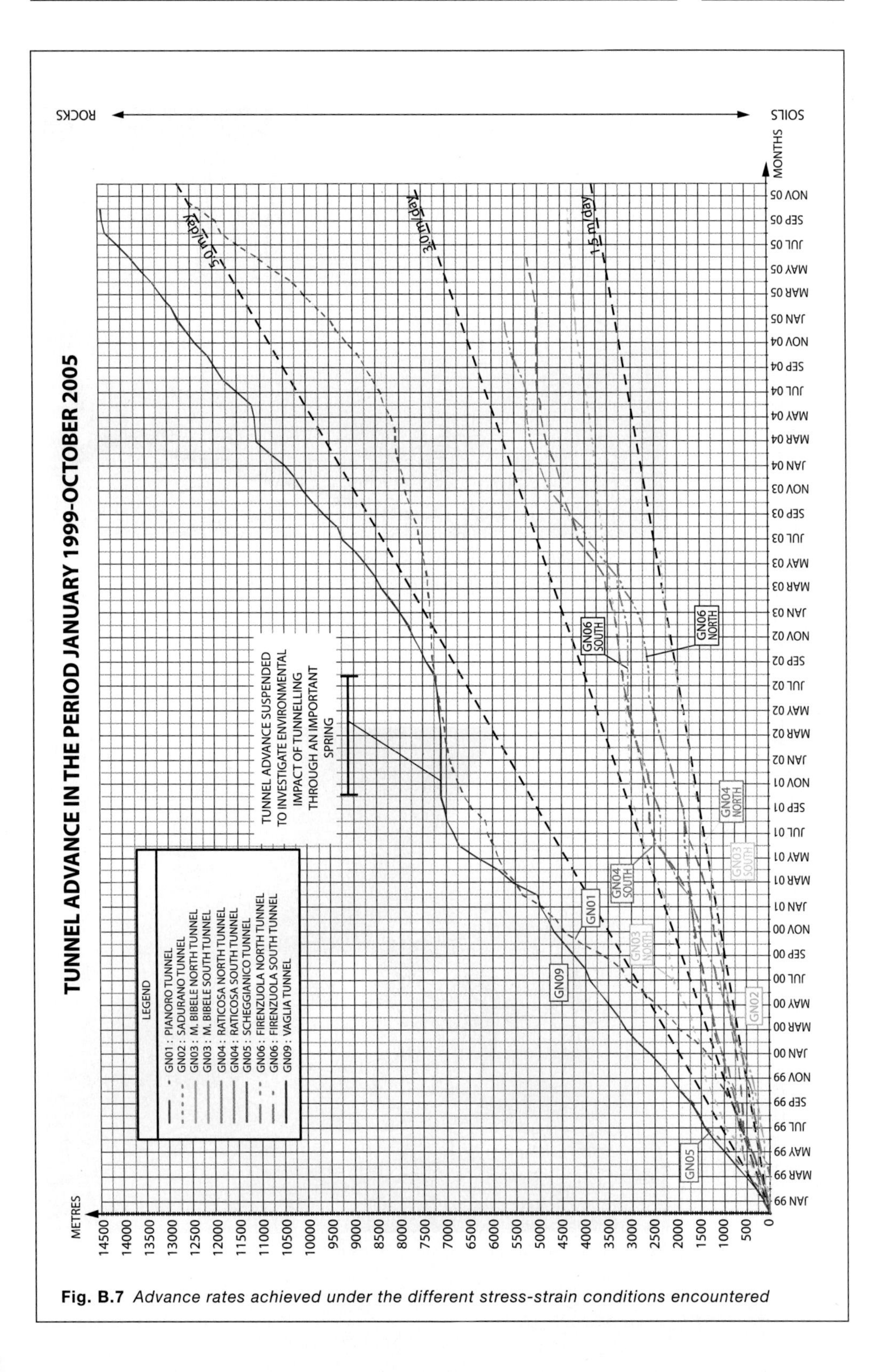

Fig. B.7 *Advance rates achieved under the different stress-strain conditions encountered*

Raticosa Tunnel. Reinforcement of the tunnel invert very close to the face (ground: scaly clays of the Complesso Caotico, overburden: ~550 m)

Raticosa Tunnel. Mobile formwork for casting the final lining in reinforced concrete

detailed design as illustrated above. It subcontracted all the various activities out to the CAVET consortium (land expropriation, design, construction, testing, etc.).

Operational phase

Immediately after the contract was awarded, the construction design of works began at the same time as excavation work (July 1996).

Additional survey data and direct observation in the field generally confirmed the validity of the detailed design specifications, while the following minor refinements were made in the construction design phase:

- to deal specifically with particularly delicate stress-strain conditions, a steel strut was introduced as a variation to section types B2 and C4 in the tunnel invert to produce much more rapid containment of deformation. This modification of the B2 section type was found to be much more versatile and appropriate even for many situations where the core-face was unstable. Use of the heavier C section types was thus limited to the more extreme stress-strain conditions;
- the effectiveness of core-face ground improvement using fibre glass structural elements was increased considerably by introducing an expansive cement mix to grout the fibre glass nails;
- as a consequence of positive results acquired during the construction of tunnels on the Rome-Naples section of the same railway line, a section type B2pr was designed for underground construction of the Sadurano, Borgo Rinzelli and Morticine tunnels originally designed as artificial tunnels;
- finally it was decided to replace section types B3 and C3 which involved the use of mechanical pre-cutting with section C2 (ground improvement in the core-face and around it with fibre-glass structural elements) better suited to the ground to be tunnelled.

The final result of construction design was the definition of the following percentages of tunnel section type:

- section type A: 20.5%;
- section type B: 57.5%;
- section type C: 22.1%.

Section type	Forecast advance rates [m per day of finished tunnel]	Actual advance rates [m per day of finished tunnel]
A	5.40	5 – 6
B0	4.30	5 – 5,5
B2	2.25	2.10 – 2.2
C1	1.40	1.40
C2	1.25	0.85
C4V	1.25	1.63

Table B.II

Approximately 73 km of running tunnel, 100% of the total, have been driven and almost all of it has been lined. Average monthly advance rates of finished tunnel were around 1,000 m of finished tunnel with peaks of 2,000 m reached in March 2001, working simultaneously on 30 faces.

Figure B.7 gives the advance rates for the different tunnels constructed. It can be seen that not only were the advance rates very high in relation to the type of ground tunnelled, but above all they were very constant, an indicator of the excellent match between the design specifications and the actual reality. Even for the Raticosa tunnel, driven under extremely difficult stress-strain conditions in the *Complesso Caotico* formation consisting of the much feared scaley clays, average advance rates were never less than 1.5 m per day. Table B.II gives a comparison of daily advance rates forecast by the detailed design specifications for some section types and the actual advance rates achieved.

Table B.III on the other hand gives the distribution of differences in section types between the detailed design specifications and the tunnel as built. Even if account is taken of the increased rigidity introduced to the B2 section type, which made it possible to employ the type in many situations typical of behaviour category C, there was nevertheless a significant reduction in the use of the more costly section types in favour of more economical solutions. This result was to a large extent due to the exceptional effectiveness of the preconfinement methods employed even under large overburdens. It was the first time they had been used with overburdens greater than 500 m.

Some consider making the core-face more rigid before tunnel advance as counter productive under large overburdens. However, if it is properly performed and the continuity of the action from ahead of the face back down into the tunnel is ensured with the placing of a steel strut in the tunnel invert, it is, as was demonstrated, extremely effective even with large overburdens and resort to the heavier section types was only necessary in the most extreme situations.

Another reason for the greater use of section type A was that there are only minor differences between it and the B0 sections types such as the thickness of the final lining or the distance from the face at which the tunnel invert is cast.

As a consequence, section type A was adopted in place of section types B0 or B0V, whenever the ground conditions allowed this to be done without taking greater risks (e.g. in long sections of the Vaglia tunnel where the presence of the adjacent service tunnel already built made the situation particularly clear). The differences found between the detailed design specifications and the tunnel as built did not reveal any sensational discrepancies, neither in terms of the overall cost of the works, which was a little less than budgeted under the detailed design specifications (~–5%), nor with regard to construction times. While the contractor will benefit from the lower cost, a reward for the greater risk run by agreeing to sign a rigorously lump sum, all-in contract, the client and citizens will benefit from the punctual observance of time schedules because they will be able to use the new transport services without intolerable delays.

The monitoring and calibration of the design during construction (monitoring phase)

The particular nature and the importance of the project required a thorough monitoring programme both during construction and for the completed tunnel in service.

HIGH SPEED/CAPACITY TRAIN - Milan to Naples Line - SECTION TYPES DISTRIBUTION

DETAILED DESIGN (SECTION TYPES DISTRIBUTION [m])

TUNNEL	Length [m]	A	B0	B0V	B1	B2	B3	B4	C1	C2	C3	C4	C4V	C5
Pianoro	10293.4		951.8			3886.4	3036		62.0	948.8	1083	310		15.5
Sadurano	3778.0	64.0	2580.8			875.0			68.0	190.3				
M. Bibele	9118.5	978.2	1094.6		4529.1	1212.2			76.0	1112.8		115.6		
Raticosa	10381.0	3043.0			972.2	758.4			40.0	786.7		4465.1		315.67
Scheggianico	3530.6	2089.9			1404.7					36.0				
Firenzuola	14311.5	3528.7			5950.4	716	412.2		227.5	511.9	2226.8	738.1		
B. Rinzelli	455.0								160		295			
Morticine	273.7								80	193.7				
Vaglia	16757.0	2017.2	3104.3	1129.8	5629.0			1151.2	692	708.5			2325.2	
TOTAL LENGTH [m]	68898.6	11721.0	7731.5	1129.8	18485.2	7447.9	3448.2	1151.2	1405.4	4488.6	3604.8	5628.8	2325.2	331.2

	A	B0	B0V	B1	B2	B3	B4	C1	C2	C3	C4	C4V	C5
SECTION TYPES DISTRIBUTION [%]	17.0	11.2	1.6	26.8	10.8	5.0	1.7	2.0	6.5	5.2	8.2	3.4	0.5

AS BUILT (Tunnelled length; SECTION TYPES DISTRIBUTION [m])

TUNNEL	Length [m]	Tunnelled length [%]	Tunnelled length [m]	A	B0	B0V	B2	B2pr	B2V	C1	C2	C4	C4V	C6
Pianoro	10710	100.0	10710.0		8167.0		682.5		63,0	3,5			1794.3	
Sadurano	3767.0	100.0	3767.0	2213.1	1408.0		85	53			8			
M. Bibele	9101.0	100.0	9101.1	2935.5	2015.7		3965.8		97,9	41	45			
Raticosa	10367.0	100.0	10367.2	3680.0	786.0		857		25	85		1468	673.4	2792.6
Scheggianico	3535.0	100.0	3535.0	3517.0	18.0									
Firenzuola	15211.0	100.0	15211.0	6833.1	3556.3		3081.8		263.5	577.1	9		125.83	43.46
B. Rinzelli	528.5	100.0	528.5					303.5			225.1			
Morticine	565.5	100.0	565.5					537.0			28.5			
Vaglia	16757.0	100.0	16757.7	5287.60	9299.50	96.60	1547.0		296.70	128.25	101.4			
TOTAL LENGTH [m]	70542.3	100.0	70542.4	24466.2	25250.5	96.6	10219.1	893.5	746.1	834.8	417.0	1468.0	2593.5	2836.1

	A	B0	B0V	B2	B2pr	B2V	C1	C2	C4	C4V	C6
SECTION TYPES DISTRIBUTION [%]	34.7	35.7	0.1	14.5	1.3	1.1	1.2	0.6	2.1	3.7	4.0

Table B.III *Percentage of section types used: comparison between the detailed design and as built*

Firenzuola Tunnel: cementation of the fibre glass reinforcement in the core-face

Borgo Rinzelli Tunnel: effect of stabilisation using artificial ground overburden (ground: Clays of the Mugello Basin, overburden: zero)

The following was monitored during construction:

- the tunnel faces by means of systematic geomechanical surveying of the ground at the face. The surveys were conducted to I.S.R.M. (International Society of Rock Mechanic) standards and gave an initial indication of the characteristics of the ground to be compared with design predictions;
- the deformation behaviour of the core-face by measuring both surface and deep core-face extrusion, with measurements taken as a function of the different behaviour categories. Systematic measurements of this kind under difficult stress-strain conditions are crucial: in fact as already stated in chapter 10 on the monitoring phase, monitoring convergence alone, which is the last stage of the deformation process, is not sufficient to prevent tunnel collapse under these conditions. Extrusion, on the other hand, is the first stage in the deformation process and if it is kept properly under control will allow time for effective action to be taken;
- the deformation behaviour of the cavity by means of systematic convergence measurements;
- the stress behaviour of the ground-lining system by placing pressure cells at the ground-lining interface and inside the linings themselves, both preliminary and final.

The results of monitoring activity guided the design engineer and the project management in deciding whether to continue with the specified section type or to modify it according to the criteria already indicated in the design, by adopting the 'variabilities' contained in it. Obviously, in the presence of particular conditions that were not detected at the survey phase and therefore not provided for by the design, it is always possible to design a new section type. This method of proceeding allowed the uncertainty connected with underground works to be managed satisfactorily even with a rigorously lump sum contract like that between FIAT and TAV.

Most of the instrumentation already used during construction, which is connected to automatic data acquisition systems, will continue to be employed for monitoring when the tunnel is in service. The automatic data acquisition system can be interrogated at any moment in the life of the works to obtain data and verify the real behaviour of the tunnels and compare it with design predictions.

The construction of the Ginori service tunnel for the Vaglia running tunnel

The construction of the Ginori service tunnel for the Vaglia running tunnel deserves particular attention.This tunnel is 9,259 km in length with an internal diameter of 5.6 m and runs parallel to the "Vaglia" running tunnel for 6,501 metres to which it is connected by small pedestrian tunnels every 250 m (Fig. B.8). The first section of 1,588 m and the final section of 1,170 m run down and up respectively with a gradient of 1.8%. As already mentioned, fully mechanised technology was employed to drive this tunnel. This decision was made on the basis of in-depth study of the geological, geomechanical and hydrogeological conditions of the ground and it fully satisfied a series of contingent needs and requirements:

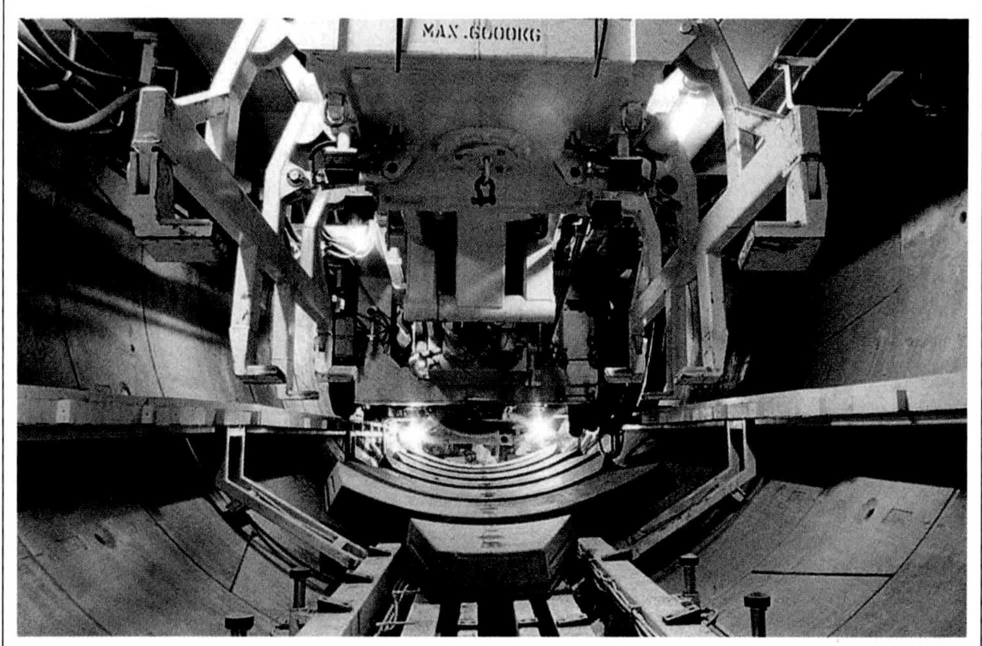

Ginori service tunnel: two views from inside the TBM. Note the conveyor line for the prefabricated concrete segments of the final lining below

- the environmental need to drive the tunnel from one single portal near Ginori;
- the need to control interference with ground water as much as possible;
- finally the requirement to complete the tunnel rapidly.

Given the success already achieved with the design and construction of the running tunnels across the Apennines, important innovations were also introduced to the design for mechanised tunnel construction. A series of criteria were developed and introduced in the design (and as a consequence in the contract) to follow during construction consisting of design and/or operational measures to be applied if 'variabilities' (differences between predicted and actual conditions) were encountered during tunnel advance.

The design of tunnel advance by TBM

The ground through which the "Ginori" tunnel passes all lies within the *Monte Morello* formation. It is certainly a complex formation with mechanical characteristics which vary greatly, ranging from argillites to compact limestones. Consequently the designer of the

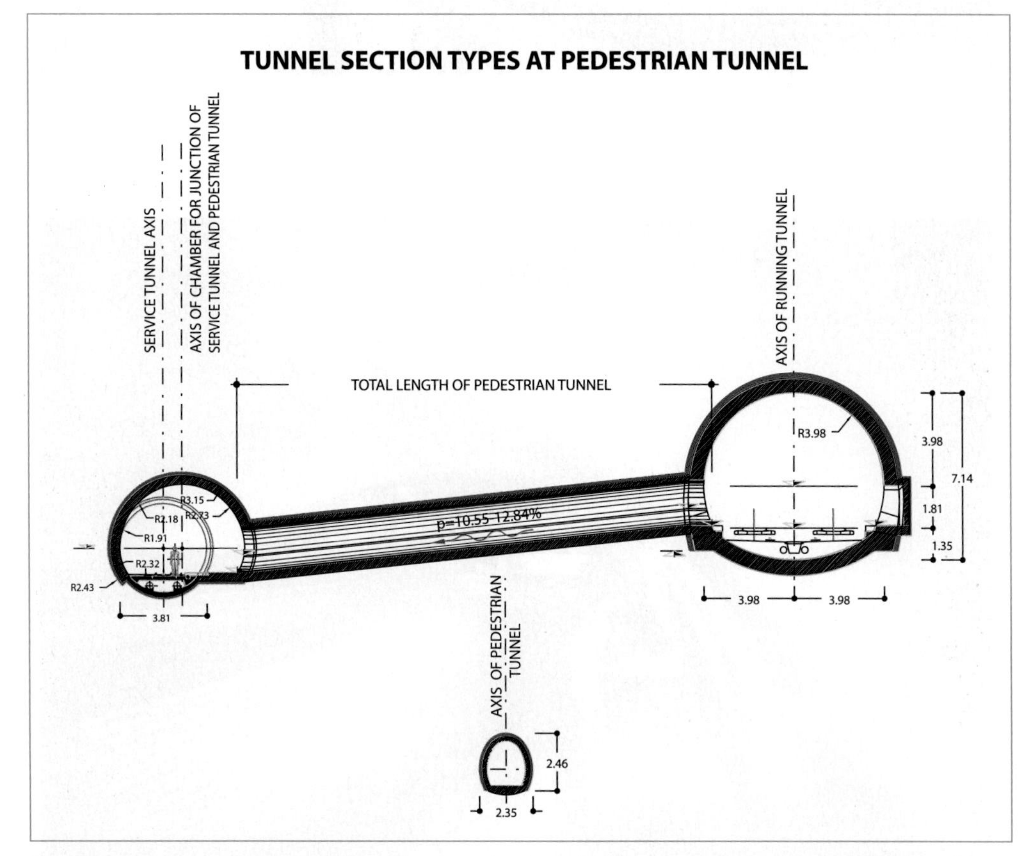

Fig. B.8 *Typical cross section of the Ginori service tunnel at the junction with a pedestrian tunnel connecting with the Vaglia tunnel*

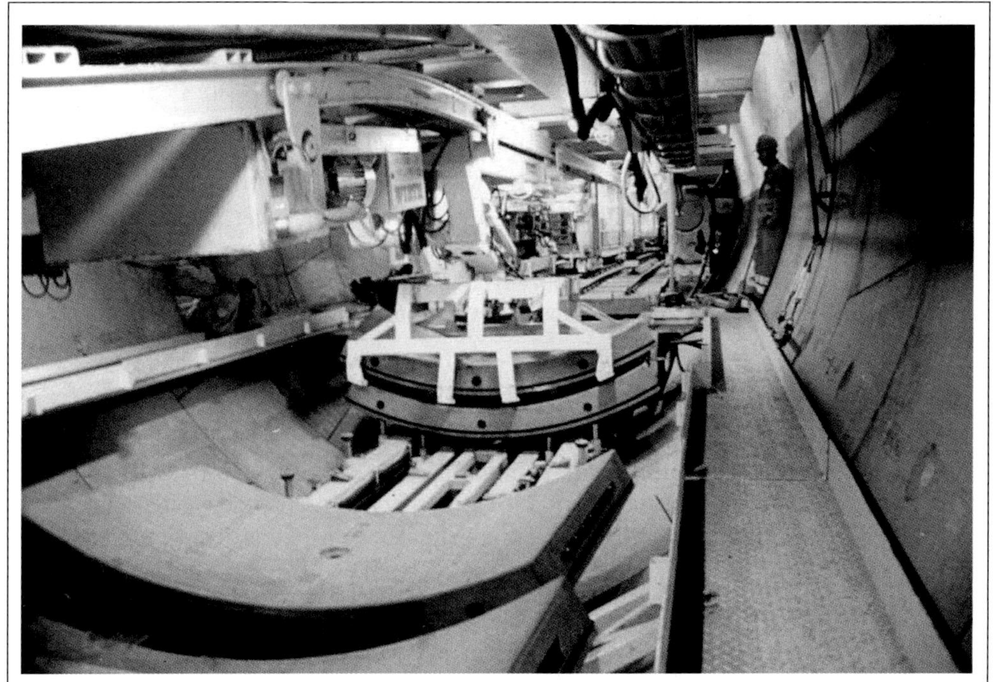

Ginori service tunnel: The mechanism for lifting prefabricated concrete segments for the final lining

Ginori service tunnel: the "tail" of the TBM inside the tunnel already lined

TBM received considerable input from the consulting civil engineer and the result was an innovative machine fitted with advanced geological surveying systems, capable of adjusting the method of advance according to the different geomechanical conditions it encounters and, where necessary, of performing advance reinforcement both in the core and around it (Fig. B.9). The ways in which it acquires cutter operating parameters and the criteria it employs to change the type of tunnel advance and to prepare the ground ahead are explained in a document appended to the design specifications entitled *Guide Lines for performing surveys in advance, for selecting the appropriate operating methods and performing ground reinforcement operations.* This document clearly defines, right at the design stage, the criteria that the TBM operator and/or design engineer must adopt to:

- confirm the current operating method;
- perform extra surveys to obtain a better understanding of the nature of the rock ahead of the face;
- change the operating method when different ground conditions are encountered;
- perform ground reinforcement operations in the face and around the tunnel that the machine is capable of doing.

To achieve it was important for the *guide lines* to give the main parameters to be followed systematically during construction with precise descriptions with quantities. These included the following:

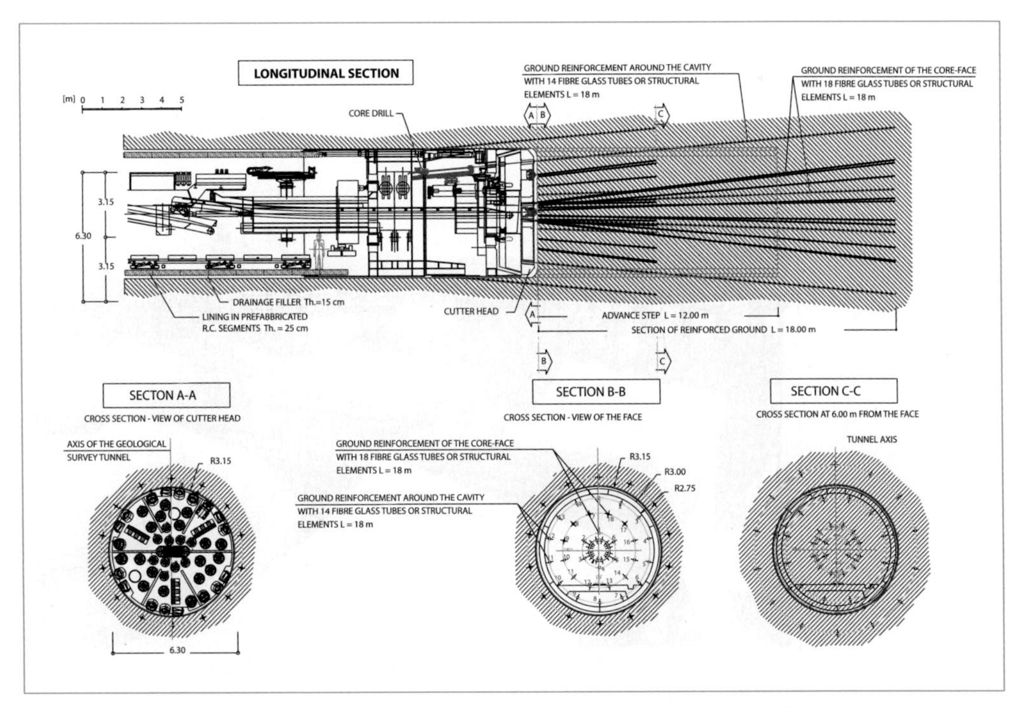

Figura B.9 *Intervention to improve the ground that can be performed from the TBM*

Ginori service tunnel: view of the head of the TBM during insertion in the tunnel

Depot for the prefabricated concrete segments for the final lining on the construction site of the Ginori service tunnel driven by TBM

- the TBM advance parameters (advance velocity, thrust, specific energy);
- the geological and geomechanical parameters acquired from systematic seismic surveying performed continuously in advance. This information may be added to by georadar tests in probe holes drilled into the face (Fig. B.10).

These parameters are useful for continuous and systematic monitoring of the characteristics of the ground and its response to excavation. They are associated with value ranges for normal functioning of the TBM. If the values actually measured differ substantially from those predicted as a result of particular unexpected local geomechanical conditions (tectonised zones, sudden lithological and structural changes, etc.), the TBM operator and/or the design engineer must take appropriate corrective or additional action as specified in the *guide lines*. Specified action may include operational changes (e.g. overbreak by the cutter head, locking the telescope mechanism, activating the tail pistons, etc.) extra surveys (e.g. core drilling) or ground reinforcement design operations. If the conditions encountered do not correspond to any of those specified in the design, then other solutions must be considered (Fig. B.11).

With this method of proceeding, action taken in a completely arbitrary manner based on no particular rules, which might have a marked effect on the stress-strain response of the rock mass, is avoided. At the same time a guide is provided for adapting to ordinary discrepancies between design predictions and the reality not of an emergency nature, with action taken based on standard design procedures.

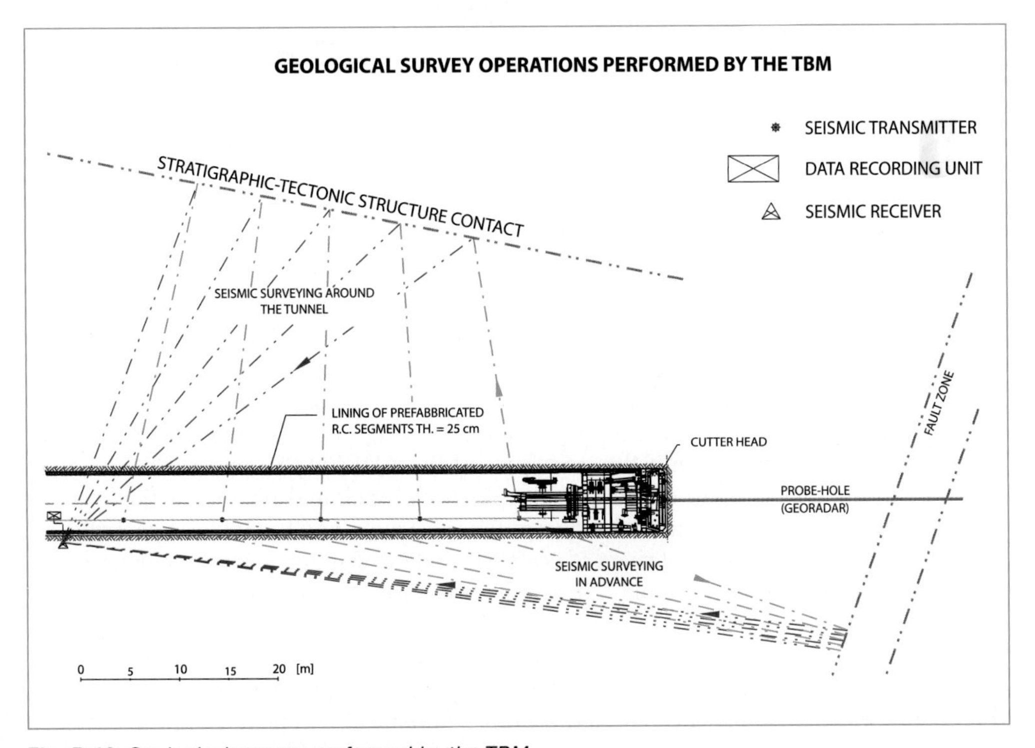

Fig. B.10 *Geological surveys performed by the TBM*

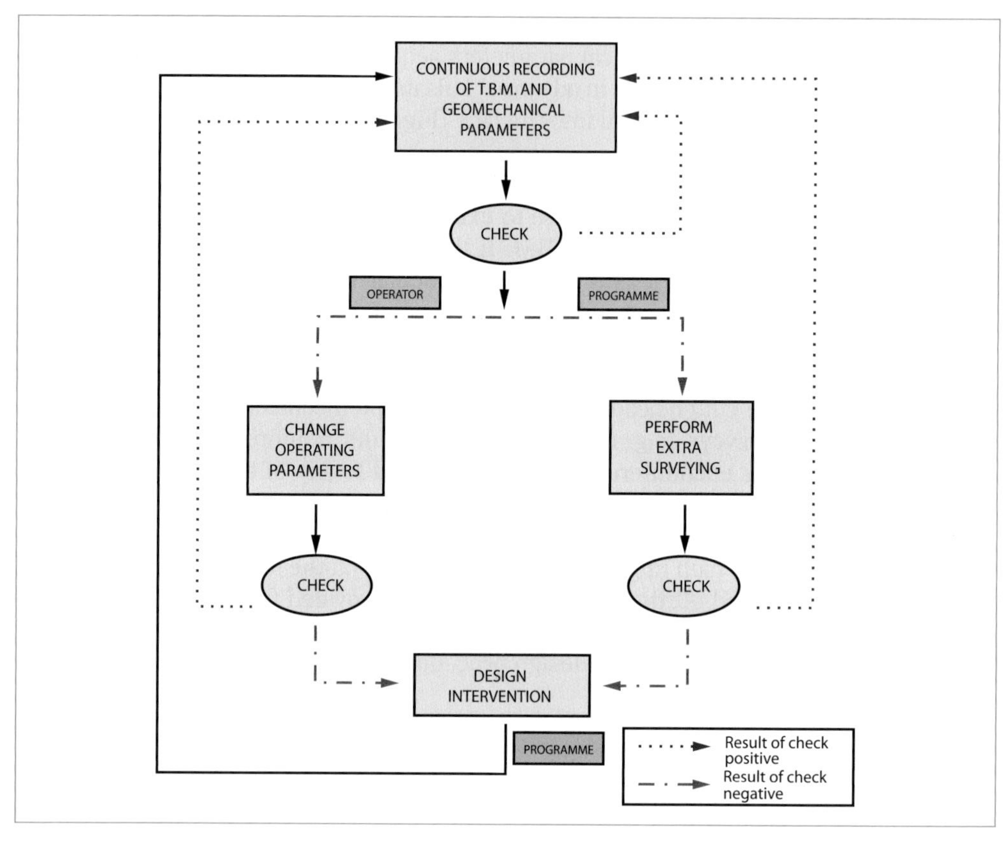

Fig. B.11 *Flow diagram for design control by TBM*

Tunnel construction

The TBM used for the excavation of the tunnel and its lining is a Wirth TB 630E/TS, double shielded telescopic rock cutter with a diameter of 6.3 m capable of generating a maximum thrust of 30,000 kN and an excavation torque of 5 kNm. As has been said, the distinguishing feature of the machine is that it was fitted with drilling equipment inside the shield capable of drilling holes in advance through special apertures in the cutter head and on the steel shell for the purpose of surveying or improving and reinforcing the ground (in accordance with the principles of the ADECO-RS approach). The spoil removal system consisted of conveyor belts. A special rack and pinion traction system supplied the TBM with the material required for excavation and tunnel lining activity.

The lining itself consisted of rings of prefabricated r.c. segments 140 cm in length and 25 cm thick, designed to withstand a hydraulic pressure of up to 5 bar. The rings, composed of six segments plus a key, had a special diagonally truncated cone shape, which made it easy to follow the theoretical alignment of the tunnel with excellent precision by simple fitting the segments for each ring in the sequence appropriate for each occasion (Fig. B.12). Mechanised excavation of the Ginori tunnel was completed without any serious problems at an average advance rate of approximately 20 m/day of finished tunnel on schedule and to budget.

Conclusions

The experience acquired in the construction on time and to budget of this project to cross the Apennines with a new high speed/capacity railway line, which was exceptional in terms of its size and the heterogeneity and difficulty of the ground, demonstrates beyond any doubt that, by using new technologies in an integrated manner, perfectly consistent with its underlying principles, the ADECO-RS approach opens up exciting new prospects for tunnelling, finally making industrialisation possible with the consequent certainty over construction times and costs previously not possible.

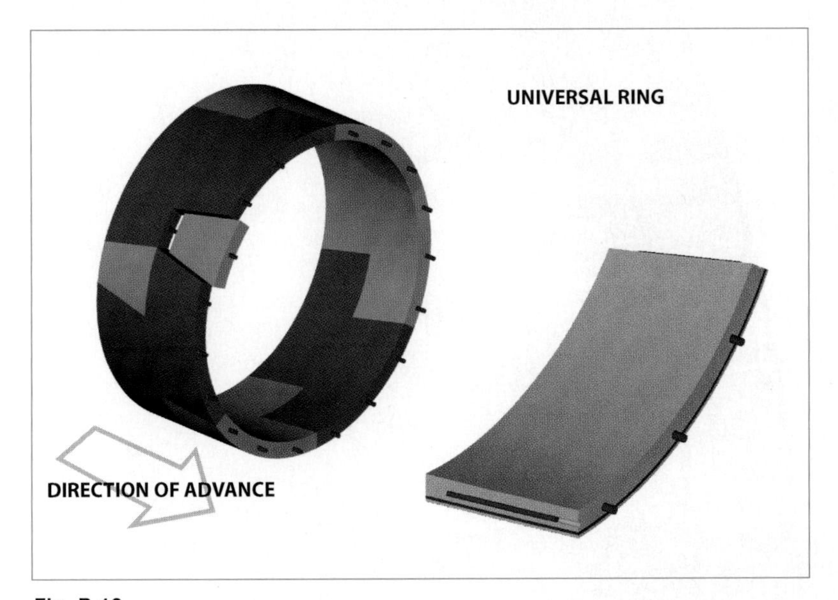

Fig. B.12

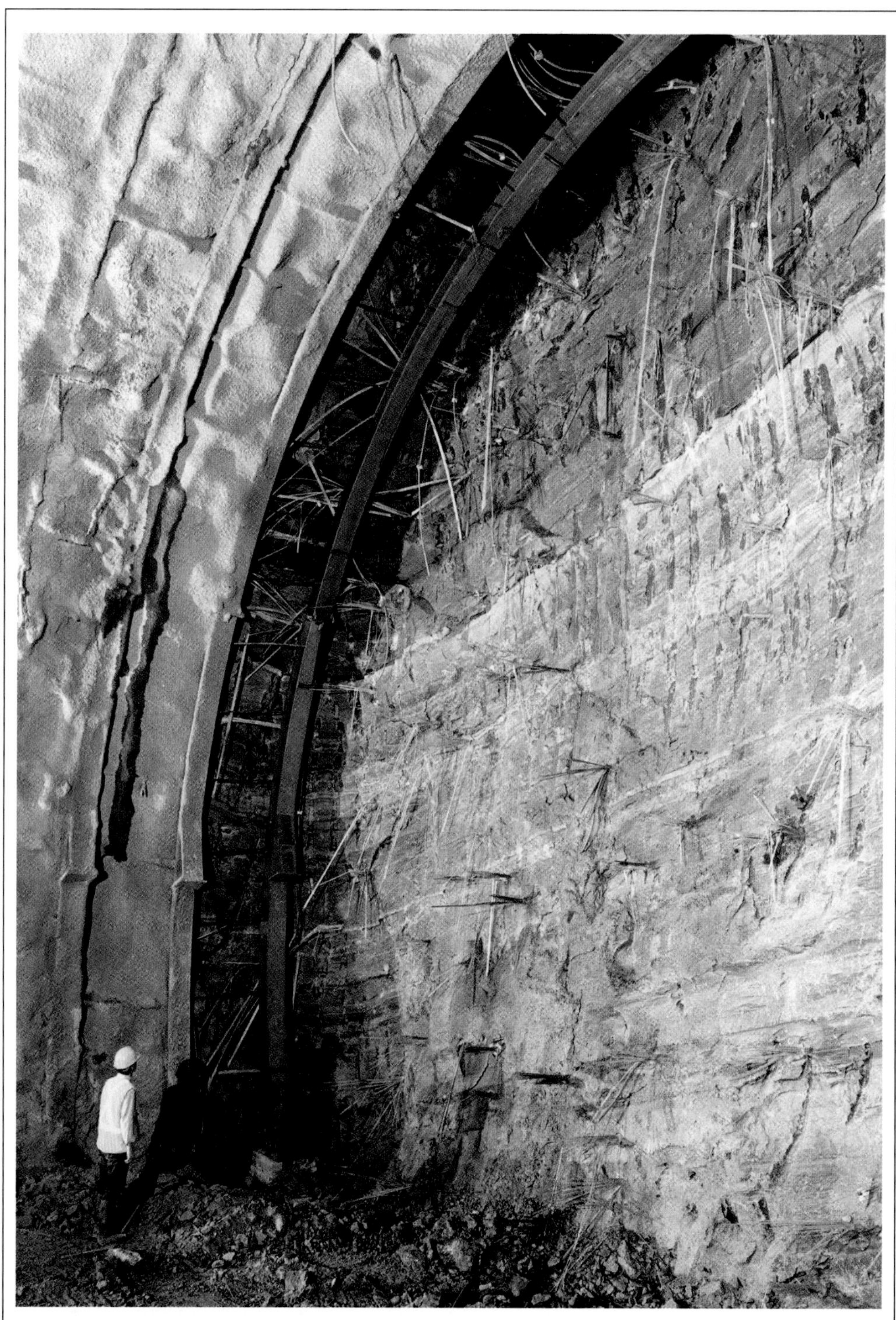

The Tartaiguille tunnel: the core-face reinforced with fibre glass elements, 1997, ground: clay, max. overburden: ~ 110 m

The Tartaiguille tunnel

The Tartaiguille tunnel, 2,330 m in length with a cross section of 180 m^2 is one of six tunnels on the TGV Méditerranée Line that connects Lyon to Marseilles on a route of approximately 250 km (Fig. C.1). The route passes through several Cretaceous formations which, from North to South, are as follows:

- upper Stampiem calcareous, fractured to a varying extent;
- lower Stampien marly clays;
- alternating Aptien blue marls and Albian sandstones.

The tunnel project described here passes through the lower Stampien formation consisting of alternating strata of marls, clays and silts with some calcareous material. The thickness of each strata varies from a few centimetres to some metres with the main discontinuities consisting of the stratification itself.

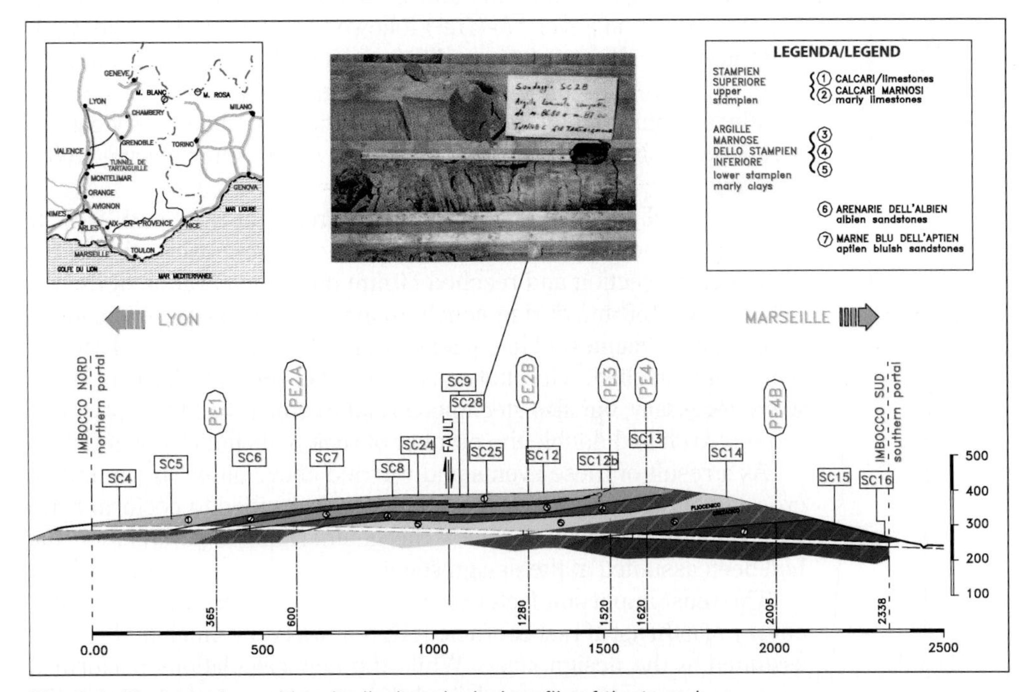

Fig. C.1 *Chorography and longitudinal geological profile of the tunnel*

This formation is extremely sensitive to the presence of water, which on contact causes a very rapid change in the strength properties of the ground. There is an immediate loss of consistency and swelling is triggered due to the high percentage of swelling material present in the clayey component (75% montmorillonite).

The maximum overburden for the whole tunnel is 137 m while that for the marly clay section is 106 m.

A brief history of the excavation

Excavation began from both portals (North and South) in February 1996.

The system of advance employed was to drive a top heading with a roadheader and bench excavation approximately 200 m behind using an excavator hammer.

The cross section types specified by the design for the long and short term stabilisation of the tunnel were of differing strength according to the geological characteristics of the ground that were encountered. They all consisted of shotcrete and steel ribs and end anchored radial rock bolts 4 m in length.

The design for crossing the 900 m stretch in the Stampien clays specified a cross section type consisting of a 25 cm layer of shotcrete, reinforced with HEB 240 ribs placed at 1.5 m intervals and micropiles beneath the base of the ribs and rock bolts after excavation of the bench (L = 4 m, one every 3–4 m^2) followed by casting of the sidewalls and the invert. When safety conditions required it the ground ahead of the face was reinforced with tubular fibre glass nails to improve stability. Final stabilisation of the tunnel was to be then obtained by casting a concrete lining in the crown (thickness 70 cm) after placing waterproofing.

At the end of September 1996, convergence at the South face increased well beyond predicted design values during excavation of the Aptien marls section and reached 60 mm during excavation of the top heading and 150 mm during bench excavation. The increase in convergence values manifested in the form of sizeable cracks in the shotcrete which also spalled. This not only made considerable barring down work necessary, but also steel mesh reinforcement had to be placed in the roof arch and double the number of rock bolts initially planned.

As a result of these events and worried above all by the imminent crossing of the marly clays, the site operator decided to perform a further geological survey to check the geomechanical parameters that had been assumed at the design stage.

The most important factor to emerge from the survey was the discovery of a thrust at rest coefficient K_0 that was very much higher than assumed at the design stage. While the new calculations performed with correct parameters explained the convergence observed in the marls, it was clear that it would be impossible to use the planned cross section type since it would have involved accepting inadmissible deformation.

In order to find the best possible solution to the problem, the SNCF (*Societè National du Chemin de Fer*) set up a study group (*"Comité de pilotage"*), at the beginning of 1996, consisting of its own experts from French Rail, experts from the consortium formed by the companies Quillery & Bard and from their consultants Coyne et Bellier, of the geotechnical consultants of the Terrasol and Simecsol Consortium and finally of experts from the CETU. This study group then turned to major European tunnelling experts, inviting them to put forward their design proposals for crossing the clayey section of the tunnel in safety and on time.

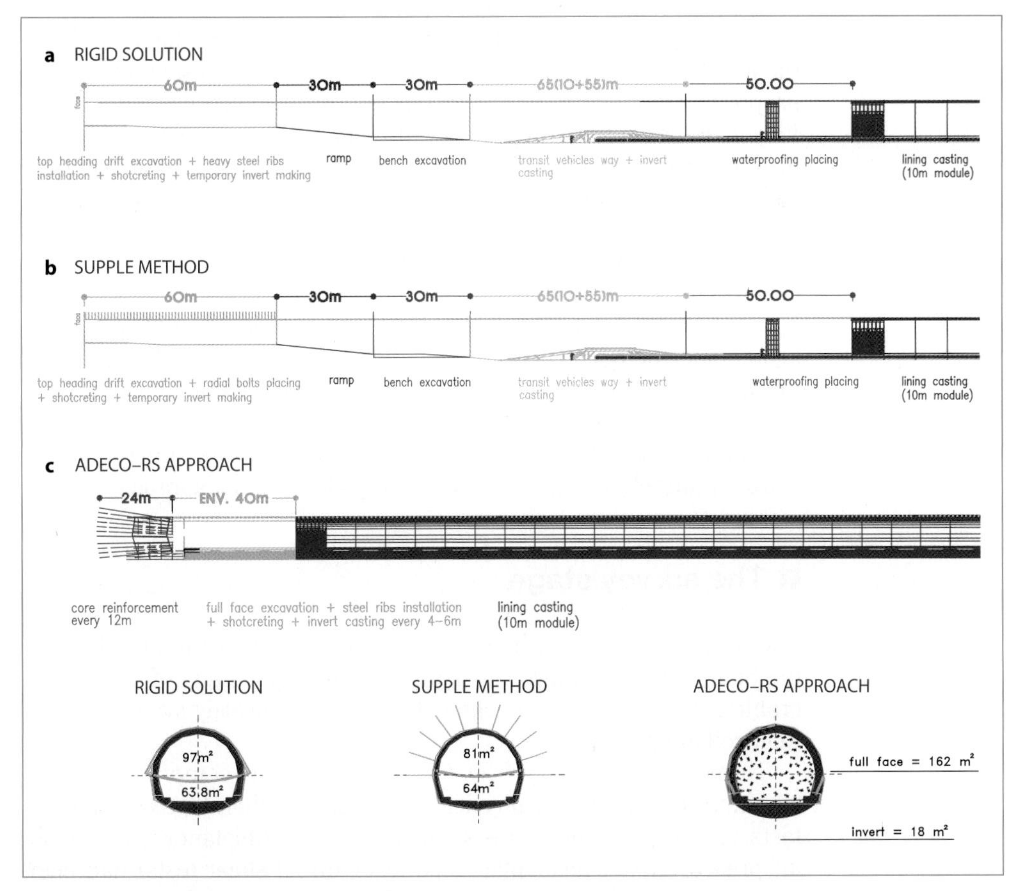

Fig. C.2 *The three solutions examined (French, Swiss, Italian)*

The first proposal (French) (Fig. C.2a) suggested half face advance with a much more rigid cross section type than was specified in the initial design, created by increasing the number of radial rock bolts placed around the excavation and by placing a stronger preliminary lining, thicker and with more reinforcement. However, this would have required very lengthy execution times and was immediately discarded.

The second proposal (Swiss) (Fig. C.2b) considered by the study group involved the use of special deformable ribs for the top heading. This would have allowed the rock mass to free itself of part of its potential energy through the development of unopposed and also considerable deformation. Reinforcement of the face using fibre glass tubes of reduced length was specified to guarantee the safety of site workers. This proposal was also discarded because of the numerous uncertainties connected with the difficulty in calculating the behaviour of the deformable ribs and due to the long time required for this type of advance (the final lining could not be cast until the strong convergence allowed by the deformable ribs had stabilised).

Before resigning itself to accepting long delays before the new line could be opened, the SNCF invited the author to give his opinion on the feasibility, and construction times and costs and to send in his own proposal. They had in fact heard of the major successes achieved in Italy in driving tunnels under very difficult conditions, by applying new design and construction principles. Consequently a third solution (Italian) (Fig. C.2c) was considered, radically different from those already described. This involved full face advance and was based on controlling deformation phenomena by, as we shall see, stiffening the advance core (ADECO-RS approach [41], [42]).

SNCF were favourably impressed by the preliminary designs presented and the construction times that were forecast and guaranteed by the consulting engineer on the basis of documentation produced for similar cases resolved using the same system. Consequently, in March 1997, the SNCF selected Rocksoil to design the remaining 860 m of tunnel to be driven in the marly clays of the Stampien.

The survey stage

The first stage in the design of the section of tunnel running through the Stampien marly clays consisted of an in-depth study of the considerable geological and geotechnical information already available when Rocksoil intervened.

Numerous laboratory (CD, CU and UU triaxial compression tests, edometer tests, tests for physical and chemical characterisation, swell tests, spectrophotometric tests) and *in situ* tests (dilatancy tests, “push-in” pressuremeter tests, plate load tests, direct shear tests) had been performed and the geomechanical characteristics of the ground had virtually been completely defined (Fig. C.3). The parameters to be used in the calculations and the principle geotechnical hypotheses to be assumed were decided jointly during a number of meetings of the study group in which consulting engineer participated directly. They are given below:

- γ = unit weight = 21.7 KN/m^3
- c_u = undrained cohesion = 1.2 MPa
- c' = drained cohesion = 0.2 MPa
- φ' = angle of friction = 27°
- E_u = undrained elastic modulus = 400 MPa
- E = drained elastic modulus = 200 MPa
- ν = Poisson modulus = 0.4
- K_0 = thrust coefficient at rest = 1.2
- p_g = swelling pressure = 0.2–0.3 MPa (below the tunnel invert)
- *fluage* of the ground: simulated as a reduction of 35% of the drained elastic modulus on a thickness equal to one radius of excavation (7.5 m) around the cavity;
- water table: approximately 25 m above the crown of the tunnel.

When the available geotechnical and geological information had been examined, the consulting engineer asked for four triaxial cell extrusion tests to be performed in order to study the response of the ground to excavation by simulating the release of stress produced in the ground by tunnel advance in the laboratory.

In this type of test a sample of the ground is inserted in a triaxial cell and put under pressure until the natural stress state of the rock mass is reached.

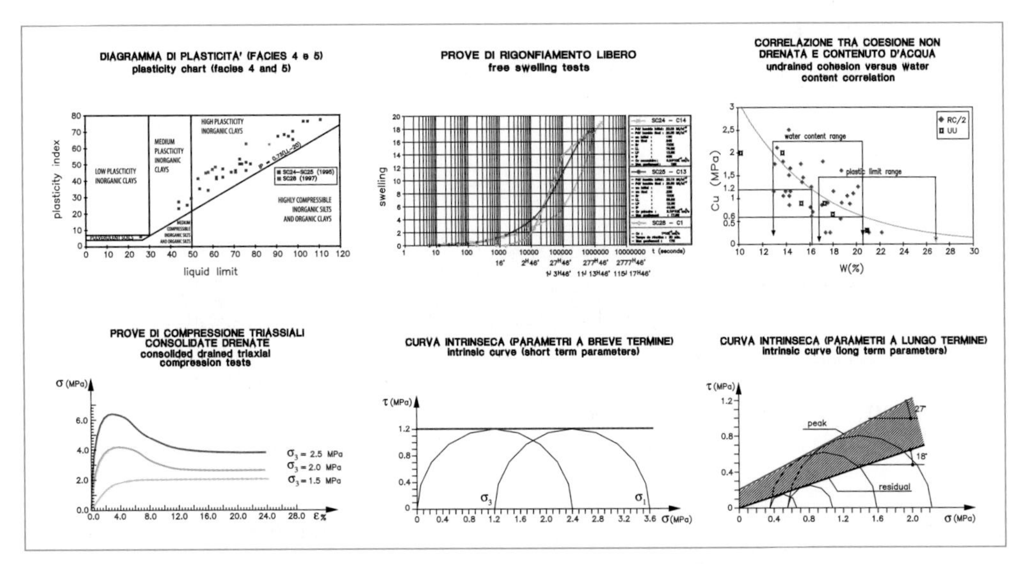

Fig. C.3 *Results of the tests performed in the laboratory on the Stampien clays*

By using a fluid under pressure, this stress state is also reproduced in a special cylindrical volume (which simulates the advance of a tunnel face) created inside the test sample and coaxial to it before the test. The stress state around the sample is maintained and the pressure of the fluid in the cylindrical volume is gradually reduced to simulate the changes in the stress state at the face induced by excavation. A prediction of the magnitude of extrusion at the face as a function of time is thereby obtained.

The use of these tests is fundamental for putting the finishing touches on the design and defining tunnel advance stages.

The diagnosis stage

In the diagnosis stage, the results of the swelling and extrusion tests (Figs. C3 and C4) and the characteristic lines at and at a distance from the face (Fig. C.5) were studied.

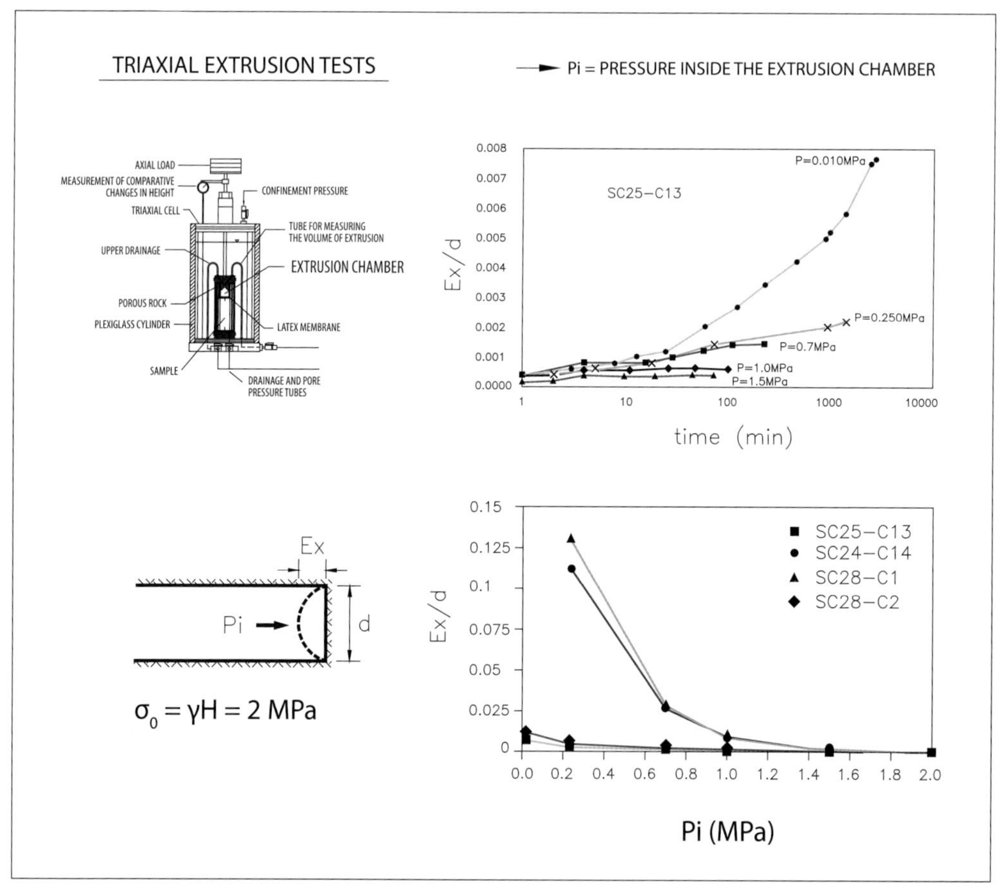

Fig. C.4 *Triaxial cell extrusion tests*

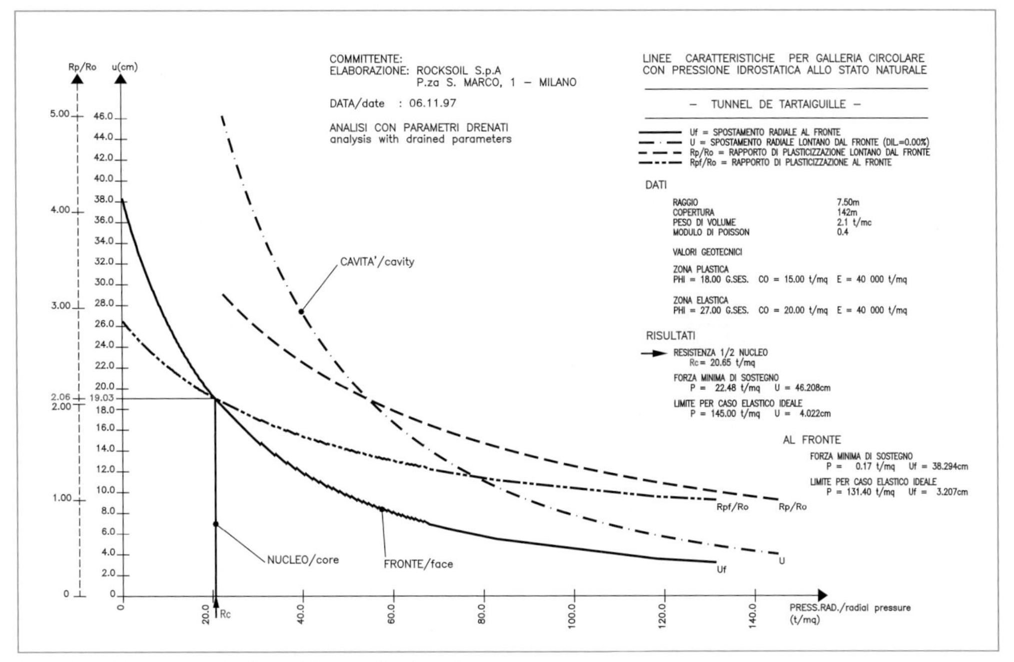

Fig. C.5 *Characteristic lines (diagnosis phase)*

A singular sudden, and rapid loss of consistency in the material was immediately seen and this placed the accent on the need to intervene appropriately ahead of the face to prevent a substantial band of plastic behaving material from developing around the cavity. Considering also the considerable presence of swelling minerals in the ground, this was felt to be the main cause of the disastrous deformation that had occurred and of the excessive loads on the linings that had been observed. Preventing this band of plastic material from developing was to be achieved by energetic preconfinement of the cavity capable of countering deformation before it is triggered.

The therapy stage

On the basis of the findings of the diagnosis stage, it was decided to cross the lower Stampien clays by adopting a cross section type that involved full face advance after first reinforcing the ground in the advance core and then almost immediately closing the preliminary lining by placing the tunnel invert.

The reinforcement of the advance core (Fig. C.6) was achieved by means of (an average of) 90 fibre glass structural elements 24 m in length, injected with an expanding cement mix. It also required the excavation and casting of the tunnel invert in steps of between 4 and 6 m behind the face and the placing of the final lining at between 20 and 40 m from the face.

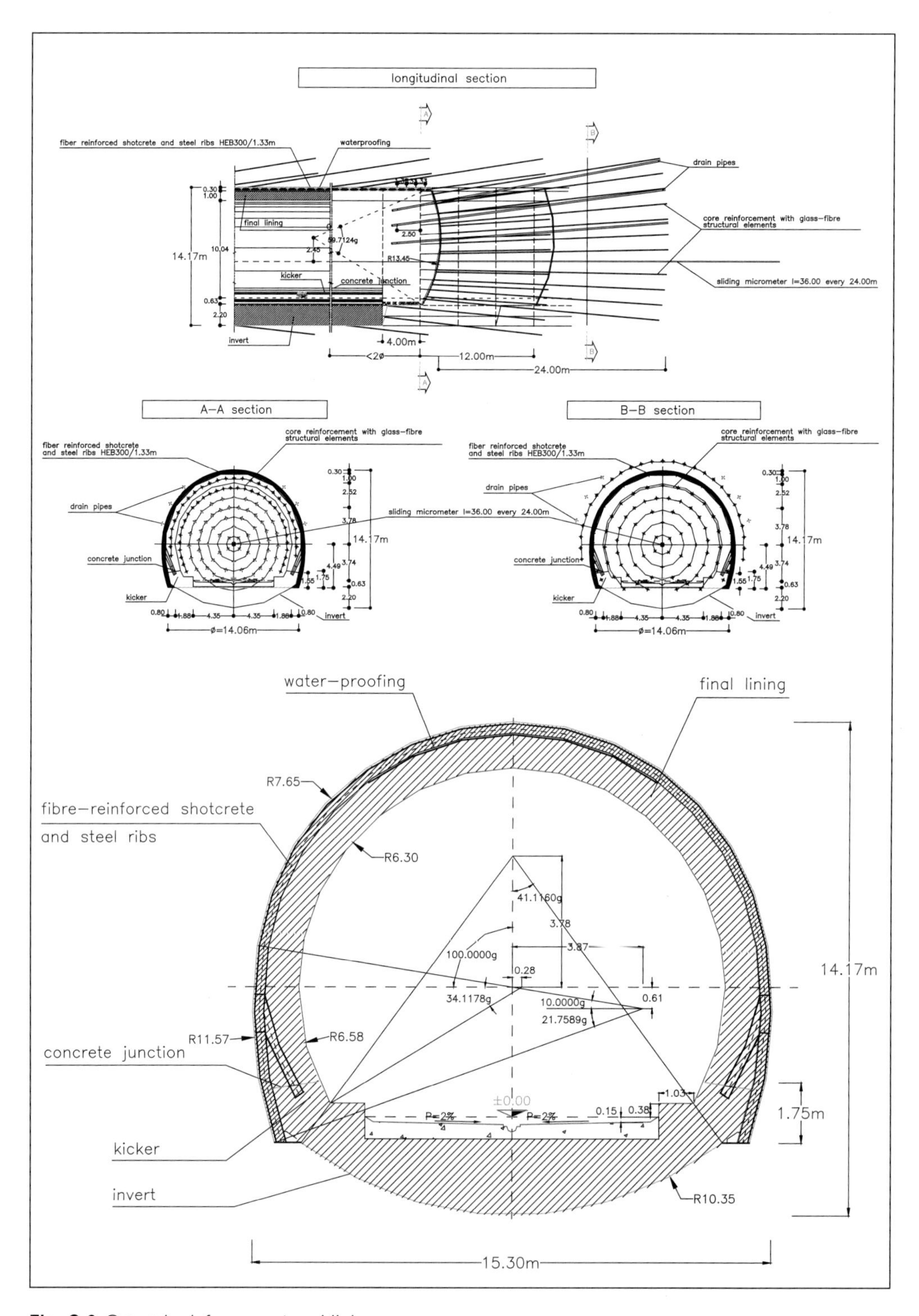

Fig. C.6 *Ground reinforcement and linings*

The preliminary lining was to have consisted of 30 cm of fibre reinforced shotcrete, also reinforced with HEB300 steel ribs placed at intervals of between 1.33 and 1.50 m. To meet the request of the contractor to maintain the formwork used for the section of tunnel already excavated (which would make it impossible to construct side walls high enough to contain the extremely high horizontal thrusts) it was finally decided to fit the steel ribs with a structural element (*jambe de force*) capable of transferring the loads from the preliminary lining onto the tunnel invert and vice versa (Fig. C.10).

Establishing the magnitude of the reinforcement of the advance core was initially performed by reprocessing the results of the extrusion tests (Fig. C.11) and by the characteristic line method. It was then tested using 2D and 3D finite element models in the elasto-plastic range (Fig. C.12) with which it was also possible to test the preliminary and final linings under the effects of complex stress-paths such as those due to *fluage* and swelling (Fig. C.13).

The operational stage during constrution

The full cross section of the "Tartaiguille" tunnel, as has been said, is approximately 180 m^2; each steel rib (HEB300) weighed approximately 5 tonne, the volume of shotcrete placed in the lining was 13.3 m^2/m and each reinforcement of the advance core required 2,160 m of fibre glass structural elements.

Despite the considerable quantities briefly described here, the advance system proposed (see diagram Fig. C.14) enabled contractual deadlines to be met and industrialisation of the various operating stages resulted in average advance rates of 1.7 metres/day (Fig. C.15), even faster than the 1.4 metres/day forecast by the consulting engineer. As a consequence, the tunnel was completed just one year after the start of full face advance (end of July 1997), a full month and a half ahead of schedule.

The monitoring phase during construction

Fine tuning of the tunnel section design in terms of:

- number of fibre glass structural elements at the face (90–150);
- length of overlap between them (9–12 m);
- length of tunnel invert steps (4–6 m);
- spacing between steel ribs (1.33–1.50 m);
- distance from the face at which the final lining was cast in the roof arch (20–40 m);

Fig. C.7 *Reinforcement of the advance core*

Fig. C.8 *Placing of steel ribs*

Fig. C.9 *Reinforcement of the core-face by means of fibre glass structural elements*

Fig. C.10 *Preparation of the tunnel invert. Note the "jambe de force"*

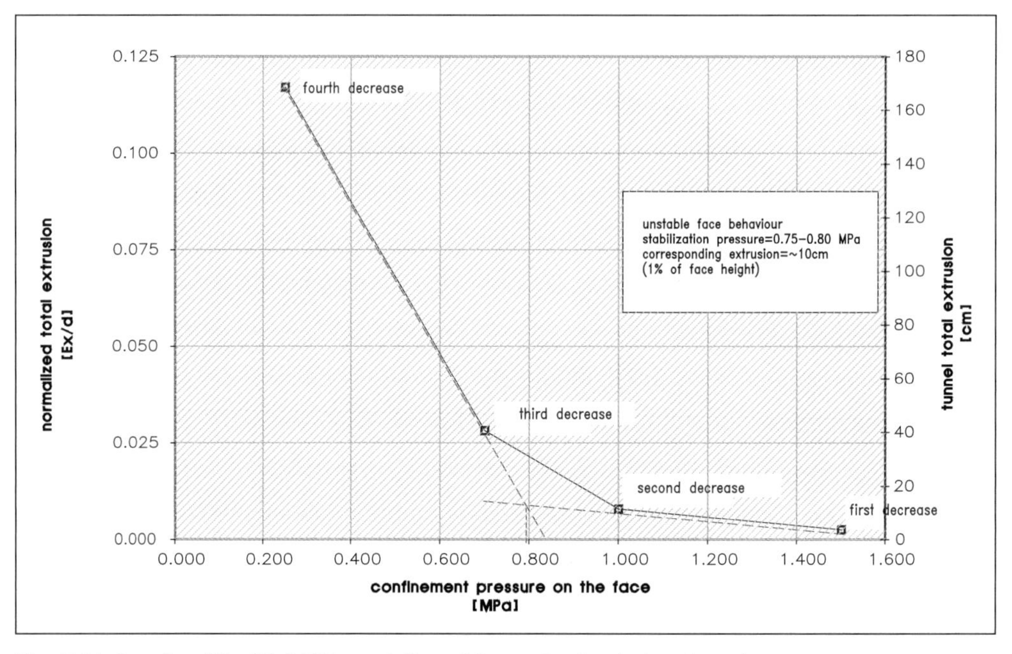

Fig. C.11 *Results of the 3D F.E.M. modelling of face extrusion (values in cm)*

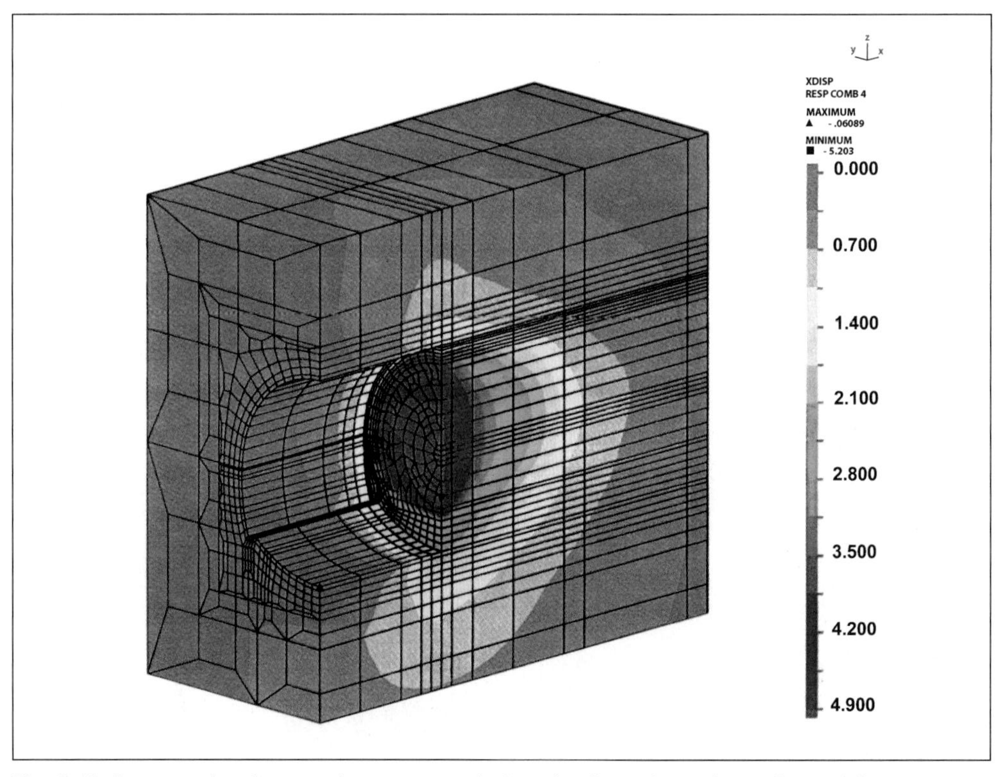

Fig. C.12 *Reprocessing the extrusion tests to calculate the dimensions of core-face reinforcement*

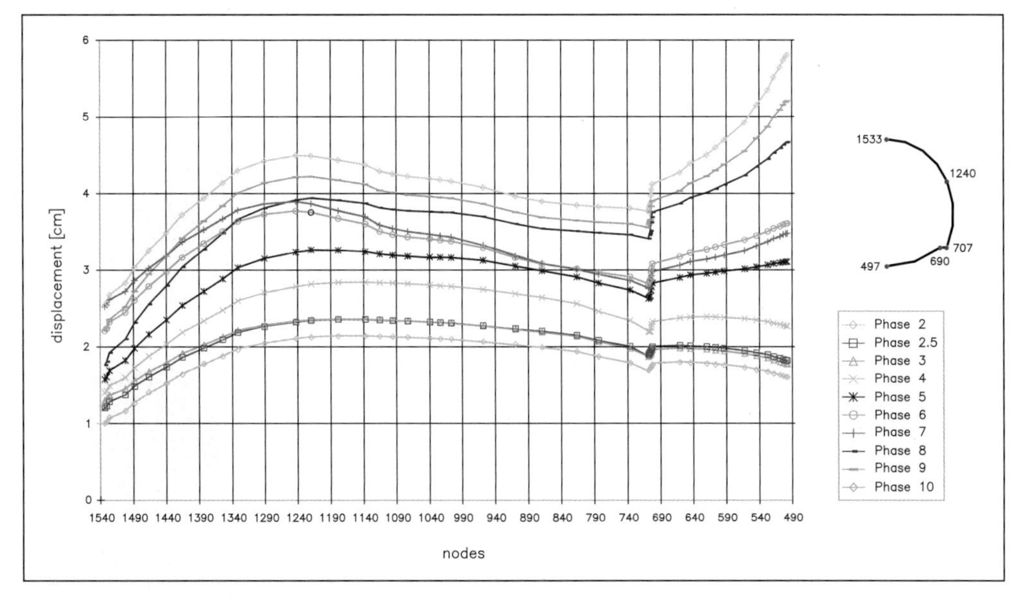

Fig. C.13 *Results of 2D modelling of radial displacements*

was performed on the basis of processing and interpretation of monitoring data and in particular on systematic measurement of extrusion of the advance core and convergence of the cavity.

With tunnel advance fully underway, average values (Figs. C.16 and C.17) for extrusion were around 3–4 cm and for diametric convergence around 5–7 cm (after casting the tunnel invert).

An examination of the extrusion curves in particular shows that the size of the zone of influence of the face varies between 8 and 13 m. This can be deduced from the appreciable change in the angle of the curves at this depth. Furthemore, it can be seen that the excavation of the last 6 m produces extrusion of 15–18 mm comparable to that of the first 6 m and that thanks to the ground improvement, the advance core behaved elastically confirming that the number of fibre glass structural elements placed had been well calibrated.

The extrusion curves obtained in the laboratory did in fact show that the ground tended to rapidly lose its strength above all in the clayey fraction and that action had to be taken to limit deformation to a minimum for excavation to occur in conditions of safety. As already stated, this is the principle that guided the design from the very beginning and was still maintained even during the stage of fine tuning the tunnel section during construction.

Convergence measurement stations were installed every 6 m with measurement points consisting of 7 optical targets inserted both directly into the shotcrete and on the steel ribs. Measurements obtained from the latter were 60–70% lower on average than those taken from the shotcrete.

FULL FACE ADVANCE AFTER REINFORCEMENT OF THE CORE-FACE

Operational stages

PHASE 1

Reinforcement of the core-face by means of fibre glass structural elements

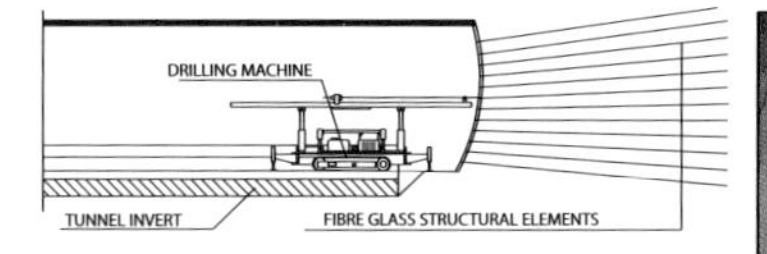

PHASE 2

Excavation step
(m 0.70 ~1.00 full face)

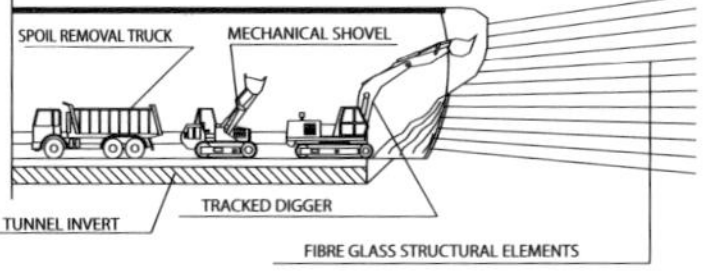

PHASE 3

Placing of shotcrete on the face and surface of the cavity to protect walls of the tunnel (thickness around 5 cm)

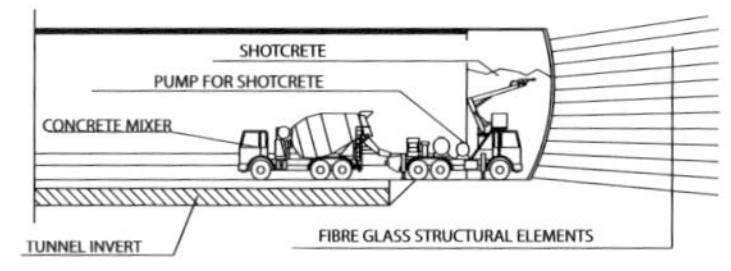

PHASE 4

Placing of the steel rib and connecting reinforcement in the section of tunnel excavated in phase 2

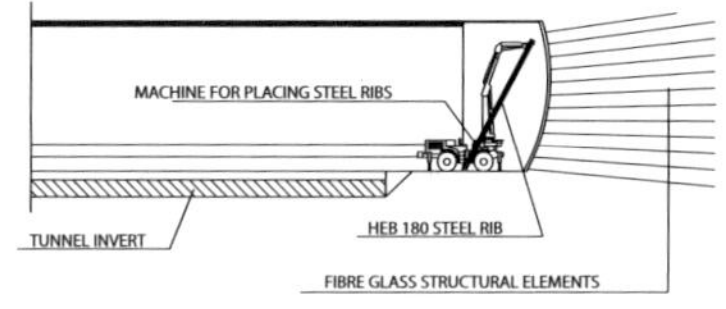

PHASE 5

Placing of the preliminary lining in fibre reinforced shotcrete (thickness around 35 cm)

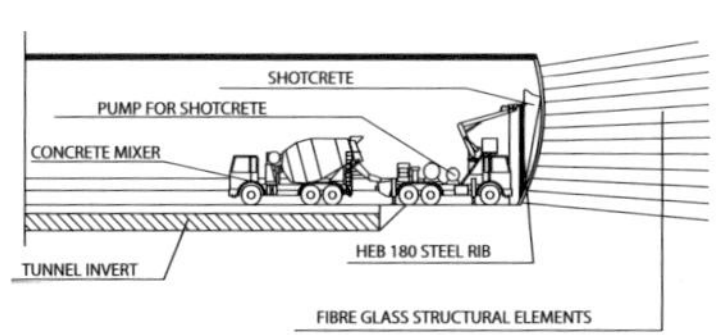

PHASE 6

Excavation and casting of side walls and tunnel invert at a distance from the face of ≤1.5 Ø

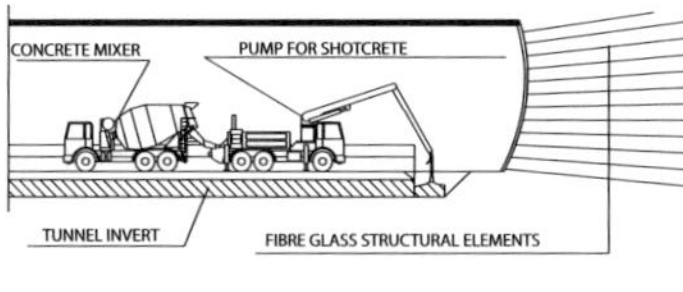

PHASE 7

Casting of the final lining at distance from the face of ≤4–5 Ø

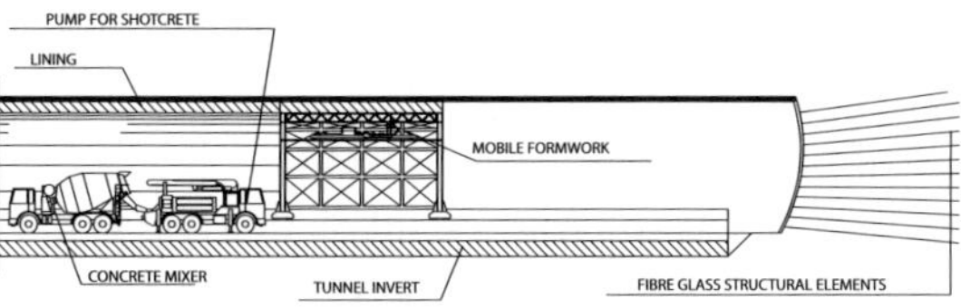

Fig. C.14 *Operating phases (with relative photographs)*

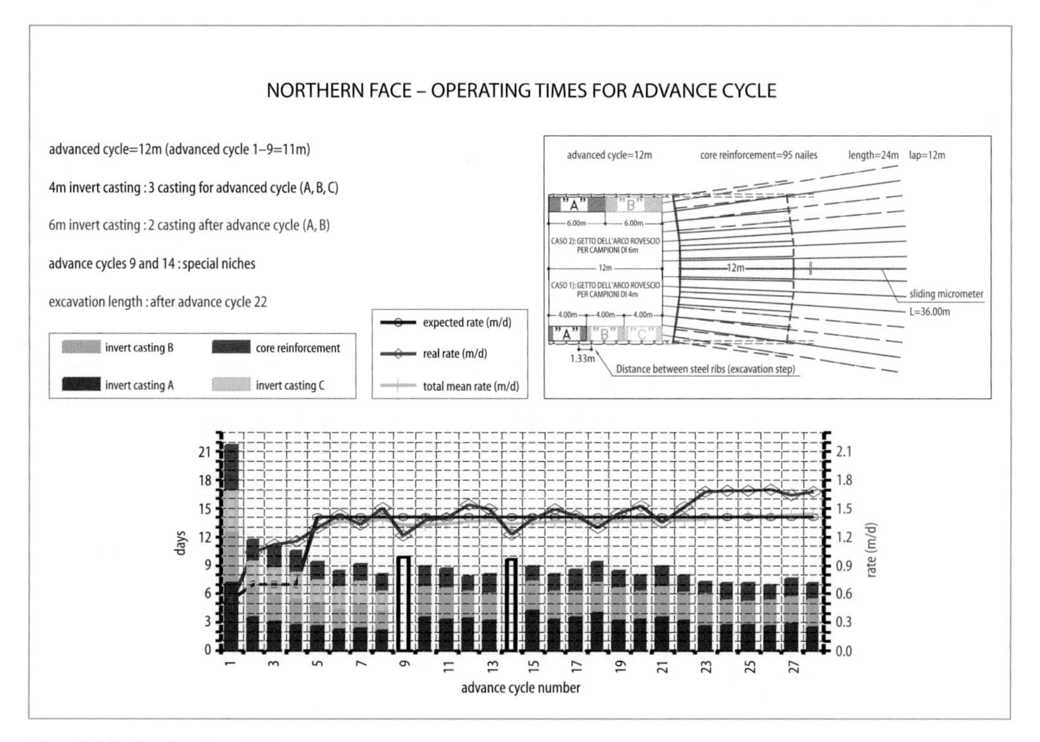

Fig. C.15 *Production data*

The measurement sections in which targets were installed before the tunnel invert was placed gave convergence readings of around 2 or 3 cm with gradients of 4–6 mm per day until the tunnel invert was cast. After the ring of the preliminary lining was closed, however, the gradient fell to 1–3 mm per day.

This data shows therefore that deformation (convergence-extrusion) was generally maintained within the elastic domain characteristic of the ground, thanks to the substantial limitations placed on the possibility of a band of ground with plastic behaviour developing around the cavity. This confirms the effectiveness of the intervention proposed and adopted both in terms of ground improvement and reinforcement and operating methods.

Each time a cross section of ground larger than the current section was excavated, as for example required for the construction of recesses in the tunnel walls, the consequent greater relaxation of the material was followed by a significant increase in deformation (up to 60%), confirming the extreme sensitivity of the ground and the importance of reducing decompression and the time taken to place linings to a minimum.

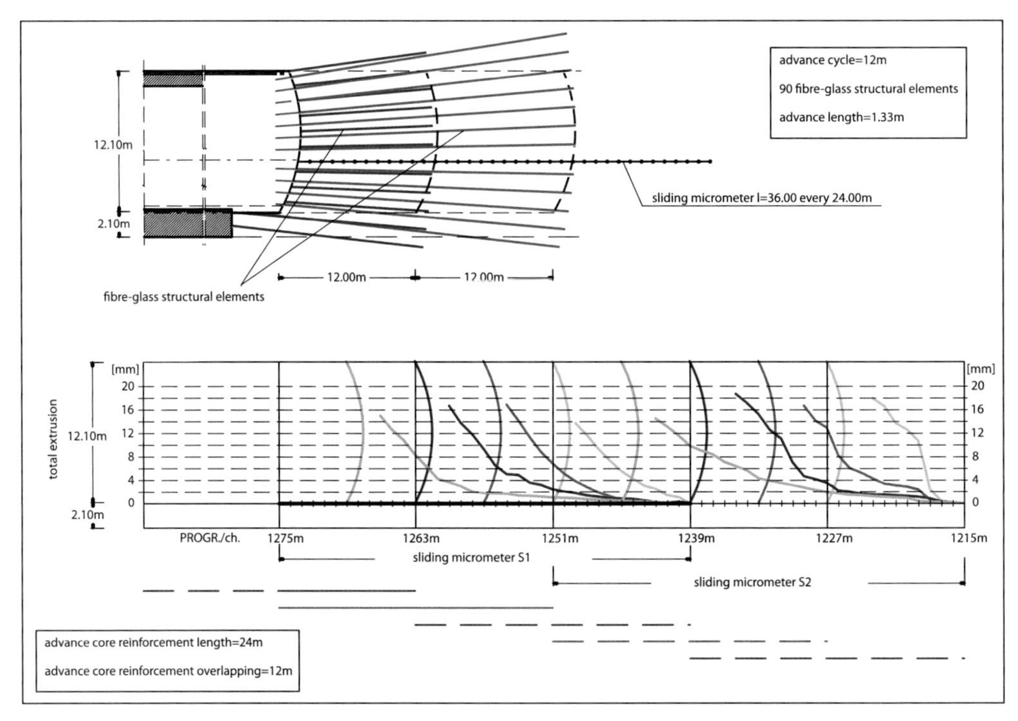

Fig. C.16 *Results of extrusion measurements*

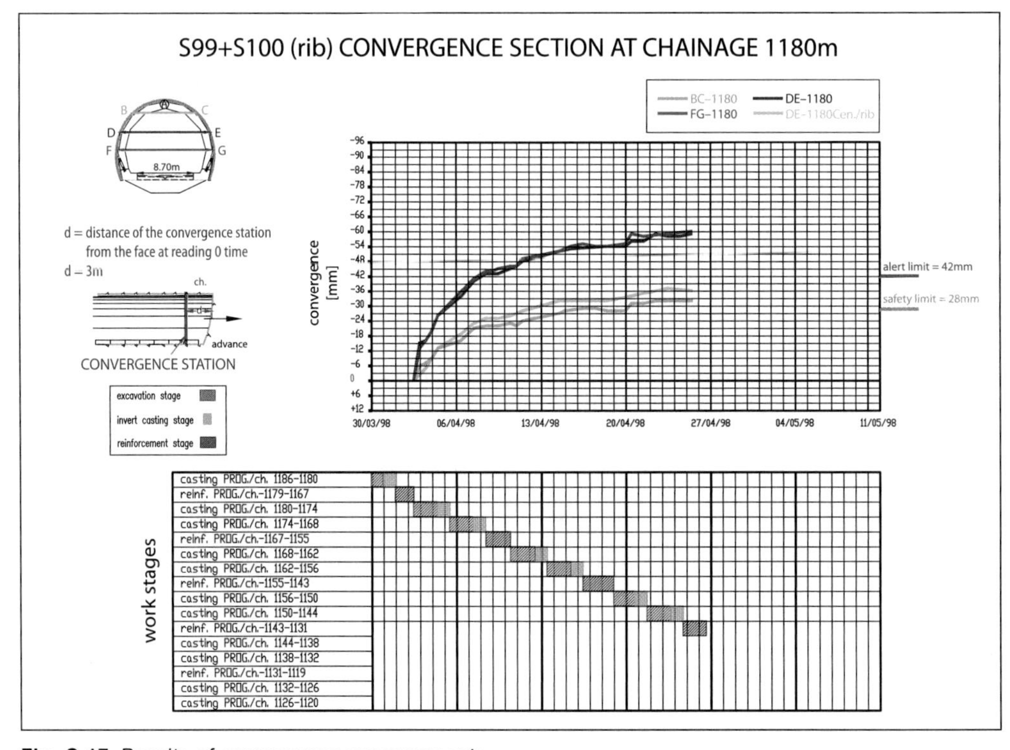

Fig. C.17 *Results of convergence measurements*

■ Conclusions

Construction of the central section of the "Tartaiguille" tunnel (180 m^2 cross section) in the clays of the Stampien formation was the cause of considerable concern to the consulting engineers and contractors involved, when the inadequacy of the design and construction systems adopted (NATM) soon became clear. They were unable to guarantee the feasibility and safety of the works with acceptable advance rates when faced with the extensive swelling phenomena characteristic of the ground concerned. Given these problems, SNCF sought a solution to the problem that would enable them to construct the tunnel in safety and on schedule without delaying the opening of the new line. They therefore set up a study group and invited the most renowned European tunnelling experts to form part of it. Various proposals were examined, but the only one that seemed able to meet all the requirements demanded was the Italian proposal suggested by the author and drawn up by Rocksoil S.p.A. of Milan. It was based on ADECO-RS principles and consisted of full face advance after first reinforcing the advance core using fibre glass structural elements. The adoption of this design, together with the on site assistance of engineers from Rocksoil S.p.A. enabled a difficult tunnel to be driven without problems, with very reduced deformation, ahead of schedule and at costs even lower than budgeted.

This considerable achievement made a great impression and aroused some degree of amazement in France, where the specialist press paid numerous tributes to all the Italian consulting engineers, technicians and construction contractors who supplied the know-how required to complete the project on schedule ["*Débuté en juillet, le chantier, qui fait travailler 200 personnes, posait principalement des difficultés liées aux pressions exercées par la montagne. Une nouvelle méthode a donc été instaurée sur l'idée d'un ingénieur italien: le percement en pleine section (plotôt qu'en demi-section) ...*" (Tunnels et ouvrages souterrains, January/February 1998), "*Lorsqu'elle en prend les moyens, l'Italie peut réaliser des travaux à faire pâlir les entreprises françaises ...*"(Le Moniteur, 20th February 1998). "*Le creusement du tunnel de Tartaiguille a été très difficile, en raison notamment de convergences inattendues du terrain qui ont nécessité un changement de méthode en cours de chantier: le professour italien Pietro Lunardi a convaincu la SNCF de travailler à la pelle en pleine section dans les argiles, en boulonnant le front sur 24 m ...*" (Le Moniteur, 7th August 1998)]. It was therefore a great success, which has established Italian leadership in the field of underground works under difficult stress-strain conditions.

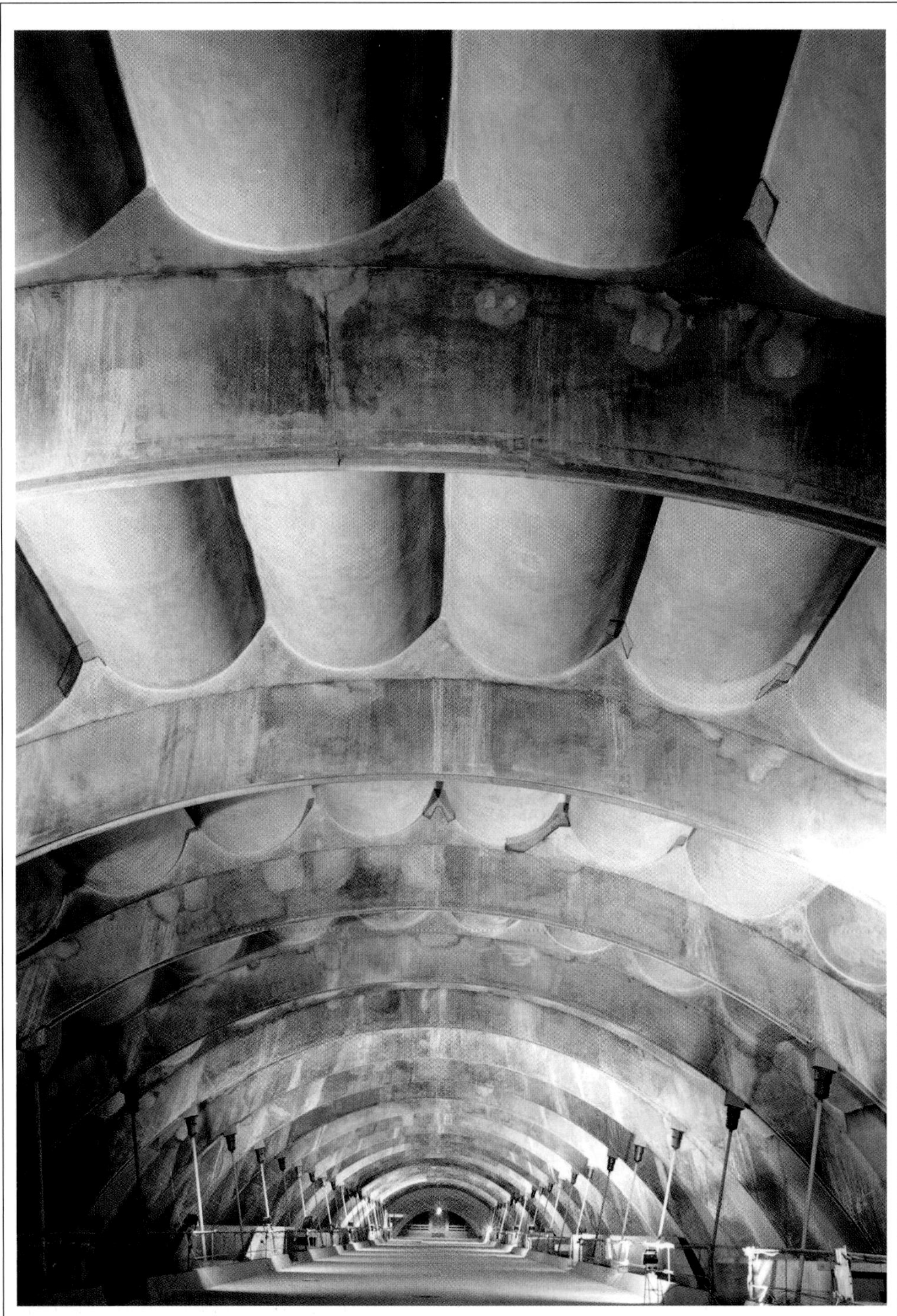

Cellular arch technology: view of the structure built for the Venezia station of the Milan urban link line (1990, ground: sand and gravel under the water table)

APPENDIX D

Cellular arch technology

The cellular arch is an innovative construction technology designed for the construction of large span tunnels in urban environments when the geotechnical and stress-strain situations, the shallow overburdens and the requirement for construction work to have negligible effects on surface constructions and activities are either not compatible with conventional tunnelling methods, or make them less reliable and competitive.

It is a composite structure with a trellis-like framework and a semicircular cross section. The longitudinal members (cells) consist of pipes (minitunnels) filled with r.c. joined together by a series of large transverse ribs (arches) (Fig. D.1). Studies performed to find the limits to its application suggest that it can be employed successfully to build shallow overburden, bored tunnel cavities with a span of more than 60 m even in loose soils under the water table, without causing any appreciable surface settlement.

The advantage of this technique over traditional methods lies in the way that the passage from the initial equilibrium condition of the still undisturbed ground to the final equilibrium condition of the finished

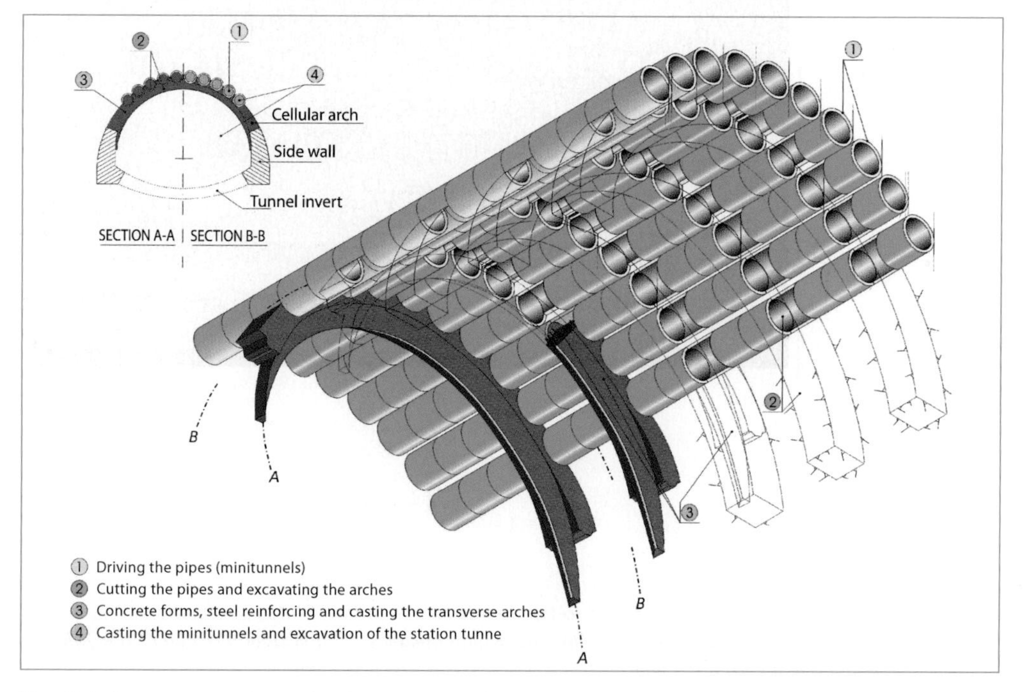

Fig. D.1 *Construction stages for the cellular arch*

The plaque awarded to the author by the United States publication Engineering News – Record, which makes a "Construction's man of the year" award each year

tunnel is controlled to prevent the onset of decompression in the material and consequently also of surface settlement.

Excavation is in fact performed when the very rigid load bearing structure has already been constructed and is able to furnish the ground with the indispensable confinement required without suffering any appreciable deformation.

To achieve this the entire construction of the cellular arch structure takes place in the following stages (Fig. D.2):

- half face excavation of side drifts for the tunnel walls after first improving the ground from a pilot tunnel and completion of ground improvement around the final tunnel;
- completion of excavation of the tunnel wall side drifts and casting of the tunnel sidewalls inside them while, in a completely independent site above, a series of r.c. pipes (minitunnels) are driven side by side (2.10 m in diameter) into the ground along the profile of the future tunnel roof from a thrust pit (by pipe jacking them);
- excavation from the minitunnels of transverse tunnels to be used as the formwork (the walls of this consisting of the ground itself) for the casting of the connecting arches in r.c. and then placing of reinforcement and casting of the arches themselves;
- the ten longitudinal minitunnels forming the tunnel arch are filled with concrete and then the tunnel for the station is excavated beneath the protection of the "cellular" arch practically already active;
- the tunnel invert is cast in steps.

This innovative construction method for which the author, who invented it, was awarded a prestigious international award was used for the first time in the world to construct the "Venezia" station of the Milan urban link railway line. If conventional systems had been used its construction would have been much more troubled and costly, if not actually impossible.

What follows is an attempt to fully illustrate the characteristics of the technology on the basis of the experience acquired during the construction of this exceptional civil engineering work with a focus also on the operational aspects of construction and the final results achieved.

The construction of "Venezia" station

Venezia station has now been in service for a number of years. It is located strategically in the commercial centre of Milan and is the largest underground construction in the whole of the regional transport system.

It is a large tunnel with an external diameter of approximately 30 metres and a length of 215 metres. It is bored tunnel that passes through cohesionless soil under the water table with an overburden

MILAN URBAN LINK RAILWAY LINE
THE CONSTRUCTION OF THE CELLULAR ARCH STRUCTURE FOR VENEZIA STATION

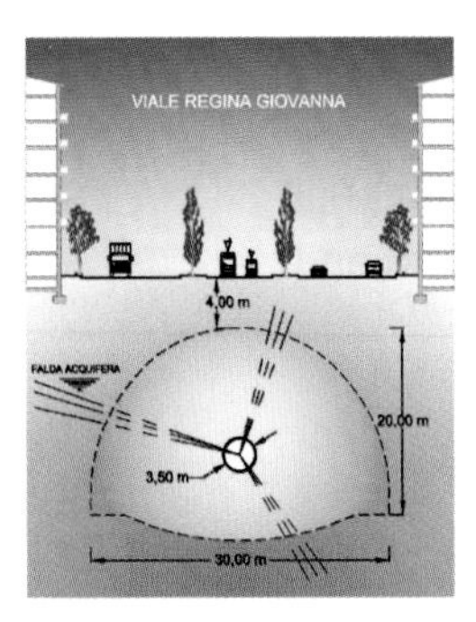

1 - Ground improvement injections from a pilot tunnel

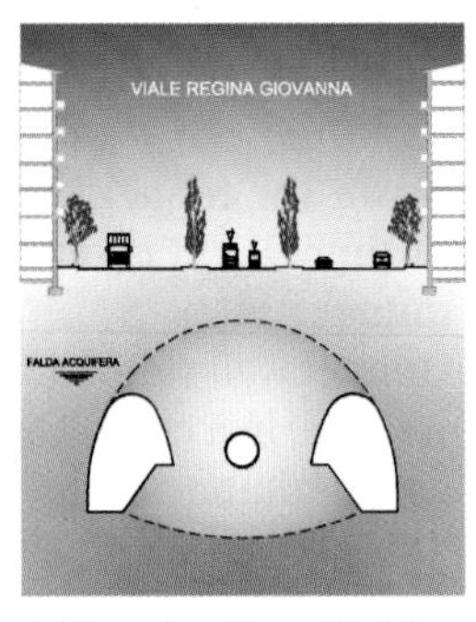

2 - Excavation of two side drifts for the sidewalls of the station tunnel

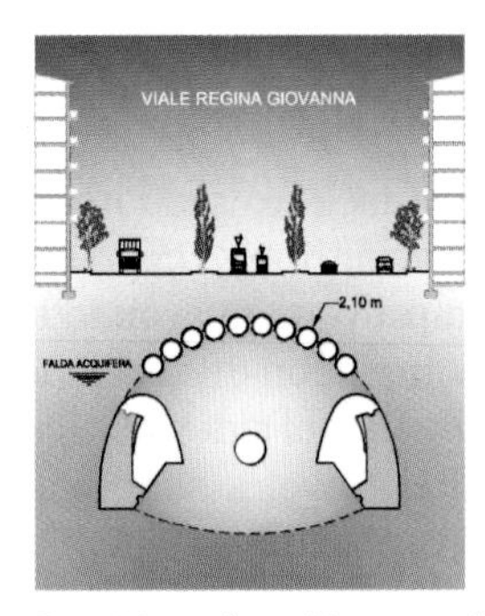

3 - Ten minitunnels are driven around the profile of the crown of the station tunnel and the sidewalls are cast

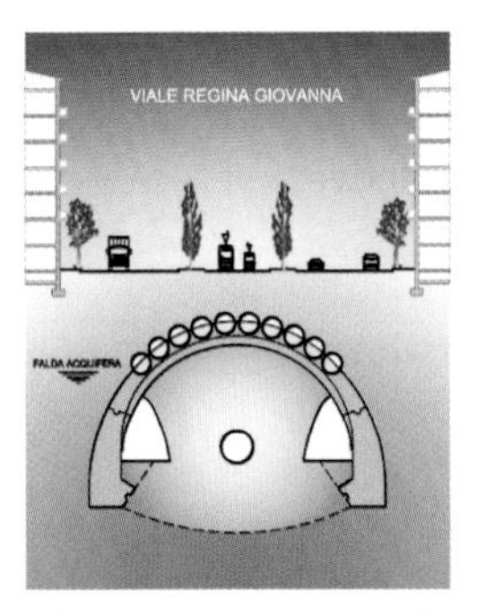

4 - The transverse arches are built working from inside the minitunnels

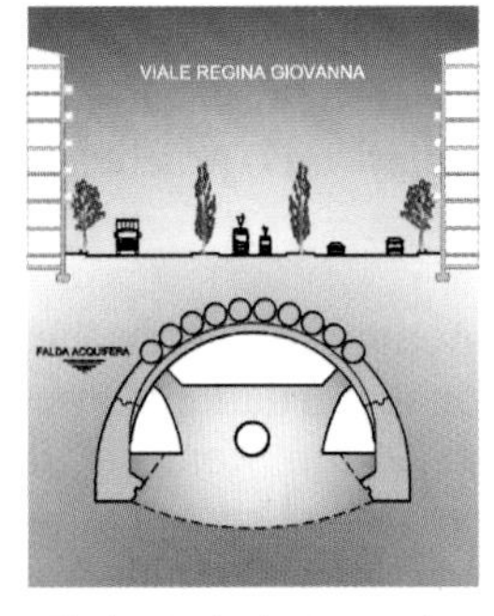

5 - The longitudinal minitunnels are cast and the station tunnel is excavated

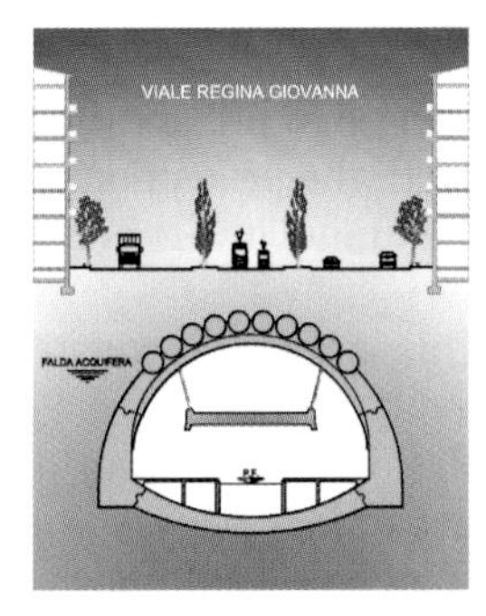

6 - The tunnel invert is cast and the mezzanine floors built with finishing works

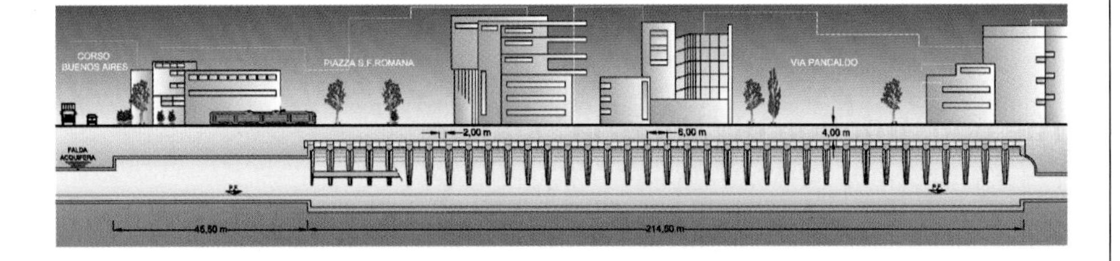

Fig. D.2

Venezia Station on the Milan urban link railway line: the construction of a sidewall in a side drift

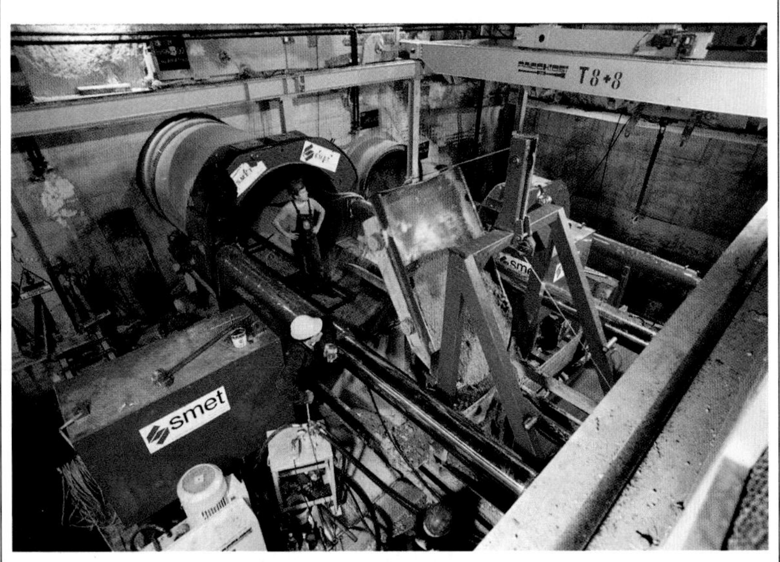

Venezia Station on the Milan urban link railway line: pipe jacking to drive minitunnels side by side around the perimeter of the roof arch of the station

Venezia Station on the Milan urban link railway line: a view of the thrust pit after the minitunnels have been driven

of only 4 metres beneath the foundations of 16th century buildings (Fig. D.3). The 440 m^2 cross section of the station tunnel is six times larger than that of normal twin track metropolitan railway running tunnels and almost twice the size of the second largest tunnel built to-date in Milan. To be able to build it with the same geometry would certainly have been impossible without the cellular arch technology with which the final lining was built before excavation began. It was driven by boring through recent and cohesionless soils. The overall diameter was approximately 30 metres and the overburden under the foundations of ancient buildings was only 4–5 metres.

Such a shallow overburden caused considerable doubt at the design stage over what the results might be if traditional methods were employed based on advance in steps after first improving the ground with sprayed concrete and steel ribs placed immediately. It would in fact have been impossible to improve the ground over the arch of the tunnel before excavation began to a degree that would have been sufficient for the large dimensions of the tunnel cross section. Numerical analyses carried out using the finite element method also showed that a conventional confinement structure of steel ribs and shotcrete would have been too deformable and would not have been able, even temporarily, to contain surface settlement within the limits required to safeguard nearby structures and utilities.

That is why the development of the cellular arch method was necessary.

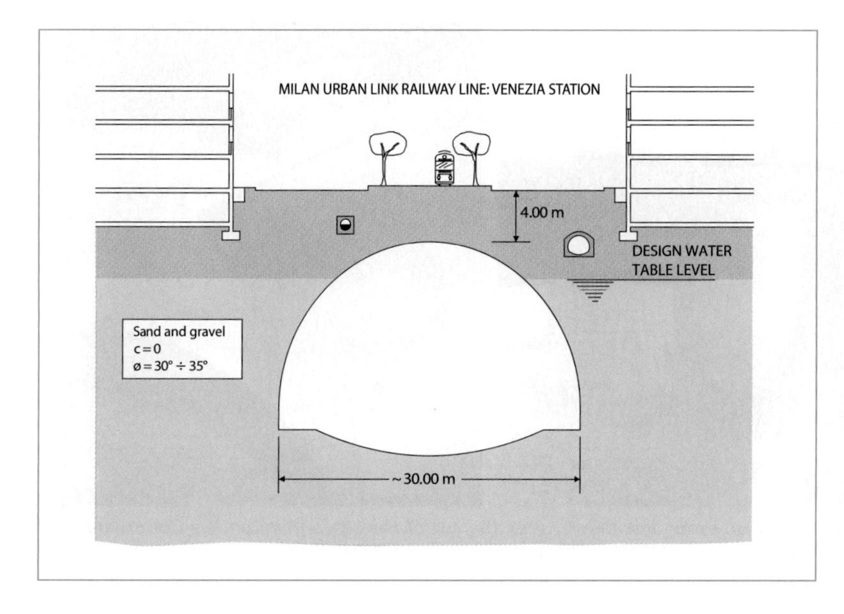

Fig. D.3
The overburden of the ground above the extrados of the tunnel was not sufficient to create a band of treated ground around it of adequate thickness

Venezia Station on the Milan urban link railway line: the cut of the pipes (minitunnels) for creating the transverse arches

The construction of the cellular arch structure for Venezia Station

From an operational point of view, the side drifts 60 m^2 (7.6 m in width and 11.0 m in height) being the same length as the future station tunnel were constructed in two stages:

1. excavation of the crown of 40 m^2 down to the water table;
2. ground improvement injections under the water table, from the floor of the first stage, under the future sidewalls of the tunnel and then subsequent deepening of the excavation down to the foot of the sidewalls.

The lining consisted of steel ribs, wire mesh reinforcement and shotcrete.

Once completed, the side walls of the future station tunnel were then cast inside them. Average production was approximately 2 metres per day of finished side wall, including the time taken for excavation work. It therefore took 11 months to complete the 430 metres of sidewall (215 m for each side), about the same time employed for pipe jacking (minitunnels) which was performed simultaneously from a completely independent site.

The design specification was for the excavation of ten minitunnels along the profile of the crown of the tunnel. This involved driving 1,080 pipes for a total length of about 2,160 m. The pipes were prefabricated, manufactured using the radial prestress system and high strength cement mixes, and had an outer diameter of 2,100 mm, an internal diameter of 1,800 mm and a length of 2 m.

They were pipe jacked into the ground from a thrust pit (Fig. D.4) using equipment consisting of a cylindrical metal shield with a diameter of 2,100 mm and a length of 7.7 m divided into two parts: the front part for cutting, which was jointed to allow the operator to adjust the vertical and horizontal movement, was fitted with a computer operated hydraulic cutter and a conveyor belt for mucking out and the rear part, 3.50 m long containing, motors, pumps and reservoirs for the hydraulic fluid. The thrust equipment included two hydraulic long stroke jacks, the indispensable load distribution structures and a hydraulic pump operating at a pressure of 600 bar.

Two sets of equipment were employed to obtain daily pipe advance rates of approximately 8–9 m per day. Topographic monitoring carried out during and after pipe jacking ensured and then confirmed that it was performed accurately with negligible deviations in direction and depth.

Once the side walls had been cast and all the pipes driven into place, construction of the load bearing cross members of the arch of the future tunnel started, undoubtedly the most characteristic part of the cellular arch technique.

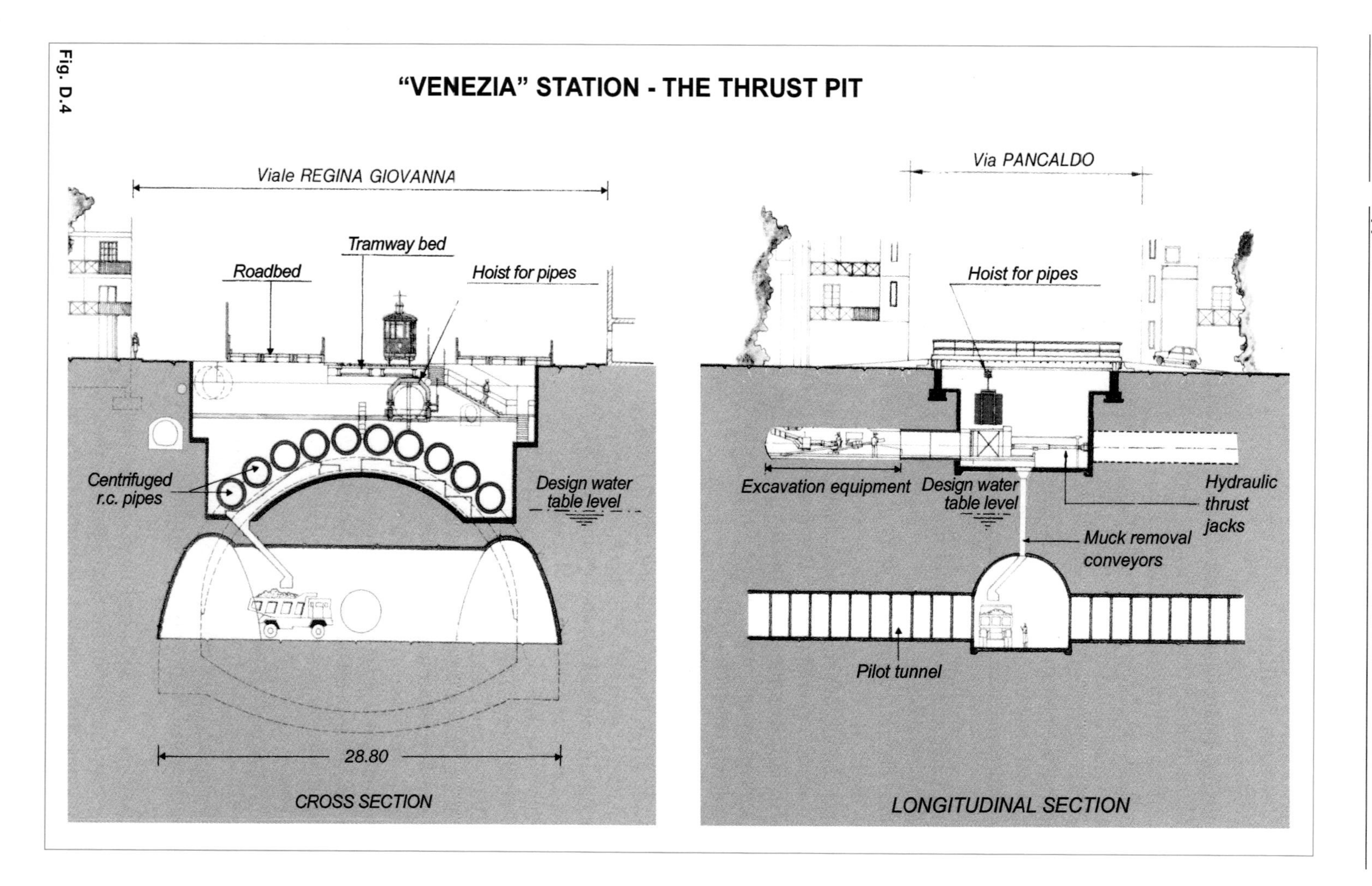
"VENEZIA" STATION - THE THRUST PIT
Viale REGINA GIOVANNA
Tramway bed
Roadbed
Hoist for pipes
Centrifuged r.c. pipes
Design water table level
28.80
CROSS SECTION
Via PANCALDO
Hoist for pipes
Excavation equipment
Design water table level
Hydraulic thrust jacks
Muck removal conveyors
Pilot tunnel
LONGITUDINAL SECTION

Fig. D.4

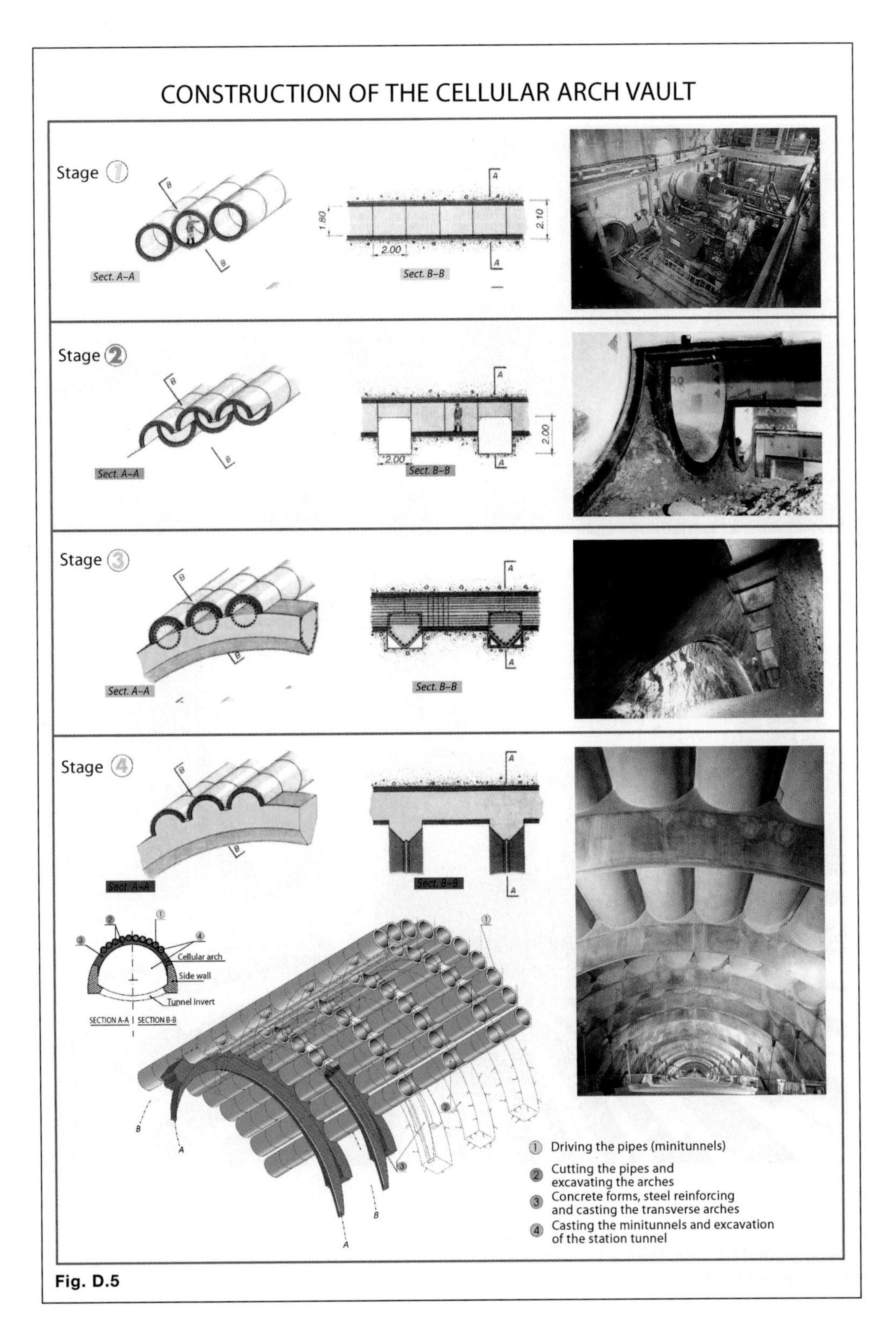
CONSTRUCTION OF THE CELLULAR ARCH VAULT
Stage 1
Sect. A–A
1.80
2.00
2.10
Sect. B–B
Stage 2
Sect. A–A
2.00
2.00
Sect. B–B
Stage 3
Sect. A–A
Sect. B–B
Stage 4
Sect. A–A
Sect. B–B
Cellular arch
Side wall
Tunnel invert
SECTION A-A | SECTION B-B
1 Driving the pipes (minitunnels)
2 Cutting the pipes and excavating the arches
3 Concrete forms, steel reinforcing and casting the transverse arches
4 Casting the minitunnels and excavation of the station tunnel

Fig. D.5

Venezia Station on the Milan urban link railway line: stages for the construction of the arches inside the transverse drifts

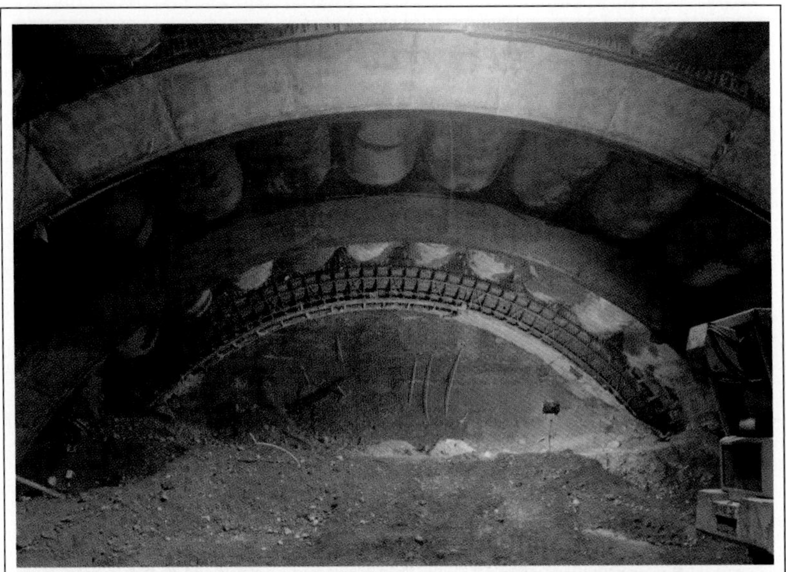

Venezia Station on the Milan urban link railway line: excavation of the station tunnel under the protection of the cellular arch already constructed. Note the presence of an arch with the formwork not yet removed

Venezia Station on the Milan urban link railway line: a view after half face excavation

Fig. D.6 *Venezia Station on the Milan urban link railway line: the full cross section*

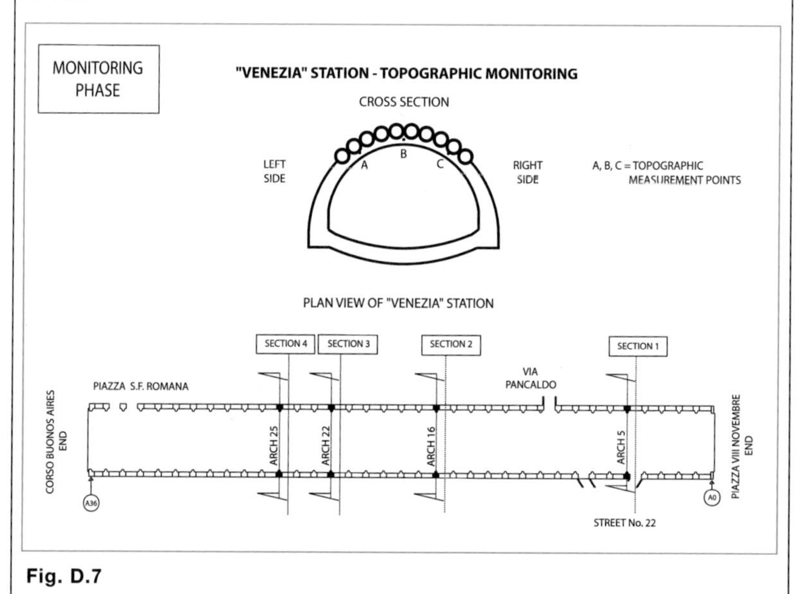

Fig. D.7

The cross members consisted of a series of 35 intermediate arches placed at intervals of 6.00 m plus the two end pieces. Construction was performed as follows (Fig. D.5):

1. cutting and removal of the part of the pipe intersected by the arches and excavation of the arches mainly by hand down to the tunnel wall side drifts;
2. assembly of the prefabricated steel forms inside the excavation, placing of the reinforcement for the cells (minitunnels) and the arches and casting of the latter on the side walls already in place;
3. casting of the cells.

Excavation in steps of the top heading in the crown and the bench of the large tunnel was then able to start in complete safety under the already active load bearing cellular arch structure. The lining was finally completed with an invert, varying from 1.5 to 2 m in thickness (Fig. D.6). It was cast in 5 m steps for a total length of 92 m with a total average cross section of 38 m^3/m. Each 5 m advance took a seven day working week to complete on average and work was co-ordinated so that the ring of the tunnel lining was left unclosed for three days only.

Finally, it is interesting to consider that the average overall production rate for the civil engineering works of Venezia station, constructed using the cellular arch technique, was 57 m^3 per day, for a final cost of approximately 516 euro/m^3, comparable to, if not less than, the current price of an ordinary one automobile garage in the centre of the city.

Monitoring system

The large dimensions of the cavity, the completely new construction method and the delicate surface constraints meant that a vast monitoring programme had to be designed and implemented to measure:

- settlement, at ground level, especially of buildings, during all phases of the works;
- deformation of the ground around the tunnels;
- stresses and strains in the final lining of the tunnel.

The programme included (Fig. D.7):

- topographic measurements;
- levelling, deflectometer and inclinometer measurements to monitor the development of deformation on existing buildings;
- convergence measurements in tunnels;
- pressure and deformation measurements on the lining structures.

Continuous recording and processing of the various measurements taken enabled the stress-strain conditions of the ground and the lining to be kept under control during the various stages of construction to provide a useful and constant comparison with both the design forecasts and the threshold limits within which existing structures will maintain their integrity.

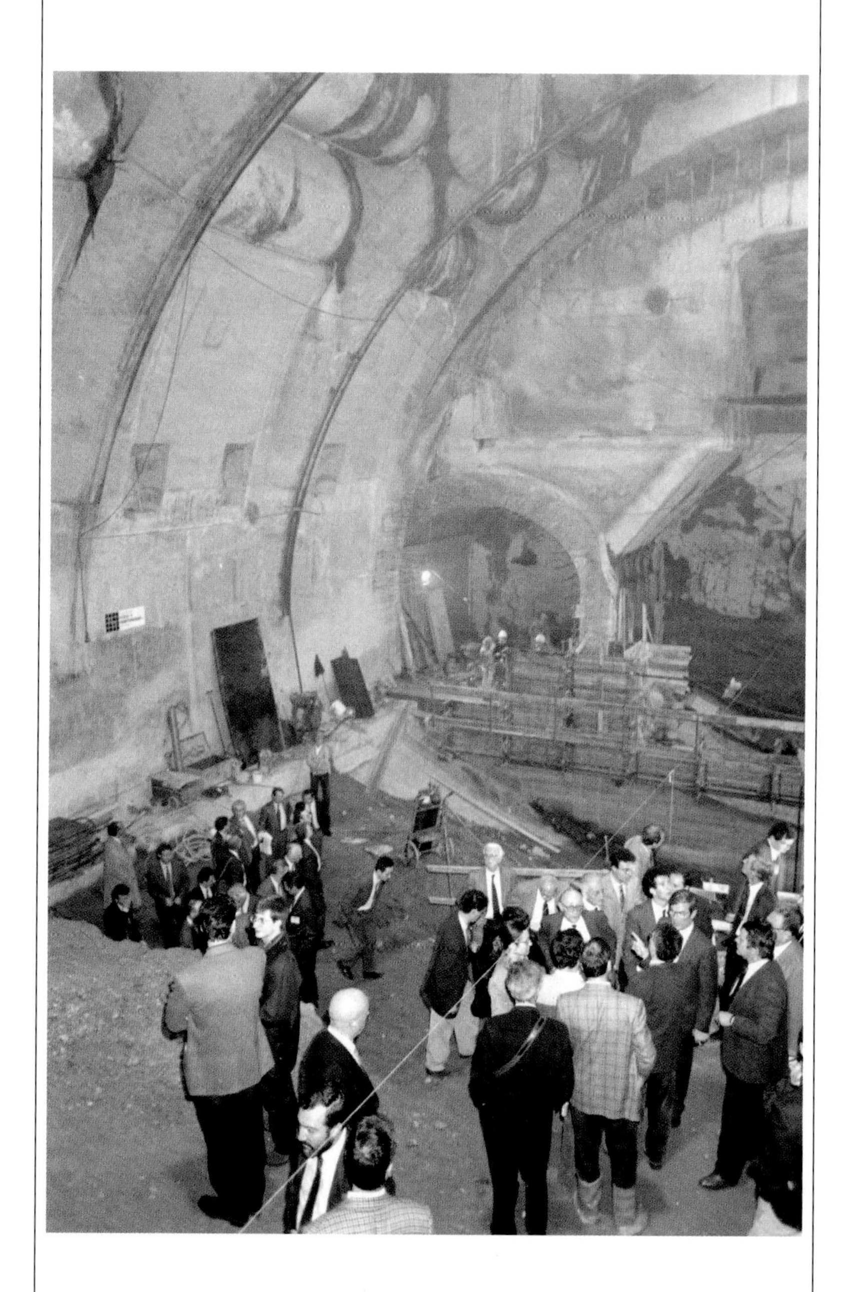

Venezia Station on the Milan urban link railway line: the construction site was visited by more than forty delegations of engineers from all over the world during construction work

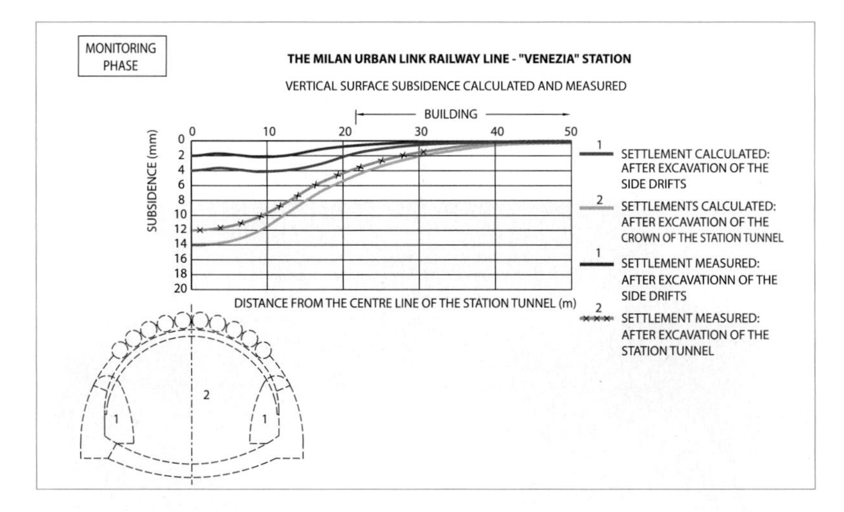

Fig. D.8

As Fig. D.8 shows, surface movements always remained below the calculated values. Of course, the greatest settlement occurred during excavation of the crown. The increase in deformation, slow at first and then more rapid as soon as the face passed the cross section measured, gradually died down as the face moved further away. This behaviour shown from datum points at street level above the route of the tunnel (Fig. D.9) was confirmed, though to a less marked extent, by those located on buildings for which maximum settlement during the passage of the face did not exceed 1–2 mm.

This monitoring system guaranteed constant surveillance over actual conditions demonstrating the compatibility between the efficiency of the construction system and the urban environment, to provide a reassuring overall picture.

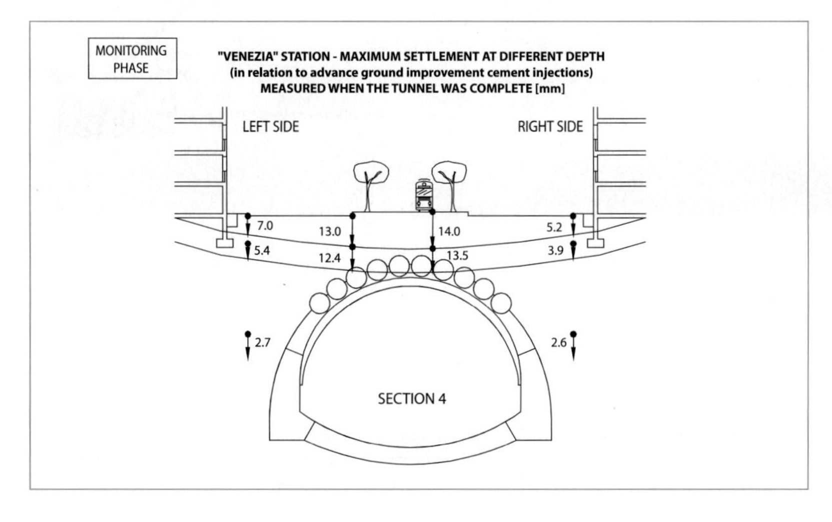

Fig. D.9

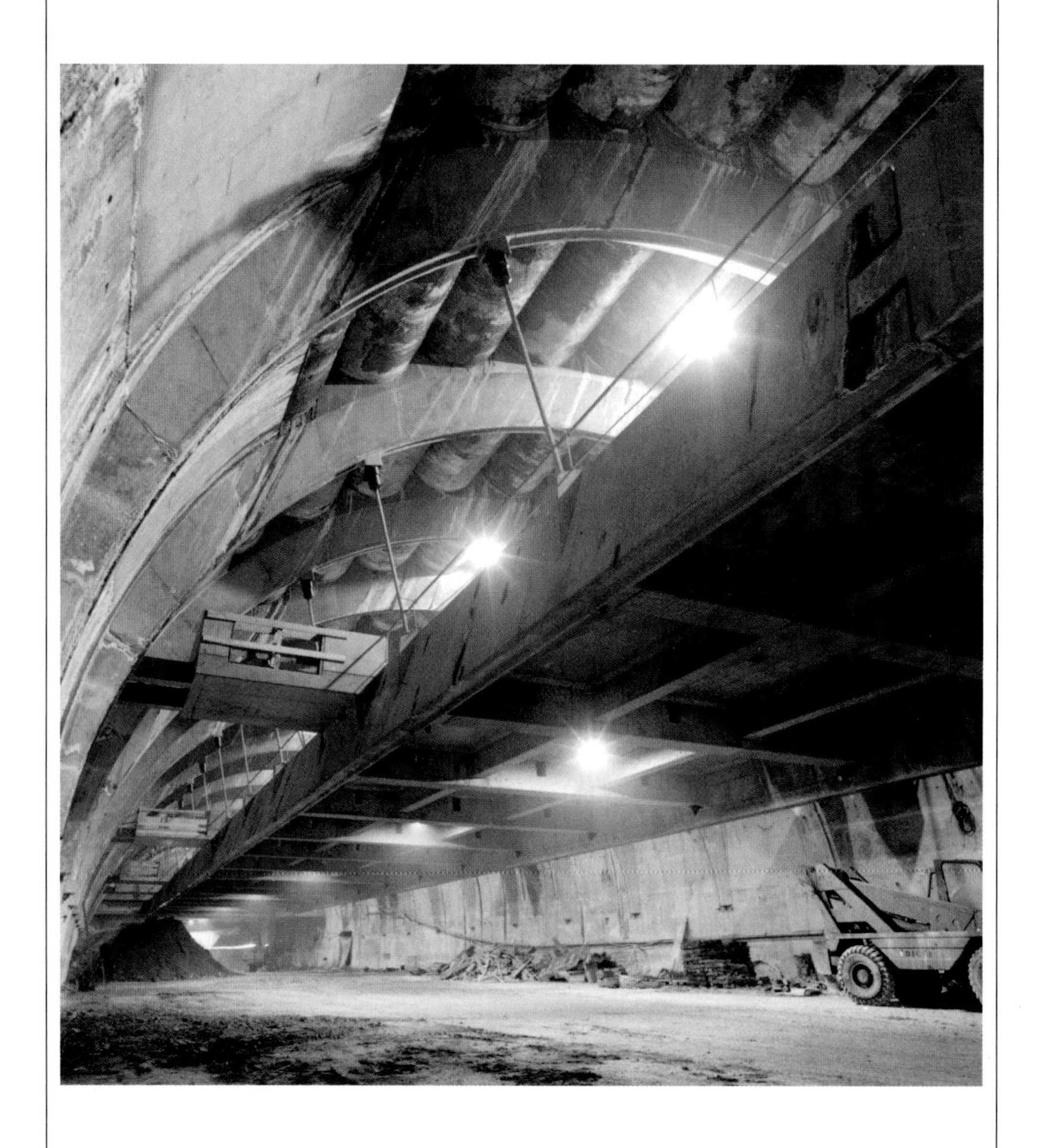

Venezia Station on the Milan urban link railway line: the mezzanine level

Possible developments of the system

Studies were performed to establish the limits of the cellular arch method for the construction of wide span tunnels in cohesionless soils, with shallow overburdens, under the water table.

Once a basic outline of the problem had been constructed and the variables determined, approximate dimensions of the main construction elements were calculated using a one dimensional finite element model to simulate the behaviour of the structure and its interaction with the ground. Three different geometries were considered with S:H ratios of 2.09, 1.73, and 1.5, with the span S varying up to 60 m.

The results of the calculation led to the production of tables giving the minimum thickness of the structural elements and surface settlement as a function of the geometry, the outer diameter, the depth of the water table and the size of the overburden (see the example in Fig. D.10). It would appear from the results that the cellular arch method could be employed with success on the bored tunnel construction of shallow tunnels with a span of over 60 m in cohesionless ground, under the water table without any significant surface settlement.

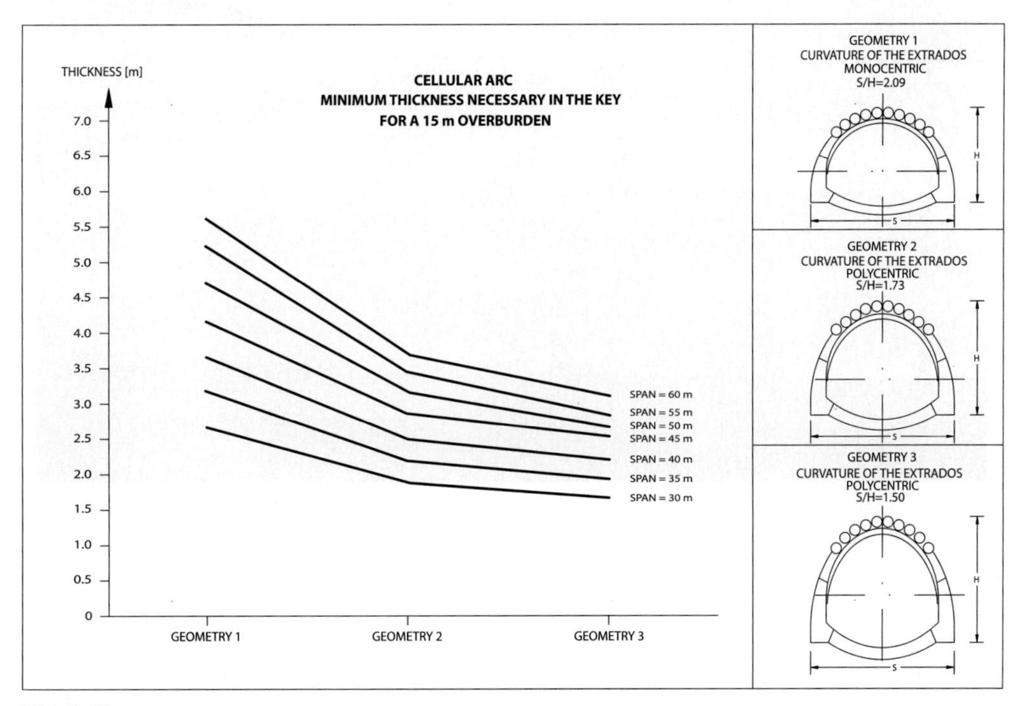

Fig. D.10

Venezia Station on the Milan urban link railway line: view of the mezzanine floor of the finished station

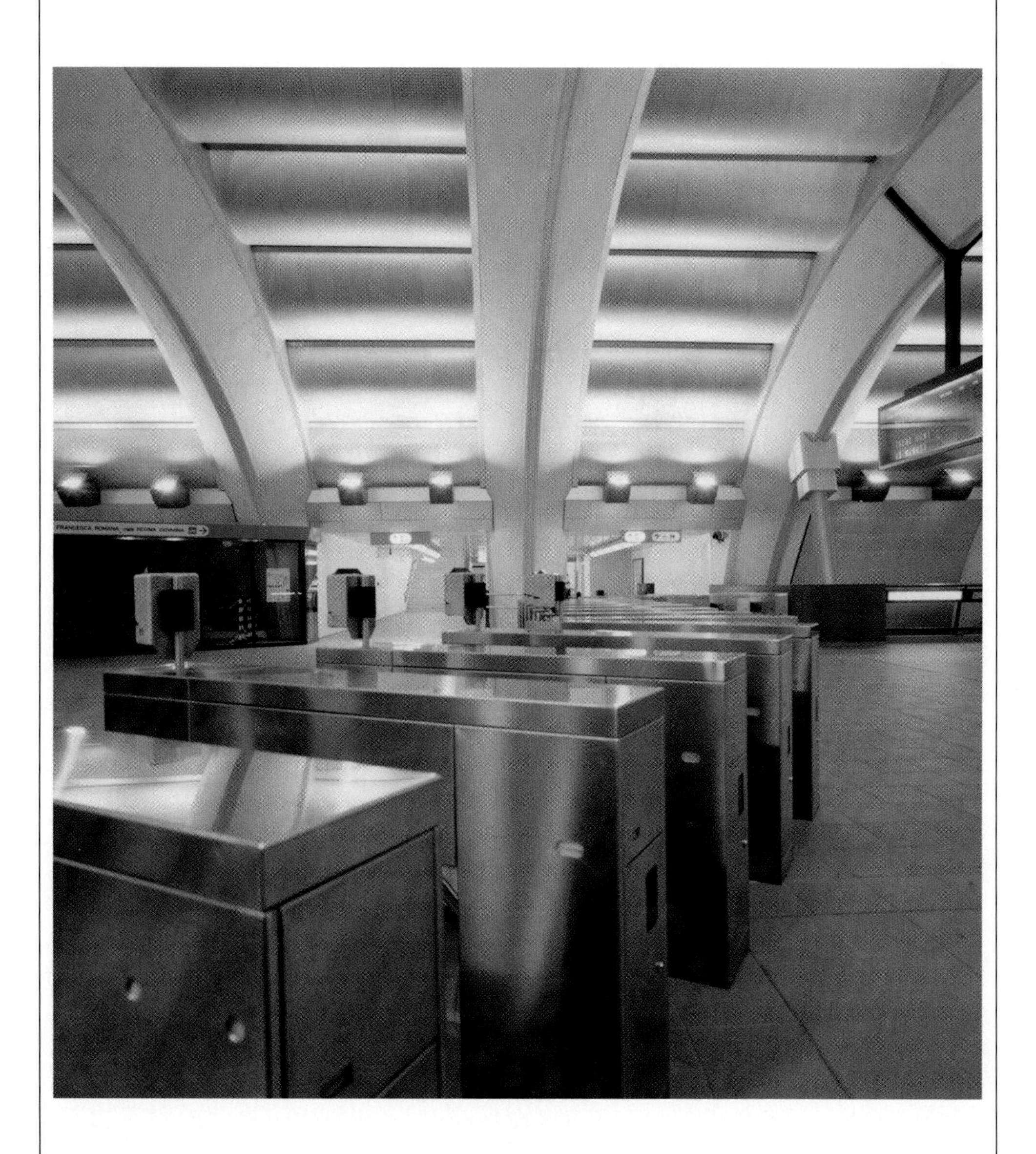

Venezia Station on the Milan urban link railway line: view of the cellular arch structure from the turnstiles

Artificial Ground Overburdens: the "Borgo Rinzelli" tunnel (high speed/high capacity railway line between Bologna and Florence), completely bored tunnel construction under A.G.O. with a very shallow overburden

Artificial Ground Overburdens (A.G.O.)

Some of the tunnels forming part of the works for the construction of the new high speed Rome-Naples railway line had very shallow overburdens because of their vertical location on the alignment and the gentle nature of the morphology. Consequently, the detailed design specified long sections of artificial tunnel. The tunnels concerned were Piccilli 1, Piccilli 2, Castagne, Santuario, Caianello and Briccelle. However, the construction of an artificial tunnel requires deep cuts to be made into the slopes to be crossed with consequent problems of:

- safety with regard to the stability of the slope itself;
- what to do with the huge volumes of material excavated;
- solving the problem of any interference there may be with existing surface structures;
- more difficult statics conditions if there is seismic activity;
- environmental and landscape impact.

These considerations, and also those of costs, led to the study of an alternative solution for the construction of these tunnels using direct bored tunnel methods, thereby avoiding all the problems which the execution of the original design would basically have involved.

Artificial Ground Overburdens

As is known the stability and therefore the long and short term existence of a tunnel is dependent on an arch effect being developed in the ground around the excavation by which the excess stresses, that are generated in the material as a result of excavation, are channelled and transmitted to the intact rock mass below it. In the case of a very shallow tunnel there is an insufficiently thick layer of ground above the crown for an arch effect to be generated naturally and consequently the design engineer must act to ensure that one is formed by taking appropriate construction measures. One conventional way of tackling the problem is, as has been said, to build an artificial tunnel. With this method, the arch effect is created by the concrete lining itself, which in the long term will be loaded with the weight of the backfill material.

Another system is to improve the ground around the cavity before driving the bored tunnel in order to give a sufficiently thick layer of it the indispensable strength it needs to channel the stresses. This path can only be taken if a minimum layer of ground exists over the crown of the tunnel to be driven on which to perform the improvement and, naturally, if the ground in question can be improved at a reasonable cost. The extremely shallow overburdens (see the longitudinal profile in Fig. E.1) in the case of the tunnels in question made this second path impossible, while the former solution was unattractive because of the difficulties mentioned in the introduction. An innovative solution was therefore studied which, by exploiting some of the particular characteristics of the pyroclastic soils in question, allowed the difficulties to be overcome, the tunnels to be driven using bored methods and the landscape and the environment to be respected.

The idea was triggered in the author's mind by considering not only the low unit weight ($\approx 1.3\,t/m^3$) and the interesting characteristics of the pyroclastic soils found on the sites concerned, but also the ease with which they can be mixed with lime to generate a product with reasonable strength. The method devised consisted of replacing the ground over the crown of the future tunnel with structural elements termed *Artificial Ground Overburdens (A.G.O.)*, obtained by using the ground itself after first treating it appropriately. Figure E.2 illustrates the various operational stages of the new construction methodology.

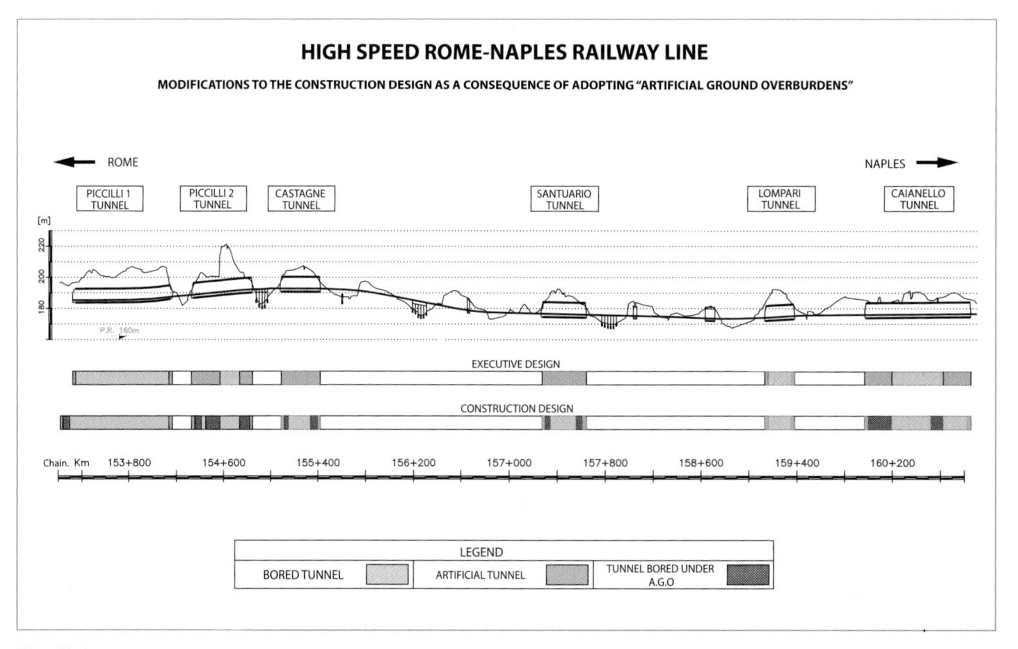

Fig. E.1

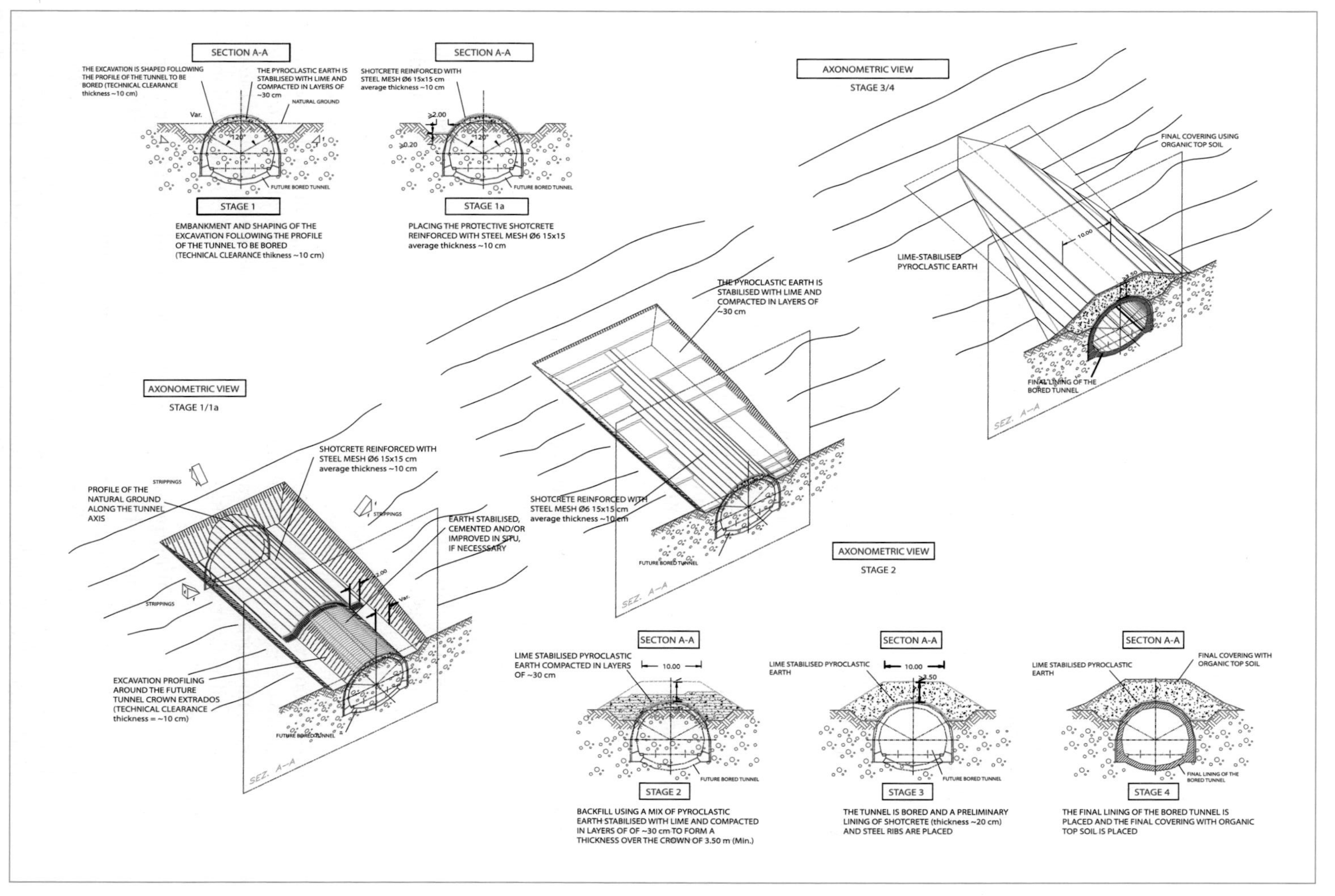

SECTION A-A
THE EXCAVATION IS SHAPED FOLLOWING THE PROFILE OF THE TUNNEL TO BE BORED (TECHNICAL CLEARANCE thickness ~10 cm)
THE PYROCLASTIC EARTH IS STABILISED WITH LIME AND COMPACTED IN LAYERS OF ~30 cm
NATURAL GROUND
Var.
FUTURE BORED TUNNEL
STAGE 1
EMBANKMENT AND SHAPING OF THE EXCAVATION FOLLOWING THE PROFILE OF THE TUNNEL TO BE BORED (TECHNICAL CLEARANCE thikness ~10 cm)
SECTION A-A
SHOTCRETE REINFORCED WITH STEEL MESH Ø6 15x15 cm average thickness ~10 cm
FUTURE BORED TUNNEL
STAGE 1a
PLACING THE PROTECTIVE SHOTCRETE REINFORCED WITH STEEL MESH Ø6 15x15 average thickness ~10 cm
AXONOMETRIC VIEW
STAGE 3/4
FINAL COVERING USING ORGANIC TOP SOIL
LIME-STABILISED PYROCLASTIC EARTH
FINAL LINING OF THE BORED TUNNEL
SEZ. A–A
THE PYROCLASTIC EARTH IS STABILISED WITH LIME AND COMPACTED IN LAYERS OF ~30 cm
AXONOMETRIC VIEW
STAGE 1/1a
SHOTCRETE REINFORCED WITH STEEL MESH Ø6 15x15 cm average thickness ~10 cm
PROFILE OF THE NATURAL GROUND ALONG THE TUNNEL AXIS
STRIPPINGS
EARTH STABILISED, CEMENTED AND/OR IMPROVED IN SITU, IF NECESSARY
EXCAVATION PROFILING AROUND THE FUTURE TUNNEL CROWN EXTRADOS (TECHNICAL CLEARANCE thickness = ~10 cm)
FUTURE BORED TUNNEL
SEZ. A–A
SHOTCRETE REINFORCED WITH STEEL MESH Ø6 15x15 cm average thickness ~10 cm
FUTURE BORED TUNNEL
SEZ. A–A
AXONOMETRIC VIEW
STAGE 2
SECTON A-A
LIME STABILISED PYROCLASTIC EARTH COMPACTED IN LAYERS OF ~30 cm
FUTURE BORED TUNNEL
STAGE 2
BACKFILL USING A MIX OF PYROCLASTIC EARTH STABILISED WITH LIME AND COMPACTED IN LAYERS OF OF ~30 cm TO FORM A THICKNESS OVER THE CROWN OF 3.50 m (Min.)
SECTON A-A
LIME STABILISED PYROCLASTIC EARTH
FUTURE BORED TUNNEL
STAGE 3
THE TUNNEL IS BORED AND A PRELIMINARY LINING OF SHOTCRETE (thickness ~20 cm) AND STEEL RIBS ARE PLACED
SECTON A-A
LIME STABILISED PYROCLASTIC EARTH
FINAL COVERING WITH ORGANIC TOP SOIL
FINAL LINING OF THE BORED TUNNEL
STAGE 4
THE FINAL LINING OF THE BORED TUNNEL IS PLACED AND THE FINAL COVERING WITH ORGANIC TOP SOIL IS PLACED

Fig. E.2

After first having removed the surface layer of ground present over the tunnel to be constructed, following the profile of the crown, with a technical clearance of 10 cm, down to the springline, according to the geometry shown in the figure, a 10 cm layer of steel mesh (6 Ø 15 × 15 cm) reinforced shotcrete is then sprayed on the floor of the excavation shaped in this way with the function of:

- shaping the future tunnel;
- distributing the future loads that will weigh on the crown.

At this point it is filled in and embanked with the same ground previously removed, after adding 3–6% of lime to increase its strength, until the whole of the crown of the future tunnel is covered in individually compacted 30 cm layers to a thickness of at least 3.5 m.

Bored tunnel excavation of the tunnel can now start.

The application of the method on the tunnels of the new Rome-Naples railway line

The application of new methods of driving tunnels always requires very thorough preliminary studies to assess whether it is genuinely feasible and compatible with the specific local conditions. In the case of the six tunnels on the new Rome-Naples railway line, these studies regarded primarily the soils found in the ground on the sites in question to test and optimise the effect of the treatment with lime with which it was intended to improve them.

The geology, the survey campaign and experimentation

From a geological viewpoint the route of the tunnels borders on the Roccamonfina volcano system and only marginally intersects the lava deposits, running almost entirely through eruption material. The various lithotypes consist of successions of stratified pyroclastites (tuffs), sub-horizontally positioned, with granulometry varying from sandy silt with inclusions of gravel to coarse sand, to give a morphology characterised by undulating hills with valley depressions originated by seasonal streams.

A geological survey campaign was conducted consisting of continuous core bore sampling and exploration trenches (one every 7–10 m) to assess the quality of the weathered material to improve before placing the A.G.O. The main geophysical characteristics of the material were measured in the laboratory in view of the subsequent study of the treatment with lime: unit weight and granulometry and modified ASSHTO compaction tests were also performed.

Samples of ground were then prepared mixed with lime in percentages varying from 3% to 5% using the optimum humidity found with the modified AASHTO test. They were then left to harden in a saturated water vapour environment.

Compressive strength tests were then performed after 3, 7 and 28 days of hardening (Figs. E.3 and E.4), which gave average results of higher than 200 t/m^2, 250 t/m^2, 320 t/m^2 (2 MPa, 2.5 MPa, 3.2 MPa) respectively. Compaction tests did not give results significantly different from those performed before the treatment.

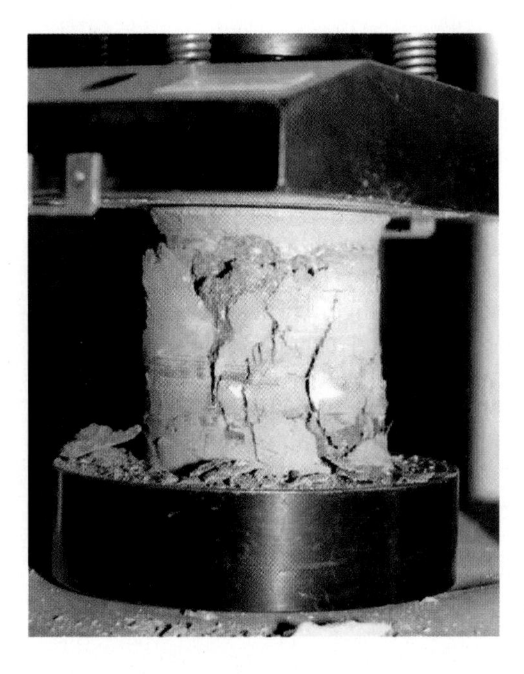

Fig. E.3
Test piece subjected to mono-axial compression test

The optimum percentage of lime identified by the experiments was 3–4%.

Full scale field trials were also performed to check the results obtained in the laboratory. The results confirmed expectations.

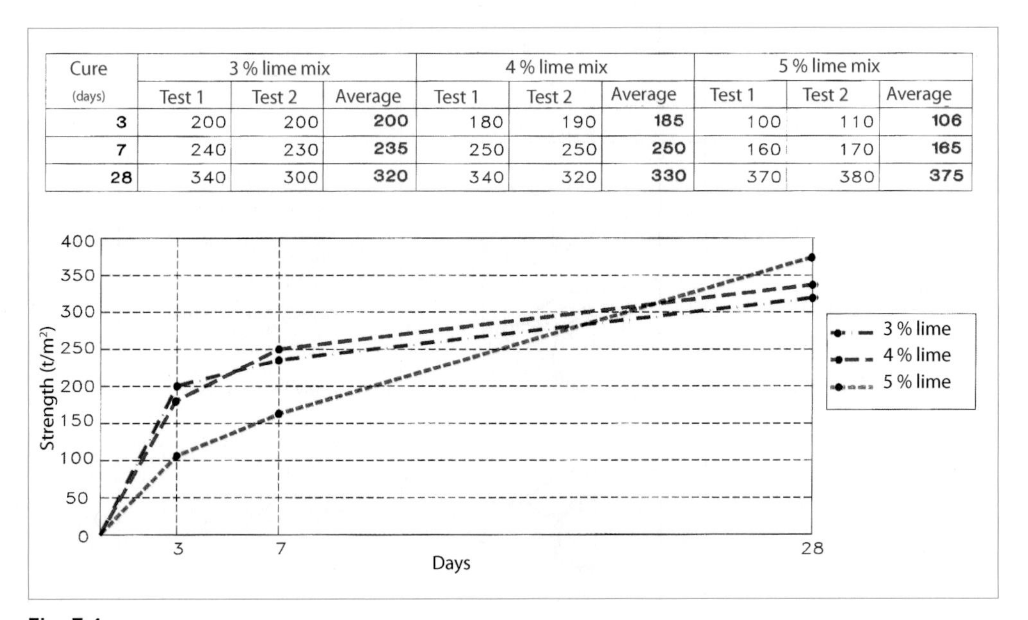

Cure (days)	3 % lime mix			4 % lime mix			5 % lime mix		
	Test 1	Test 2	Average	Test 1	Test 2	Average	Test 1	Test 2	Average
3	200	200	**200**	180	190	**185**	100	110	**106**
7	240	230	**235**	250	250	**250**	160	170	**165**
28	340	300	**320**	340	320	**330**	370	380	**375**

Fig. E.4

Statics tests

The feasibility of the new solution and the reliability of the statics were tested by means of finite element numerical analysis conducted in the non linear field using version 6.0 of the ADINA software application.

A typical cross section of the running tunnel was considered for this purpose (with final lining in r.c., 80 cm thick in the crown and 90 cm in the invert), constructed with isoparametric, plane strain, four node elements.

An elasto-plastic stress-strain response model according to Drucker-Prager was chosen for those representing both the natural and the treated ground, while a linear elastic model was chosen for those representing the preliminary and final linings.

As the analysis was intended to study the stress-strain behaviour of the entire ground-structure system at the different stages of construction and after commissioning, seven calculation 'times' were performed (see Fig. E.5) to model the succession of those stages and the implementation of stabilisation intervention during construction work as realistically as possible.

STAGES OF THE FINITE ELEMENT CALCULATION

TIME	MODEL	DESCRIPTION
1	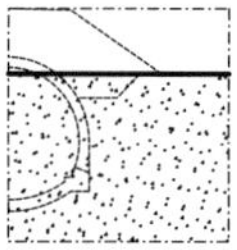	CALCULATION OF THE GEOSTATICS
2	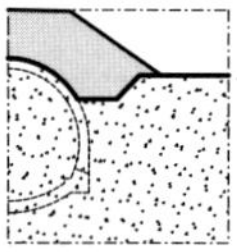	IMPROVEMENT AND BACKFILL WITH PYROCLASTITES TREATED WITH LIME
3	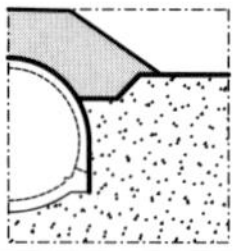	THE TUNNEL IS DRIVEN FULL FACE AND THE PRELIMINARY LINING OF SHOTCRETE AND STEEL RIBS IS PLACED RELAXATION OF THE CORE 70%
4	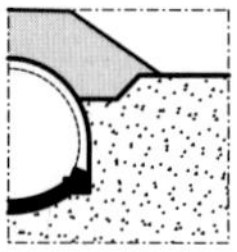	THE TUNNEL INVERT AND KICKERS ARE CAST RELAXATION OF THE CORE: 80%
5	...	CURING OF THE TUNNEL INVERT RELAXATION OF THE CORE: 90%
6	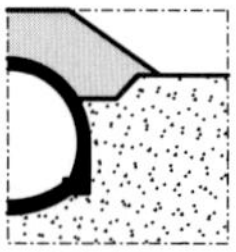	THE TUNNEL ARCH IS CAST AND THE GEOMECHANICAL CHARACTERISTICS OF THE GROUND TREATED WITH LIME FALL RELAXATION OF THE CORE: 100%
7	...	SEISMIC ACTION

Fig. E.5

The progressive demolition of the face due to tunnel advance was simulated in the model by progressively reducing the strength and deformability characteristics of the elements representing the material excavated to zero.

The calculation parameters adopted for the natural and the treated ground are summarised in table E.I. Extremely low long term (service phase) values for the strength and deformability parameters of the treated ground were chosen to err on the side of safety, while 2.00 t/m³ was assumed for unit weight of both (natural ground and treated ground) was assumed. Finally an extra live load of 2.00 t/m² was considered on the summit of the model.

Table E.I

SOILS	γ (t/m³)	E (t/m²)	ν	c (t/m²)	φ (°)
Weathered surface ground	1.8	7,000	0.30	2.00	30
Pyroclastites	1.8	20,000	0.30	10.00	30
Lime treated Pyroclastites (*time* 2–5)	2.0	40,000	0.30	20.00	30
Lime treated Pyroclastites (*time* 6–7)	2.0	7,000	0.30	0	30

Results of the statics tests

The results of the calculation confirmed that bored tunnel full face advance was possible under the protection of the artificial ground treated with lime, by employing the geometry and construction methods illustrated.

From the viewpoint of deformation, they gave very low values for convergence (less than 2 mm), while the role played by the treatment in the crown of the tunnel before excavation started was very important for stress. Thanks to its arched shape the improved ground was subject to contained compressive stress action only (σ_{max} = 12.9 t/m² at *time* 5), which were appropriately channelled around the tunnel and transmitted to the natural ground on the sides of the tunnel (Fig. E.6).

Similarly the preliminary lining in shotcrete was found to be fully compressed ($\sigma_{average} \leq 210$ t/m²). The final lining in r.c. with the ring closed with a tunnel invert tested well even without reinforcement (which was placed anyway, in case of potential asymmetry or local anomalies) since it was subject to extremely low tensile stress (≤ 18 t/m²) in the crown and at the joint with the tunnel invert and to maximum compressive strength of less than 220 t/m².

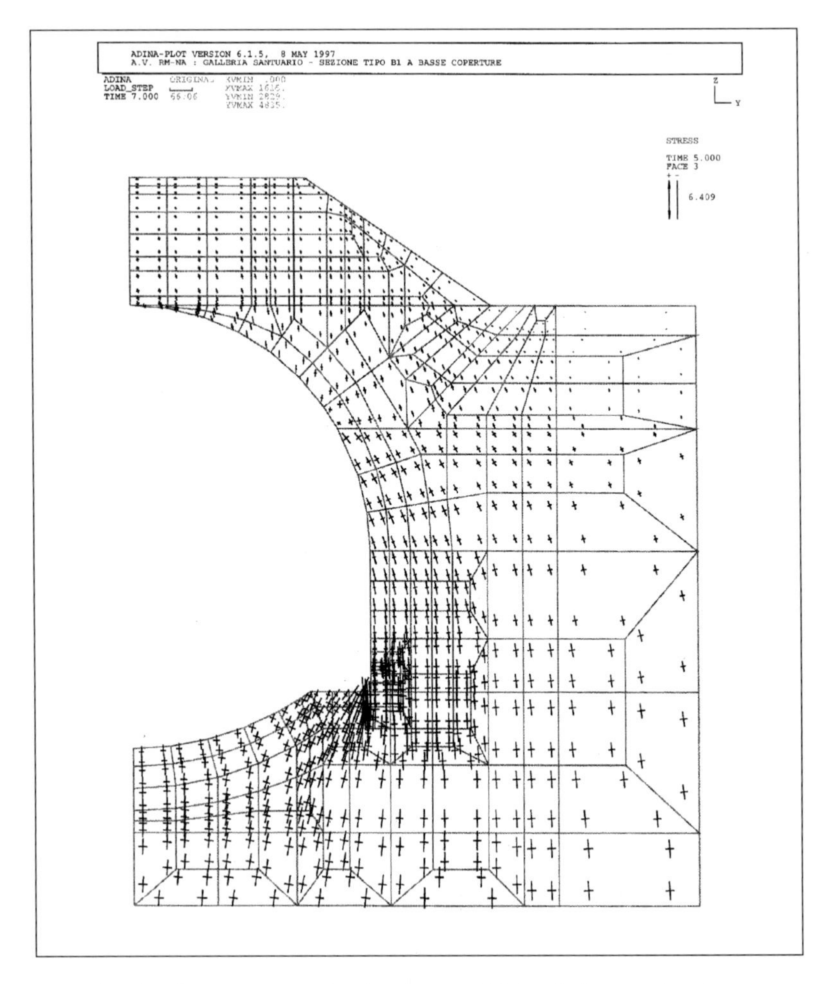

Fig. E.6 *High Speed Rome-Naples: Santuario Tunnel – Section type B1 for shallow overburdens*

Construction

Once it had been ascertained that the solution studied was feasible and that the statics were reliable, experimental implementation began following the procedures and phases already described and illustrated in figures E.2 and E.6.

Work started from the Picilli 2 tunnel where bored tunnel advance was performed under the protection of the A.G.O. for more than 250 m. The photographs in Figs. E.7 and E.8 illustrate some of the different stages of the work.

Two different procedures were employed on site to mix the tuff with lime:

- a plant with a hopper was used to store the lime (Fig. E.9) and the mix was prepared at the plant; it was then transported to the construction site, spread in layers of not more than 30 cm and rolled at optimum humidity, or
- the pozzolana spread and covered with an adequate quantity of lime was milled and then rolled as above.

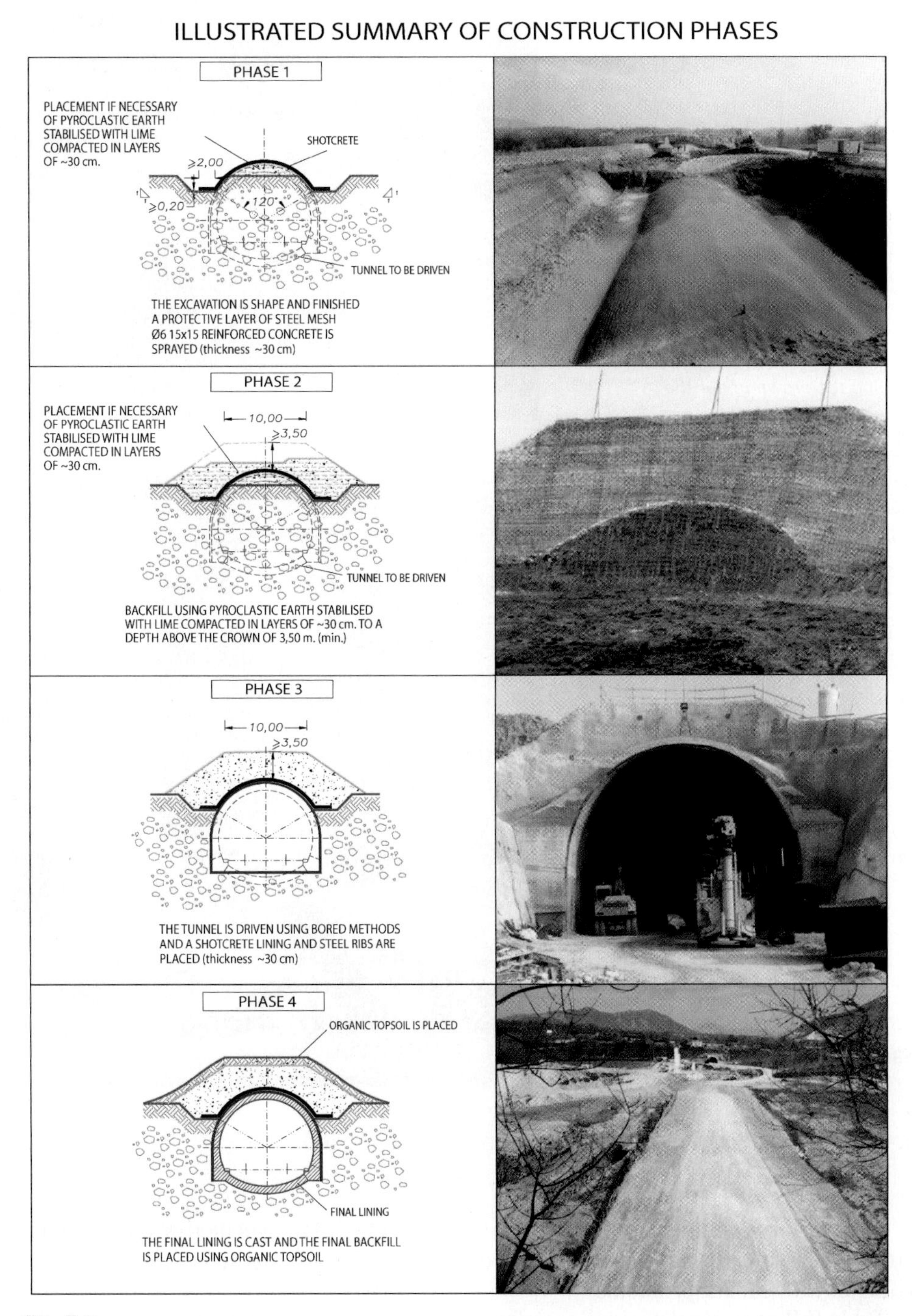
ILLUSTRATED SUMMARY OF CONSTRUCTION PHASES
PHASE 1
PLACEMENT IF NECESSARY OF PYROCLASTIC EARTH STABILISED WITH LIME COMPACTED IN LAYERS OF ~30 cm.
SHOTCRETE
≥2,00
≥0,20
120°
TUNNEL TO BE DRIVEN
THE EXCAVATION IS SHAPE AND FINISHED A PROTECTIVE LAYER OF STEEL MESH Ø6 15x15 REINFORCED CONCRETE IS SPRAYED (thickness ~30 cm)
PHASE 2
PLACEMENT IF NECESSARY OF PYROCLASTIC EARTH STABILISED WITH LIME COMPACTED IN LAYERS OF ~30 cm.
10,00
≥3,50
TUNNEL TO BE DRIVEN
BACKFILL USING PYROCLASTIC EARTH STABILISED WITH LIME COMPACTED IN LAYERS OF ~30 cm. TO A DEPTH ABOVE THE CROWN OF 3,50 m. (min.)
PHASE 3
10,00
≥3,50
THE TUNNEL IS DRIVEN USING BORED METHODS AND A SHOTCRETE LINING AND STEEL RIBS ARE PLACED (thickness ~30 cm)
PHASE 4
ORGANIC TOPSOIL IS PLACED
FINAL LINING
THE FINAL LINING IS CAST AND THE FINAL BACKFILL IS PLACED USING ORGANIC TOPSOIL

Fig. E.7

Fig. E.8 *Bored tunnel advance: shaping the crown under the protection of the A.G.O.*

Fig. E.9 *Plant for mixing the pyroclastites ground with lime*

The work continued smoothly for the whole of the 251 m specified without any problems at all arising. It was therefore decided to extend the new method to the other tunnels of the lot with similar problems.

Table E.II contains data on the application of the A.G.O. method. It can be seen that a total of 1,346 metres of tunnel have been driven to advantage using bored tunnel methods since the first experimentation thanks to this system.

Table E.II

TUNNEL	TOTAL LENGTH OF THE TUNNEL [m]	LENGTH OF TUNNEL DRIVEN UNDER A.G.O. [m]
Piccilli 1 (State Railway, High Speed Rome-Naples)	907	58
Piccilli 2 (State Railway, High Speed Rome-Naples)	485	251
Castagne (State Railway, High Speed Rome-Naples)	289	73
Santuario (State Railway, High Speed Rome-Naples)	322	80
Caianello (State Railway, High Speed Rome-Naples)	832	363
Sadurano (State Railway, High Speed Bologna-Florence)	3767	68
Borgo Rinzelli (State Railway, High Speed Bologna-Florence)	528	73
Morticine (State Railway, High Speed Bologna-Florence)	654	380
Total length driver under A.G.O.		**1346**

Monitoring and measurements during construction

Design of the A.G.O. required a series of measurements to be taken during construction to ensure that its performance matched the design predictions. More specifically:

- for the pozzolana-lime mix:
  - average compressive strength after 7 days of hardening must not be less than $80\,t/m^2$;
  - the *in situ* density achieved must not fall below 95% of the maximum density achieved in the laboratory;
- for the shotcrete:
  - average compressive strength after 48 hours must not be less than $1300\,t/m^2$;
  - after 28 days it must not be less than $2000\,t/m^2$.

In the case of the Piccilli 2 tunnel samples were taken systematically during construction, which always gave results above the minimum specified limits when tested. Average strength of $300\,t/m^2$ was measured for the pozzolana-lime mix, similar to the results obtained in the laboratory.

In addition to the tests for quality control of the materials, various load measurements were taken by positioning load cells at the level of the removed ground at around the height of the springline of the future tunnel and also at the foot of the steel ribs.

Piccilli 1 Tunnel: treatment of the ground with lime

Piccilli 1 Tunnel: shaping the extrados of the future tunnel

The former gave average readings of $2\,t/m^2$, taken, however, before driving the bored tunnel and therefore of little significance, while the latter gave average readings of $30\,t/m^2$.

Tunnel convergence was also monitored systematically: the average readings obtained from the five-nail stations fluctuated between 1 and 2 mm, and were therefore very close to those predicted by the FEM calculations.

Final considerations

The construction of artificial tunnels for sections of tunnel without the necessary overburden above the future crown of a tunnel needed for bored tunnel technology, inevitably requires deep cuts to be made in the slopes to be crossed with consequent problems of safety, environmental impact, etc.

The alternative solution, illustrated here, using A.G.O. allows bored tunnel technology to be employed, thereby avoiding all the problems connected with the construction of artificial tunnels.

This new solution, initially studied and implemented for the construction of tunnels on the new Rome-Naples high speed railway line, was found to be extremely practical, safe and also advantageous economically (Fig. E.10), as well as from an environmental and landscape viewpoint. It has been applied with considerable success on various other tunnels suffering from similar problems due to insufficient overburdens.

COMPARISON OF CONSTRUCTION COSTS AND TIMES FOR BORED TUNNELS DRIVEN THROUGH ARTIFICIAL GROUND OVERBURDEN (A.G.O.) AND FOR THE SAME TUNNELS CONSTRUCTED ARTIFICIALLY

WORK*	ARTIFICIAL TUNNEL €/m	BORED TUNNEL UNDER A.G.O. €/m	DIFFERENCE
EARTH REMOVAL, GROUND IMPROVEMENT, TUNNEL EXCAVATION	2,464.00	4,450.00	-1,986.00
BACKFILL	915.00	0.00	915.00
LINING OF CROWN, KICKERS AND SIDEWALLS	3,421.00	2,383.00	1,038.00
TUNNEL INVERT	1,664.00	790.00	874.00
TOTALS	8,464.00	7,623.00	841.00

*ONLY THOSE ITEMS OF THE TWO CONSTRUCTION METHODS WHICH ARE DIFFERENT

TIME EMPLOYED FOR THE CAIANELLO TUNNEL (FOR 365 m) WITH THE SAME MEANS EMPLOYED

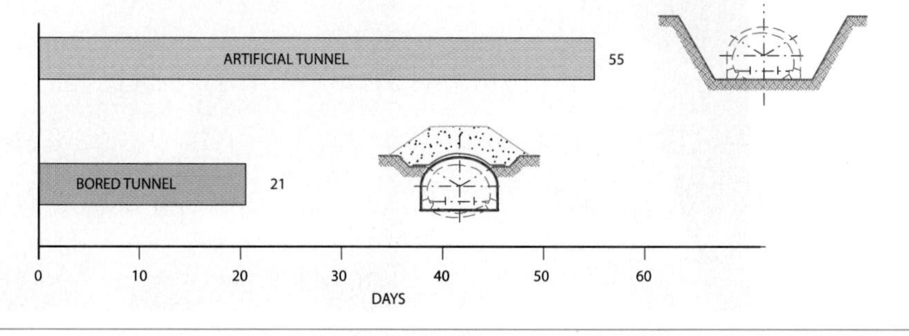

Fig. E.10

Sadurano Tunnel (high speed/capacity Bologna-Florence railway line): view of the 'sprayed' crown of the future bored tunnel

Piccilli 2 Tunnel: backfill with stabilised ground and compaction

Piccilli Tunnel 1 (above) and 2 (below): two stages in the construction of the A.G.O.

Portals in difficult grounds: aerial view of a portal constructed using vertical jet-grouting technology for a tunnel on the Sibari-Cosenza railway line (1985, ground: clay)

Portals in difficult ground

The construction of tunnel portals often involves solving problems that are closely connected with the morphology of the slope to be entered, with the existence of nearby constructions, with the geometry of the structures to be constructed and the type of material involved. The preparation of a portal site and of the wall to be excavated, very frequently requires substantial cuttings which are of no particular concern when working in rock, but are very problematic when working in soft soils, especially if they are loose.

If the decompression caused by excavation in ground with little cohesion is not adequately confined, there is a risk that it will easily and rapidly propagate into the medium with serious effects for the whole slope. The only way the entrance to the portal of a tunnel can be excavated without decompressing the ground is clearly by creating a structure to confine (or better preconfine) the ground in advance ahead of the future excavation which is capable of conserving the existing natural equilibriums. Until just a few years ago, the only systems available for achieving this goal were to place the following structures at the entrance of the bored tunnel:

- large diameter pile walls, anchored if necessary;
- "Berlin" or soldier pile walls;
- r.c. diaphragm walls.

The construction of large diameter piles on slopes tending towards instability is, however, often difficult, if not impossible, because, it is often the case that the morphology of the slopes does not allow the use of the heavy operating machinery required.

On the other hand the lability of Berlin type structures themselves, which rely to a large extent on anchors which reach into stable zones of the ground for their effectiveness, mean that these systems are not always sufficiently reliable.

Not even r.c. diaphragm walls are sufficient for the most delicate situations: earth removal and the introduction of water have the effect of reducing the shear strength of the ground with consequences for which there is no remedy in some situations.

These problems were then made worse because the lack of adequate techniques to improve the ground in advance and of suitable operating systems, required bored tunnels to be driven with overburdens measurable in terms of tunnel radii. This required huge portions of the ground to be removed with the risk of triggering ground decompression that is very difficult to confine, as has been demonstrated by the

A very deep cut is made into the slope for a tunnel portal in semi-cohesive soils using conventional systems (anchored Berlin wall; overburden: twice the diameter of the tunnel)

The same portal as the one above, after it had collapsed because of the decompression induced in the slope

Fig. F.1

many cases of failure seen in the past (see Fig. F.1 and the photographs on the page opposite).

The availability of new jet-grouting systems for improving ground, introduced in Italy around twenty years ago, allowed the author to invent and experiment an innovative solution to create preconfinement structures in loose or poorly cohesive soils with properties that would overcome the problems that have been described.

The basic concepts are explained below, together with a few of the most significant case histories.

Shells of improved ground formed by means of jet-grouting

The idea consists of creating a confinement shell, before excavation for the portal commences, consisting of rows of columns of ground improved by means of jet-grouting, which geometrically enfold the section of artificial tunnel (see Figs. F.2 and F.3).

The magnitude and distribution of the treatment are decided on the basis of each specific operating and geotechnical context. A top beam in r.c. joins the tops of the columns to help make it a single rigid structure.

TUNNEL PORTALS WITH SHELLS OF IMPROVED GROUND BY MEANS OF VERTICAL JET-GROUTING

Construction stages

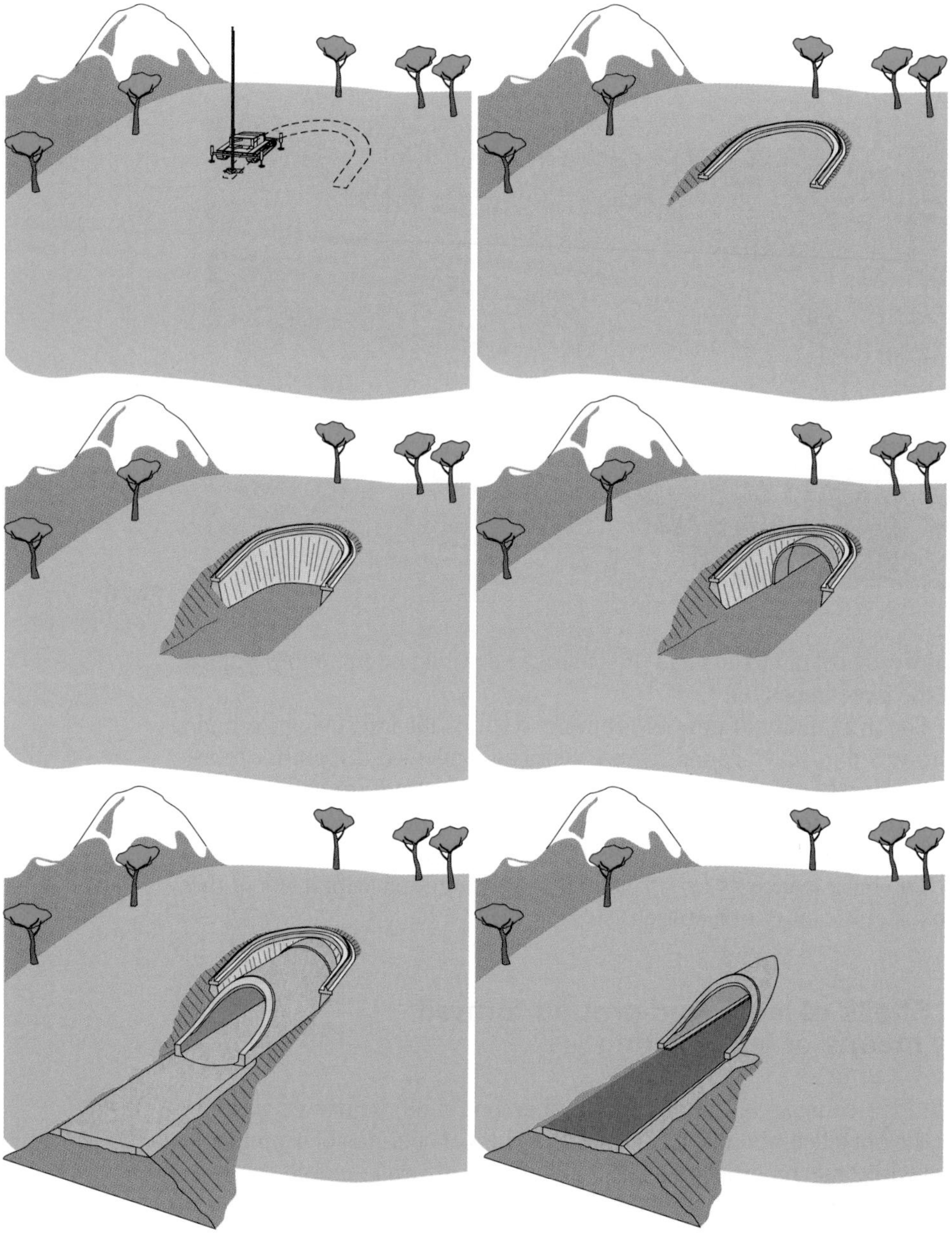

Fig. F.2

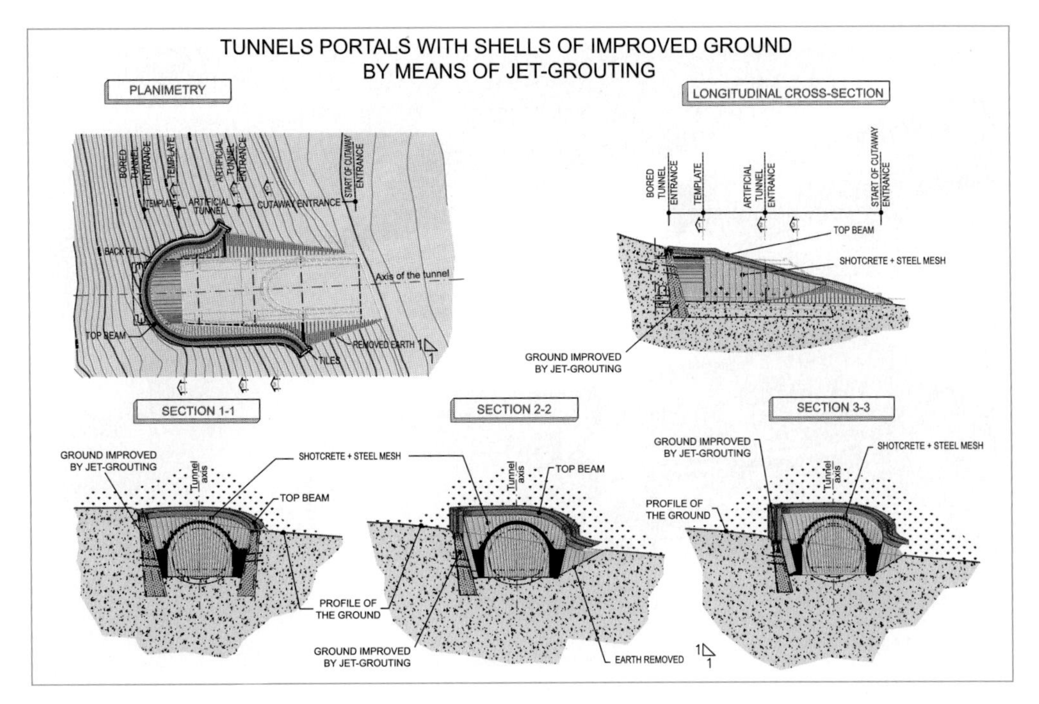

Fig. F.3

Once the earth has been removed for the entrance to the portal, the work is completed with:

- a layer of shotcrete sprayed on the whole surface of the exposed wall;
- drainage pipes placed sub-horizontally through the improved ground to prevent the formation of heads of water behind it.

The structure thus built functions by means of the sub-horizontal arches and is subject mainly to compressive and shear stress.

The creation of shells of this type is strictly dependent on the subsequent tunnel being driven using horizontal jet-grouting methods. Thanks to the characteristics of this ground preconfinement technology, which requires only minimum overburdens, bored tunnels can be driven with extremely low cover, with many important advantages, including the way it fits unobtrusively into the environment (see the photographs in this appendix).

Applications

The first time that sub-vertical jet-grouting to preconfine excavation was experimented was in 1980 at Sesto San Giovanni (Milan), in an area destined to contain a ventilation chamber 9.80 m deep on a section of line 1 of the Milan metro then under construction. The first application for the construction of tunnel portals occurred five years later for the T1 portal on the Pontebba side of the S. Leopoldo tunnel on the Pontebba-Tarvisio. railway line.

A shell of columns of ground improved by vertical jet-grouting constructed for the portal of the S. Leopoldo tunnel under the motorway embankment

Fig. F.4 *The portal of the same tunnel during excavation works*

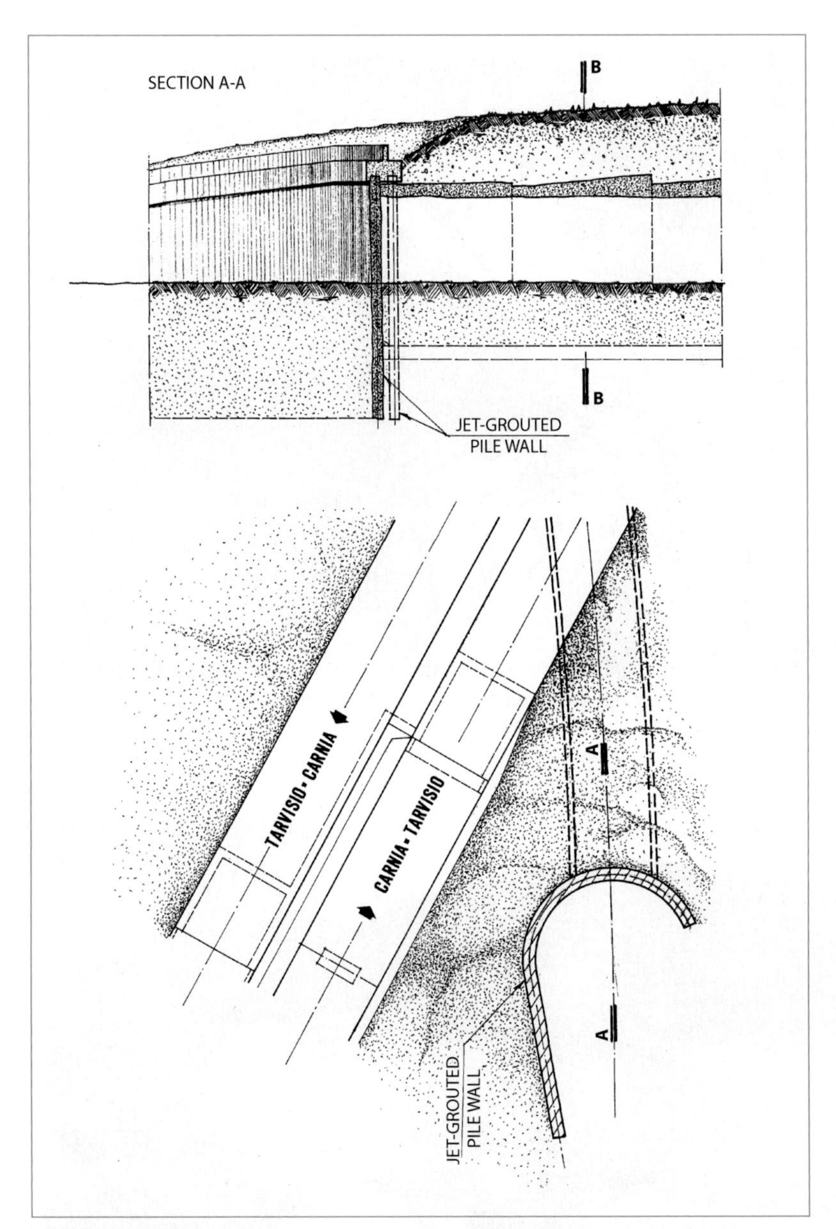

Fig. F.5 *Plan view and vertical cross section of the T1 portal of the S. Leopoldo tunnel*

The design problem consisted of the simultaneous presence of both the embankment for the Carnia-Tarvisio motorway under which the tunnel was to pass and the shoulder of the motorway viaduct almost immediately on top of the point of the chainage for the portal, which meant that the tunnel had to advance without causing deformation in the ground (see Figs. F.4 and F.5). Since this seemed impossible using conventional techniques, it was decided to employ jet-grouting technology.

Aerial view of the construction site for the portal on the Assergi side of the Gran Sasso Tunnel

The Gran Sasso Tunnel, portal on the Assergi side: the depth of the cut made in the slope using conventional methods for the portal of the tunnel certainly makes an impression

The study of the statics, conducted from the outset with the assistance of computers on a finite element model, confirmed the feasibility of the intervention and provided useful indications for the final fine tuning of the design in order to obtain a structure stressed mainly by compressive action. It was therefore decided to create a shell with three rows of sub-vertical jet-grouted columns arranged according to the geometry shown in Fig. F.5 and to then employ horizontal jet-grouting to start and drive the tunnel. It was possible in this manner to start driving a bored tunnel with an overburden of just two metres.

The works were performed without any particular difficulties and without any damage at all to either the embankment or to the shoulder of the motorway viaduct. The success achieved gave rise to numerous new applications of the same type performed in terrains of various kinds.

These included the portal constructed for the S. Elia tunnel, illustrated below (lot 33 *bis* on the Messina-Palermo motorway) positioned close to the provincial Cefalù-Gibilmanna road, which cut across the West slope of the S. Elia valley in conditions of precarious stability, given the non cohesive nature of the soil.

Geological and geotechnical picture

The morphology of the area affected by excavation was characterised by the presence of slopes of varying steepness depending on the lithology of the outcropping terrain and the structural conditions. The overall picture was typical of a young, little developed morphology and this was confirmed by the almost rectilinear geometry of the main water courses and by the almost total absence of secondary incisions.

The zone was investigated by carrying out an intense geological survey campaign in which the following were performed:

- 3 boreholes (S2, S3, S4) with a total of 56,5 m drilled;
- 60 standard penetration tests;
- 4 tests with an excavator down to 5 m from ground level (Sa3, Sa4, Sa5, Sa6);
- 1 sample was taken in the proximity of the trench Sa3 and was reassembled and reconstituted in the laboratory (C4).

Thorough examination of all the documentation collected established the following:

- the portal zone of the tunnel was located in a substantial layer of detritus, 12–19 m thick, consisting of light brown coloured silty-clayey soils;
- this layer included a high percentage of sharp grey-greenish arenaceous elements with dimensions of between 0.1 and 1 m chaotically dispersed;

Pianoro Tunnel (new high speed/capacity railway line between Bologna and Florence): the progression of the works for the South portal built in semi-cohesive ground using jet-grouting technology

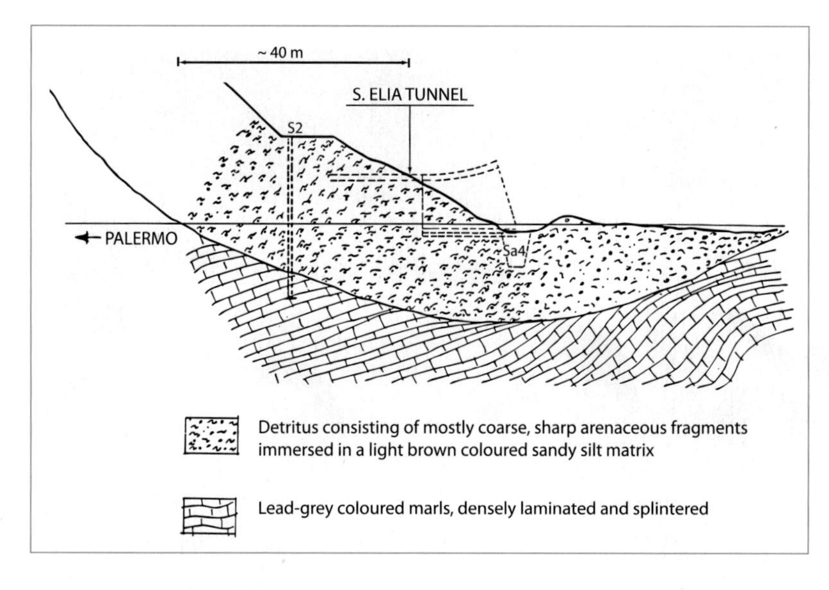

Fig. F.6

- as is shown in the geological cross section in Fig. F.6, it was resting on a densely stratified and fractured substrate of clayey marls belonging to the Polizzi formation;
- a free water table located at–3.14 m from ground level, but subject to variations in level as a function of precipitation on the slope;
- as concerns granulometry, the clayey fraction of the detritus was around 20%.

The presence in the ground of some tensile fractures also revealed the state of decompression of the existing material, a symptom of precarious stability.

From a geotechnical viewpoint the standard penetration tests performed in the zone surveyed gave average values of more than 30 blows per foot. This result seemed, however, excessively optimistic when compared with the other data acquired and was certainly influenced by the abundant presence of coarse rocky material dispersed in the matrix.

The following parameters were therefore attributed on initial analysis to the soils in question for the necessary design and verification calculations:

- unit weight 1.8–2 t/m^3 (18.0–20.0 kN/m^3)
- effective angle of friction 30°
- effective cohesion 0 Mpa

S. Elia Tunnel (Messina-Palermo motorway): portal on the Messina side under construction under the provincial Cefalù-Gibilmanna road

MESSINA - PALERMO MOTORWAY
S. ELIA TUNNEL - PORTAL ON THE MESSINA SIDE

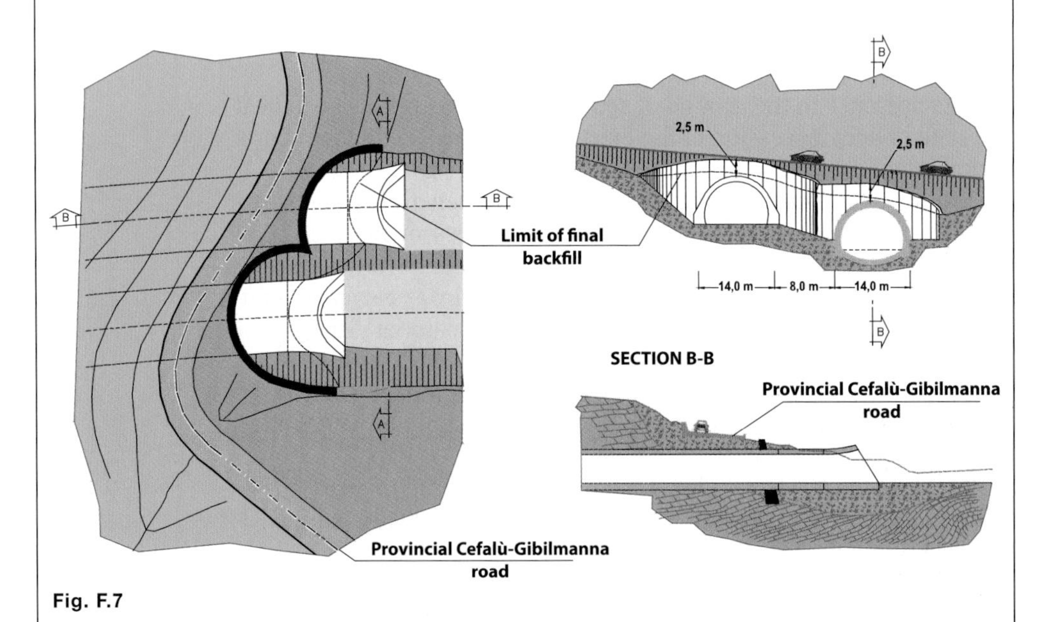

Fig. F.7

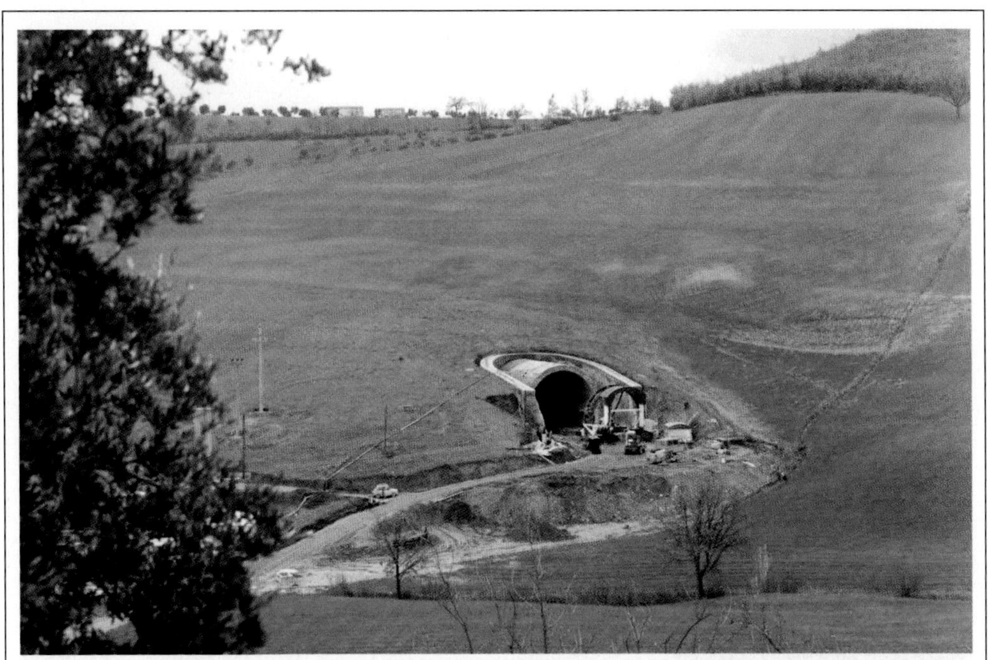

Tunnel No. 2 (Sibari-Cosenza railway line): the use of jet-grouting technology enables tunnel portals to be built in difficult grounds with full respect for natural equilibriums and for the landscape

The same portal as above: notice the extremely shallow overburden of the tunnel and the way it fits perfectly into the landscape of the environment

Malenchini Tunnel (Livorno-Civitavecchia motorway): North portal built by first placing a shell of ground improved by means of jet-grouting

The same portal as above, as it is now

General stability of the slope

As shown in Fig. F.6, the tunnel portals were to lie in the detritus deposits for a length of approximately 40 m from the start of the bored tunnels. Given the heterogeneous and non cohesive nature of the ground to be excavated and tunnelled, it was clearly going to be rather difficult both because of the presence of the provincial road over which the tunnels were to pass precisely in the vicinity of the portals and also because any land slip/slide would have immediately affected the stability of a small villa located just ahead of the portals.

As can be seen from the cross section through the axis of the tunnel shown in Fig. F.7, the overburden below the provincial road did not exceed 4 m, while the distance from the edge of it was around 7 m.

The original design solved the problem by moving the road further ahead of the portals and then making a deep cut into the slope.

However after direct *in situ* observation and examination of the results of the geological survey, it was decided in agreement with the project management to discard this solution. The reasons were as follows:

- moving the provincial road farther up the slope would have required cutting substantially into the layer of detritus, which was already in a precarious condition and would therefore have required adequate works to improve the ground and protect the excavation;
- deep cuts into the slope to construct the artificial tunnels, as specified in the design, would have caused dangerous decompressions in the layer of detritus and would have drawn water into the excavation. This would have resulted in serious danger for the stability of the residential structures above;
- it was impossible to make cuts with escarpments with a gradient of 1:1 even if only temporary, as would have been required with the presence of nearby buildings;
- the length of the construction times, without considering the risks of triggering uncontainable deformation phenomena in the slope.

In consideration of the reasons listed above, changes had to be made to construction plans to enable the tunnel portals to be built without dangerously affecting the pre-existing natural equilibriums given the already precarious stability of the slope. Obviously this could only be achieved by not starting excavation until the future perimeter walls had been stabilised. Systematic ground improvement by jet-grouting it around the portals of the bored tunnel and around the foundations and portals of the artificial tunnels seemed to be the best means of achieving this and it would also eliminate the need to move the provincial road and halt traffic on it.

The geometry of the ground improvement

Following similar procedures to those already successfully employed experimentally for the portal on the Pontebba side of the San Leopoldo tunnel, a systematic campaign of ground improvement was performed along three main lines (Fig. F.8):

- for the faces for driving the two bores: ground improvement and reinforcement by creating a pile wall of vertical jet-grouted columns, which would confine the excavation required to prepare the tunnel access portals and fully prevent any decompression of the slope alongside the provincial road;
- for the perimeter of the two bores: ground improvement using horizontal jet-grouting in the band of ground around the extrados of the theoretical profile of the tunnel in order to eliminate convergence of the future tunnel and, as a consequence, prevent even the slightest settlement of the structure of the road;
- for the foundations of the artificial tunnels and portals: vertical jet-grouting treatment with a column approximately every $3\,m^2$, designed to increase the shear strength of the ground locally to eliminate the need to dig deep foundations for the structures.

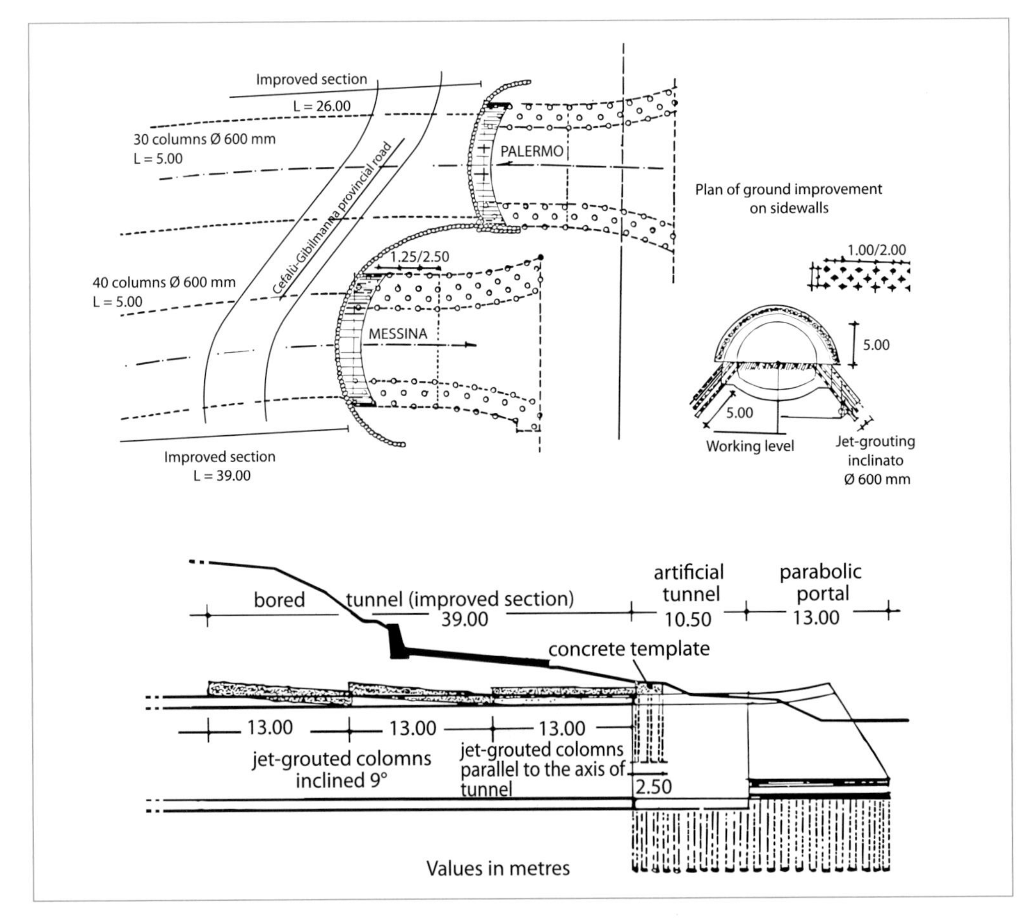

Fig. F.8

Ground improvement and reinforcement

Determination of the operating parameters

It is absolutely critical to the successful outcome of intervention when using jet-grouting to improve ground to use operating parameters that are appropriate to the geotechnical context and to obtain the desired results. It was therefore decided to perform six test treatments with varying combinations of the dosage of the mix, the injection steps, pressure and injection speed. It was decided that the zone most appropriate for performing the tests was the portal on the Messina side of the escarpment between the two alignments at the depth of the working level of the higher bore and that is at a depth of approximately 5.00 m from the intrados of the lining of the bored tunnel. The operational procedures for implementing the test columns, positioned geometrically as indicated in Fig. F.9, are summarised in table F.I below.

Column No.	Depth m.	Pressure Atm.	Quantity		Ratio H_2O/cm^3 cement mix	Extraction time in seconds per 4 cm	Total volume of back-flow	Notes
			Cement quintal/m	Mixture L./m.				
1		300	2.5	333		15	450	Fissures open parallel to the escarpment
2		500	2.5	333		11	700	It is noticed that fissures open parallel to the escarpment during injections
3	5.00	400	1.5	200	100 $\gamma = 1.5$	8	200	Drilling water is absorbed into the ground
4		400	1.5	200		8	300	Drilling water flows out along all the hole
5		400	1.5	200		8	nil	Drilling water is absorbed
6		400	2.5	333		13	500	Fissures open parallel to the escarpment

Table F.I *Operational procedures for jet-grouting column tests*

Selection of optimum parameters

The improved columns of ground were laid bare after they had hardened for three days. The random presence of boulders was found which, combined with the dishomogeneity of the ground, led to continuous zones of ground in which the cement mixture had mixed with and consolidated the clay component and wrapped around the blocks present in it to form blades of hydrofracturing (*claquage*), which extended for more than a metre from the axis of the columns themselves. The following was decided on the basis of the examination of the columns:

Malenchini Tunnel (Livorno-Civitavecchia motorway): South portal built by first placing a shell of ground improved by means of jet-grouting

The same portal as above, as it is now

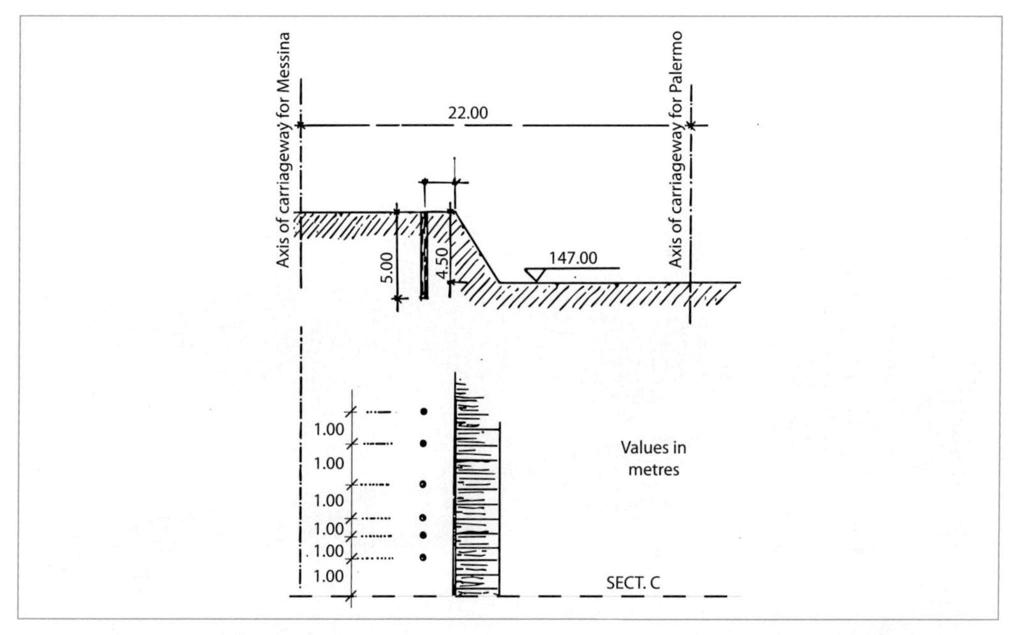

Fig. F.9

- to adopt the operating parameters used for columns 3, 4 and 5 for the ground treatment;
- to reduce the distance between centres of the vertical columns to 0.45 m and to set that distance at 0.40 m for the horizontal columns to take account of the dishomogeneity of the ground, the presence of erratic boulders and the possibility of potentially large deviations during drilling.

Geotechnical results

Table F.II below reports the main results of the comparative tests performed on soils before and after ground improvement. The values for c (cohesion) and φ (internal angle of friction) measured on samples of natural ground were obtained using undrained anisotropic triaxial tests on samples from the preliminary core drilling, S2 and the trench Sa4 (see Fig. F.6) and from samples taken during the preparation of the portals, where only one sample was undisturbed.

NATURAL GROUND			IMROVED GROUND		
Sample No.	c [Mpa]	φ [°]	Sample No.	c [Mpa]	φ [°]
21	0.015	17	31	1.2	39
22	0.02	16	32	1.5	41
23	0.019	17	33	1.3	39
24	0.02	18	34	1.5	40
25	0.016	15	35	1.2	38

Table F.II

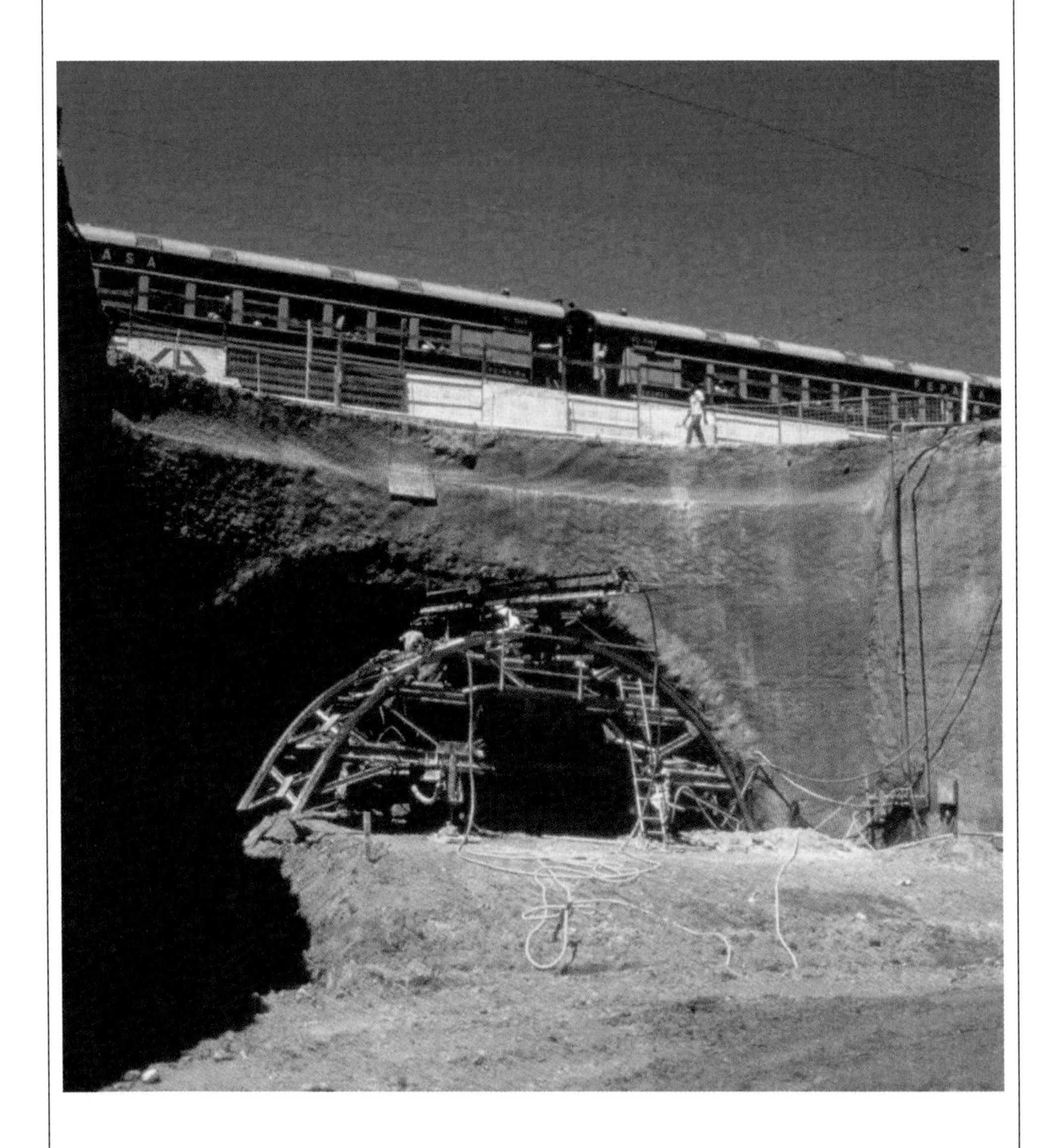

View of the portal constructed using a shell of ground improved by means of vertical jet-grouting under the railway station of the city of Campinas (Brazil)

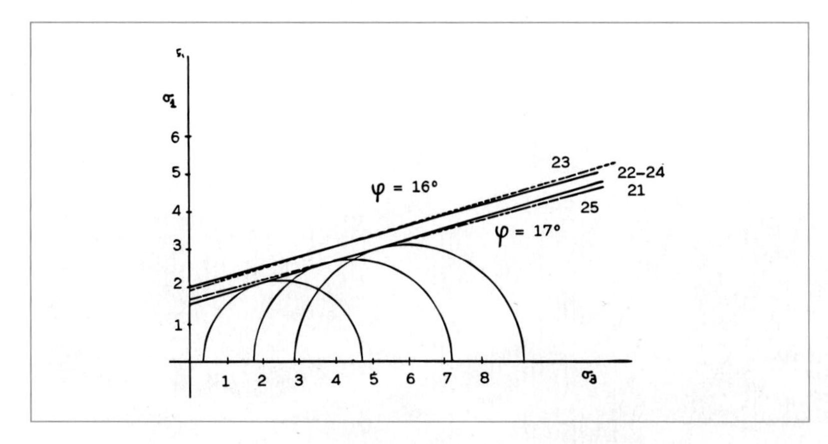

Fig. F.10
Undrained anisotropic triaxial cell test (ground not improved)

The samples of improved ground, however, were taken as cubes and then shaped into cores with a diameter of 50.4 mm and length of 100.8 mm in the laboratory. The triaxial tests on the latter were performed on a 70 MPa Hoek cell.

Each sample was composed of three test pieces, each of which was given a σ_3 which was a multiple of the previous one to give three points on the same straight line (see Figs. F.10, F.11). The piezometric pressure was calculated analytically for the first group of samples, but not for the triaxial Hoek cell test on the second group.

The outcome of the ground improvement was considered satisfactory given the results of these tests, which registered an increase in the average cohesion value from 0.018 to 1.34 MPa and in the average internal angle of friction from around 17° to around 40°. Three direct shear tests (Hoek type) performed on the improved ground confirmed this opinion with average values of 1,12 MPa for cohesion and 36° for the internal angle of friction.

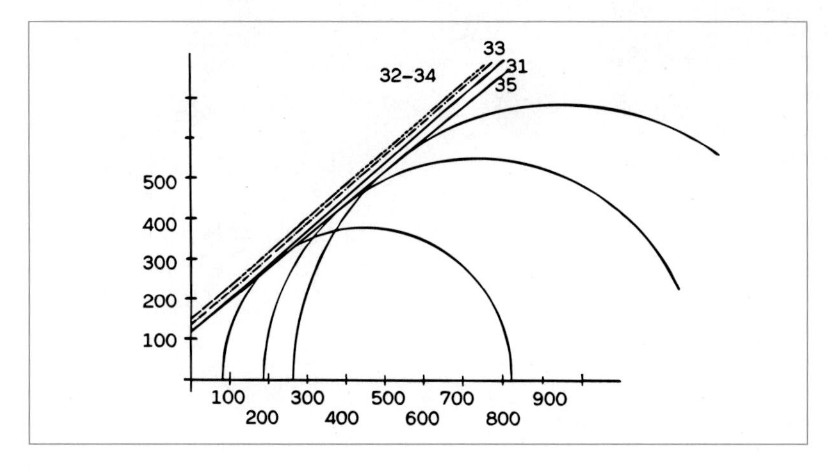

Fig. F.11
Triaxial Hoek cell test (improved ground)

Tunnel No.1 (Sibari-Cosenza railway line): jet-grouting technology enabled tunnel portals to be built without damage to the environment and landscape even in grounds with the poorest geotechnical characteristics

The same portal as above, seen from the front

Technical and operational considerations

Given the satisfactory results and having taken account of the full scale tests in the field, work commenced on the construction of the portal according to the design that has been illustrated.

In actual practice the work to improve the ground before starting to drive the tunnel was performed in the following six distinct stages starting with the lower bore and then on the upper bore, alternating between the two (Fig. F.7).

Stage 1: removal of the earth from the approach after having first stabilised the future walls of the cut with vertical jet-grouting.

Stage 2: the creation of a ring of drainage pipes, if necessary, and horizontal jet-grouting treatment in the crown of the future tunnel for more than 13 m ahead of the face. The telescoped horizontal columns of improved ground were performed with a distance between centres of 40–45 cm in the crown and of 50 cm in the zone of the springline.

Stage 3: half face advance[1], for approximately 10 m, positioning of the steel ribs and immediate placing of the steel mesh reinforced concrete for a thickness of approximately 10–15 cm.

Stage 4: a new series of drainage pipes are created (if necessary) from the new face and another section of sub horizontal columns of treated ground is created (13 m approx.), positioned with a distance between centres of 45–50 cm around the whole perimeter of the tunnel with a maximum inclination of 9% in the crown with respect to the axis of the tunnel; ground treatment is performed for the sidewalls of the rear section already excavated.

Stage 5: excavation of the bench in the section already fully treated with ground improvement and the application of shotcrete on the walls; excavation for the tunnel invert in the section of tunnel already open and subsequent casting of the sidewalls and the tunnel invert.

Stage 6: the final lining is cast in concrete reinforced with INP 200 steel ribs.

1 It is interesting to note that in the first tunnels driven using the new jet-grouting technology, tunnel advance was generally half face, while today, with much greater familiarity with this technology and as has been affirmed many times in this book, full face advance is always preferred, naturally using adequate designs and operational procedures. The advantages are considerable in terms of safety, production rates and the long and short term stability of the tunnels.

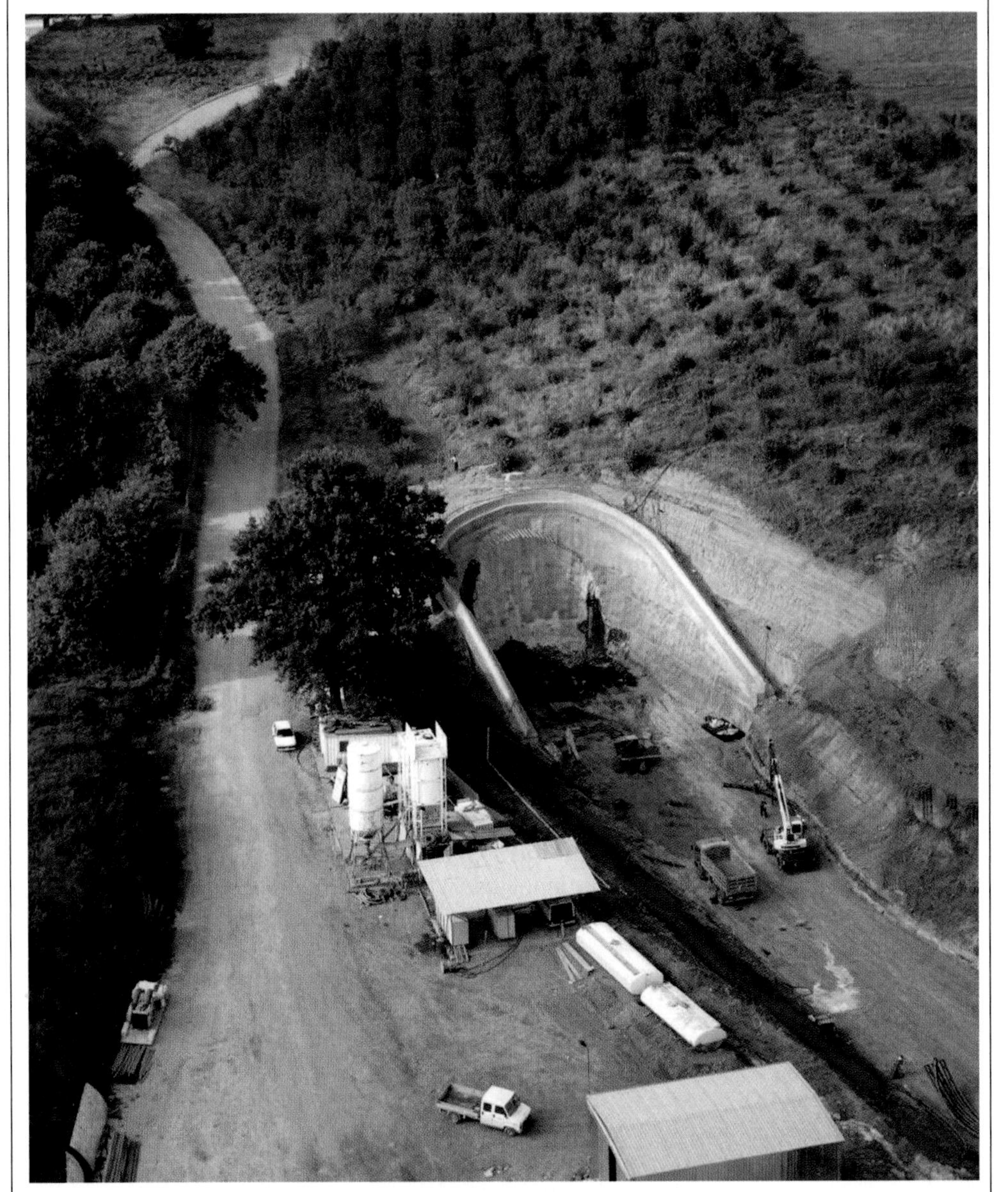

Colli Albani Tunnel (new Rome-Naples high speed railway line): aerial view of the works for the construction of the South portal of the tunnel

Apart from the reasonable certainty of obtaining good technical results from the use of jet-grouting for ground improvement, one important factor for the contractor was the relative fast construction times and the fact that it interfered little with construction schedules, favoured undoubtedly by the possibility of working on two adjacent bores at the same time. As concerns the latter aspect, it was objectively imperative to create an arch effect in advance as it was indispensable to the stability of the excavation and to the absence of even the slightest convergence. From this viewpoint:

- the main function of the band of jet-grout improved ground is not only to prevent immediate loosening of the ground, which would otherwise have direct effects on the surface, with serious damage to the structure of the road and the building above. but also to distribute asymmetric thrusts which manifest in the long term in soils like those of the S. Elia tunnel, redistributing them on the final lining which therefore need not be so thick. Given the reduced overburden between the crown of the future tunnel and the provincial road, as well as the permanent loads and live loads due to the passage of vehicles, the band of improved ground around the extrados of the S. Elia tunnel could be expected to be subject to compressive stress of around 6.00 MPa and tangential stresses of around 0.30 MPa, values compatible, temporarily, with the strength of the ground treated. Radial convergence measurable in millimetres was therefore to be expected given those stresses;
- thanks to its excellent strength and deformability properties, the layer of steel mesh reinforced shotcrete placed in direct contact with the treated ground acts as a protective mantle capable of developing confinement action of greater than 0.20 Mpa and it therefore increases the strength of the jet-grouted columns considerably. It also plays an important role as an element of transition in the overall economy of the intervention and of intermediate rigidity between the improved ground and the final lining: in the event of an earth quake, the wave of the shear stress on the final lining would be cushioned greatly so that it could guarantee the integrity of the tunnel even if immersed in non cohesive material;
- finally the task of the concrete structure reinforced with steel ribs and closed with the tunnel invert is to prevent long term settlement, which might result not only from viscous phenomena, but also from heavily asymmetrical individual thrusts, which are not improbable in non cohesive soils with extremely heterogeneous granulometry like those of the S. Elia tunnel, which can also be easily affected by external meteorological events, given the very shallow overburdens.

Vertical jet-grouting for a portal shell

Final considerations

Experiences to-date in the application of vertical jet-grouting to solve the problems of constructing tunnel portals in soft grounds have demonstrated that it can be used to start tunnels in slopes consisting of non cohesive or poorly cohesive soils without causing practically any movement in the surrounding ground.

When used in conjunction with horizontal jet-grouting, it allows bored tunnel excavation to start under extremely shallow overburdens of a little more than one metre, thereby minimising the problems of harmonising new constructions set in existing environments. Other advantages can also be gained from the application of these methods:

- great flexibility in adapting to any morphological situation;
- fast construction times;
- competitive costs.

Although jet-grouting technology does require considerable expenditure, it nevertheless guarantees an excellent ratio between costs and the quality and safety of the results (obviously provided the necessary morphological conditions exist).

Quite apart from any considerations concerning traffic, there was also a modest saving on costs in the construction of the portal for the S. Elia tunnel, as is shown by the summary figures given in table F.III below.

• Ground improvement work using jet-grouting to pass under the provincial road	Lump sum	610	
• Extension of the embankment wall	m 20	25	
• Additional bored tunnel to replace the artificial tunnel	m 39	275 910	910 +
• Cost of artificial tunnels eliminated	m 59	575	
• Savings for not having to change the route of the provincial road	a corpo	156	
• Savings for not having to improve the ground of the slope	a corpo	219	
• Savings for not having to expropriate property	a corpo	69 1019	1019 –
Total savings			**109**

Table F.III
Comparison of construction costs for the two alternatives (in millions of lire)

Preparation for casting the upper r.c. beam over a shell of ground improved by vertical jet-grouting

Detail of the wall of a shell of columns of ground improved by means of vertical jet-grouting

Pianoro Tunnel (new high speed/capacity railway line between Bologna and Florence): work in progress on the southern portal of the tunnel

S. Giovanni Tunnel (Catanzaro ring road East): tunnel portal with diameter 14 m approx., excavation under urban buildings

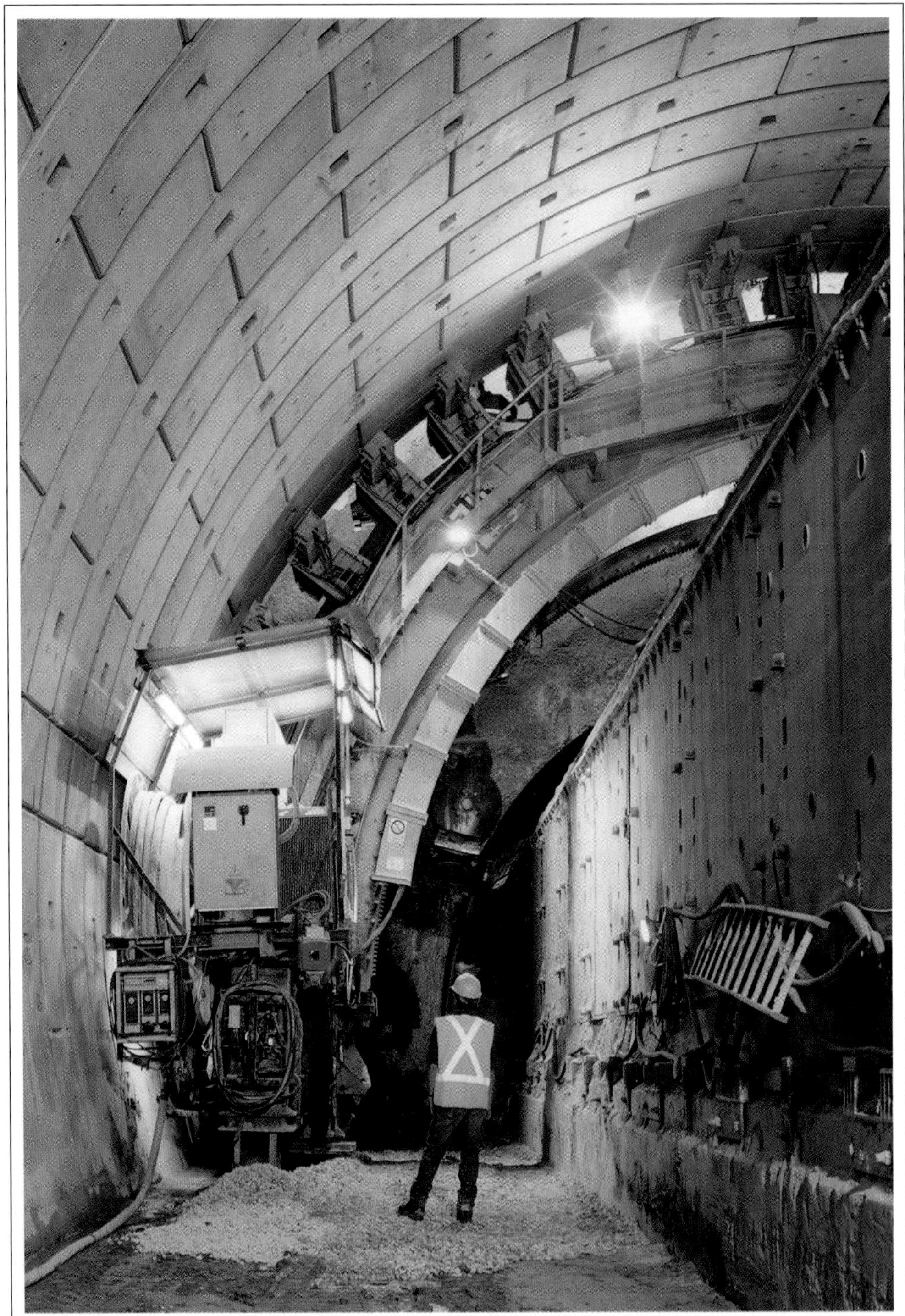

Widening of tunnels without interrupting traffic: Milan-Rome motorway, Nazzano tunnel, placing an arch of the lining constructed using prefabricated segments (2005)

Widening road, motorway and railway tunnels without interrupting use

As traffic increases, there is an increasing need to widen roads, motorways or railways to increase their capacity. Meeting these needs is easy if the routes run completely on the surface, however it is very complicated when they run through stretches of tunnel, because it is indispensable in these cases to resort to costly alternative routes to construct the new tunnels in addition to those that already exist.

And then a tunnel can only be widened if it is possible to:

- guarantee the necessary safety of users and limit inconvenience within an acceptable threshold;
- solve the technical and operational problems connected with driving the face to widen the tunnel in ground that has already been disturbed by the previous excavation;
- construct the new load bearing structure at the same time as the old one is demolished and deal adequately with any stress-strain conditions, even unexpected conditions, which might be met during construction, without any danger to tunnel users and to any human activity there may be on the surface.

Currently as part of the work on the widening of the "Nazzano" tunnel, on the A1 Milan-Naples motorway, from two to three lanes, plus an emergency lane in each direction, a new innovative construction method is being used, proposed by the author, which will allow the requirments listed above to be met and road, motorway and rail tunnels to be widened without interrupting traffic during construction work (Fig. G.1). This appendix first discusses the major problems that had to be solved and then illustrates the basic concepts of this new technology with references to some of the first results of its application.

■ The indispensable requirements for the technique to be employed in the presence of traffic

If a technique to widen a tunnel in service without interrupting traffic is to be really possible, as mentioned in the introduction, it must solve at least two types of problem satisfactorily:

- the problem of performing the works required to excavate and construct the lining of the widened tunnel and demolish the existing

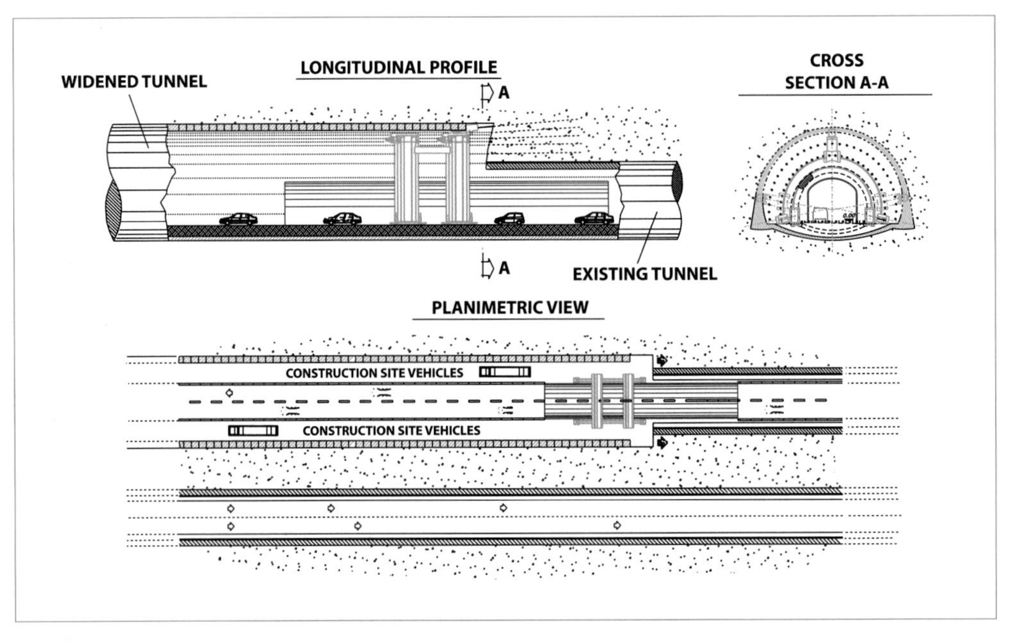

Fig. G.1

tunnel, while ensuring the safety of tunnel users and minimising inconvenience;

- the problem of adapting the technique for use in any type of ground and stress-strain conditions it might encounter, minimising the effects on the ground surrounding the tunnel and ensuring safety constantly during operations.

Clearly the development of a specific construction approach is required to solve these problems. It must allow all types of ground improvement in advance that may be required at the face and around the tunnel to be performed without any danger to traffic as well as the placing and activation of the final lining at a very short distance from the widening face.

It is only by operating in this way that it will be possible to:

- control the effects of the probable presence around the existing cavity of a band of ground that has already been plasticised and must not be subject to further disturbance;
- widen the cross section of the tunnel without triggering harmful deformation in the ground which would translate into huge thrusts on the lining of the final widened tunnel and differential settlement at the surface, dangerous for any constructions that might be present;
- ensure at the design stage that construction schedules are observed independently of the type of ground and stress-strain condi-

tions to be addressed, with costs contained and with reliable planning of construction costs and time schedules in order to minimise route deviations for traffic and inconvenience for users.

How the idea developed

The conviction that it was possible to develop a construction method suitable for these purposes, capable, that is, of satisfying the requirements listed above without interfering with normal motorway traffic, first began to take shape in the mind of the author at the time of the construction of the "Baldo degli Ubaldi" underground station in Rome. The large dimensions (span of 21.5m and a height of 16m) and the severe constraints on surface settlement required the design of an innovative construction method. Using that method the tunnel for the station was constructed in four main phases (Fig. G.2):

1a. two side tunnels were driven 5m wide and 9m high from two access shafts, which would house the future side walls of station tunnel, after first reinforcing the ground ahead of the face with fibre glass structural elements and lining the excavation with fibre reinforced shotcrete and steel ribs equipped with invert struts;

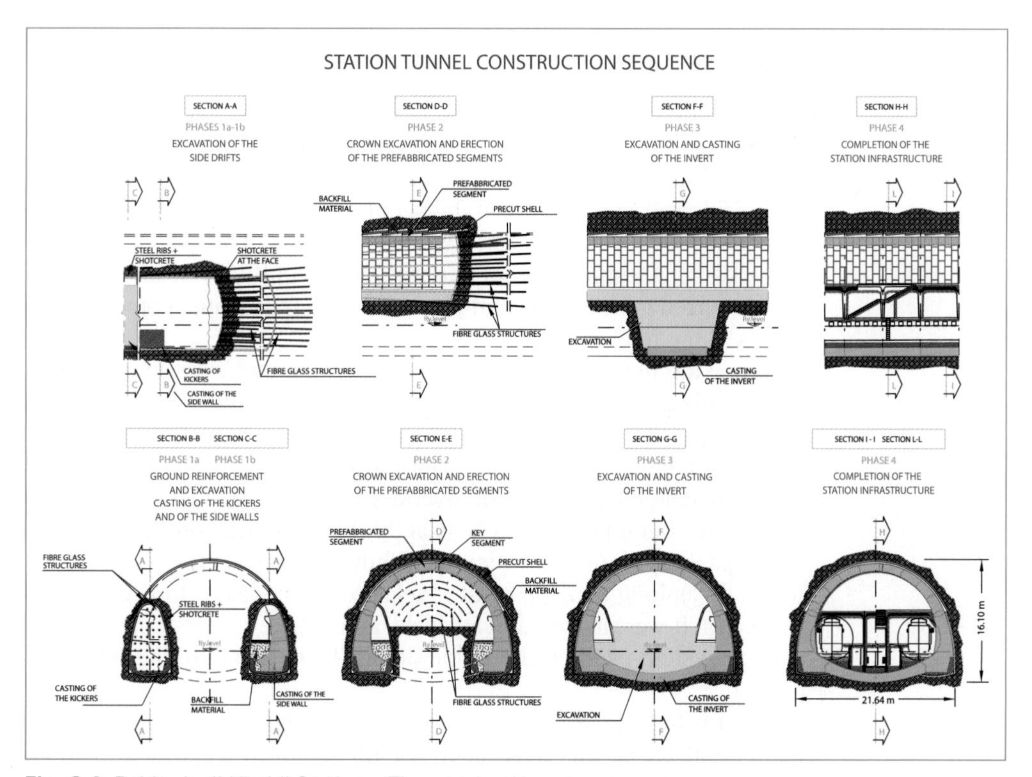

Fig. G.2 *Baldo degli Ubaldi Station – The construction stages*

1b. casting of the side walls with reinforced concrete;
2. driving the tunnel for the crown of the station tunnel (21.5 m span, 8.5 m high with a cross section of 125 m^2) after first reinforcing the ground ahead of the face with fibre glass structural elements and protecting it with a strong shell created using the mechanical precutting method, and then immediately placing the lining of the crown with an "active arch" of prefabricated concrete segments;
3. excavation downwards of the station tunnel (cross section of 90 m^2) and immediate casting of the tunnel inverts in steps after the construction of the crown;
4. completion of the station infrastructure.

The new construction system contained essentially two new elements:

- one was ground improvement ground ahead of the face with fibre glass structural elements and the mechanical precutting technology (employed for the first time in the world on a span of 21.5 m) combined with the "active arch" principle;
- the other was the extremely high level of industrialisation achieved with intense use of machinery.

In fact a special machine was designed, developed and constructed jointly with STAC S.p.A. of Mozzate (Como) to combine the technologies employed, all fairly recent, in a single construction system It consisted (Fig. G.3) of a large metal portal, with the same geometry as the profile of the crown of the station tunnel, which rested on the inside of the sidewall tunnels with stabilisers positioned on its side members to enable it to travel backwards and forwards. The portal not only contained the equipment for mechanical precutting but also housed that required for handling and placing the prefrabricated concrete segments of the final lining.

Once the machine and its accessory equipment were in operation, the author noticed during the construction operations that the area consisting of the bench of the future station tunnel, with a cross section of similar size to that of a normal motorway or rail tunnel, was not used at all for construction operations (Fig. G.4). These operations could have been performed in exactly the same way on the extrados of an existing tunnel, to widen it, without having to close it to traffic, naturally as long as appropriate measures were taken to protect tunnel users. It was, in the final analysis, a question of extending the half section system used on the Baldo degli Ubaldi tunnel to the full cross section.

That is how the idea of a technique was born using machinery and equipment based on those used for the Baldo degli Ubaldi station, which would be capable of widening a tunnel without putting it out of service during construction work (Fig. G.1).

Fig. G.3

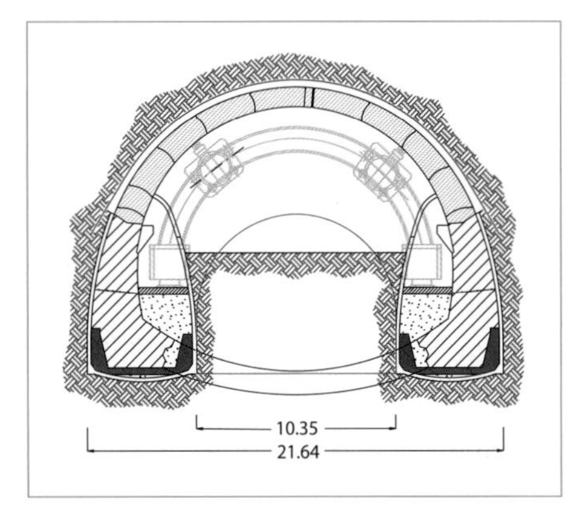

Fig. G.4

Illustration of the technique

The idea on which to work basically consisted of the following (Figs. G.5 and G.6)

- *the first stage*: operations, if necessary, to reinforce the widening face and/or for preconfinement of the cavity, and then excavation of the ground in steps between the theoretical profile of the future widened tunnel and that of the old existing tunnel;
- *the second stage*: the construction immediately behind the face of the final lining with the placing of one or more arches of prefabricated concrete segments according to the "active arch" principle;
- *the third phase*: the construction of the foundation (tunnel invert), if necessary.

During the first two stages, which should be performed in very regular cycles inside the profile of the old tunnel, there would be in place a *steel traffic protection shell* and all the machinery used in the operations would move and work above this. The hollow space between this steel protection and the lining of the existing tunnel would be filled with sound proofing and anti-shock material. The steel shell would be at least four times longer than the diameter of the tunnel to be widened and would occupy a relatively small space within it and allow construction work without interrupting traffic in the existing lanes. When tunnel widening advances to the point where the distance between the face and the front end of the shell reaches what is considered the minimum safety limit, the shell must be moved forward and the various stages repeated in cycles until the entire tunnel has been widened.

The paragraphs below give a detailed description from an operating viewpoint of each stage of the technique.

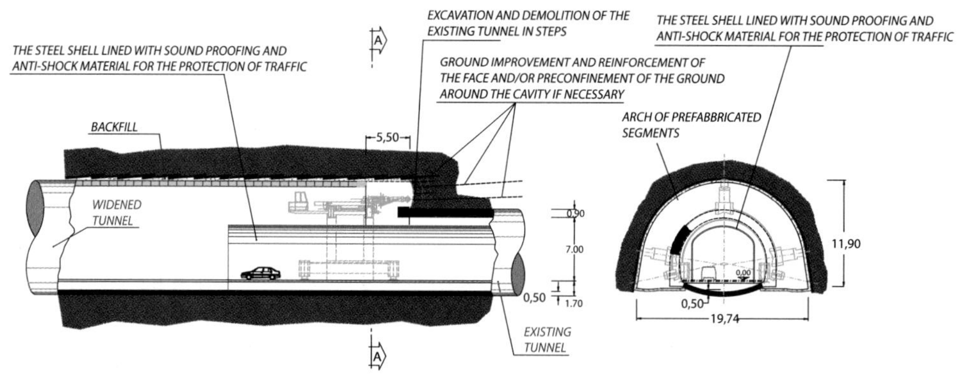

Fig. G.5 *Excavation of the ground at the widening face and demolition of the existing tunnel*

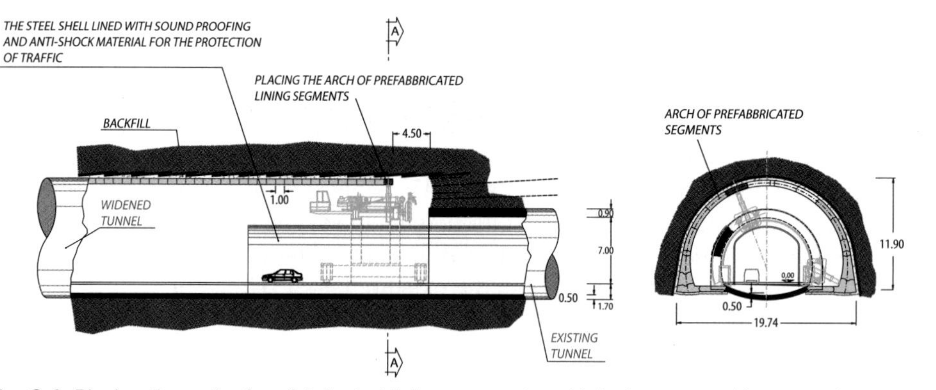

Fig. G.6 *Placing the arch of prefabricated lining segments, which alternates with excavation of the widening face*

The first operating stage

The first stage of the work, as has been said, involves first of all the ground improvement operations ahead of the face, if necessary, and then the excavation in steps (how large they are depends on the characteristics of the ground concerned) of the ground between the design profile of the future widened tunnel and that of the old existing tunnel.

The ground improvement operations ahead of the face performed on the basis of the geological and geotechnical context may consist of reinforcement in the widening face and/or preconfinement of the cavity, such as: horizontal jet-grouting, mechanical precutting or improvement using valved and injected fibre glass structural elements around the future widening face. They may be placed in advance or radially, working from inside the existing tunnel but in any case always above the "steel traffic protection shell".

After the ground improvement ahead of the face has been performed, work starts on driving the widening face (see Fig. G.5). This is done by excavating the ground and demolishing the lining of the existing tunnel in small steps [from 60 cm to 200 cm depending on the stress-strain conditions of the material being tunnelled and the size of the prefabricated segments designed for the final lining (see the section: *The second construction stage*)]. If the stress-strain conditions allow, tunnel advance may proceed in steps that are as long as several lining segment lengths.

The machinery used for ground reinforcement and improvement in advance and for lining the tunnel operates completely above the steel traffic protection shell and is equipped with all the equipment needed for the necessary ground improvement operations.

The second construction stage

The second construction stage entails placing the final lining of the widened tunnel consisting of prefabricated concrete segments.

Excavation in steps with the final lining consisting of prefabricated concrete segments placed immediately according to the "active arch" principle constitutes the key factor. It halts any possible deformation phenomena before it starts at a short distance from the face and overcomes all the problems of the deformation response of the rock mass. This is the most characteristic feature of the technique presented here.

The work involved the following operations (Fig. G.6):

a. transport of the concrete segments to the face using conveyor belts and fork lift trucks positioned on one side of the widened tunnel;

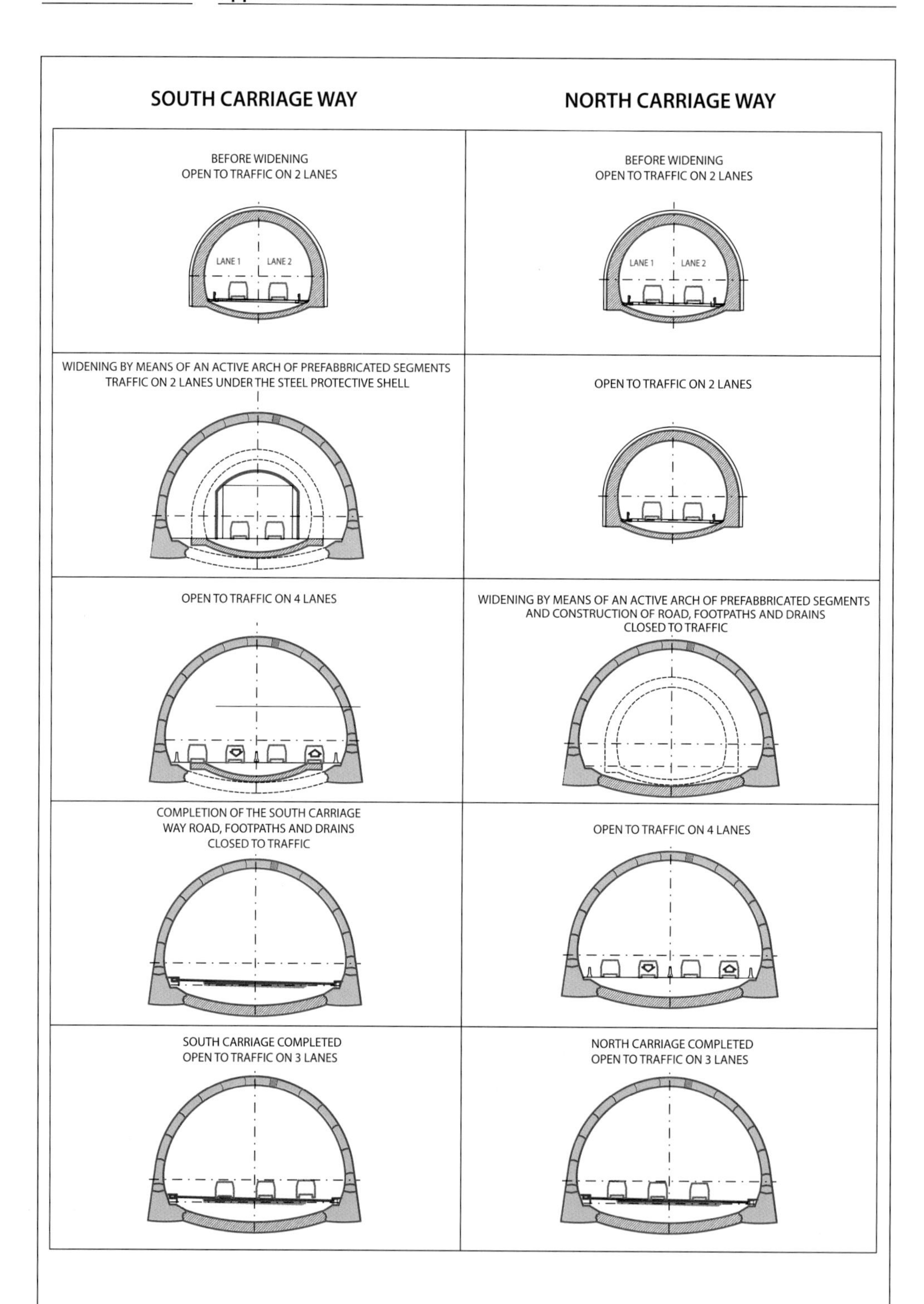

Fig. G.7 *The traffic regulation plan during widening of the carriage way*

b. the application of slow setting epoxy resins on the two sides of the segment to be placed and on the front end that will be in contact with the arch of the lining already placed;
c. raising and positioning of the segments using a special erector machine which places the lower segments first on both sides of the tunnel and then the upper segments until the arch is completed with the key segment in the roof;
d. mortar is introduced between the extrados of the prefabricated segments and the wall of the excavation behind it;
e. the pressure jack in the key segment is then activated to bring the segments into firm contact with each other and immediately produce the confinement pressure required on the ground around the tunnel according to the "active arch" principle.

The third construction stage

If a foundation structure is needed, traffic is first appropriately deviated, as we will see, and the lining of the new widened tunnel is simply joined to the tunnel invert of the old tunnel or a genuine new tunnel invert is cast.

Resolution of particular problems to maintain traffic flow during construction work

While the first two construction stages do not pose any particular problems to the maintenance of traffic flow since everything takes place above the steel traffic protection shell, with the third stage there are two distinct cases.

Rail tunnels

Once the tunnel has been widened, rail traffic will have to be interrupted to change the layout of the tracks in the new situation. The structure to join the final lining of the widened tunnel with the existing tunnel invert or, alternatively (if the statics conditions require it), the casting of a new tunnel invert can be performed in this interval of time.

Road (single bore) or motorway (twin bore) with 2 lanes in each direction

Traffic in road tunnels can always be kept flowing on at least one lane in each direction by appropriate organisation of the works to construct the foundations and to widen the road itself. Similarly two lanes in each direction can always be kept open with two bores by switching the works between the two tunnels and deviating traffic flow accordingly onto lanes as they become free (as illustrated in the example in Fig. G.7).

Application of the system for the widening of the "Nazzano" tunnel

The new technology is being applied experimentally for the first time in the world on the "Nazzano" tunnel located on the A1 Rome-Naples motorway betweem Orte and Fiano Romano, between chainage km 522 + 00 and km 523 + 200 (Fig. G.8).
The route of this tunnel is completely rectilinear and lies at an altitude of 166 m a.s.l.. It is 337 m long and runs under an overburden of 45 m From a geological viewpoint, the tunnel runs through sandy and silty-clayey ground of the Plio-Pleistocene series on which the town of Nazzano is located (Fig. G.9).
Given the type of ground to be tackled, the design first specifies the creation of a shell of fibre reinforced shotcrete around the tunnel using mechanical precutting before starting excavation to widen the tunnel.
Tunnel widening will therefore take place in the following four main stages (Figs. G.10 and G.11):

1. creation of a mechanically precut shell around the future tunnel (19.74 m span) 5.5 m in length and 35 cm thick and ground improvement ahead of the widening face if necessary;
2. demolition of the old lining in steps under the protection of the previously improved ground and excavation of the ground until the design profile of the widened tunnel is reached;
3. immediate erection of the final lining behind the face (4.5–6.5 m max.) by placing an arch of prefabricated concrete segments, using the "active arch" principle;
4. construction of the foundation structure (new tunnel invert).

All the operations for the first three stages will be performed protecting the road with a self propelled steel traffic protection shell under which vehicles may continue to pass in safety. The shield is designed with a length of 60 m and will extend for a length of approximately 40 m ahead of the widening face. It consists of a modular steel structure and is equipped with runner guides, anchors, motors, sound proof and anti shock panels to absorb the shock of falling blocks of material during excavation and demolition of the existing tunnel including any ground that falls accidentally.

All the machinery for performing the various operations will move and operate above the shell. When tunnel widening advances to the point where the distance between the face and the front end of the shell reaches what is considered the minimum safety limit for vehicle traffic, the shell will be moved forward and the various stages repeated in cycles until the whole tunnel has been widened.

The machine and its equipment

The design and construction of the machine prototype and its equipment required particular effort because a series of operating functions had to be optimised to work in a very limited space between the finished tunnel and the shield: precutting at the face, excavation, placing of the segments, various grouting operations, demolition of the existing tunnel.

The problems were solved by using innovative technology and the result was a versatile and compact design, highly computerised capable of performing all the functions required with movements and therefore also operating times reduced to a minimum.

Basically it consisted of a robust double arch steel structure (Fig. G.12) connected at the bottom by telescopic beams which gave rapid and precise longitudinal movement forwards and backwards.

Centring in the cross section and correct positioning of the height are achieved by hydraulic control systems.

A particularly sophisticated carriage is fitted on the arch at the face which carries the precutting blade and the cutter for excavation and demolition or alternatively a demolition hammer.

The circular movement of the carriage around the arch is obtained by gear reduction motors and a rack and pinion and the single and

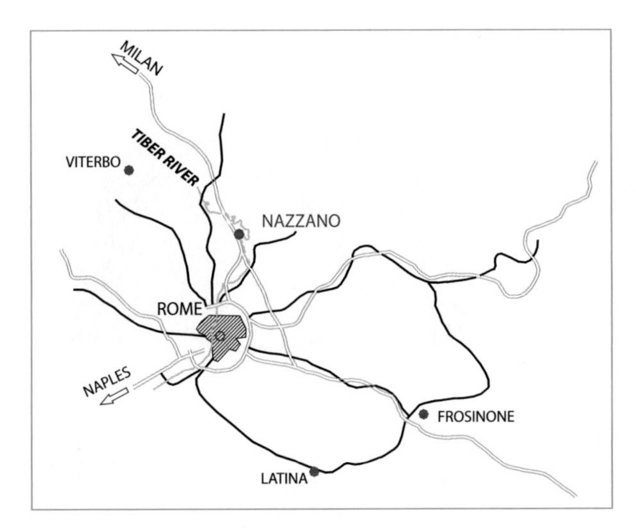

Fig. G.8 *Chorography*

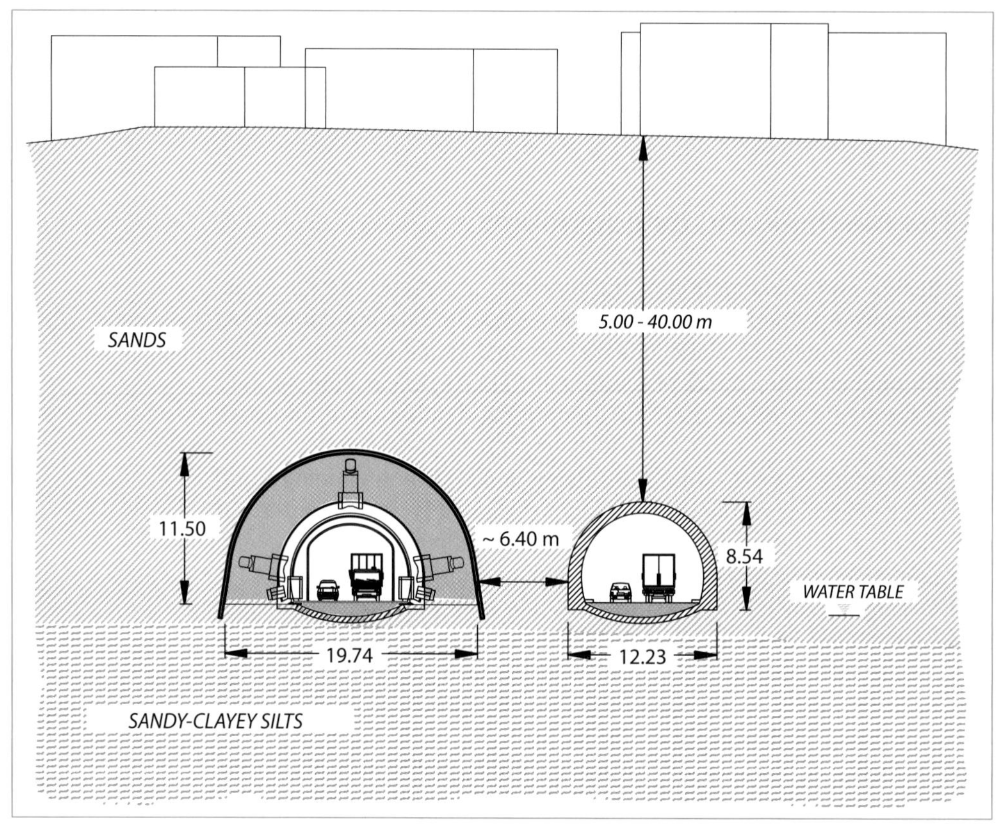

Fig. G.9

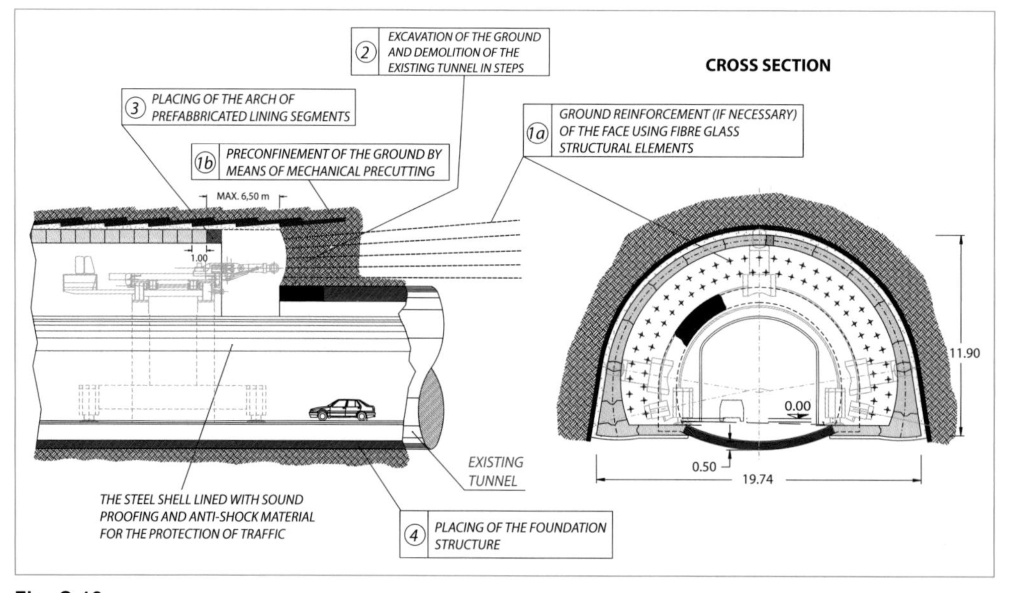

Fig. G.10

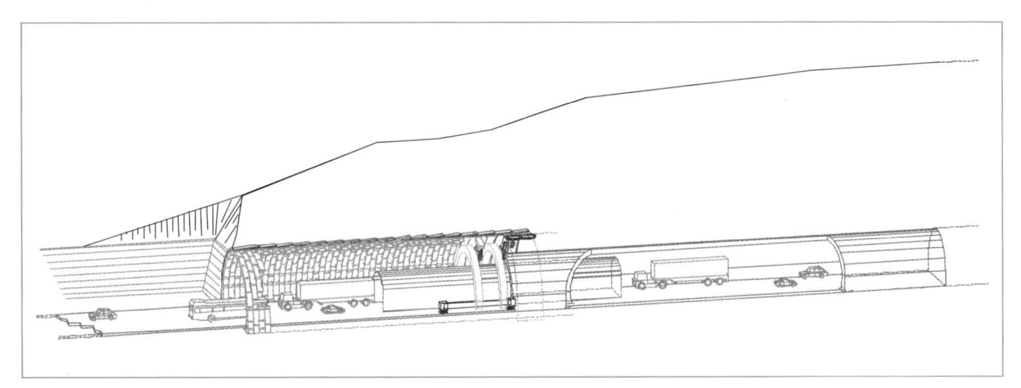

Fig. G.11 *Three dimensional view*

complex movements of the different parts allow the different operations specified in the design to be performed.

A dual system is also appropriately positioned on the same arch to manage the tubing used for filling the cut made with the precutting blade with mortar and the space between the segments and the walls of the excavation.

The rear arch was designed for placing the concrete segments. A carriage runs on it equipped with an *erector* capable of "grabbing" the segments and placing them. The movement of the erector is totally powered by electricity and hydraulics and it is controlled from a panel

Fig. G.12 *View of the machine with the large chain saw for making the precut in the foreground*

equipped with a display which gives information on manoeuvres and errors that may have been committed.

Before the key segment of the arch is placed and it consequently becomes self supporting, the segments rest on special telescopic structures anchored to the arch. They are equipped with sensors which allow the different manoeuvres to be made in safety.

The structure is equipped with various service gangways to allow personnel to work with a clear view of operations.

The various functions of the equipment are controlled by a PLC *(Programmable Logic Controller)*, which recognises commands it receives and sends information to the display for correct and safe control of the equipment.

Table G.I below contains the main technical specifications of the machine in question.

Table G.I

TECHNICAL SPECIFICATIONS OF THE MACHINE	
Cutting capacity of the blade	L = 550 cm; th = 30 cm
Erection of the segments	max load = 7 ton at 10.70 m
Rated power	214 KW
Means of power	electricity-hydraulics
Movement	hydraulic controlled by a *Programmable Logic Controller*

The progress of works

Once a series of affairs had been resolved, attributable exclusively to contractual and financial problems, which delayed the start of widening work several times and after solving a series of problems connected with passing through the portal zone with a very shallow overburden, which was more heavily affected by the work already performed for the construction of the existing tunnel, regular advance rates were finally achieved for tunnel widening after the system was fine tuned.

On 17th November 2005, widening work was completed on the bore with North moving vehicles in the full presence of traffic. Average advance rates achieved were between 0.7 and 0.8 m/day.

Optimisation of the system and the advance cycle mainly concerned the adoption of a longer cutting tool capable of making a "precut" 5.5 m long and 35 cm thick (compared to the initial tool of 4.5 m and 30 cm).

The precut was then followed by two stages of excavation, the demolition of the existing tunnel and the placement of shotcrete at the face, for an advance of 2 m each, interposed by the erection of two consecutive "active arches" in the crown with a length of 1 m each.

Fig. G.13 *The work to widen the tunnel was performed with the motorway in service*

5. Conclusions

The first results of the experiment indicate that the technique illustrated solves the specific problem of widening a road, motorway or rail tunnel while allowing traffic to continue to flow during construction work.

Its main features are:

1. the adoption of a final lining consisting of prefabricated concrete segments to stabilise the widened tunnel placed in short steps according to the "active arch" principle, which therefore comes into operation at a very short distance from the widening face (4.5–5.6 m max). As a consequence passive stabilisation operations such as shotcrete and steel ribs are avoided;
2. the final lining can be put under pressure by using jacks in the key segment designed to recentre asymmetrical loads should there be bending moments sufficient to make the resisting section of the arch of prefabricated segments act partially;
3. the ability to perform ground improvement ahead of the face, if required, to contain or even completely eliminate deformation of the

Fig. G.14 *The steel shell to protect traffic. The tracks on which the machine moves can be seen at the bottom of the photo*

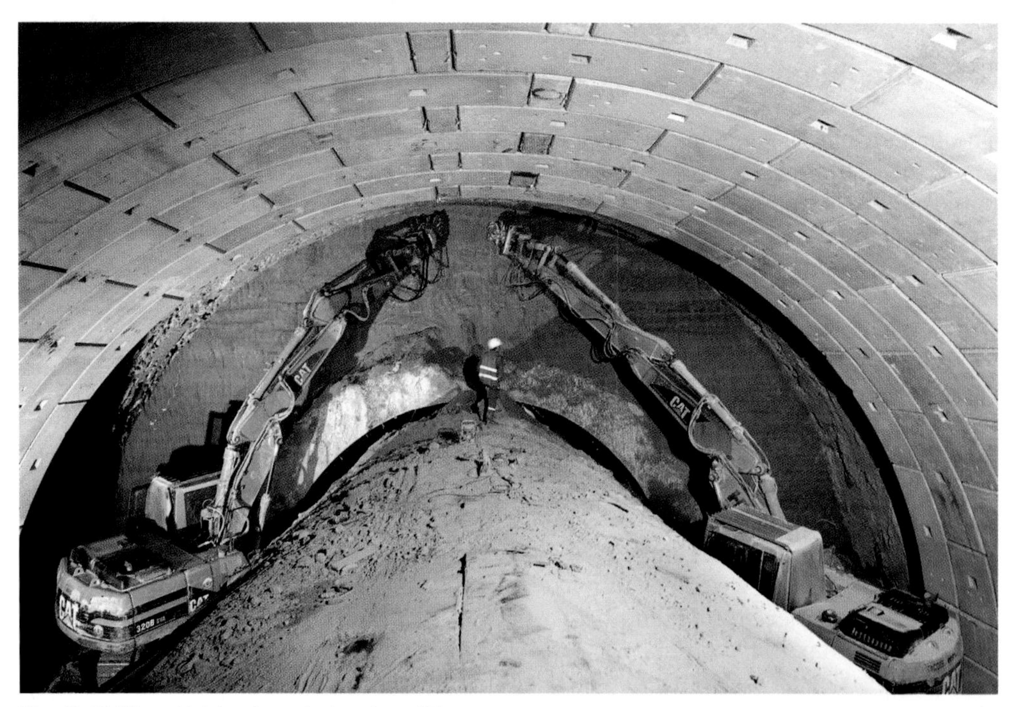

Fig. G.15 *The widening face during demolition*

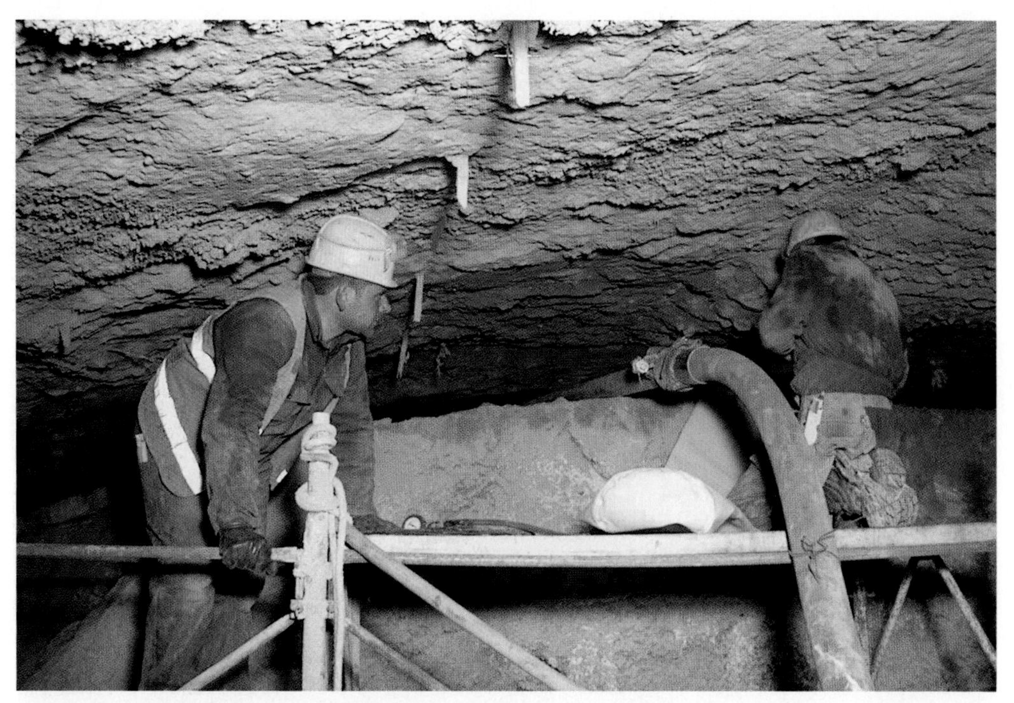

Fig. G.16 *Backfill of the precut with pumped mortar*

Fig. G.17 *Erection of the prefabricated lining segments in pretressed r.c.*

Fig. G.18 *View of the widened tunnel*

face and of the cavity and therefore avoid the uncontrolled loosening of the rock mass and thereby ensure operational safety;

4. intense mechanisation of the various construction stages, including the operations for ground improvement ahead of the face if required, with consequent *regular advance rates* and shorter construction times, all factors that have advantageous repercussions for construction economics and the production rates that can be achieved;
5. the extremely linear production rates obtainable (industrialised tunnelling), which it is predicted will be around 0.6–1.2 m/day of finished tunnel;
6. the ability to perform all construction operation while protecting the road with a "steel shell" under which traffic can continue to flow in safety;
7. The extreme versatility of the machine used, which is able to operate under extremely varied ground and stress-strain conditions.

After a significant period spent fine tuning the system, connected with the fact that it was the first time that this technology had ever been used to solve the problem of widening a tunnel with traffic running, the experimentation in progress demonstrated that the following can be achieved with this technology:

- controlling the effects of the probable presence around the existing cavity of a band of ground that has already been subjected to plasticisation and must not be disturbed any further;
- widening the cross section of the tunnel without causing damaging deformation of the ground and therefore preventing substantial thrusts on the lining of the widened tunnel from developing with differential surface settlement dangerous for existing structures;
- ensuring that construction occurs on schedule as specified in the design, independently of the type of ground and the stress-strain conditions, with construction times and costs contained and planned in order to reduce traffic deviations and inconvenience to users to a minimum.

Glossary

Acquired consistency: the consistency assumed by a material, which varies as a function of the magnitude and anisotropy of the stress tensor acting on it.

Arch effect: a phenomenon of *channelling stresses* around an excavation, which can be performed either by natural or artificial means, inside the mass of the ground, as a result of which underground cavities are able to exist and can be created.

Back-analysis: a reverse calculation based on measurements performed during the monitoring phase (extrusion, convergence, pressure on linings, etc.), used to calculate the strength and deformability parameters of the ground with greater accuracy (dependent also on the calculation models employed). It is therefore an important tool for testing the accuracy of the design hypotheses and procedures.

Barring down: an operation that is performed for safety purposes after each round of shots when blasting, which consists of either manually, or using mechanical equipment, knocking down fragments of rock which remain partially attached to the crown or walls of the tunnel.

Behaviour Category: a general concept with which tunnels to be constructed or which are being constructed can be interpreted and classified according to the universal criterion of the stress-strain response of the core-face in the absence of stabilisation intervention. The ADECO-RS approach identifies three distinct stress-strain behaviour categories: category A: stable core-face; category B: stable core-face in the short term; category C: unstable core-face, with the observation that all known cases of tunnels built fall within these three categories. The design engineer defines the action which must be produced to achieve the complete stabilisation of a cavity in the long and short term and as a consequence the most appropriate stabilisation intervention to employ on the basis of the expected behaviour category of the ground's response to excavation.

Bench: the intermediate part of the cross section of a tunnel between the crown and the tunnel invert.

Cavity: a space artificially hollowed out of a mass of ground, the existence of which in time depends on the formation in the mass around the cavity of an arch effect which, if it is not created by natural means, can be created by artificial means, by intervening to stabilise the ground.

Channelling of stresses: a phenomenon which is produced in ground subject to natural stresses after a disturbance is generated in the stress equilibrium by the creation of a cavity, generally consisting of one or more series of parallel flow lines, where the space between the lines depends on the intensity of the field. If the geomechanical characteristics of the medium are sufficient to support the intensity, then the continuity of the stress flow is conserved by natural means by deviating the lines intercepted by the creation of a cavity in an outwards direction. This deviation of the stress flow lines away from the cavity and the consequent concentration of the lines in a narrow space around the perimeter of the cavity is identified by this term the *channelling of stresses.*

Characteristic lines (method of the): calculation method based on plotting curves which place the confinement pressure σ_3 applied on the perimeter of a tunnel in relation to its *radial convergence.*

Confinement pressure: radial pressure that acts on the perimeter of the excavation. It is given by the principal minor stress σ_3.

Confinement: action exerted on the walls and roof and/or on the face of a tunnel to prevent the principal minor stress σ_3 in the surrounding ground from falling to zero. This action can be produced either directly or indirectly with either active of passive action. As long as it is not produced in statics conditions that have already been compromised, confinement action will reduce the magnitude of the plasticisation which normally affects the ground around a tunnel when the material is stressed beyond its elastic limit (thickness of the plasticised ring). It therefore favours the formation of an arch effect, which is indispensable for the stability of a tunnel in the long and short term.

Convergence: reciprocal movement between two points located on the perimeter of a cavity which is produced as a consequence of the deformation response of the rock mass to the action of excavation. Convergence can develop in time either instantaneously or in a deferred manner, depending on whether the deformation response develops in the elastic or the elasto-plastic range, and on the rheological characteristics of the material.

Core-face: the system consisting of the face and the advance core.

Cut: in conventional excavation by blasting this is a hole or cavity created in the face, generally in the centre to create a free surface in the direction of which the surrounding rock can be demolished with excavation blasts.

Decompression (of the ground, of the medium): decay, by the decrease or disappearance of the principal minor stress σ_3, of the geomechanical characteristics of the ground (of its cohesion in particular) to residual values, which can occur in the band of ground surrounding the excavation in the absence of adequate intervention when the material is stressed beyond its peak strength.

Deformation phenomena: see the *deformation response*.

Deformation response: the reaction of the medium to the action of excavation. It manifests in the form of extrusion, preconvergence and convergence.

Diagnosis phase: that phase in the *design stage,* during which the design engineer uses the information collected during the survey phase to divide the tunnel to be bored into sections with uniform stress-strain behaviour (categories A, B, C), defining how deformation will develop and the types of load that will be mobilised for each of them;

Disturbance in the medium: modification of the natural field of stresses that is produced in the ground following tunnel advance. This field, which can be described as a network of flow lines, is deviated by the presence of the excavation and is concentrated in its proximity (*channelling of stresses*), producing an increase in the deviator of the stresses.

Extrusion: movement of the ground that forms the core-face of a tunnel into the tunnel itself in a longitudinal direction along the axis of the tunnel. It is a primary component of the deformation response of the medium to the action of excavation, which develops mainly inside the advance core as a function of the strength and the deformability of the core and of the original field of stresses it is subject to. It manifests at the surface of the tunnel face in a longitudinal direction along the axis of the tunnel, either with more or less axial symmetrical deformation geometry (bellying of the face) or as gravitational turning (rotation of the face). It is generally measured as a function of time (face halted) or of tunnel advance.

Face: the front wall of a tunnel which is being driven, which closes its foremost extremity in the mass of the ground.
Its position moves progressively forwards as the medium (ground) is excavated.

Failure range: this is identified in the three dimensional space of the principal stresses as the dominion of stresses that are incompatible with the strength capacity of the material.

Ground improvement: action performed on the ground to conserve or improve its natural strength, deformability and permeability properties.

Ground improvement in advance: intervention to stabilise the ground ahead of the face, by improving or conserving the geomechanical characteristics of the ground in the advance core.

Interpretation of measurements: processing and critical analysis of measurement readings taken in the monitoring phase in order to obtain a sufficiently accurate and detailed understanding of how the stress-strain conditions in the ground are evolving.

Intrinsic curve: line of failure of the ground or an earth mass which is normally obtained as the envelope of the circles of principal failure stresses, according to the criterion employed (Mohr-Coulomb, Hoeck-Brown, Drucker-Prager etc.).

Measurement station: section of tunnel in which instruments are installed, either before or after the passage of the face, to take measurements of stress and/or deformation and/or hydraulics (e.g.: extensometers, pressure and load cells, piezometers, distometer nails, dilatometers etc.).

Muck/spoil removal: the operation which consists of removing loading and transporting excavated earth to the surface.

Natural consistency: the consistency typical of a material which depends exclusively on its intrinsic physical and mechanical properties.

Pilot tunnel: small diameter tunnel driven using a full face TBM, before full tunnel excavation commences, located near or along its axis with the primary purpose of performing an in-depth survey of the geomechanical and structural characteristics of the ground concerned (see also *RS-Method*), but sometimes also to perform advance ground improvement which may be necessary for the construction of the full tunnel. If it is driven outside the cross section of the full tunnel to be driven, it may also serve as a permanent service and safety tunnel (ventilation, drainage, emergency exits and first aid access, etc.) in addition to its other temporary purposes.

Plasticised ring: zone of ground with plastic behaviour which is formed around a tunnel when the stresses induced by excavation exceed the peak strength of the ground which is tunnelled. The thickness of the zone depends on the peak and residual strength parameters of the ground, the natural stress state and, when tunnel advance takes place under the water table in hydrodynamic conditions, on the hydraulic load and the permeability of the material.

Preconfinement: stabilisation intervention which acts ahead of the face on the ground around the future cavity to either prevent the principal minor stress from falling or even to increase it. It uses the advance core to achieve this, by protecting it from excess stress and/or by conserving or even by improving its rigidity.

Preliminary lining: general term which includes all those structural stabilisation instruments employed ahead of the face and inside the tunnel, to eliminate or contain the deformation response of the medium to the action of excavation in the long and short term.

Presupport: a stabilisation instrument placed ahead of the face, but which does not produce any arch effect ahead of the face. Presupports (e.g. forepoles) can sometimes be employed in fractured rocky ground when tunnelling under elastic conditions to protect against concentrated falls of rock and to prevent the localised fall of material.

Radial convergence: radial movement in relation to the axis of a tunnel of a generic point located on the perimeter of the excavation or inside the mass of the ground which occurs as a consequence of the deformation response of the ground to the action of excavation. Radial convergence can develop in time either instantaneously or in a deferred manner, depending on whether the deformation response develops in the elastic or the elasto-plastic range. Surface or depth (radial) convergence is spoken of according to whether it refers to a point on the walls of the tunnel or a point inside the mass of the ground.

Radius of influence of the face R_f: the size of the disturbed zone surrounding the face. It depends on the radius of the tunnel, on the geomechanical characteristics of the medium, on the rate and means of excavation and on the stabilisation intervention implemented.

Relaxation: decrease or disappearance of the principal minor stress σ_3 with a consequent decrease in the shear strength of the medium.

Rigidity (of the advance core, of the core-face): a characteristic of the advance core or the core-face, which determines its greater or lesser tendency to extrude when stressed in the elasto-plastic range, or the magnitude of the cavity preconfinement action that it is able to exert.

Round of shots: simultaneous or predetermined sequence of explosive detonations specifically placed to give a determined result in terms of face excavation.

Rock mass: this term has been widely used in the book to translate the Italian word *ammasso,* which really means just the “mass” of the ground generally without specifying whether it is rock or soil. As a consequence the term "rock mass" has often been used generically to refer to the ground whether as a rock or soil mass.

RS Method: survey instrument which uses a TBM as a huge penetrometer, measuring its operating parameters by means of special sensors (thrust, advance velocity, energy consumed, etc.). It is thus possible to determine the specific energy employed per m^3 of material excavated and, by means of appropriate correlations, also the strength of the rock mass.

Scaling: see *Barring down*.

Section with uniform stress-strain behaviour: portion of the underground alignment in which it is reasonable to hypothesise that the deformation response, understood as the reaction of the ground to the action of excavation, will be qualitatively constant for the whole of its length.

Solid load: mass of ground potentially unstable due to the effect of gravity, which, if not adequately treated, will produce instability sooner or later or which would weigh, as a dead weight, on the lining of the tunnel.

Spoil: material detached and demolished by the blast of explosions or by any other procedure for excavating the earth.

Survey phase: that phase in the *design stage*, during the which the design engineer, proceeds to characterise the terrain, or the medium, through which the tunnel will pass in terms of rock or soil mechanics. It is indispensable for analysing the pre-existing natural equilibriums and for proper conduct of the subsequent diagnosis phase.

The advance core: the volume of ground which lies ahead of the face, practically cylindrical in shape with a cross section and length of around the size of one tunnel diameter.

The monitoring phase: that phase of *the construction stage* in which deformation phenomena (which constitute the response of the medium to the action of deformation) are read and interpreted during the construction of a tunnel to monitor the accuracy of the predictions made during the diagnosis and therapy phases in order to perfect the design by balancing stabilisation instruments between the face and the cavity. The monitoring phase does not finish when a tunnel is completed, but continues for the whole course of its life in order to constantly monitor its safety in service.

The operational phase: that phase in the *construction stage* during which stabilisation intervention is implemented in compliance with the design specifications. It is adapted, in terms of preconfinement and confinement, to fit the real deformation response of the ground and monitored according to pre-established quality control plans.

The therapy phase: that phase in the *design stage* during which the design engineer decides the action to exert (*preconfinement* or simple *confinement*) and the intervention required to achieve the complete stabilisation of the tunnel in the long and short term, on the basis of the predictions made in the diagnosis phase. He then decides the composition of the longitudinal and cross section types and their dimensions and tests their effectiveness using mathematical tools, specifying the criteria for their application and possible variations to be made as a function of the stress-strain behaviour of the ground that will actually be observed during excavation. He then finally draws up the monitoring plan.

Tunnel: a civil engineering work designed to provide a continuous path for communication or for a watercourse through a mass of earth which it is not possible or not worthwhile creating by other means. It is complete from a structural point of view but does not include the finishings characteristic of the particular use to which it is destined.

Variability: a possible variation in a given tunnel section type of the intensity and/or geometry of stabilisation intervention, to be applied when statistically probable conditions, clearly described in the design, occur, but for which the precise location could not be predicted on the basis of the data available at the design stage.

Zero reading: this is the first reading taken after a measurement instrument is installed, against which all the subsequent readings performed on the same instrument are compared.

Bibliography

[1] KÁRMÁN T., "Festigkeitsversuche unter allseitigem Druck", Berlin 1912

[2] BIENIAWSKI Z.T., "Rock Mechanics design in mining and tunneling", Balkema 1984

[3] BARTON N.R., LUNDE J., "Engineering classification of rock masses for the design of tunnel support", International Journal of Rock Mechanics and Mining Sciences & Geomech. Abst. 25, No. 13–13, 1974

[4] BIENIAWSKI Z.T., "Rock mass classification as a design aid in tunnelling", Tunnels & Tunnelling, July 1988

[5] RABCEWICZ L., "The New Austrian Tunnelling Method", Water Power, November 1964, December 1964, January 1965

[6] THE INSTITUTION OF CIVIL ENGINEERS, "Sprayed concrete linings (NATM) for tunnels in soft ground", Telford 1996

[7] KOVÁRI K., "On the Existence of NATM, Erroneous Concepts behind NATM", Tunnel, No. 1, 1994 (English and German), or Gallerie e Grandi Opere Sotterranee, No. 44, December 1994 (Italian and English)

[8] KOVÁRI K., "Il controllo della "risposta dell'ammasso" in galleria: pietre miliari fino agli anni 1960", Quarry and Construction, July–August 2001

[9] LUNARDI P., "Applicazions de la Mecanique des Roches aux Tunnels Autoroutiers. Example des tunnels du Frèjus (côte Italie) et du Gran Sasso", Revue Francaise de Geotechnique No. 12, 1979

[10] KASTNER H., "Statik des Tunnel- und Stollenbaues", Sprinter, Berlin-Göttingen-Heidelberg, 1962

[11] PANET M., "Le calcul des tunnels par la méthode convergence-confinement", Ponts et chaussées, 1995

[12] LOMBARDI G., AMBERG W.A., „Une méthode de calcul élasto-plastique de l'état de tension et de déformation autour d'une cavité souterraine", Congresso Internazionale ISRM, Denver, 1974

[13] LUNARDI P., "La costruzione della galleria S. Elia lungo l'autostrada Messina-Palermo", Costruzioni, December 1988

[14] LUNARDI P., "Aspetti progettuali e costruttivi nella realizzazione di gallerie in situazioni difficili: interventi di precontenimento del cavo", International Convention on «Il consolidamento del suolo e delle rocce nelle realizzazioni in sotterraneo» – Milan 18–20 March 1991

[15] LUNARDI P., FOCARACCI A., "Aspetti progettuali e costruttivi della galleria Vasto", Quarry and Construction, August 1997

[16] GOLINELLI G., LUNARDI P., PERELLI CIPPO A., "La prima applicazione del jet-grouting in orizzontale come precontenimento delle scavo di gallerie in terreni incoerenti", International Convention on "Il consolidamento del suolo e delle rocce nelle realizzazioni in sotterraneo" – Milan 18–20 March 1991

[17] ARSENA F.P., FOCARACCI A., LUNARDI P., VOLPE A., "La prima applicazione in Italia del pretaglio meccanico", International Convention on «Il consolidamento del suolo e delle rocce nelle realizzazioni in sotterraneo» – Milan 18–20 March 1991

[18] LUNARDI P., BINDI R., "The evolution of reinforcement of the advance core using fibre glass elements for short and long term stability of tunnels under difficult stress-strain conditions: design, technologies and operating methods", Proceedings of the International Congress on "Progress in Tunnelling after 2000", Milan, 10–13 June 2001

[19] LUNARDI P., BINDI R., FOCARACCI A., "Nouvelles orientation pour le projet et la construction des tunnels dans des terrains meubles. Études et expériences sur le préconfinement de la cavité et la préconsolidation du noyau au front", Colloque International "Tunnels et micro-tunnels en terrain meuble", Paris 7–10 February 1989

[20] LUNARDI P. et al., "Tunnel face reinforcement in soft ground design and controls during excavation", International Convention on "Towards New Worlds in Tunnelling" – Acapulco 16–20 May 1992

[21] LUNARDI P., "Fibre-glass tubes to stabilize the face of tunnels in difficult cohesive soils SAIE: Seminar on "The application of fiber Reinforced Plastics (FRP) in civil structural engineering" – Bologna, 22 October 1993

[22] LUNARDI P., "La stabilité du front de taille dans les ouvrages souterraines en terrain meuble: etudes et experiences sur le renforcement du noyau d'avancement", Symposium international "Renforcement des sols: experimentations en vraie grandeur des annes 80", Paris, 18 November 1993

[23] LUNARDI P., "Avanza la galleria meccanica" – Le Strade, May 1996

[24] LUNARDI P., "The influence of the rigidity of the advance core on the safety of tunnel excavation", Gallerie e grandi opere sotterranee, No. 52, 1997 (Italian and English)

[25] BROMS B.B., BENNERMARK H., "Stability of clay at vertical openings", Journal of the Soil Mechanics and Foundation Division, 1967

[26] ATTEWELL P.B., BODEN J.B., "Development of the stability ratios for tunnels driven in clay", Tunnels and Tunnelling, 1971

[27] PANET M., GUENOT A., "Analysis of convergence behind the face of a tunnel", Tunnelling '82, The Institute of Mining and Metallurgy, London 1982, pages 197–204

[28] TAMEZ E., "Estabilidad de tuneles exavados en suelos", Work presented upon joining the Mexican Engineering Academy, Mexico 1984
[29] ELLSTEIN A.R., "Heading failure of lined tunnels in soft soils", Tunnel and Tunnelling 1986
[30] CORNEJO L., "El fenómeno de la instabilidad del frente de excavación y su repercusión en la construcción de túneles", International Congress on "Tunnels and Water", Madrid 1988
[31] PANET M, LECA E ., "Application du calcul à la rupture à la stabilité du front de taille d'un tunnel", Geotecnique 1988
[32] CHAMBON P., CORTÉ J.F., "Stabilité du front de faille d'un tunnel faiblement enterré: modélisation en centrifugeuse", Colloque International "Tunnels et micro-tunnels en terrain meuble", Paris, 7–10 February 1989
[33] CHAMBON P., CORTÉ J.F., "Stabilité du front de faille d'un tunnel dans milieu frottant. Approche cinematique en calcul à la rupture", Revue Française de Geotechnique 1990
[34] LOMBARDI G., "Dimensioning of tunnel linings", Tunnels and Tunnelling 1973, pages 340–351
[35] LOMBARDI G., "Nuovi concetti sulla statica delle gallerie", La Rivista della Strada, 406, 1975, pages 465–482
[36] LUNARDI P., "Convergence-confinement ou extrusion-préconfinement?", Colloque "Mécanique et Géotechnique " (Giubileo scientifico di Pierre Habib), Laboratoire de Mécanuque des Solides, École Polytechnique, Paris 19 May 1998
[37] GALFO R., "Studio della stabilità del fronte di scavo e metodi di stabilizzazione", Tesi di laurea, Università degli Studi di Firenze, Facoltà di Ingegneria Civile, 1990
[38] FUKUSHIMA S., MOCHIZUCHI Y., KAGAWA K., YOKOYAMA A., "Model study of pre-reinforcement method by bolts for shallow tunnel in sandy ground", International Congress on "Progress and Innovation in Tunnelling", Toronto 9–14 September 1989
[39] ANTHOINE A., "Une méthode pour le dimensionnement à la rupture des ouvrages en sols renforcés", Revue Française de Geotechnique, 50, 1990
[40] LUNARDI P., "Conception et exécution des tunnels d'après l'analyse des déformations contrôlées dans les roches et dans les sols", (Italian and French), Article in three parts: Quarry and Construction, March 1994, March 1995, April 1996 or Revue Française de Geotechnique, No. 80, 1997, No. 84, 1998, No. 86, 1999
[41] LUNARDI P., "Design & constructing tunnels – ADECO-RS approach", Tunnels and Tunnelling International, special supplement, May 2000
[42] LUNARDI P., Progetto e costruzione di gallerie – Approccio ADECO-RS", Quarry and Construction, supplement, May 2001

[43] LUNARDI P., "La stabilite du front de taille dans les ouvrages souterraines en terrain meuble: etudes et experiences sur le renforcement du noyau d'avancement", Symposium international "Renforcement des sols: experimentations en vraie grandeur des annes 80", Paris, 18 November 1993
[44] LUNARDI P., "Lo scavo delle gallerie mediante cunicolo pilota", Politecnico di Torino, Primo ciclo di conferenze di meccanica e ingegneria delle rocce – Turin, 25–26 November 1986
[45] CAMPANA M., LUNARDI P., PAPINI M., "Dealing with unexpected geological conditions in underground construction: the pilot tunnel technique", Acts of 6th European Forum on "Cost Engineering" – Università Bocconi, Milan, 13–14 May 1993, Vol. 1
[46] GOODMAN R.E., SHI GEN-HUA, "Block theory and its application to rock engineering", Prentice Hall, New Jersey, 1985
[47] FAZIO A.L., RIBACCHI R., "Stato di sforzo e di deformazione intorno a una galleria sotto falda", Gallerie e grandi opere sotterranee, 17, 1983
[48] HUERGO P.J., NAKHLE A., "L'influence de l'eau sur la stabilité d'un tunnel profond", International Congress on "Tunnels and Water", Madrid 1988
[49] LUNARDI P., BINDI R., La teoria dell'Area Indice nel dimensionamento dell'interasse tra cavità adiacenti. Applicazioni al caso del tunnel sotto La Manica", International Congress on "Grandi Opere Sotterranee", Florence 8–11 June 1986
[50] LUNARDI P., FROLDI P., "Influence des conditions géostructurelles et géomécaniques sur les phénomènes de rupture dans la calotte du tunnel, VII International IAEG Congress – Lisbon 5–9 September 1994 – vol. VI
[51] HOEK E., BRAY J.W., "Rock slope engineering", Institution of Mining and Metallurgy, London 1981
[52] LUNARDI P., "Il progetto dello scavo nella realizzazione di opere in sotterraneo", Atti della 4° sessione delle Conferenze Permanenti Alta Velocità su "Lo scavo: metodi, tecniche ed attrezzature nella progettazione e costruzione della tratta Bologna-Florence", Scarperia (FI) 31 March 2000
[53] Recommendations and Guidelines for Tunnel Boring Machines (TBM), International Tunnelling Association, Working Group No. 14, September 2000
[54] MAIDL B., HERRENKNECHT M., ANHEUSER L., "Mechanised shield tunnelling", Ernst & Sohn, Berlin 1996
[55] ARNOLDI L., Lo scavo meccanizzato, nella nostra esperienza, con TBM del tipo aperto", Convention on "Scavo meccanizzato integrale di gallerie: esperienze ed esigenze di committenti, fabbricanti di macchine, imprese e progettisti", Rome 13–15 May 1997
[56] PANCIERA A., PICCOLO G., "TBM scudate. Blocco in rocce spingenti", Convention on "Scavo meccanizzato integrale di gallerie:

esperienze ed esigenze di committenti, fabbricanti di macchine, imprese e progettisti", Rome 13–15 May 1997
[57] LUNARDI P. et al., "Soft ground tunnelling in the Milan Metro and Milan Railway Link. Case histories", Soft Ground Tunnelling Course – Institution of Civil Engineers – London, 10–12 July 1990
[58] LUNARDI P., "Evolution des technologies d'escavation en souterrain dans des terrains meubles", Comité Marocain des Grands Barrages" - Rabat, 30 September 1993
[59] LUNARDI P., "The construction of large-span stations for underground railways", Tunnels, 8, December 2000
[60] LUNARDI P:, FOCARACCI A., RICCI D., VALENTE A., "Una soluzione innovativa per la realizzazione di gallerie naturali senza copertura", Quarry and Construction, May 1997
[61] LUNARDI P. et al., "Quality Assurance in the Design and Construction of Underground Works", International Congress on "Underground Construction in Modern Infrastructure", Stockholm 7–9 June 1998
[62] TERZAGHI K., "Mécanique théorique des sols", Dunod, Paris 1951
[63] FALCHI DELITALA G., "Calcolo dei rivestimenti delle gallerie", Vitali e Ghianda, Genoa 1971
[64] COLONNA P., "Su alcuni problemi di stabilità dei fronti di scavo in galleria, con particolare riferimento agli ammassi interposti tra due fronti in progressivo avvicinamento", Quarry and Construction, January 1997
[65] CUNDALL P.A., "Numerical modeling of jointed and faulted rock", Mechanics of Jointed and Faulted Rock, A. A. Balkema, Rotterdam 1990, pages 11–18
[66] BERRY P., DANTINI E.M., LUNARDI P., "Pressioni in aria e sismi indotti in opere sotterranee da volate in galleria", Strade & Autostrade, 5, 1999
[67] BERTA G., "L'esplosivo strumento di lavoro", Italesplosivi, Milan 1989
[68] LUNARDI P., "Pretunnel advance system", Tunnels and Tunnelling International, October 1997
[69] ANDRE D., DARDARD B., BOUVARD A., CARMES J., "La traversée des argiles du tunnel de Tartaiguille", Tunnels et ouvrages souterrains, No. 153, May–June 1999
[70] LUNARDI. P., "The "Tartaiguille" tunnel, or the use of the ADECO.-RS approach in the costruction of an "impossible" tunnel", Gallerie e grandi opere sotterranee, No. 58, August 1999 (Italian and English)
[71] MARTEL J., ROUJON M., MICHEL D., "TGV Méditerranèe – Tunnel de Tartaiguille: méthode pleine section", Proceedings of the International Conference on "Underground works: ambitions and realities", Paris, 25th–28th October 1999

[72] "Florence to Bologna at high speed", Tunnels and Tunnelling International, April 1999
[73] "Progress report on the underground works on the Bologna to Florence high-speed railway line", Tunnel, No. 8, 2000
[74] LUNARDI P., BINDI R., CASSANI G., "Prime evidenze e risultati dell'impiego dell'approccio ADECO-RS per la realizzazione di oltre 73 Km di gallerie di linea", Strade e Autostrade, No.1 2006
[75] LUNARDI P., "Tunnelling under the Via Appia Antica in Rome", Tunnels & Tunnelling International, April 2000
[76] LUNARDI P., "Une methode de construction innovante pour elargir les tunnels routiers, autoroutiers et ferroviaires sans interrompre la circulation; son application au tunnel de Nazzano sur l'autoroute A1 Milan-Naples", Proceedings of the Conference on "Instandsetzung von Tunneln", Olten, 21 October 1999

Contents of the special focus boxes

Chapter 1

Chapter 2

Chapter 3

Chapter 4

Chapter 5

Chapter 6

Chapter 7

Chapter 8

Chapter 9

Chapter 10